CONTENTS

FOREWORD *Koïchiro Matsuura*	13
PREFACE In praise of UNESCO and its principles *M. S. Swaminathan*	15
A PERSONAL NOTE The history of the History *Walter Erdelen*	18
GENERAL INTRODUCTION *Editorial Committee*	21

PART I: SETTING THE SCENE, 1945–1965 — 27
INTRODUCTION — 29
VISIONS AND REVISIONS
Defining UNESCO's scientific culture, 1945–1965
Patrick Petitjean

CONCEPTION AND CREATION OF UNESCO — 35
AN UNSUNG ANCESTOR — 35
The International Institute of Intellectual Cooperation (IIIC) and science
Jean-Jacques Renoliet

HOW THE 'S' CAME TO BE IN UNESCO — 36
Gail Archibald

BRAVE NEW ORGANIZATION — 40
Julian Huxley's philosophy
John Toye and Richard Toye

BLAZING THE TRAIL — 43
Needham and UNESCO: perspectives and realizations
Patrick Petitjean

INTERNATIONAL SCIENTIFIC COOPERATION 48
FINDING A FOOTING 48
The sciences within the United Nations system
Patrick Petitjean

GIVING SCIENCE FOR PEACE A CHANCE 52
The post-war international laboratory projects
Patrick Petitjean

COOL HEADS IN THE COLD WAR 57
Pierre Auger and the founding of CERN
Patrick Petitjean

OF MOLECULES AND MEN 60
UNESCO and life sciences during the Cold War
Bruno J. Strasser

NOURISHING SCIENCE, ON LAND AND SEA 65
FAO collaboration with UNESCO and the creation of the IOC, 1955–1962
Ray Griffiths

PIERCING THE IRON CURTAIN 68
UNESCO, marine science, and the legacy of the International Geophysical Year
Jacob Darwin Hamblin

THE 'PERIPHERY PRINCIPLE' 71
CROSSING BORDERS 71
Contributing to the development of science in Latin America
Patrick Petitjean

GOING GLOBAL 72
UNESCO Field Science Offices
Jürgen Hillig

MEANWHILE IN THE MOTHERLAND 75
Regional Office for Science and Technology in Africa (ROSTA)
Robert H. Maybury

SCIENTIFIC NGOS AND UNESCO 77
A SPECIAL RELATIONSHIP 77
The early years of the UNESCO-ICSU partnership
Patrick Petitjean

A FAILED PARTNERSHIP 78
The WFSW and UNESCO in the late 1940s
Patrick Petitjean

SOCIAL ASPECTS OF SCIENCE 81
MEETINGS OF MINDS 81
UNESCO and the creation of the International Union of History of Science
Patrick Petitjean

SIXTY YEARS OF SCIENCE AT UNESCO 1945-2005

The ideas and opinions expressed in this publication
are those of the authors and are not necessarily those
of UNESCO and do not commit the Organization.

The designations employed and the presentation of
material throughout this publication do not imply the
expression of any opinion whatsoever on the part of
UNESCO concerning the legal status of any country,
territory, city or area or of its authorities or concerning
the delimitation of its frontiers or boundaries.

Published by the United Nations Educational,
Scientific and Cultural Organization
7, place de Fontenoy, 75352 Paris 07 SP, France

Typeset by UNESCO Publishing/Roberto Rossi
Printed by Imprimerie Barnéoud, Bonchamps-lès-Laval

Cover design based on Science and Technology UNESCO
poster by Shoichi Hasegawa © UNESCO 1986

ISBN 10: 92-3-104005-7 (softcover)
ISBN 13: 978-92-3-104005-4 (softcover)

ISBN 10: 92-3-104047-2 (hardcover)
ISBN 13: 978-92-3-104047-4 (hardcover)

© UNESCO 2006

All rights reserved

Printed in France

SIXTY YEARS OF SCIENCE AT UNESCO 1945–2005

UNESCO Publishing

Suggested citation: Petitjean, P., Zharov, V., Glaser, G., Richardson, J., de Padirac, B. and Archibald, G. (eds.). 2006. *Sixty Years of Science at UNESCO 1945–2005*. Paris, UNESCO.

Editorial Committee
General Editor: Jake Lamar
Assistant Editor: Brian Smith
Project Director: Bruno de Padirac
Historical Coordinator: Gail Archibald

Section Coordinators
Part I: Setting the Scene, 1945–1965: Patrick Petitjean
Part II: Basic Sciences and Engineering: Vladimir Zharov
Part III: Environmental Sciences: Gisbert Glaser
Part IV: Science and Society: Jacques Richardson
Under the overall authority of: Walter Erdelen
Assisted by: L. Anathea Brooks

CONTENTS

ON THE ROAD   83
UNESCO's travelling science exhibitions
Alain Gille

THE ULTIMATE ODYSSEY   85
The birth of the *Scientific and Cultural History of Mankind* project
Patrick Petitjean

SPREADING THE NEWS   88
The natural sciences in the *UNESCO Courier*, 1947–1965
Lotta Nuotio

PART II: BASIC SCIENCES AND ENGINEERING   93
INTRODUCTION   95
AT THE HEART OF IT ALL
The basic and engineering sciences as the key to civilization
Vladimir Zharov

MATHEMATICS, PHYSICS AND CHEMISTRY   104
CALCULATED RISKS   104
Initiatives in pure and applied mathematics
Franck Dufour

QUANTUM LEAPS FOR PEACE   107
Physics at UNESCO
Franck Dufour

TAMING THE ATOMIC TIGER   110
The creation of the International Centre for Theoretical Physics
André M. Hamende

WARM RELATIONS AFTER THE COLD WAR   114
The Physics Action Council, 1993–1999
Irving A. Lerch

OPENING SESAME   116
A landmark of scientific cooperation in the Middle East
Clarissa Formosa Gauci

FINDING THE RIGHT CHEMISTRY   120
The International Union of Pure and Applied Chemistry
(IUPAC) and UNESCO
Mohammed Shamsul Alam

INFORMATICS, INFORMATION AND
COMMUNICATION TECHNOLOGIES   126
MAKING EVERYTHING COMPUTE   126
ICTs at the crossroads of applied mathematics, physics and chemistry
Rene Paul Cluzel

ORGANIZING INFORMATION The origins and development of UNISIST *Jacques Tocatlian*	129
THE WAY IT WAS UNESCO and informatics, a memoir *Sidney Passman*	131

BIOLOGICAL SCIENCES — 134

TO BE OR NOT TO BE UNESCO's biological and microbiological programmes *Franck Dufour, Julia Hasler, Lucy Hoareau*	134
MAKING CONNECTIONS International Biosciences Networks (IBN) *Julia Marton-Lefèvre*	143

ENGINEERING SCIENCES — 147

THE SPIRIT OF INNOVATION Engineering and technology at UNESCO *Tony Marjoram*	147
APPENDIXES: PERSONAL REFLECTIONS ON ENGINEERING	158
James McDivitt	158
Charles Gottschalk	161
Edward Beresowski	163
Benjamin Ntim	165
FUEL FOR THOUGHT UNESCO initiatives on renewable energy sources *Osman Benchikh*	167

SCIENCE EDUCATION — 176

LIVE AND LEARN The early days of science education at UNESCO *Albert Baez*	176
WORKING KNOWLEDGE Symbiosis of programmes in science teaching, environmental education, and technical and vocational education *Saif R. Samady*	182

PART III: ENVIRONMENTAL SCIENCES — 193

INTRODUCTION — 195

EARTH MATTERS
The environment and sustainable development focus in UNESCO's science programmes
Gisbert Glaser

CONTENTS

NATURE TO THE FORE   201
The early years of UNESCO's environmental programme, 1945–1965
Malcolm Hadley

THE ESSENCE OF LIFE   233
UNESCO initiatives in the water sciences
Sorin Dumitrescu

A PRACTICAL ECOLOGY   260
The Man and the Biosphere (MAB) Programme
Malcolm Hadley

ROCKY ROAD TO SUCCESS   297
A new history of the International Geoscience Programme (IGCP)
Susan Turner

THE FINAL FRONTIER   315
UNESCO in outer space
Robert Missotten

SAFETY FIRST   320
UNESCO's initiatives in natural disaster reduction
Badaoui Rouhban

OBSERVING AND UNDERSTANDING PLANET OCEAN   332
A history of the Intergovernmental Oceanographic Commission (IOC)
Geoffrey Holland

BUILDING BLOCKS FOR MARINE SCIENCE   354
A history of UNESCO's Marine Sciences Division (OCE)
*Dale C. Krause, Selim Morcos, Marc Steyaert and Gary D. Wright,
in consultation with Alexei Suzyumov and Dirk G. Troost*

BREAKING DOWN BARRIERS, BUILDING BRIDGES   371
Coastal Regions and Small Islands (CSI) Platform
Dirk Troost and Malcolm Hadley

THREADS OF LEARNING AND EXPERIENCE   383
Recording and valorizing traditional ecological knowledge
Malcolm Hadley

LINKING BIOLOGICAL AND CULTURAL DIVERSITY   385
Local and Indigenous Knowledge Systems (LINKS) project
Douglas Nakashima and Annette Nilsson

A GIFT FROM THE PAST TO THE FUTURE   389
Natural and cultural world heritage
Bernd von Droste zu Hülshoff

ACTING TOGETHER   401
Promoting integrated approaches, and the paradigm shift from environmental policy to sustainable development from Stockholm 1972 to Johannesburg 2002
Gisbert Glaser

PART IV: SCIENCE AND SOCIETY 429
INTRODUCTION 431
THE SHOCK OF THE NEW
The contemporary age begins
Jacques Richardson

HELPING HANDS, GUIDING PRINCIPLES 434
Science and technology policies
Jürgen Hillig

PUSHING AND PULLING 452
Scientific research and industrial applications
Jacques Richardson

CRUNCHING NUMBERS 453
Science and technology statistics at UNESCO
Ernesto Fernández Polcuch

CLOSING THE CULTURAL GAP 460
Science and society
Jacques Richardson

TAPPING AT THE GLASS CEILING 465
Women, natural sciences and UNESCO
Renée Clair

QUESTIONING AUTHORITY 474
Science and ethics
Jacques Richardson

HARD TALK 476
The controversy surrounding UNESCO's contribution to
the management of the scientific enterprise, 1946–2005
Bruno de Padirac

WHAT IS TO BE DONE? 482
A few conclusions
Jacques Richardson

PART V: OVERVIEWS AND ANALYSES 485
IMPRESSIONS: FORMER ASSISTANT
DIRECTORS-GENERAL FOR NATURAL SCIENCES 487
A TREMENDOUS PRIVILEGE 487
Abdul-Razzak Kaddoura

GREAT THINGS 490
Adnan Badran

IN THE SERVICE OF MEMBER STATES 491
Maurizio Iaccarino

UNESCO AND ICSU: SIXTY YEARS OF COOPERATION	496
PARTNERSHIP IN SCIENCE	507

Cross-cutting issues in UNESCO's natural sciences programmes
Malcolm Hadley and Lotta Nuotio

OVERVIEW 571
The Natural Sciences Sector, 2005
Walter Erdelen

PART VI: LOOKING AHEAD 591
THOUGHTS ON THE NATURAL SCIENCES AT UNESCO 593
Beyond 2005
Walter Erdelen

ANNEXES 639
1. ACRONYMS 641
2. UNESCO SCIENCE MILESTONES, 1945–2005 648
3. HEADS OF NATURAL SCIENCES AT UNESCO 660
4. A TRIBUTE TO TWO OF OUR OWN:
   MICHEL BATISSE AND YVAN DE HEMPTINNE 671
5. CHRONOLOGY OF UNESCO INTERNATIONAL SCIENCE PRIZES 675
6. CONTRIBUTORS 677

INDEX 683

16 November 2005, Celebration of UNESCO's 60th Anniversary, the Director-Generals (in the forefront, from left to right): Mr Koïchiro Matsuura (Japan, Director General since 1999), Mr Amadou-Mahtar M'Bow (Senegal, 1974-1987), and Mr Federico Mayor (Spain, 1987-1999). In the background, from left to right: Mr Jaime Torres Bodet (Mexico, 1948-1952), Mr Julian Huxley (United Kingdom, 1946-1948), Mr René Maheu (France, 1961-1974), Mr Luther Evans (USA, 1953-1958), and Mr Vittorino Veronese (Italy, 1958-1960).

FOREWORD

Koïchiro Matsuura, Director-General of UNESCO

SCIENTIFIC knowledge has led to remarkable innovations that have been of great benefit to humankind. Life expectancy has increased strikingly, and cures have been discovered for many diseases. Agricultural output has risen significantly in many parts of the world to meet growing population needs. Technological developments and the use of new energy sources have created the opportunity for freeing humankind from arduous labour. These developments have also facilitated the generation of an expanding and complex range of industrial products and processes. Technologies based on new methods of communication, information handling and computation have brought unprecedented opportunities for scientific endeavour and for society at large. At the same time, however, the applications of scientific advances and the development and expansion of human activity have also led to environmental degradation, technological disasters and a range of adverse societal effects.

During its six decades of existence, UNESCO's endeavours in the field of science have taken place within a context marked by this dual aspect of the face of modern scientific development. Throughout this period, UNESCO has been fortified by the clear purpose and functions of the Organization laid down in Article I of its Constitution:

> The purpose of the Organization is to contribute to peace and security by promoting collaboration among the nations through education, science and culture in order to further universal respect for justice, for the rule of law and for the human rights and fundamental freedoms which are affirmed for the peoples of the world, without distinction of race, sex, language or religion, by the Charter of the United Nations.

UNESCO is the only organization in the United Nations family which has science inscribed in its name. Its first Director-General, Julian Huxley, was a zoologist, committed to the popularization of science. From the beginning, UNESCO knew the importance of making official contact with the international scientific societies, of promoting scientific cooperation across the boundaries of nation, ideology and culture, and of bringing together scientists and others concerned with the uses of science. UNESCO has also been at the forefront of scientific trends; for example, before the world gave a name to ecology and the environment, one of the Organization's first large-scale science projects was in the Amazon rainforest. Today, UNESCO is well known, and justly so, for its literacy campaigns and World Heritage sites; it is time to tell the story of its achievements over the past sixty years in the field of science.

This volume is not an official history but, by drawing upon the personal reflections, memories and views of staff who have worked for the Organization, and others who have worked with it, this collection captures much of the spirit of science at UNESCO. From inside UNESCO, former and current staff members, and from outside the Organization, scientists and historians, all thought it important to retrace the history of natural sciences at UNESCO. It is a history of international scientific cooperation, a political history, a history of people and, of course, an essential part of the history of the United Nations as a whole.

Through their signed contributions, the authors have been encouraged to speak freely, to criticize past actions (though never those involved), and to highlight shortcomings and failings as well as accomplishments. Just as UNESCO staff could not do their work or carry out their programmes without the help of other scientists around the world, this book could not have been written without this same team spirit. The contributors deserve our thanks for sharing their thoughts and experiences; they are united by their affection for UNESCO and by their enduring commitment to its ideals and values.

UNESCO remains well placed to proclaim the message that access to scientific and technological progress is a human right and that human beings must be at the centre of science's priorities. UNESCO's commitment, now as ever, is to put science and technology at the service of the welfare of people everywhere, to eradicate poverty, and to ensure sustainable development for all.

Koïchiro Matsuura

PREFACE
In praise of UNESCO and its principles
M. S. Swaminathan[1]

MY association with UNESCO started in 1949, when I was awarded a UNESCO–Netherlands Government fellowship for postgraduate studies at the Agricultural University at Wageningen (the Netherlands) in the field of genetics. I then spent a week at UNESCO Headquarters in Paris, learning about the goals and programmes of this unique organization from the late Malcolm Adiseshiah and other UNESCO staff members. Over the past sixty years, UNESCO's path-breaking role in the areas of science and technology has been one of integration of the principles of ecology, equality and ethics into the scientific enterprise. In addition, the Organization has helped to advance the frontiers of scientific knowledge. Its guiding principles in science have been akin to those articulated by Bertrand Russell and Albert Einstein when they helped to establish the Pugwash Conferences on Science and World Affairs in 1955: 'We appeal, as human beings, to human beings: *Remember your humanity and forget the rest*. If you can do so, the way is open to a new paradise; if you cannot, there lies before you the risk of universal death'.

Allow me to cite a few examples to illustrate the catalytic role played by UNESCO, during these sixty years, in mainstreaming the ecological, egalitarian and ethical dimensions into the development and dissemination of science-based technologies.

In the area of ecology, the first Director-General of UNESCO, Sir Julian Huxley, played a key role in the founding of the International Union for the Conservation of Nature and Natural Resources (IUCN, now the World Conservation Union) at Fontainebleau, France, in 1948. IUCN came into existence with the help of UNESCO and the Swiss League of Nature. This initiative ultimately led to the concept of environmentally sustainable development. Initiatives with far-reaching implications for sustainable human security and well-being – such as the Man and the Biosphere Programme (MAB), the World Heritage sites and the International Hydrological Programme – were launched in the first thirty years of UNESCO's history.

---

1   M. S. Swaminathan holds the UNESCO Chair in Ecotechnology at the M. S. Swaminathan Research Foundation, in Chennai (Madras), India, and is President of the Pugwash Conferences on Science and World Affairs. His long association with UNESCO began in 1949, and with the Man and the Biosphere Programme (MAB), in the 1970s.

UNESCO's foresight in launching programmes to help save coastal mangrove wetlands worldwide gained recognition in the aftermath of the immense tsunami that caused such great loss of life and of livelihoods in India, Indonesia, Malaysia, Myanmar, Seychelles, Somalia, South Africa, Sri Lanka, Thailand, and the United Republic of Tanzania, on 26 December 2004. Mangrove forests acted as a speed-breaker in areas hit by the tsunami and thus saved numerous lives. Today, UNESCO's coastal bio-shield concept has become an important component of the disaster management strategy in coastal areas under conditions of cyclonic storms, tidal waves, and rising sea levels resulting from global warming. UNESCO's pioneering work in coastal and marine ecosystems has helped to combine the livelihood security of coastal communities with the ecological security of coastal areas.

Thanks to the foundation laid by Joseph Needham, the first Head of Natural Sciences at UNESCO, principles of social and gender equality have guided the programmes and policies of the Organization and have led to an ever-increasing awareness of the symbiotic links between science and society. The latest manifestation of this concern that science help promote sustained and sustainable human security in all its dimensions is UNESCO's imaginative programme for bridging the digital divide, an initiative based on the principle of social inclusion in access to technologies.

The growing rich–poor divide started with the Industrial Revolution in Europe; and the current technology divide has led to more than a billion members of the human family suffering from unacceptable poverty and deprivation, while another billion lead unsustainable lifestyles. The growing violence in the human heart, as evidenced by the almost daily occurrence of suicide bombings resulting in the loss of innocent lives, is a serious threat to human destiny. It is here that UNESCO's efforts in including the excluded and giving voice to the voiceless – through bringing the tools of information and communication technologies (ICTs) to the economically and educationally deprived – are assuming increasing relevance. Bridging the digital divide is also proving to be a powerful method of bridging the gender divide in rural areas. UNESCO's initiatives in fostering distance education, blending traditional wisdom with frontier technologies through eco-technologies, and its efforts in mobilizing science for achieving the UN Millennium Development Goals, deserve widespread support.

Finally, UNESCO must be commended for underlining the need to pay attention to the ethical dimensions of science. With the rapid growth of genomics, proteomics, recombinant DNA technology and nanotechnology, the need for internalizing ethical considerations both in the design of experiments and in the application of their results, is growing. UNESCO has been wise in setting up both the International and the Intergovernmental Bioethics Committees and in developing Universal Declarations on the Human Genome and Human Rights, and on Bioethics and Human Rights.

Recognition of the ethical responsibility of scientists for the consequences of their work is also expanding. Not only does the nuclear peril persist even sixty years after Hiroshima and Nagasaki, but today there is the frightening possibility of nuclear-armed individuals and groups emerging – in addition to the existing and potential Nuclear Weapon States – due to the availability of enriched uranium. Similarly, bio-perils are increasing. The recent outbreak of the avian flu, caused by the H5N1 strain of the virus, serves as a wake-up call. In my view, UNESCO should further strengthen its leadership role in pointing out to heads of states and governments that bioethics is fundamental to biosecurity.

During the last sixty years, UNESCO has become the flagship of the humanistic science movement. By integrating the fundamentals of ecology, equality and ethics in scientific endeavour, UNESCO has played a pivotal role in ensuring that science becomes an instrument of human happiness and well-being. The multidimensional role of science in our day-to-day life has been captured beautifully in the inscription in the dome of the US National Academy of Sciences building in Washington DC: 'To Science, pilot of industry, conqueror of disease, multiplier of the harvest, explorer of the universe, revealer of nature's laws, eternal guide to truth'.

Sometimes it is said that UNESCO has not done enough in promoting excellence and relevance in scientific research throughout the world. I am reminded in this context of what Mother Teresa once said, when someone told her that her work is like a drop in the ocean. 'Yes, my work is like a drop in the ocean,' she replied, 'but the ocean will be less without that drop'. It can be said with confidence that UNESCO's drop has, over the past six decades, developed into a steady stream, flowing into the ocean of science.

I would like, on behalf of the global family of scientists, to pay tribute and express gratitude to the visionary directors-general of UNESCO, starting with Sir Julian Huxley and continuing up to the present Director-General, Mr Koïchiro Matsuura, for their catalytic contributions to fostering humanistic science and scientific humanism in our troubled world. All the past and present directors and staff members of UNESCO's Natural Sciences Sector have earned our gratitude for their labour of love for science and society. Our sincere thanks go also to all those who helped produce this outstanding volume, which promises to be of lasting value.

21 November 2005, Chennai (Madras)

A PERSONAL NOTE
The history of the History
Walter Erdelen, Assistant Director-General for Natural Sciences at UNESCO

As this book tells the history of the natural sciences at UNESCO, I feel it opportune to highlight the history of this book. When we were approaching the sixtieth anniversary of both the United Nations (in 2005) and, in particular, UNESCO (in 2005–06), I thought this would be a good occasion to look into what had happened during the past sixty years in the field of natural sciences at UNESCO. Inspired by the booklet *UNESCO: Why the S?* and similar publications, I felt it would be useful to explore the history of science in the Organization, with a particular focus on the processes that shaped the past, and on what we can learn from the past for our present and future. An idea was born; the next question was how to realize it.

Very soon our colleagues Gail Archibald and Bruno de Padirac expressed their interest in cooperating, and they soon felt at home in their roles of coordinators for this ambitious initiative. For me, as a comparatively new Assistant Director-General for Natural Sciences (I joined UNESCO in 2001), it was now of the essence to look into who could contribute to our undertaking. We started with a lot of good ideas but virtually no financial resources. We all felt that the only way of realizing our plan was to call on the good will of a group of idealistic and enthusiastic former colleagues, along with other scientists and historians, who would possess the institutional memory required for a history of natural sciences at UNESCO, joined by current staff. The Association of Former UNESCO Staff Members (AFUS) was therefore very helpful in many aspects of the production of this history.

Having convened the first group of volunteers, we began to consider ways of bringing our idea to fruition. The timeline was clear: UNESCO was going to celebrate its sixtieth anniversary over a period linking two essential dates – the anniversary of its birth, with the signature of its Constitution (16 November 2005), and that of the coming into force of the same (4 November 2006).[1]

Our meetings on the 'History Project' – as we used to call it – began in May 2004 (there were nine in all, the last in September 2005), and it was clear from

---

1   The Constitution of UNESCO, signed on 16 November 1945, came into force on 4 November 1946, after ratification by twenty countries.

the first that we were embarking on a very interesting journey. Early discussions already centred on the question of a title for the book. One of the first suggestions was 'La vie tumultueuse du "S" dans l'UNESCO'. At this first meeting, a lively debate took place on how to organize the contents of the history; several participants expressed the need for criticism of past UNESCO activities and to include opposing points of view, and so on. In short: the debate was already in full swing. An excellent starting point!

Also at the May 2004 meeting, a draft schedule was set up, and the contents started evolving. During a second and a third meeting (June and September 2004), it was agreed to put four major parts together, with a coordinator for each, namely: Part I: 'Setting the Scene, 1945–1965' (coordinator Patrick Petitjean), Part II: 'Basic Sciences and Engineering' (Vladimir Zharov), Part III: 'Environmental Sciences' (Gisbert Glaser) and Part IV: 'Science and Society' (Jacques Richardson), plus Annexes (Gail Archibald and Bruno de Padirac). Later, it was decided to include overviews and information on cross-cutting themes (Part V), as well as a concluding section looking towards the future (Part VI).

Over sixty authors contributed to this history – including many former and current staff members, scientists, and historians of science and of international organizations, both inside and outside UNESCO. I wish to thank them all very much indeed. I would also especially like to express my appreciation to my predecessors, former Assistant Directors-General for Natural Sciences Abdul-Razzak Kaddoura, Adnan Badran and Maurizio Iaccarino, for their support of this enterprise and for their contributions.

An undertaking such as this publication shows the dedication of UNESCO's former and current staff and its collaborators to the Organization, and their commitment to working together to both face its challenges and meet its obligations. UNESCO is considered an intellectual organization. I would add that it is also an idealistic one, as shown not only in its mandate and ideals but particularly in the engagement and idealism of our staff. As this history clearly demonstrates, UNESCO has been through – and continues to experience – difficult times, times during which this engagement and idealism are *sine qua non*.

Just as the United Nations itself is going through a major reform process, so is UNESCO. Our reform process has been one of the priorities of our Director-General, Mr Koïchiro Matsuura. It has been ongoing since Mr Matsuura took office, in 1999, with the most recent discussions in our Executive Board centring on such important issues as a long-term staff policy and decentralization.

I do hope that our 'history' – which eventually became *Sixty Years of Science at UNESCO 1945–2005* – will be a useful tool for all of us who are directly involved in the changes at UNESCO, but also for those who wish to learn something about the history of science at a truly fascinating organization, whose mandate (or should I say, mandates) has become increasingly important and relevant since its beginning.

BIBLIOGRAPHY

UNESCO. 1985. *UNESCO:Why the S?* Paris, UNESCO, 63 pp.

GENERAL INTRODUCTION
Editorial Committee

THE United Nations Educational, Scientific and Cultural Organization (UNESCO) is the only United Nations agency dealing specifically with science. And its history reflects the evolution of international scientific cooperation. The Organization's early priorities in the sciences were not essentially national priorities, but rather precursors of what would later become mainstream concerns. Programmes concerning the environment, renewable energy, science and ethics, and informatics were among the UNESCO initiatives that were notably ahead of their time. From the very beginning, its scientific programmes were characterized by the determination to share information between what Joseph Needham called the 'bright areas' (technologically advanced nations) and the 'periphery' (underdeveloped nations). Yet the decisions and directions the Organization has taken over the past sixty years concerning international scientific cooperation were rarely arrived at without a struggle. On the contrary, differences and debates, trial and error, hard-won successes and fortuity all played their part in this story.

The UNESCO Constitution[1] was finalized at the London Conference in November 1945, where it was at last decided to add the 'S' – for Scientific – to the Organization's name. UNESCO is a specialized agency within the United Nations system and is governed by its Member States; its Constitution came into force, after ratification by twenty States, on 4 November 1946. This text clearly sets out UNESCO's goals as being international peace and the common welfare of humankind:

> In consequence whereof they do hereby create the United Nations Educational, Scientific and Cultural Organization for the purpose of advancing, through the educational and scientific and cultural relations of the peoples of the world, the objectives of international peace and of the common welfare of mankind for which the United Nations Organization was established and which its Charter proclaims (Preamble).

Yet the text then goes on to state: 'The purpose of the Organization is to contribute to peace and security by promoting collaboration among the nations

---

1 Available at: http://unesdoc.unesco.org/images/0013/001337/133729e.pdf#page=7.

through education, science and culture' (Article 1: Purpose and Functions). And it is to do so by disseminating knowledge and encouraging intellectual activity:

> By encouraging cooperation among the nations in all branches of intellectual activity, including the international exchange of persons active in the fields of education, science and culture and the exchange of publications, objects of artistic and scientific interest and other materials of information (Article 1: Purpose and Functions: [c] Maintain, increase and diffuse knowledge).

Hence the dilemma of UNESCO's scientific activities: Should their effectiveness be judged on the basis of their direct contribution to peace in the world? Or, rather, should the Organization be judged by the success of its international efforts and exchange within the scientific fields, which might indirectly contribute to peace? Implicit in the Constitution is the ageless tension between 'intellectual activity' and concrete action, between reflection and results. Over the past six decades, UNESCO's role has had to evolve in the continually changing worlds of politics and science. How has it managed its precarious balancing act between scientific integrity and intergovernmental necessity?

A need has long been felt for an analytic overview of the history of the natural science programmes at UNESCO, one which would address several questions. How interested are governments in multilateral scientific cooperation? How did UNESCO's programmes influence intellectual evolution in various scientific fields? How impermeable (or open) was the Organization to what was going on in the scientific world at large? Much depended on a few dedicated men (and women, very few women), who developed their concept of what science and UNESCO should do and pushed its programmes into the future. Who were they? What drove them? And how did they cope with the politics of science within an intergovernmental organization? In order to move forward, the Organization has to understand what it has done in the past.

Most of those who were at UNESCO in the very beginning are no longer with us; their contributions must be interpreted by historians. Those who remember these founding fathers and have built on their legacy are thus valuable witnesses; it is time to share their experiences, knowledge and understanding of the Organization. Contributors to this multi-author book therefore include historians, past and present international civil servants, and those who worked on or with UNESCO's scientific programmes, both within and outside the Organization. Some authors have taken a critical approach; others have preferred more objective chronologies of events. This volume is not an 'official' history but rather a collection of invaluable insights and points of view. Contributions are signed by their authors.

GENESIS OF THE HISTORY

The *Sixty Years of Science at UNESCO, 1945–2005* project was initiated in November 2003 at the request of UNESCO's Assistant Director-General for the Natural Sciences, Walter Erdelen. On 30 April 2004, the Director-General launched the 'UNESCO History Project', to support research on the Organization's history, scheduled to end in 2010. The science history project, while autonomous, is part of this endeavour to trace UNESCO's past. It is also part of an overall effort within the United Nations system to make its history available in print form, an initiative which most notably includes the United Nations Intellectual History Project.[2] Thanks must go to the Association of Former UNESCO Staff Members (AFUS) History Club, whose efforts put history back on the official UNESCO map, and whose patient work has been an inspiration.

The present project really got off the ground in May 2004, with a meeting at UNESCO Headquarters attended by both retired and active UNESCO staff having extensive experience in the Organization, together with a historian of science. This group brought to the project the vast wealth of their own previous scientific and international experience, together with a worldwide network of contacts with scientists and historians of science. The discussions that ensued covered many subjects; hence, so do the contributions to this book.

OVERVIEW OF THE VOLUME

This volume is divided into six main parts, concluding with a presentation of the Natural Sciences Sector as it is today – its major programme activities, and staff structure – and a look at what the future may hold.[3]

PART I: SETTING THE SCENE, 1945–1965

In the 1920s and 1930s, the League of Nations' International Institute for Intellectual Cooperation modestly launched scientific publications, collaborated with the International Council of Scientific Unions (ICSU, now the International Council for Science) to organize scientific meetings, and lobbied for governmental involvement in linking scientific advancement with social transformation.

---

[2] The United Nations Intellectual History Project (UNIHP) began in mid-1999, when its secretariat was established at the Ralph Bunche Institute for International Studies of the Graduate Center of the City University of New York. The project has two main components: a series of books on specific topics, and oral histories.

[3] From 1946 to 1948, scientific activities were carried out within the Natural Sciences Section, which in July 1948 became the Natural Sciences Department, remaining so until 1964, when UNESCO was divided into Sectors. The Natural Sciences Sector, like the other programme sectors, is headed by an Assistant Director-General. It is important to note, however, that throughout its history, UNESCO has regularly carried out scientific activities outside of the Natural Sciences Sector.

Following the Second World War, the reconstruction of war-torn countries became a major concern, which included assisting countries in their efforts to catch up on lagging research and the education and training of a new generation of scientists, engineers and technicians.

In its overview of UNESCO's science programmes in its first two decades, Part I of this volume focuses on the Organization's role in, and influence on, international scientific cooperation. It introduces Julian Huxley, UNESCO's first Director-General, and Joseph Needham, first Head of Natural Sciences, whose ideas on what the 'S' in UNESCO should represent already included environmental programmes. Particular attention is given to scientific collaboration within the UN system and the creation of new partners (both intergovernmental and non-governmental organizations), and UNESCO's Regional Offices for Science and Technology. Part I was coordinated by Patrick Petitjean, a historian of science with the French National Centre for Scientific Research (CNRS) and University of Paris 7.

PART II: BASIC SCIENCES AND ENGINEERING

UNESCO has always worked closely with international scientific unions through ICSU. It has also supported the creation of non-governmental organizations (NGOs) and developed partnerships with existing NGOs to achieve its goals. Setting up centres of excellence, institutes and networks in order to propagate both the teaching and learning of science and engineering has also always been a main (and often successful) objective of the Organization.

Part II encompasses the history of UNESCO programmes in mathematics, physics and chemistry, the biological sciences and engineering, and the regional dimensions of these programmes. It was coordinated by Vladimir Zharov, professor of Physical Chemistry, who was Director of the Division of Basic Sciences at UNESCO from 1984 to 1998.

PART III: ENVIRONMENTAL SCIENCES

The second UNESCO General Conference (Mexico City, Mexico, 1947) identified the role of environmental sciences in underpinning the protection of nature as a priority of the Organization's scientific activities. Subsequently, UNESCO's environmental sciences programmes became pioneering examples of intergovernmental cooperation on environmental protection issues and contributed to the shaping of today's concern for the protection of the environment. A project promoting the ecological research of tropical forests, known as the Hylean Amazon Project, was launched in 1947. Twenty years later, another pioneering project, the Man and the Biosphere Programme (MAB) initiated a world network of biosphere reserves. It was the first international scientific programme linking the natural and social sciences and charted the way for the concept of sustainable development.

Part III also covers the history of UNESCO's programmes in the hydrological, ecological, Earth and ocean sciences, and the Organization's contribution to environmental sciences and policy in general, as well as more recent activities in the area of sustainable development. It explores the history of the natural heritage part of the Convention concerning the Protection of the World Cultural and Natural Heritage, adopted by UNESCO in 1972. Part III was coordinated by Gisbert Glaser, a geographer with a long career at UNESCO, who was in charge of coordinating the organization's enviromental programmes, and served as Assistant Director-General for the Natural Sciences a.i., from March 2000 to February 2001.

PART IV: SCIENCE AND SOCIETY

The 'S' was included in UNESCO's name because of the post-war necessity to link science to society and to the humanities; implicit in this was the need to promote social responsibility among scientists. The relationship between science and society has been interpreted in various ways over UNESCO's history. In the early 1960s, the term 'science policy' was introduced into its programme, which led to important regional conferences and the review of national and regional policies in the scientific domain. The information-gathering process may have been more effective than its follow-up, and the programme, as such, was discontinued in the 1990s, when the issue of scientific ethics came to the fore.

UNESCO's programmes on science and technology policy, science ethics, and the role of women in science at UNESCO are analysed in this section, which was coordinated by Jacques Richardson, Head of UNESCO's Science and Society Section from 1972 to 1985, and former editor of the UNESCO journal *Impact of Science on Society*.

PART V: OVERVIEWS AND ANALYSES

This section includes contributions from three former assistant directors-general for Natural Sciences, a survey of the long relationship between UNESCO and ICSU, and a detailed analysis of cross-cutting issues, which considers capacity-building, institutional relations, interdisciplinarity, governance and the political dimensions of the science programme, partnerships with outside communities, and internal issues such as science staff and budget evolution. It concludes with an overview of UNESCO's Natural Sciences Sector as it is today, written by the current Assistant Director-General (since March 2001), Walter Erdelen.

PART VI AND ANNEXES

Mr Erdelen then looks towards the future in Part VI, with some ideas on what the 'S' in UNESCO might imply in the years to come. Finally, the Annexes include a chronology, brief biographies of Heads of Science at UNESCO, and a list of acronyms.

Sixty Years of Science at UNESCO 1945–2005 is an ambitious project that contributes to the historical record of UNESCO and of the United Nations system. It documents the personal observations of key actors and historians, rather than being an 'official' institutional history. We hope and believe that it will respond to a need within the political science and history of science communities for a readable account of the beginnings – and the interlinkages – of many of today's most influential international science bodies. UNESCO has played a remarkable role in science since 1945, yet one which is not often fully appreciated. We hope that this publication will give rise to others like it, for the edification of specialists and the general public alike.

Editorial Committee

PART I
SETTING THE SCENE, 1945–1965

PART I: SETTING THE SCENE, 1945-1965

INTRODUCTION
VISIONS AND REVISIONS
Defining UNESCO's scientific culture, 1945–1965

Patrick Petitjean[1]

BORN of the cataclysm of the Second World War, battered by the storms of the Cold War, transformed by the end of colonialism, UNESCO (United Nations Educational, Scientific and Cultural Organization) – during the first two decades of its history – reflected much of the tumultuous change that defined the mid-twentieth century. The Organization's Natural Sciences Sector found itself engaged in some of the most significant issues of the era. The story of how this most idealistic of endeavours – trying to better the world and further the cause of peace through international scientific cooperation – developed, between 1945 and 1965, makes for an intriguing intellectual and political history. Part I of this book examines the priorities and principles that defined UNESCO's scientific programme from its inception to its early maturity.

It was in 1942 that the Allies began discussing post-war cooperation and laying the groundwork for new international, intergovernmental organizations. During the late war years, from 1943 to 1945, a continuing collaboration between the key players in the anti-Fascist alliance seemed assured. When the war ended with mushroom clouds over Hiroshima and Nagasaki, it was blindingly obvious that scientific development would play a critical role in the future of nations. Scientists had played an essential role in the war effort; now many hoped to do the same for keeping the peace. Science was understood to be neutral while at the same time promoting progress. For many scientists, international cooperation represented a means of not only continuing their anti-Fascist commitment but also preventing the destructive use of science in the post-war era.

Nevertheless, the subject of science had been neglected by the former international cultural and educational organizations. Before the war, the International Institute for Intellectual Co-operation (IIIC) took steps in that direction but had little time to achieve much. Moreover, this scientific cooperation was between persons, not governments. While there was some government participation in

---

1   Patrick Petitjean is a historian of science with the REHSEIS team (Recherches Épistémologiques et Historiques sur les Sciences Exactes et les Institutions Scientifiques) of the French National Centre for Scientific Research (CNRS) and University of Paris 7.

the International Council of Scientific Unions (ICSU, since then renamed the International Council for Science), its activities too were cut short by the war.

Even after the 'S' was added (at the last moment) to the Organization's name, at the conference which established UNESCO in November 1945, the place of science in the Organization – as in the United Nations system as a whole – remained ill-defined. Thanks to the determination of UNESCO's first scientific staff, however, within a few years significant programmes were launched and successfully developed, often in the face of setbacks and formidable opposition from some Member States. In the process, leaders such as Julian Huxley, UNESCO's first Director-General, and Joseph Needham, its first Head of the Natural Sciences Section,[2] promoted their compelling vision of the world, as well as ideas on the social and international role of science.

SETTING UP THE 'S', 1946–1950

UNESCO's first science programmes were focused on the reconstruction of countries ravaged by the war. The funding for these programmes (less than 10 per cent of UNESCO's total budget) was divided into three main fields:

- support to ICSU and the creation of new scientific unions;
- establishment of UNESCO's Regional Scientific Offices and conferences; and
- creation of new forms of scientific cooperation, such as the Amazonian Institute project (initiated at the first session of UNESCO's General Conference in Paris in 1946), the Arid Zone Institute project, and the International Computation Centre (both initiated at the General Conference in Beirut, Lebanon in 1948).

Other initiatives concerned the social aspects of science and nurtured the creation of films, publications, exhibitions and educational projects addressing a range of issues. There was also the establishment of the World Centre for Scientific Liaisons, which was engaged in exchange programmes, travel facilitation, the standardizing of analytical reports and other publications, and the creation of an international directory of scientists.

But the freezing winds of the Cold War had a profound effect on UNESCO. The Soviet Union refused to join the Organization, wary of the West's prominent

---

2   The names of the Sector have been, from 1946 to 1948: the Natural Sciences Section; from July 1948 to 1964: Natural Sciences Department; and from September 1964 to the present day: Natural Sciences Sector.

role in its creation. Meanwhile, Anglo-Saxon Member States accused UNESCO of being pro-Communist. They also applied pressure to reduce its scientific activities; their reasons varied: programmes were too disparate, bureaucracy was being expanded without results, the financial crisis in Europe required limits on funding.

In the spring of 1948, finding a successor to Joseph Needham as Head of Natural Sciences at UNESCO proved difficult. The United States vetoed candidates suggested by Needham, and Director-General Julian Huxley refused candidates suggested by the United States. Pierre Auger – liberal but less left-wing than Needham, and a man with close ties to the French Government – was a compromise choice. Succeeding Needham, Auger held the post of Director of the Natural Sciences Department until December 1958.

UNESCO also had difficulties fitting into the United Nations system as a whole. The scientific mandate was shared by several UN agencies and by the UN Economic and Social Council (ECOSOC). The responsibility for nuclear power was reserved for the United Nations Security Council. While scientific development had become an essential political issue during the war, a divergence soon became apparent between scientists and diplomats. Unlike ECOSOC, UNESCO was a hybrid organization, intergovernmental while also acknowledging the importance of intellectual personalities. The negotiation, timing and execution of a number of projects reflected the way scientists did things, which was largely incompatible with the rhythm and ways of diplomats and the necessities imposed by intergovernmental consultations.

The scientific–diplomatic culture clash was a main factor in the failure of the Amazonian Institute. Because of the persistent hostility of the United States to costly projects, the Arid Zone Institute was reduced to a simple 'consultative committee'. Nevertheless, it would go on to become a resounding success.

LOOKING FOR A BALANCED PROGRAMME, 1950–1954

Jaime Torres Bodet succeeded Julian Huxley after the Beirut General Conference (December 1948), with only the last-minute endorsement of the United States. Writer and poet, diplomat and ex-minister of education, less opinionated than Huxley and more government orientated, Torres Bodet hoped that UNESCO would receive the funding necessary to fulfil its pacifist mission. He wanted UNESCO to bridge the gap between East and West by promoting contacts between intellectuals. At Florence (Italy, May–June 1950, fifth session of the General Conference), the United States prevented both: the Director-General was denied the budget he hoped for, and UNESCO maintained a bias towards the West. Membership was denied to the People's Republic of China, and the Organization voiced its support for American intervention in Korea. A major conference between Eastern and Western intellectuals was refused. Torres Bodet resigned, then changed his mind and stayed two more years as Director-General. He again resigned, this time for good, in November 1952 (seventh session of the General Conference, Paris).

Within the scientific field, dire changes took place at Florence. The place reserved for science was reduced to only one out of the ten programme priorities. Needham's 'periphery principle' – which favoured the inclusion of developing nations in the scientific progress pioneered by advanced countries – was undermined by the launching of the European Organization for Nuclear Research (CERN) project. Finally, the World Federation of Scientific Workers was judged pro-Communist and scratched from the list of non-governmental organizations (NGOs) benefiting from an official relationship with UNESCO.

Despite resistance from Anglo-Saxon countries, Torres Bodet, during his tenure, succeeded in imposing the idea that scientific cooperation undertaken by UNESCO should help countries rather than just individual scientists, notably in the formation of national policy. He reaffirmed that UNESCO should 'favour progress and the applications of science for the benefit of all'. This idea would be taken up again during the seventh session of the General Conference (Paris, 1952). There it was decided that international scientific cooperation should be based on a new type of social contract in which Western nations share with others the benefits of modern science. This represented, in effect, a renewed commitment to the periphery principle. Furthermore, the Paris Conference launched an assistance programme to underdeveloped countries for the creation of national research centres.

There were some prominent American voices calling for more aid to developing countries, if only as part of a strategy of waging the Cold War. The United States President Harry S. Truman, in his inaugural address on 20 January 1949, proposed his Point Four foreign-aid programme, which was approved by the US Congress in June 1950. The programme called for Technical Assistance in the effort to improve living standards in underdeveloped countries. Funds administered by several US agencies and the United Nations were used to provide industrial and agricultural equipment, as well as to teach useful skills to people in need. Despite its altruistic dimension, Technical Assistance was also intended as a bulwark against Communism.

UNESCO struggled to find its place within the context of the dramatic changes that scientific research was undergoing during the early 1950s. This research was now completely different from that preceding UNESCO's creation. With the advent of the Cold War, powerful countries invested massively in research and nationally organized development. 'Big Science', especially in physics, became the maxim. The armed forces dominated in sensitive areas. Many scientists worked closely with their governments and were considered a purely national source of 'wealth'. A large part of research was therefore excluded from international exchange and suffered from limited circulation of individuals and results. The Cold War inevitably complicated the relationships between scientists and governments and diminished the role of international organizations.

CONSOLIDATION, 1954–1965

The eighth session of the General Conference (Montevideo, Uruguay, 1954) marked the beginning of UNESCO's consolidation phase, and the gradual thawing of Cold War hostilities. The most important indicator of change was the conference on the peaceful uses of atomic energy (Geneva, Switzerland, August 1955). The Union of Soviet Socialist Republics (USSR) finally became a Member State in 1954, and the Russian Victor A. Kovda replaced Pierre Auger as Director of the Natural Sciences Department in January 1959. International scientific cooperation was revived by this détente. There was also a sense of peaceful competition, exemplified by Polar expeditions and the first International Geophysical Year (1957–1958).

In 1954, the scientific programme's objective to improve living conditions for humankind was reconfirmed and was divided into four equal budget chapters funding international scientific cooperation, contributions to research, the teaching and diffusion of science, and the spread of UNESCO's Regional Offices for Science and Technology (ROSTs). UNESCO's determination to encourage intergovernmental cooperation was reiterated at its tenth session of the General Conference (Paris, 1958), which contributed to the participation of newly independent countries in international scientific cooperation during the 1960s.

The success of the Arid Zones project led to a proliferation of new ventures during UNESCO's second decade of existence. A similar project for Humid Tropics was launched in 1955, as was a consultative committee on marine sciences. In 1960, the Intergovernmental Oceanographic Commission (IOC) of UNESCO was created. In 1961, the International Computation Centre (ICC, in Rome, Italy) became, at last, operational. That same year, the FAO/UNESCO project for a world soil map was devised. And in 1965, the UNESCO magazine *Nature and Resources* was launched, and the International Hydrological Decade began.

The end of colonialism represented a major turning point for science at UNESCO. The sector benefited from a notable budget increase and greater involvement in policy issues. A United Nations conference (Geneva, February 1963) was organized around the 'Application of Science and Technology for the Benefit of the Less Developed Areas'. The thirteenth session of the General Conference (Paris, 1964) then decided to raise science to the same high-priority level as education. During the second half of the 1960s, posts in the Natural

Sciences Sector doubled, and the portion of UNESCO's budget allocated for science rose, from an average of less than 10 per cent, to 15 per cent. UNESCO organized a first regional Conference on the Applications of Science and Technology (CAST) in Chile in 1965. More conferences followed in other parts of the world with the help of the ROSTs. The first volume in the series Science Policy Studies and Documents was published in June 1965.

TOWARDS THE NEW CENTURY

By the mid-1960s, UNESCO had mastered its own original approach to international scientific cooperation. The importance accorded to the environment, the emphasis placed on the social aspects of science, and the priority given to developing countries gave science in UNESCO its particular culture and identity. In the closing decades of the twentieth century, UNESCO's scientists were well prepared for the challenges ahead.

BIBLIOGRAPHY

Baker, F. W. G. 1986. *ICSU-UNESCO: Forty Years of Cooperation*. Paris, ICSU.
Florkin, M. 1956. Dix ans de sciences à l'UNESCO [Ten years of science at UNESCO], *Impact of Science on Society*, Vol. VII, No. 3, pp. 133–59.
King, A. 1953. La Coopération scientifique internationale: ses possibilités et ses limites [International scientific cooperation: its possibilities and limits], *Impact of Science on Society*, Vol. IV, No. 4, pp. 195–231.

CONCEPTION AND CREATION OF UNESCO
AN UNSUNG ANCESTOR
The International Institute of Intellectual Cooperation (IIIC) and science
Jean-Jacques Renoliet[3]

Although the Pact of the League of Nations did not provide for the creation of a technical body for intellectual cooperation, the League established the International Committee on Intellectual Cooperation (ICIC), a political body, in 1922. The Intellectual Cooperation Organization (ICO), which covered all of the League's intellectual activities, was active from 1922 to 1946, backed by the International Institute of Intellectual Cooperation (IIIC). The Institute's activities in the field of the exact and natural sciences, preceding those of UNESCO, are reviewed by Jean-Jacques Renoliet.

THE NATURAL SCIENCES

IN the field of the exact and natural sciences, the IIIC conducted research and worked in cooperation with international scientific organizations. The Institute published a *Bulletin on International Scientific Relations* and studied various questions including the conservation of manuscripts and printed matter, the standardization of scientific terminology, the coordination of scientific bibliographies, and collaboration among science museums (which was the subject of a publication). Starting in 1931, the Institute devoted all its efforts to the conclusion of a cooperation agreement with the various international scientific Unions (notably physics and chemistry unions) and their International Council. Following the convening of several committees of experts by the Institute, an agreement, signed in July 1937,[4] established the Council of Unions as a consultative organ of the ICO for scientific matters: the Council was to consult the

---

[3] Jean-Jacques Renoliet is the author of *L'UNESCO oubliée: La Société des Nations et la coopération intellectuelle (1919–1946)* [The Forgotten UNESCO: the League of Nations and intellectual cooperation], from which this excerpt is extracted and translated (Paris, Publications de la Sorbonne, p. 309).
[4] UNESCO CICI C.327.M220.1937.XII (Appendix 4), p. 503, 510.

Organization on all international questions relating to the organization of scientific work, while the Institute would provide the secretariat of the commissions set up by the Council. In the framework of the agreement, the Institute held, between 1937 and 1939, study meetings bringing together scientists for the purpose of studying a precise theme, some of which led to publications. The Institute also worked to ensure the dissemination of scientific work to the public at large. To that end, in February 1939, it convened a committee of scientists, which recommended the creation of an international centre for scientific documentation and dissemination. However, despite the financial backing of the Rockefeller Foundation, the implementation of the project was hindered by the war.

HOW THE 'S' CAME TO BE IN UNESCO
Gail Archibald[5]

IT was only on the sixth day of the United Nations Conference for the Establishment of an Educational and Cultural Organization, in 1945, that the reference to science was added to the new organization's name.[6] It had taken nearly three years for the 'S' to make its way into the acronym of an intergovernmental organization that was initially conceived as focusing on education, and then on culture and education. The following is a brief account of how science came to be part and parcel of UNESCO.

During the 1920s, international scientific cooperation had been rekindled with the restoration of peace after the First World War. The League of Nations' International Institute of Intellectual Cooperation, founded in Paris in 1925, included a section devoted to Scientific Information and Scientific Relations. Among the activities of the International Bureau of Education, established in Geneva (Switzerland) the same year, was scientific research. In the non-governmental sector, the International Council of Scientific Unions (ICSU) would be founded in Brussels (Belgium) in 1931.

Then the Second World War broke out. By 1943, however, Allied victories encouraged politicians to turn their attention to post-war planning. On both sides of the Atlantic, non-governmental projects proposed the creation of an international organization for education. The same year, Joseph Needham

---
5   A UNESCO staff member since 1981, Gail Archibald is the author of *Les Etats-Unis et l'UNESCO 1944–1963* [The United States and UNESCO, 1944–1963] (Paris, Publications de la Sorbonne, 1993).
6   Before it was agreed to add 'Scientific' to the name, the original proposal was for a 'United Nations Educational and Cultural Organization' (thus, 'UNECO').

launched a campaign from China to develop post-war international scientific cooperation in the form of a World Science Cooperation Service. A biochemist at Cambridge University in the UK (1920–42), Needham was also a socialist. In 1937, he had received a cultural shock when three Chinese students arrived in Cambridge to work with him and his wife. He found their company exhilarating, learned Chinese and, much later on, married one of them.

The British Government sent Needham to China in February 1943 as a representative of the Royal Society to consolidate Anglo–Chinese cultural and scientific relations. In December of the same year, Needham wrote to China's Foreign Minister elaborating his idea of international scientific cooperation: 'The time has gone by when enough can be done by scientists working as individuals or even in groups organized as universities, within individual countries … Science and technology are now playing, and will increasingly play, so predominant a part in human civilization that some means whereby science can effectively transcend national boundaries is urgently necessary'. Needham's immediate goal was the transfer of advanced basic and applied science from highly industrialized Western countries to the less industrialized ones, 'but', he assured, 'there would be plenty of scope for traffic in the opposite direction too'.

Meanwhile, in London (UK), at the Conference of Allied Ministers of Education (CAME), representatives – many exiles from Nazi-occupied countries – met between 1942 and 1945 to discuss and plan post-war educational reconstruction. Among the conference's various activities was the creation of the Commission on Scientific and Laboratory Equipment. During the war, the Nazis had sabotaged scientific laboratories, and ransacked and closed down universities and institutions, in an effort to halt scientific activity in occupied countries. The commission took charge of assessing these post-war reconstruction needs and the appropriate measures to meet them.

In a declaration to the press in March 1944, the US Secretary of State Cordell Hull explained the rationale for US participation in emergency educational and cultural reconstruction of war-torn countries: 'Teachers, students and scientists have been singled out for special persecution. Many have been imprisoned, deported or killed, particularly those refusing to collaborate with the enemy. In fact, the enemy is deliberately depriving the victims of those tools of intellectual life without which their recovery is impossible'.

A month later, in London, an American delegation presented the CAME with a 'Suggestion for the development of the CAME into the United Nations Organization for Educational and Cultural Reconstruction'. A modified version of this text referred to 'science' four times. Reparation of damage through theft of scientific apparatus was mentioned twice and the restoration of scientific laboratories once, as was 'including scientific research' in the 'interchange between nations bearing upon educational and cultural problems'.

Joseph Needham pursued his campaign in China by sending out the first of three memoranda to scientists, politicians and diplomats in Allied countries on the creation of an International Science Cooperation Service (July 1944). He explained that the Service would have permanent representatives in all countries or regions, advise governments and assist international organizations on scientific matters. After discussions with colleagues in the British Council and Royal Society, he sent out his second memorandum from London, 'Measures for the organization of international cooperation in science in the post-war period'.

On a journey to Washington DC (USA) in February 1945, Needham was astonished to find that one of the main topics of conversation was the creation of an organization for culture and education. Surprised at how far the project had come, he concluded that it would be more reasonable to incorporate scientific cooperation into this organization, on the condition that the word 'science' be included in its name. Needham's influence could be seen in the March version of the American project, which contained multiple references to scientific cooperation as a contribution to peace and security. However, 'science' was still missing from the name, which remained 'the International Organization for Education and Cultural Cooperation'. This new project was presented to the CAME in April 1945, whose drafting committee turned the American project into a CAME document. Responding to the suggestion that 'science' be included in the organization's name, a member of the US delegation explained that, for the American public, the word 'culture' covered 'science'.

Needham sent a third memorandum from China in April 1945 to important scientific officers in several Allied countries. He insisted that, if they wanted scientists to be interested and involved in the organization, it must be evident that the organization was interested in them. Needham also requested that 'science' include applied sciences, in other words technology, which the word 'culture' would not cover. For Needham, the new organization's principal role would be to promote exchanges between industrially advanced countries – which he called 'the bright zone' – and the less advanced ones, nations 'on the periphery'. Needham supposed that the organization would not transfer commercial secrets from technologically advanced countries to less developed ones, but rather encourage industries to introduce the use of new technologies in the 'periphery.'

Delegations to the San Francisco Conference (which elaborated the United Nations Charter, from 25 April to 26 June 1945, in San Francisco, USA) agreed upon a French recommendation to convene a conference to establish an international organization of intellectual cooperation. At San Francisco, the American astronomer Harlow Shapley was for including 'science' in the name of the proposed organization, but other members of the American delegation felt this would make for a wordy name.

For Joseph Needham, it was a June 1945 trip from Tehran (Iran) to Moscow (USSR) that was the turning point. Tehran airport turned out to be the meeting place for national scientific delegations on their way to Moscow to celebrate the 220th anniversary of the Russian Academy of Sciences. Needham's third memorandum was distributed to the American, Indian and Chinese delegations, all of whom displayed strong interest. Other delegations were given copies in Moscow; only the Soviet delegation proved unresponsive. The American scientists promised to undertake an important campaign to push Needham's point of view.

UNESCO's first General Conference, Paris, 1946. From left to right: Jean Thomas (profile); Julian Huxley, Director-General of the Organization (standing); Léon Blum, President of the General Conference (centre).

The United Nations Conference for the Establishment of an Educational and Cultural Organization was held in London from 1 to 16 November 1945. Ellen Wilkinson, British minister of education and president of the conference, announced in a plenary session that, although 'science' was not part of the original title of the organization, the British would put forward a proposal for it to be included. 'In these days', said Wilkinson, 'when we are all wondering, perhaps apprehensively, what scientists will do to us next, it is important that they should be linked closely with the humanities and should feel that they have a responsibility to mankind for the results of their labour'.

On 5 November, the conference divided itself into Commissions. The First Commission was charged with drafting the Title, Preamble and Aims and Functions of the new organization. It was the American delegate who proposed that it be called the United Nations Educational, Scientific and Cultural Organization. After hesitating for twenty-four hours, the commission decided in favour of the UNESCO title, which simultaneously served as an instruction to insert the word 'science' in the text of the Constitution wherever indicated. For example, 'The purpose of the Organization is to contribute to peace and security by promoting collaboration among the nations through education, science and culture'.

In its concluding report, the First Commission felt it necessary to explain that the inclusion of 'scientific' in the title and elsewhere in the text implied inclusion in the Organization's activities of the philosophy of science and not its applications (science as touching on military security would be dealt with by the disarmament conference). It was vital that scientists be in touch with those who saw the world in 'human' terms. On the afternoon of 16 November 1945, the heads of thirty-seven delegations signed UNESCO's Constitution.

That UNECO should have become UNESCO is proof that the need for such an organization was greater than any mistrust prevailing at the time. The delay in including science in the Organization's mandate, on the other hand, underlines the multiple, delicate and difficult relationships between science and governments in those turbulent years. This in turn would influence the various definitions attributed to the term 'international scientific cooperation'. But that is another story.

BRAVE NEW ORGANIZATION
Julian Huxley's philosophy
John Toye and Richard Toye[7]

JULIAN Huxley (1887–1975) served as the first Director-General of UNESCO from 1946 to 1948. Feeling the need to clarify his ideas about the Organization's role, he took two weeks to write a substantial pamphlet. This was published on 15 September 1946, with the title *UNESCO: Its Purpose and Its Philosophy* (Huxley, 1946).[8]

---

7   John Toye, a professor at the University of Oxford (UK), is former Director of the Globalization Division of the UN Conference on Trade and Development (UNCTAD) (1998–2000).
Richard Toye is a lecturer in History at Homerton College, University of Cambridge (UK).
8   Published in book form the following year (Huxley, 1947).

PART I: SETTING THE SCENE, 1945-1965

The eminent zoologist and popularizer of science, Julian Huxley (1887–1975), first Director-General of UNESCO, 1946–48.

Huxley wanted not only to clarify and elaborate on UNESCO's constitution, but also to provide the Organization with 'a working philosophy ... concerning human existence' (ibid., p. 6), to guide its approach to the issues with which it had to deal. He noted that UNESCO was clearly debarred from endorsing the viewpoint of any one of the world's religions, and from espousing capitalism or Marxism, or indeed any other political, social, economic and spiritual approach that he called 'sectarian'. However, Huxley deduced from UNESCO's concern with peace, security and human welfare that 'its outlook must, it seems, be based on some form of humanism' (p. 7). Moreover, this humanism needed to be 'scientific' but not 'materialistic'; and furthermore, 'it must be an evolutionary as opposed to a static or ideal humanism' (p. 7). It is striking that Huxley does not seem to have appreciated, at the point he was writing, that such an approach was likely to be problematic for many Member States – almost as much so, in fact, as the numerous philosophies that he recognized it was politically impossible for UNESCO to adopt.

Huxley viewed the term 'evolution', in its broadest sense, as denoting 'all the historical processes of change and development at work in the universe' (p. 8). He believed that humankind could guide these processes consciously, to achieve further world progress. As human societies could benefit from 'cumulative tradition' or 'social heredity', natural selection was being replaced by conscious selection as the motor of evolution, the possible rate of which was therefore 'enormously speeded up' (p. 9). Huxley believed that UNESCO had a significant role to play in 'constructing a unified pool of tradition' for the human species.

Manifestly, his conception of the Organization's purpose was extremely bold and ambitious. He argued that 'the more united man's tradition becomes, the more rapid will be the possibility of progress'; and that 'the best and only certain way of securing this will be through political unification' (p. 13). While conceding that such an ideal was remote and that it fell outside the field of the Organization's competence, Huxley argued that there was much that UNESCO could do to lay the foundations of world political unity.

He presented examples of activities that UNESCO could undertake to achieve this, in the fields of education, the natural and social sciences, and culture and the arts. The topics that he championed were widely esteemed in the 1940s, and many had a basis in psychology. They included the classification of psycho-physical types, IQ testing, applied psychoanalysis, human resources planning, parapsychology, yoga and the history of the rise of individuality. Perhaps most significantly for the pamphlet's reception, it emphasized the need for UNESCO to promote population control and the study of 'the eugenic problem' (pp. 10, 12, 21, 37–8 and 45).

Understandably, Huxley's ideas proved controversial. The pamphlet had already been presented to UNESCO's Preparatory Commission and been ordered to be printed as an official document, when Sir Ernest Barker, one of the commission's members, took exception to it. Barker was an historian, political theorist, and convinced Anglican. According to Huxley, 'he argued forcibly against UNESCO's adopting what he called an atheist attitude disguised as humanism' (Huxley, 1978). The commission's Executive Committee was therefore agreed that, when the document was circulated, a slip of paper should be inserted into it, saying that the essay was a statement of Huxley's 'personal attitude', and that it was 'in no way an official expression of the views of the Preparatory Commission'.[9]

In his memoirs, Huxley conceded that Barker had been right to object. 'Though UNESCO has in fact pursued humanistic aims, it would have

---

9   UNESCO Document Misc./72, 6 December 1946.

been unfortunate to lay down any doctrine as a basis for its work', he wrote. 'Further, a purely humanist tone would have antagonized the world's major religious groups, including the Russians with their pseudo-religion of dialectical materialism' (Huxley, 1978). He did not admit, however, that deriving moral principles and policies from the science of evolution was in any sense a doubtful philosophical procedure.

BIBLIOGRAPHY

Huxley, J. 1946. UNESCO: Its Purpose and Its Philosophy. UNESCO/C/6. Paris, UNESCO, available at: http://unesdoc.unesco.org/images/0006/000681/068197eo.pdf
——. 1947. *UNESCO: Its Purpose and Its Philosophy*. Washington DC, Public Affairs Press.
——. 1978. *Memories II*. Harmondsworth, Middlesex, UK, Penguin, p. 12.

BLAZING THE TRAIL
Needham and UNESCO: perspectives and realizations

Patrick Petitjean

THE biochemist and historian of science Joseph Needham served a brief tenure as the first Head of the Natural Sciences Section – a mere two years. However, with the support of the first Director-General, Julian Huxley, Needham largely defined the role of 'science' in UNESCO, guaranteeing that his influence on the Organization would endure far beyond his limited mandate. In three important memoranda composed during the Second World War, Needham outlined an ambitious project for international scientific cooperation. They were to become the basis of the first scientific programme that he proposed in June 1946 and which was presented to the first session of the General Conference (Paris, November 1946).

Born in London, England, in 1900, Needham studied medicine and biochemistry at Cambridge University (UK), but he also had a keen interest in religion and philosophy. His political commitment was forged during the Great Depression. The massive unemployment resulting from the economic crisis that began in 1929 led many people to criticize the role of science and its applications to industry. It also brought about a reduction in both finances and employment within the field of scientific research. Needham joined the International Council of Scientific Unions (ICSU) and, throughout the 1930s, benefited from his experiences with 'movements for social relations in science'.[10] Needham was part of an idealistic generation of scientists who wanted to use discoveries and their applications to improve living conditions for all and to develop democracy.

The war did not interrupt this commitment; quite the contrary. Needham, like most of his peers, was horrified by the way the Nazis deformed and used science to justify the racist ideology that led to the Holocaust. In the democratic countries, many scientists, including Germans in exile, participated directly in the struggle against Nazism. Even during the war, several conferences were

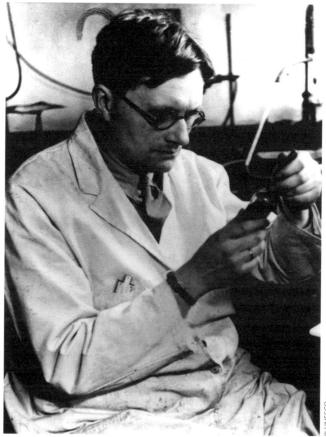

Joseph Needham (1900–1995), biochemist, Head of Natural Sciences Section, 1946–48.

---

10   Including the 'Division for the Social and International Relations of Science' of the British Association for the Advancement of Science (BAAS), the 'Committee on Science and its Social Relations' of ICSU, the British Association of Scientific Workers (AScW) and other similar groups. The reference book for these groups was *The Social Function of Science*, by John Desmond Bernal (1939); 'science will come to be recognized as the chief factor in fundamental social change' was their leitmotiv.

organized in London by the British Association for the Advancement of Science (BAAS) and the British Association of Scientific Workers (AScW) to discuss the post-war role of science. Participants were determined that science and its applications be used for the well-being of all. The importance of international scientific cooperation would be paramount. In February 1945, several foreign delegations took part in the 'Science for Peace' conference, during which the creation of international scientific associations was notably discussed. From 1946 onward, these same scientists quite naturally met up again at UNESCO, ICSU or the World Federation of Scientific Workers to put into practice their ideas and projects.

NEEDHAM'S ORIENTATIONS

In 1942, Needham, who had learned Chinese, travelled to China to head the Sino-British Science Cooperation Office, one of several scientific liaison offices that were created during the war. The Anglo-Chinese bureau, according to Needham, devoted one-third of its activities to 'war science', another third to 'pure science' and the final third to scientific applications for agriculture and industry. In spite of the war, the office exchanged a large amount of equipment, information and research with the West. In April 1946, Huxley summoned Needham back from China to join the UNESCO Secretariat.

Needham's ideas for UNESCO were inspired by both his war and his peacetime experiences. He was familiar with both scientific unions, which covered one subject and several countries, and scientific liaison offices, which covered all subjects but were bilateral. The scientific unions Needham had joined in peacetime were independent bodies, but – often lacking financial and administrative resources – they could be inefficient. The scientific liaison offices he had come to know during the war were better financed but were subject from time to time to bureaucratic controls. Thus, he concluded:

> What we need today is fundamentally a system which will combine the methods which have spontaneously grown up for assuring international relations in time of peace, with those which the nations have had to work out under the stress of war. None of the machinery ought to be scrapped. The problem is to weld it into a satisfactory functioning system (Needham, 1946, p. 6).

One of Needham's most original concepts was the 'periphery principle'. He believed that the most scientifically advanced nations must share their knowledge and resources with less developed countries – that is, countries 'on the periphery' – in order to reduce disparities between the different regions of the world. This principle was Needham's personal brainchild and represented a radical break from the past. At the creation of UNESCO, the majority of scientists were Eurocentric and did not think this way. Needham criticized 'the parochial theory of the "laissez faire" school', according to which everybody in the scientific world knew each other and therefore projects got done spontaneously; he pointed out that 'the picture of world science looks very different when seen from Romania, Peru, Java, Iran or China' (ibid., p. 7–8).

Needham believed that the 'social function of sciences' had to be part and parcel of UNESCO's science programmes. The Organization would need to address the history of science, scientific education and the social consequences of scientific development. Behind Needham's thinking was the idea of the universality of science and its subsequent internationalism. In his report to UNESCO's Preparatory Commission, July 1946, Needham defined the aims of the Natural Sciences Section:

> UNESCO is an agency for peace through active international cooperation. In the field of scientific cooperation and service, we have one of the immediately effective means of accomplishing this. This is partly because scientific research is essentially and traditionally international and cooperative, and also because the applications of scientific knowledge to human welfare, if properly made, can be one of the most effective methods of removing some of the causes of war (UNESCO, 1946).

CONCLUSION

Joseph Needham met with two main difficulties in trying to realize his objectives. Little by little, UNESCO became a hostage of the Cold War. Even though the Soviet Union only joined the Organization in 1954, political bickering among the principal contributors (France, the UK and the USA) stymied commitments and projects formulated at UNESCO's founding. As early as 1947, budgets were limited, for science as for all UNESCO programmes.

Moreover, Needham and his left-wing friends were very marginal in the scientific world. UNESCO's support for ICSU was unanimously endorsed, but this was not the case for the 'periphery principle', nor for Needham's ideas about the social relations of science. Support for underdeveloped countries by scientists did not really get off the ground until the massive decolonization of the 1960s.

Pierre Auger, a French physicist, replaced Needham in April 1948, and Julian Huxley quit the post of Director-General at the end of the same year. The political climate led UNESCO (as well as the rest of the United Nations specialized agencies) towards a system of cooperation based on Technical

Assistance, an objective announced by US President Harry S. Truman in his January 1949 inaugural address. Needham's idealistic aims were replaced by a more utilitarian concept of the 'social and international functions of science', based on a Western liberal model for the economic development of societies. However, the 'periphery principle' had pointed UNESCO in a direction that it would later resume and continues to follow.

After leaving UNESCO, Needham expressed his bitterness concerning scientific colleagues from the 'bright zone' of developed nations:

> I am frankly rather tired of the people who sit in their laboratories and never give a thought for their colleagues at the other end of the world who are working in difficult conditions and even desperate need. If they were to travel about the world and visit the places which are really remote, those are the conditions they will find. There must be an end of parochialism among scientific men themselves (Needham, 1949, p. 29).

Joseph Needham lived until 1995, long enough to see that much of his original vision for UNESCO came to be realized by the Organization.

BIBLIOGRAPHY

Malina, F. J. 1950. L'UNESCO et les sciences exactes et naturelles: bilan de trois ans d'efforts [UNESCO and the exact and natural sciences: record after three years of effort]. *Bulletin of the Atomic Scientists*, Vol. VI, No. 4.

Needham, J. 1946. Science and UNESCO: international scientific cooperation – tasks and functions of the secretariat's division of natural sciences. UNESCO/Prep.Com./Nat.Sci.Com./12. Paris, UNESCO.

——. 1949. *Science and International Relations* (Fiftieth Robert Boyle Lecture, Oxford, 1 June 1948). Oxford, UK, Blackwell Scientific Publications.

UNESCO. 1946. UNESCO/Prepcom/51, 3 July 1946. Paris, UNESCO.

INTERNATIONAL SCIENTIFIC COOPERATION
FINDING A FOOTING
The sciences within the United Nations system
Patrick Petitjean

THE San Francisco Conference adopted the Charter of the United Nations on 26 June 1945.[11] While science was not explicitly mentioned in the Charter, the vast programme of international action in the economic, social, cultural and humanitarian fields (Articles 13 and 57 of the Charter) encompassed nearly the entire field of human activity, thus including science and its applications. The implementation of the programme was entrusted to the main bodies of the United Nations, notably the Economic and Social Council (ECOSOC), and to the different specialized agencies. The Secretariat was responsible for conducting studies and certain operations.

The nuclear attacks on Hiroshima and Nagasaki (Japan) occurred after the San Francisco Conference but before the London Conference (UK, November 1945), which established UNESCO. Hiroshima had made governments more aware of the social and political implications of science. One consequence would be the reference to science in the name 'UNESCO'. UNESCO and ECOSOC would therefore have joint competence in that domain.

The sharing of competencies was complicated by differing cultural approaches to science. The French traditionally viewed science as an intellectual activity (pure science) separate from its applications, which were seen as being more closely tied to the economic sphere. The Anglo-Saxon tradition took the opposite view. Thus, for example, Joseph Needham considered that UNESCO should deal with science as a whole, even if it meant overlapping with the Food and Agriculture Organization (FAO) and the World Health Organization (WHO), and limiting ECOSOC to what could be called a general orientation role. In contrast, Henri Laugier, in line with his responsibilities as deputy secretary-general of the United Nations in charge of ECOSOC, wished to give ECOSOC an active role in scientific development, while confining UNESCO to basic science.

---

11   The United Nations Conference on International Organization, which established the United Nations, met in San Francisco, California (USA), from 25 April to 26 June 1945.

The United Nations had a more explicitly political role since it was composed of career diplomats appointed by their governments. The United Nations was the first choice of the Union of Soviet Socialist Republics (USSR), which refused to join UNESCO on the grounds that the United Nations and its bodies (the Security Council, where it had a veto, and ECOSOC) were sufficient. From ECOSOC's point of view, UNESCO was above all a bridge to the scientific communities and non-governmental organizations (NGOs).

THE EARLY INITIATIVES

The first three initiatives taken by the United Nations in the field of science illustrate the initial division of responsibilities: the United Nations Atomic Energy Commission (UNAEC), the United Nations Scientific Conference on Conservation and Utilization of Resources (UNSCCUR), and the international scientific laboratories dossier. It would take more than ten years to create a structure – the Scientific Advisory Committee (SAC), reporting directly to the secretary-general – that was responsible for harmonizing the United Nations' efforts in the field of science. More ambitious in scope, the United Nations Office for Science and Technology (OST) was established at the end of the 1960s.

In January 1946, the United Nations General Assembly decided to create the United Nations Atomic Energy Commission, entrusting direct responsibility for it to the Security Council. The entire field of nuclear energy and its applications was therefore removed for several years from the scope of UNESCO and ECOSOC. The aim of the Atomic Energy Commission was to seek ways of controlling existing nuclear weapons and preventing the further development of such weapons. Confronted with contradictory proposals with regard to arms control, the commission found itself at an impasse after only several months. The Soviet's first nuclear test explosion, on 29 August 1949, made the commission obsolete, and the Cold War prevented new negotiations for many years. The commission was formally dissolved in 1952 by the United Nations General Assembly.

In May 1946, on behalf of ECOSOC, Henri Laugier proposed the creation of an international body to coordinate research and the development of international scientific laboratories. The matter would be handled by the studies and research division of the Department of Social Affairs. This decision led to several months of debate between UNESCO and ECOSOC, with regard to their mutual competencies. In 1950, ECOSOC entrusted the entire dossier to UNESCO.

The United Nations Scientific Conference on Conservation and Utilization of Resources, inspired by an idea of US President Franklin D. Roosevelt's, had been officially proposed by the United States in September 1946 at a meeting of ECOSOC, even before UNESCO was created. The difficulties in obtaining raw materials during the war had revealed the crucial nature of these problems; and post-war reconstruction and development encountered the same difficulties. The economic affairs department of ECOSOC was mandated to organize the conference. After consultation with governments, and exclusion of nuclear issues from the scope of the conference, the project was definitively adopted by ECOSOC in March 1947. The conference was held in August 1949. The United States had insisted that it should be very technical in nature: a meeting of experts that would be devoted 'solely'[12] to an exchange of findings and information in those fields, and to the determination of the economic costs and benefits of the different technical possibilities, but without going so far as to make recommendations to governments.

The Food and Agricultural Organization (FAO) was the principal agency concerned and participated actively in the conference. UNESCO was more reluctant because it approached the problems rather differently: Julian Huxley favoured an approach emphasizing the protection of nature. The second session of the General Conference, held in Mexico City (Mexico) in November 1947, therefore made a twofold decision: (1) to participate directly in the UNSCCUR conference and organize regional conferences for the protection of nature, sponsored by UNESCO, including one at Fontainebleau, France (September 1948) that would give rise to the International Union for the Protection of Nature (IUPN); and (2) to convene a technical meeting for the purposes of synthesis, to be held at the same time as the UNSCCUR, while emphasizing the cultural and scientific dimensions and social implications rather than the economic dimension underlined by ECOSOC.

SUBSEQUENT CONFERENCES

With the era of détente, nuclear scientific cooperation was back on the international agenda (for instance, the 'Atoms for Peace' speech by US President Dwight D. Eisenhower to the General Assembly of the United Nations in 1953). A plan for creating a special agency for the peaceful applications of nuclear power took shape in 1955. Early in 1956, Brazil, Czechoslovakia, India and the USSR joined with the Western countries to form a preparatory committee. Chaired by Homi Bhabha, the first United Nations Conference on the Peaceful Uses of Atomic Energy was held in Geneva (Switzerland) in August 1955. More than 1,500 delegates participated in the conference, which represented the first international scientific and technical exchanges in the field in more than

---

12   The term 'solely' appears in Resolution 32(IV) of ECOSOC (28 March 1947), as does a refusal to draw up 'international agreements' and to 'lay down political principles'.

fifteen years. Those exchanges confirmed that scientists from the two blocs had independently arrived at the same level of knowledge. The International Atomic Energy Agency (IAEA) was formally established on 29 July 1957, following the ratification by twenty-six states of the convention creating it. The Agency was initially devoted solely to peaceful uses of nuclear energy.

The development of Technical Assistance at the beginning of the 1950s (US President Harry S. Truman's 'Point Four' plan) ensured at one and the same time the United Nations' role in the definition of scientific policy, and the primacy of economics over science. It placed the question of the use of science and technology for development on the United Nations agenda, which became the second major area (together with nuclear energy) of direct intervention by the United Nations in scientific fields.

This led directly to the convening of the first United Nations Conference on the Application of Science and Technology in Developing Countries (UNCAST),[13] which was held in Geneva in February 1963, in the early days of decolonization. The conference was organized by the Scientific Advisory Committee (SAC), whose secretary-general was the Brazilian biochemist Carlos Chagas, chairman (since 1956) of the United Nations Scientific Committee on the Effects of Atomic Radiation. UNESCO actively participated in the preparations for the conference and in the conference itself. One of the intellectual bases of the conference was the report on 'Current Trends in Scientific Research', prepared for UNESCO by Pierre Auger, former Director of the Natural Sciences Department.

It was a conference involving 1,665 participants, 96 governments, 1,839 communications and 250 documentary films ... but with only 16 per cent of participants from the developing countries. Like UNSCCUR, UNCAST was not supposed to make recommendations or take decisions, but rather to assess and summarize knowledge. The conference nevertheless called for the establishment of national research and technology systems and, simultaneously, for planning scientific development. Robert Oppenheimer, one of the participants, called the conference a pointless exercise, whereas Carlos Chagas considered that it had successfully launched several initiatives. Assessments of the conference were therefore mixed (see Chagas Filho, 2000).

One outcome of the conference was the establishment of the Advisory Committee for the Application of Science and Technology (ACAST), which gave the United Nations greater responsibility for international scientific cooperation through coordination of the various agencies. In the division of labour, UNESCO was given the task of developing national science and technology policies and

---

13   This paragraph is based on a working document by K.-H. Standke, 'The interaction between the United Nations and UNESCO in the field of science and technology: An account'.

of organizing regional conferences in that field. The Advisory Committee was asked to draw up a 'programme of international cooperation in science and technology in which the developed and developing countries could join in a drive on problems of importance for the developing countries'. The programme was published in 1971. The major scientific conferences initiated by the United Nations became widespread beginning in the 1970s.

BIBLIOGRAPHY

Chagas Filho, C. 2000. *Um Aprendiz de Ciencia* [A science apprentice]. Rio de Janeiro, Brazil, Editora Nova Fronteira e Editora Fiocruz.

Standke, K.-H. The interaction between the United Nations and UNESCO in the field of science and technology: an account. Working document.

GIVING SCIENCE FOR PEACE A CHANCE
The post-war international laboratory projects
Patrick Petitjean

AFTER the devastation of the Second World War, many people looked to science as an antidote to chaos. It was hoped that scientific advances would usher in an era of peace and social progress. In the most idealistic vision, science would help eradicate war and Fascism. International scientific cooperation became an ideological and political objective. The peaceful and rational application of science would allow new challenges to be met: such as combating hunger in the world, tackling the population boom, the curing of diseases, battling desertification, conserving natural resources, and improving living conditions in the tropics and at high altitudes. All these challenges were dependent on scientific progress.

Many international initiatives were aimed in this hopeful direction in 1945 and 1946: the creation of the Atomic Energy Commission under the United Nations Security Council; the inclusion of the 'S' in UNESCO, with a Natural Sciences Section; and an American proposal for a 'Conference on the Conservation and Use of Natural Resources'.

Following a decision at the San Francisco Conference (USA, April–June 1945), a United Nations Economic and Social Council (ECOSOC) was created in January 1946,[14] composed of eighteen countries. Its function was defined by Article 62 of the Charter of the United Nations: 'The Economic and Social Council may make or initiate studies and reports with respect to international economic, social, cultural, educational, health, and related matters and may make recommendations with respect to any such matters to the General Assembly, to the Members of the United Nations, and to the Specialized Agencies concerned.'

ECOSOC was explicitly required to coordinate the activities of the different agencies. Two assistant secretaries-general of the United Nations were put in charge of ECOSOC, one responsible for economic affairs, and the other for social affairs. The latter, Henri Laugier,[15] was responsible for, among other things: human rights, the fight against drugs, the right to work, child protection, women's rights, education, culture, health and science. Laugier surrounded himself with an 'intellectual secretariat' made up of high-level scientists.[16] With this arrangement, he hoped that the United Nations would steer all aspects – from the debates to the coordination – of scientific research at the international level. His objective was to create what would, in essence, be an international equivalent of France's Centre National de la Recherche Scientifique (CNRS), with laboratories and an International Research Council (IRC).

THE INTERNATIONAL LABORATORY PROJECTS OF ECOSOC

In June 1946, Laugier presented his proposals for the United Nations to set up international research laboratories, speaking for the first time about establishing an IRC. On 19 June 1946, the *New York Times* devoted a front-page article to the United Nations programme on science. Its first sentence declared: 'The UN's Secretariat is ready to marshal the world's scientists for peace as they were for war'. The article explained that many of the 'most famous' scientists had already been consulted. In peacetime, research should focus above all on tuberculosis, cancer, soil erosion, urbanization and astronomy; social problems should also be treated as a priority. According to Laugier, the 'dream of many scientists' was the existence of an international research authority, linked to the United Nations, for the purpose of solving social problems. In his opinion, the coordination of international research should come directly within the purview of the United Nations, and not only of UNESCO.

---

14 First General Assembly of the United Nations.
15 Henri Laugier: French physiologist, first director (1939) of the CNRS (French National Centre for Scientific Research).
16 Jean Gottman, geographer, was the key mover behind the international laboratory projects. Alongside him worked Alfred Métraux, ethnologist, and Louis Gros, historian, among others.

Laugier's argument was based on the social and economic function of science and the weaknesses of national research systems (pre-war international cooperation was not particularly effective):

> This work of liaison, information, coordination between national scientific activities must be continued and considerably amplified by the United Nations. It will be one of the main tasks of the Department of Economic, Social and Cultural Affairs in the United Nations Secretariat and the specialized agency, UNESCO, that is being created ... However, we may think today that in the general interest of humanity, we must go further than just coordination and we may well ask whether certain areas of scientific research should not be taken directly under the wing of the central bodies of the United Nations or of the specialized agencies.

Laugier gave examples of the areas in which research could be carried out 'rationally, efficiently and selflessly'. He concluded: 'A significant amount of research in various fields would be far more effective if it was carried out with all the necessary means in perfectly equipped international laboratories, for which the United Nations would have financial and intellectual responsibility.'

A resolution was adopted by ECOSOC in October 1946, in which the prospect of an IRC had disappeared. The resolution confined itself to inviting 'the Secretariat to consult UNESCO and the other specialized agencies concerned and to submit to ECOSOC, if possible during the next session, a global report on the matter of the creation of United Nations research laboratories'.[17]

... AND THOSE OF UNESCO

At the same time, in June and July 1946, the 'science' sub-commission, and then the Preparatory Commission for UNESCO, looked into the matter. The initial draft programme put forward by Needham did not include specific proposals for laboratories, but was complemented in June with suggestions from delegations: the Amazon Hylean project (proposed by Brazil); a computation centre for applied mathematics (by France); nutrition institutes (by Brazil, France, and the USA) since food problems were global; health institutes for parasitology and immunology (by Mexico, Brazil and France); astronomical observatories (by the USA and the International Astronomical Union); and a meteorological laboratory (by the USA).

Unlike Laugier, Needham did not include in his document any idea for centralizing the laboratories: his approach was more pragmatic than political or

---

17   ECOSOC Draft Resolution E/147.

ideological. The key words used were 'to facilitate' cooperation, to 'not replace existing arrangements', to investigate whether the work proposed was not already being done in an institute. The idea was to start with relatively modest, pilot projects. The proposal for an International Institute of the Hylean Amazon (IIHA) was ideal for that purpose and was selected as one of four priorities for all of UNESCO in 1947.

Needham was very sceptical about Laugier's projects, which he feared might separate the basic sciences (for UNESCO) and applied sciences (for ECOSOC). Needham prevented ECOSOC from coordinating international research. However, he did accept the ECOSOC resolution, which stressed consultation with UNESCO.

THE NEEDHAM REPORT

On 20 February 1947, Needham submitted a report[18] underscoring two principles. First: 'the natural sciences [are] the most international of all human activities ... Race, colour, creed, or geographical location, have demonstrably nothing whatever to do with the plausibility of a hypothesis ... Scientific men understand one another at once, from whatever quarter of the world they come to meet together'. Secondly, there was the 'necessity of [a] concerted attack on the great problems of natural phenomena which still elude us.'

Needham remained very prudent regarding the selection criteria for laboratories. He rejected laboratories in the traditional university disciplines (general physics, botany, physiology, and so on) because what was being done in national laboratories was sufficient. And he did not select those in which commercial interests were at stake, since there, too, research already had sufficient support. Certain criteria were spelled out: no duplication; a problem to be examined had to have reached scientific 'maturity'; a laboratory must be installed someplace where there were problems to solve, particularly in regions that had not yet been sufficiently studied. Needham had a vision of laboratories 'without borders.'

On the basis of these principles and criteria, Needham identified nine priority areas: astronomical observatories; laboratories of nutritional science and food technology; meteorological centres and stations; applied mathematics laboratories; medical and biological research institutes; a centre for study of tropical life and resources; projects for international ornithological observatories; international oceanographic laboratories; and stockrooms for type-collections and standards.

---

18   UNESCO Document Nat.Sci./24/1947.

Of the fifteen proposals made in these fields, Needham narrowed the final priority list down to four:[19]
1. an institute for the study of the chemistry and biology of self-reproducing substances, including cancer research;
2. a chain of laboratories and field teams in nutritional science and food technology (a) in China, (b) in the arid and humid tropical zone, and (c) in the humid equatorial zone;
3. study of life and resources in the humid tropical zone, beginning with an institute of the Hylean Amazon and expanding into a chain of equatorial zone stations;
4. one or more institutes of oceanography or fisheries in Asia, their work to be correlated with that of the nutritional laboratories.

ECOSOC GIVES UP

The Needham Report, for the most part, went unheeded. For a year it was kept in a drawer, before being included as a principal contribution to a more general document[20] of the United Nations, presented to the seventh session of ECOSOC (July 1948), which referred it once more to a group of experts.

The meeting of this group of experts was held in August 1949 in Paris.[21] Needham, Levi-Strauss, Shapley and Ozorio de Almeida were invited and Auger (UNESCO) and Laugier (ECOSOC) were also present. They produced a lengthy document with three main priorities (a computation centre, a brain institute and a social sciences institute) and four secondary priorities (meteorology, arid zones, cancer and astronomy). This document too remained a dead letter: the eleventh session of ECOSOC (August 1950) once more referred the projects to a commission of experts and definitively entrusted UNESCO with the matter.

In parallel to the ECOSOC debates, UNESCO launched the creation of the IIHA at the end of 1946 and discussed the possibility of an International Institute of the Arid Zone (IIAZ). At the fourth session of UNESCO's General Conference, held in Florence (Italy, June 1950), the principle of two laboratories was decided upon: the international computation centre and – an American proposal – an international laboratory for particle physics, to become CERN

---

19 IBID. pp. 59–60. The second priorities: astronomical observatory in the Southern Hemisphere, a tuberculosis research institute, a computing laboratory, a high-altitude station in the Himalayas, laboratories for human biological and genetic analysis, an institute for the study of human evolution in Africa, an Arctic research institute, arid zone institutes, and an individual and social psychology institute.
20 ECOSOC Document E/620.
21 Minutes of the meeting of experts: ECOSOC Documents E/Conf/PC/SR1 to SR11. Final report: ECOSOC Documents E/1694 and E/1694 Add (UNESCO Archives).

(European Organization for Nuclear Research).[22] The latter was far removed in concept from the initial proposals: it would be in Europe and not a country of the South; and it concerned basic physics, which was not very likely to meet the immediate needs of ordinary people.

The projects debated between 1946 and 1950 about establishing, if not a United Nations version of France's CNRS, then at least a significant group of international research laboratories in several disciplines, did not lead to much. Despite their differences, Laugier and Needham both wanted to provide new bases for international scientific relations: establishing the primacy of 'thinking internationally' and the overall interests of humanity over national agendas and the spontaneous 'laissez-faire' of academics. Set against these objectives, the failure was obvious. Laugier's desire to use ECOSOC rather than UNESCO, and his goal of strong centralization on the CNRS model, were more of a disservice than an aid to the laboratory projects developed more pragmatically by Needham and UNESCO.

COOL HEADS IN THE COLD WAR
Pierre Auger and the founding of CERN
Patrick Petitjean

THE year 2004 marked the fiftieth anniversary of the founding, near Geneva (Switzerland), of the European Organization for Nuclear Research (CERN),[23] a particle physics laboratory established under the auspices of UNESCO.

The war had profoundly affected developments in nuclear physics, sharply accelerating the pace of research. On the other hand, countries such as Denmark, France, Germany and Italy – which had been at the cutting edge in the 1930s – were excluded from those developments by the war. The stakes surrounding the atomic bomb were such that the United States and the United Kingdom monopolized – and maintained a veil of secrecy around – nuclear research, especially after the bombing of Hiroshima and Nagasaki. When the United Nations was established, nuclear issues (including basic research) were assigned directly to the Security Council, and not to UNESCO or the Economic and Social Council (ECOSOC).

---

22   CERN was founded in 1954 near Geneva (Switzerland).
23   Its provisional name had been the Conseil Européen pour la Recherche Nucléaire (European Council for Nuclear Research), thus the acronym, which was kept when the organization was formally established (in 1954), with its present name.

From 1949, European physicists, in particular French and Italian (notably Edoardo Amaldi, the mainstay of the project), joined forces to bridge the gap and to draw up a plan for a European laboratory. Their aim was to attain a level of research equivalent to that of the United States. The project was examined in Geneva in December 1949 at the European Cultural Conference, organized by the European Movement. It was endorsed by scientists, diplomats and science officials, but had not yet won governmental support.

The situation changed completely when the physicist Isidore Rabi took up the idea on behalf of the American delegation to the fifth session of UNESCO's

At UNESCO's Headquarters, signature of the Convention establishing the European Organization for Nuclear Research (CERN), 19 July 1953.

General Conference, held in Florence (Italy) in June 1950. Rabi announced that, following the test explosion of an atomic bomb by the Union of Soviet Socialist Republics (USSR) in August 1949, the United States could no longer maintain absolute secrecy and was prepared to assist the Europeans in rebuilding their nuclear physics research capacity. Through UNESCO, he gave intergovernmental legitimacy to the project. But, in the minds of some, 'intergovernmental' equalled 'cumbersome machinery'. Moreover, nuclear physics had not yet been given priority in the discussions on international laboratories. To implement the

PART I: SETTING THE SCENE, 1945-1965

decision made in Florence, Pierre Auger, Director of UNESCO's Natural Sciences Department, drew on his scientific and administrative contacts to build up the content of what was still only a general idea.

Auger, who had taken part in nuclear research in Canada during the war, had the advantage of still being closely involved in the world of physics. Before his appointment to UNESCO in 1948, he had participated in the UN Atomic Energy Commission, together with Isidore Rabi and Edoardo Amaldi. In France, he had also been a director of the Atomic Energy Commission (CEA) and director of higher education in the Ministry of Education. He could therefore

UNESCO organizing the European Constitutive Conference on Nuclear Research, 1952. Photo 1, from left to right: Miss Thorneycroft, J. Nielsen, Pierre Auger (Director, Natural Sciences Section/Department). Photo 2, from left to right: A. Picot, Niels Bohr, P. Scherrer.

draw on his network of relations to move the project forward rapidly.

In December 1950, a group of physicists and the European Culture Centre in Geneva put forward a more broadly based project, opting for a particle accelerator rather than a nuclear reactor. The project was further refined at a meeting of consultants at UNESCO in May 1951, but was opposed by the United Kingdom,[24] which was in favour of a much more modest laboratory, merely an annex to Niels Bohr's laboratory in Copenhagen (Denmark). The General Conference, at its sixth session in July 1951, expressed its preference for

---

24   The British did not participate in the 1949 and 1950 meetings. Nor did they sign the provisional agreement of February 1952. They considered that they were ahead of the other European countries in particle physics and were sceptical about the persistent scientific vagueness of the Auger project. A British physicist, Skinner, even said that it was one of those 'high-flown and crazy ideas which emanate from UNESCO'. The British adopted and became fully involved in the project in 1954.

the Auger project,[25] but the Anglo-Danish project was regarded as an alternative at two subsequent meetings of consultants (October and December 1951). In December 1951, UNESCO convened an intergovernmental conference, which set up a Council of Representatives of States and working groups to implement the Auger project, while proposing that experiments should be conducted in Copenhagen in the meantime. At the second intergovernmental conference (February 1952), the interim agreement was finalized, and it came into force in May 1952, after being ratified by five countries (the Federal Republic of Germany, France, the Netherlands, Sweden and Yugoslavia).

The official agreement establishing CERN and locating it near Geneva was signed in June 1954 and came into force in September of that year. In due course, the document was signed by twelve countries,[26] and the first stone of the CERN building was laid in June 1955. The rest is history: from the early 1970s, CERN has compared favourably with its American competitors in the field of particle physics, and scientists working in its laboratories have won Nobel Prizes for their major discoveries.

UNESCO played to perfection its role as initiator of an international cooperation project. The period of time between decision and implementation was unusually short – five years – for an intergovernmental organization. Pierre Auger's personal role, through numerous networks, in conjunction with the European Movement, contributed significantly to that achievement.

BIBLIOGRAPHY

CERN. Working papers in the Studies in CERN History series, published in preparation for the three-volume *History of CERN*. Geneva, CERN.

Hermann, A., Krige, J., Mersits, U., Pestre, D., et al. 1987. *History of CERN*. 3 vols. Amsterdam, North Holland publishers.

OF MOLECULES AND MEN
UNESCO and life sciences during the Cold War
Bruno J. Strasser[27]

---

25   Ronald Fraser, liaison officer between the International Council of Scientific Unions (ICSU) and UNESCO, endorsed the Anglo-Danish project.
26   Belgium, Denmark, France, the Federal Republic of Germany, Greece, Italy, the Netherlands, Norway, Sweden, Switzerland, the United Kingdom and Yugoslavia. The countries of Eastern Europe had been explicitly excluded since the meeting held in Geneva in December 1950.
27   Bruno J. Strasser, historian of science, is currently a visiting Fellow at Princeton University (USA).

PART I: SETTING THE SCENE, 1945-1965

INTRODUCTION

THE nuclear explosions in Hiroshima and Nagasaki, Japan, in 1945 inaugurated the Atomic Age. The launching of the Soviet satellite Sputnik in 1957 ushered in the Space Age. They were defining moments of the Cold War, an era in which international scientific cooperation developed in an unprecedented manner. High energy physics and atomic weaponry, space and missile research mobilized scientists, the military and government administrations around ambitious projects. The creation of the European Organization for Nuclear Research (CERN) in 1954 and the European Space Research Organization (ESRO) in 1961 resulted from and contributed to this new scientific agenda.

No single event – no Hiroshima or Sputnik – influenced international involvement in the life sciences in the post-war era. However, at the end of the 1950s, several indices suggested that a profound transformation was taking place in this field. In European countries, as in the United States, new institutions were created to develop research on aspects of living cells under the new heading 'molecular biology'. In 1962, two Nobel Prizes were awarded for work in this field. UNESCO helped make molecular biology one of the political priorities of European policy-makers. Its action in this field highlighted what was at stake in international scientific cooperation during this period.

FIRST UNESCO LIFE SCIENCES INITIATIVES

In 1954, UNESCO had already proposed to stimulate coordination of research in order to improve basic knowledge of cell growth (UNESCO General Conference, 1954). However, UNESCO only foresaw limited action, for example financial support for the organization of international conferences. In 1962, the Organization took a bigger step towards a policy for life sciences with the creation of the International Cell Research Organization (ICRO), which included representatives from Western and Eastern Europe, the United States, the Union of Soviet Socialist Republics (USSR), India and Israel (UNESCO General Conference, 1962). UNESCO justified this action in favour of cellular biology by the fact that it was interdisciplinary. In addition, the previous year, Pierre Auger, former Director of UNESCO's Natural Sciences Department, had published a report on scientific research underlining the importance of developing this category of research. ICRO became UNESCO's advisor for all activities concerning life sciences.

In September 1963, a group of scientists founded the European Molecular Biology Organization (EMBO). The aim was to create an international laboratory on the CERN model and a fellowship system to help scientists increase their mobility among European laboratories. EMBO soon became a private organization under Swiss law. From then on, its committee looked for an international organization or a government that would enable and ensure financing for its projects by European countries. As UNESCO had played this role for CERN ten years earlier, it was an ideal candidate for the EMBO Committee.

The crystallographer John Kendrew solicited support for EMBO from UNESCO. The Organization reacted very positively because its involvement with CERN had been such a gratifying one. UNESCO envisaged similar involvement if EMBO became intergovernmental. Meanwhile, the Belgian Government proposed that UNESCO study a project for an International Life Sciences Institute (ILSI), which would be based in Belgium and bring together 100 scientists and 300 technicians to study molecular biology, biophysics, cellular physiology, human genetics and cardiovascular illnesses.

In May 1964, EMBO and ILSI representatives elaborated a joint proposition and concluded that molecular biology required a new kind of international scientific cooperation, and that UNESCO was the right organization to help the scientific community to establish it. In its January 1966 report, the UNESCO expert commission concluded that it was necessary to support the creation of a laboratory and a foundation. In particular, the commission recommended that both institutions be based in Europe and, first and foremost, serve European needs. However there was strong opposition to a centralized laboratory, especially from the United Kingdom and the USSR.

COMPETITION FROM THE SWISS INITIATIVE

UNESCO was not the only institution to be interested in scientific cooperation in molecular biology. In 1964, the Council of Europe was also interested in EMBO. The following year, the Council proclaimed that the development of research in the field of molecular biology was a vital question, a point of view it communicated to all governments of its member states. However, it was reticent about the creation of an international laboratory. A year later, the Council of Europe went further and decided to organize an intergovernmental conference on molecular biology.

Being informed of similar initiatives by UNESCO and by the Council of Europe in this field, the Swiss proceeded with caution. In March 1966, the Swiss Federal authorities consulted CERN's twelve member states, as well as the Council of Europe, UNESCO and the Organization for Economic Co-operation and Development (OECD) on their intentions concerning participation in an intergovernmental conference. Contrary to UNESCO, Switzerland envisioned international scientific cooperation within the framework of Western Europe only and did not foresee an association with either Eastern Europe or the United States. UNESCO's Director-General, René Maheu, criticized the Swiss authorities for their actions, particularly the fact that only CERN member states had been consulted. As for the Council of Europe, it decided however to participate in the Swiss initiative.

During the October 1966 UNESCO General Conference, the Swiss authorities found themselves in a delicate position. They presented their initiative, underlining that the limitation to CERN member states was a first step – other countries would be allowed to join later. The Swiss also challenged CERN as an example: even if the creation of CERN was first proposed at UNESCO's 1950 General Conference, they argued, subsequently it was governmental actions that were the most effective in its realization. They proposed the same itinerary for EMBO. While CERN member states were favourable to the Swiss proposition, other nations pressed UNESCO to maintain its own initiative. Finally UNESCO decided to continue with its plans. Afraid of an openly hostile attitude from UNESCO, the Swiss authorities were relieved by the way things finally turned out.

The intergovernmental conference to be organized by UNESCO never took place. In February 1968, following various conferences organized by the Swiss authorities to give EMBO an intergovernmental basis, the twelve European States agreed on the creation of a European Molecular Biology Conference (EMBC). Within EMBC, little by little, those who supported EMBO convinced the European governments of the importance of setting up a European laboratory. So much so that, in 1974, EMBC drew up an agreement to found the European Molecular Biology Laboratory (EMBL), inaugurated in Heidelberg (Germany) in 1978.

HELPING EUROPE CATCH UP WITH THE UNITED STATES

What were the reasons this joint initiative by EMBO and the Swiss authorities succeeded in obtaining the necessary political consensus for the creation of a European laboratory, when other institutions – such as UNESCO, the Council of Europe, the World Health Organization (WHO) and the North Atlantic Treaty Organization (NATO) – had failed? This successful outcome was all the more improbable because – contrary to particle physics on space research – with molecular biology, there were no military benefits and only a limited economic impact.

The fact that the Swiss initiative was limited to Western Europe certainly contributed to its success. In effect, the other international initiatives of the era were based on different geopolitical configurations: European but including Eastern Europe (UNESCO), Atlantic (NATO), or worldwide (WHO). Basing its project on scientific cooperation in Western Europe only, EMBO benefited from general momentum to reinforce political bonds among Western European countries. European states' support for cooperation within the field of molecular biology could be analysed as political – a means of building up common interests, as for example the Treaty establishing the European Coal and Steel Community (Paris, 18 April 1951). This strategy, obvious for European Community countries, also came into play for neutral countries such as Switzerland, which saw scientific cooperation as a way of becoming closer to their European neighbours. Increased Cold War tensions during the 1960s had made including Eastern countries unattractive to Western governments.

Finally, molecular biology was sometimes presented as an 'American' science. Support for molecular biology in Europe, then, was considered a way of standing up to the 'American challenge'. Several European politicians were very pleased to reinforce capacities in this field and thus contribute to the rebuilding of European science, which had fallen behind the United States after the Second World War. The history of international scientific cooperation in the field of molecular biology reminds us that scientific development cannot escape the political and social dynamics that transformed post-war societies.

ACKNOWLEDGEMENTS

This chapter is based on documents from the UNESCO Archives (Paris), EMBO Archives (Heidelberg) and the Swiss Federal Archives (Bern), which through lack of space cannot be cited individually. I would like to thank the archivists for their collaboration, in particular Jens Boel (UNESCO), as well as the National Swiss Foundation for Scientific Research (project no. 3151-068174) for its support.

BIBLIOGRAPHY

Auger, P. 1961. *Current Trends in Scientific Research*. Paris, UNESCO.
Dahan, A. and Dominique P. (eds). 2004. *Les sciences pour la guerre 1940–1960* [Science for War 1940–1960]. Paris, Editions de l'Ecole des Hautes Etudes en Sciences Sociales.
Gaudillière, J-P. 2002. *Inventer la biomédecine: la France, l'Amérique et la production des savoirs du vivant: 1945–1965* [Inventing Biomedicine: France, America and the production of biomedical knowledge in the life sciences]. Paris, La Découverte.
Krige, J. 2002. The birth of EMBO and the difficult road to EMBL. *Studies in the History and Philosophy of Biological and Biomedical Sciences*, Vol. 33, Nos 547–64.
——. 2003. The politics of European scientific collaboration. In: J. Krige and D. Pestre (eds), *Science in the Twentieth Century*. Amsterdam, Harwood Academic Publishers.
Morange, M. 2000. *A History of Molecular Biology*. Cambridge, Mass., Harvard University Press.
Servan-Schreiber, J-J. 1967. *Le défi américain* [The American Challenge]. Paris, Denoël.
Strasser, B. J. 2003. The transformation of the biological sciences in post-war Europe. *EMBO Reports*, Vol. 4, No. 6, pp. 540–43.
Strasser, B. J. and Joye, F. 2005a. L'atome, l'espace et les molécules : La coopération scientifique internationale comme nouvel outil de la diplomatie helvétique (1951–1969) [The Atom, Space and Molecules: international scientific cooperation as a new tool of Swiss

diplomacy]. *Relations Internationales*, Vol. 121, pp. 59–72.
———. 2005b. Une science 'neutre' dans la Guerre froide? La Suisse et la coopération scientifique européene (1951–1969) [A Neutral Science in the Cold War? Switzerland and European scientific cooperation (1951–1969)]. *Revue Suisse d'Histoire*, Vol. 55, No. 1, pp. 95–112.
Strasser, B. J. and de Chadarevian, S. (eds). 2002. Molecular biology in post-war Europe. *Studies in History and Philosophy of Biological and Biomedical Sciences*, Vol. 33.

NOURISHING SCIENCE, ON LAND AND SEA
FAO collaboration with UNESCO and the creation of the IOC, 1955–1962
Ray Griffiths[28]

UNESCO and the Food and Agriculture Organization of the United Nations (FAO), as specialized agencies of the United Nations, are committed to collaborating in fields of mutual concern. As regards the natural sciences, such collaboration has occurred mainly in the fields of marine science and the rational management of marine resources, living and non-living. Cooperation among such agencies is, for most purposes, decided by their respective governing bodies and implemented through their secretariats, sometimes directly, sometimes through joint groups of experts, workshops, task teams, and so on. However, the science itself is, necessarily, carried out by the relevant institutions and the scientists of the member states themselves. In some cases, however, there is cooperation between the 'client' institutions of two different UN organizations in the implementation of a joint programme. UNESCO-FAO collaboration led to the creation of the Intergovernmental Oceanographic Commission (IOC).

UNESCO declared its specific interest in oceanography at the third session of the UNESCO General Conference, in 1948, which authorized the Director-General to promote the coordination of research on scientific problems in various fields, including oceanography and marine biology. FAO, established in 1945, by virtue of its wide mandate was concerned with all the principal natural sciences insofar as they related to living resources, especially those of interest to agriculture, forestry and fisheries.

The first specific instance of UNESCO-FAO collaboration was in 1955: the International Advisory Committee on Marine Sciences (IACOMS) was

---
28 Ray Griffiths was, for sixteen years, assigned by FAO to the Intergovernmental Oceanographic Commission (IOC) of UNESCO (1972–88).

appointed by UNESCO's Director-General from an international panel of honorary consultants set up by UNESCO in consultation with FAO. The main purpose of IACOMS was to advise the Director-General of UNESCO on the promotion of international collaboration in marine science in the preparation and execution of marine research projects, taking into account the related programmes of the United Nations and its specialized agencies. IACOMS built up close collaboration with the Special (later, Scientific) Committee on Oceanic Research (SCOR), created in 1957 by the International Council of Scientific Unions (ICSU, now the International Council for Science).

One of SCOR's first undertakings was the organization of the International Indian Ocean Expedition (IIOE, 1959–65), co-sponsored by UNESCO. Such were the scale and implications of this multinational expedition that the UNESCO General Conference, at its tenth session (in 1958), decided that not only its scientific aspects, but also the intergovernmental aspects, required the greatest

Two UNESCO Technical Assistance projects. 1. At the Natural Sciences Faculty of Caracas, Venezuela, a young parasitologist and her professor, 1957.

2. A UNESCO expert and a UNESCO scholar performing hydrogeological research in arid zones, Brazil, 1957.

consideration. It therefore decided to convene an international conference on oceanographic research, which took place in Copenhagen (Denmark) in 1960. Its organization was shared by FAO, as well as by the World Meteorological Organization (WMO), the International Atomic Energy Agency (IAEA) and the United Nations. This conference recommended, among other things, that UNESCO establish an Intergovernmental Oceanographic Commission (IOC) within the UNESCO Secretariat, with the main task of promoting the concerted action of the Member States in oceanographic research. The next session of UNESCO's General Conference (1960) accepted this recommendation, adopted the statutes of the newborn IOC and established an Office of Oceanography to serve as the secretariat of the IOC.

In 1962, the decision was taken to pass the intergovernmental coordination of the IIOE from SCOR to the IOC. The experience acquired in the organization and coordination of the IIOE led the IOC to initiate several regional cooperative investigations, which particularly marked the latter's first decade of activity. Scientifically speaking, these were broadly based and included multi-ship, multinational resource surveys, of particular interest to FAO in respect of living marine resources.

PIERCING THE IRON CURTAIN
UNESCO, marine science, and the legacy of the International Geophysical Year
Jacob Darwin Hamblin[29]

THE International Geophysical Year (1957–1958) was the first major scientific undertaking that reached across the political and ideological borders of the Cold War. Both the United States and the Soviet Union took part, along with over sixty other nations. In some ways, the International Geophysical Year (IGY) served as a model for future cooperation across international lines; but in other respects, the events of the IGY shattered the notion that science could rise above politics. One reason was the launch of Sputnik, the first artificial Earth-orbiting satellite, in October 1957, as part of the Soviet Union's IGY scientific programme. Combined with major Antarctic projects and an unprecedented Soviet commitment to oceanography, Sputnik set the stage for a period of intense scientific and technological competition between the superpowers. For organizations like UNESCO, the continued promotion of peaceful cooperation for the good of humanity became an enormous challenge.

The idea for the IGY came from American and British scientists, such as Lloyd V. Berkner and Sidney Chapman, who wished to create a third 'Polar Year' to follow up similar international cooperative ventures of 1882–83 and 1932–33. To attract more interest, they turned it into a 'Geophysical Year', which included a wider array of disciplines than had the earlier endeavours. It was not a UNESCO programme but instead was administered by an international coordinating group, the Special Committee of the International Geophysical Year.

Planning for the IGY stimulated efforts to build new cooperative scientific communities. While marine scientists met in the mid-1950s to plan for the IGY,

---

29   Jacob Darwin Hamblin teaches history at Clemson University (USA). He is the author of *Oceanographers and the Cold War: disciples of marine science* (Seattle, University of Washington Press, 2005).

they also designed a marine science programme under UNESCO. In 1955, Roger Revelle (United States), Lev Zenkevich (Soviet Union), George Deacon (United Kingdom), Anton Bruun (Denmark), Marc Eyriès (France), Koji Hidaka (Japan), and others became founding members of the International Advisory Committee on Marine Science (IACOMS), a new UNESCO body. IACOMS was conceived as a way to cultivate interest in the marine sciences in areas where it received little support, and to promote cooperation. Bruun, for example, imagined a utopian community in Asia where all countries might pool their resources and together make a genuine contribution to science. Many of the communities in South Asia had been stronger prior to the Second World War; they had not yet recovered from the destruction of libraries, the loss of lives, and the loss of national support. IACOMS hoped that, with help from UNESCO, these communities might be revived and encouraged to cooperate with each other.

Although it was not meant to do so, IACOMS weakened the role of the International Council of Scientific Unions (ICSU). ICSU's International Association of Physical Oceanography (IAPO) thus far had suppressed efforts to create new bodies that stretched the disciplinary meaning of oceanography. It preferred, for example, to keep marine biologists within the International Union of Biological Sciences, and IAPO under the International Union of Geodesy and Geophysics. Now, IACOMS saw oceanography as 'marine science', including physical, biological and chemical aspects. However, the older organization could not complain because IACOMS was directed at the developing world, whereas IAPO was designed to coordinate the activities of leading scientists.

The differences between the two outlooks – one supporting a broad definition and targeting the developing world, and the other supporting a narrower definition and coordinating top-rank scientific work – provoked serious discord among scientists. Inspired by the worldwide interest in the IGY, Roger Revelle wanted to create a new coordinating body that would establish a broad meaning of marine science, across many disciplines, and connect scientific problems to those of society (fishing, waste disposal, weather forecasting, etc.). Many physical oceanographers, such as George Deacon, resisted this, wanting to keep physical oceanographers, marine biologists, and fisheries scientists separate. His view was that the best scientists ought to coordinate work with each other, and not have to coordinate with scientists-in-training or attempt to find social relevance for research.

Revelle and the Soviet biologist Lev Zenkevich agreed that the broad definition fostered at UNESCO was a wiser course. It would strengthen their collective voice, making it easier to solicit funding from governments. By casting scientific problems as social ones, they also could gain support from UNESCO. They convinced sceptics such as Deacon that the payoff for the scientific community would offset any negative effects. Keeping walls between disciplines strong would do harm to science. In 1957, they created a new body under ICSU that incorporated all of these ideas, the Scientific Committee for Oceanic Research (SCOR),[30] and Revelle became its first president.

The attitudes of UNESCO influenced most post-IGY projects. The first of these, the International Indian Ocean Expedition (IIOE), focused on the developing world. Originally modelled on the IGY, the expedition soon became a very different kind of project. The IIOE was a loosely coordinated assemblage of national expeditions that did not necessarily work together. Focusing on the developing world, it lasted several years (1959–65), in order to help spread out the financial burden. The IIOE helped to build or resuscitate scientific institutions. For example, India's National Institute of Oceanography owes its existence to the work of the IIOE. Unlike the IGY, which scientists vaguely hoped would ease global tensions, the IIOE was sold explicitly as a project to help humanity solve its problems of food and climate.

The efforts of IACOMS seemed superfluous after the creation of SCOR. IACOMS organized marine science training courses in the late 1950s, in locations such as Bombay, India, and Nhatrang, Vietnam. The committee also co-sponsored (with SCOR) the first International Congress on Oceanography, held in 1959 at the United Nations headquarters in New York. But because of the ascendancy of SCOR as an advisory body to UNESCO, IACOMS was dissolved in 1960. Since SCOR was not equipped or staffed to coordinate the large-scale projects it promoted, it helped UNESCO create (in 1960) the Intergovernmental Oceanographic Commission (IOC). SCOR continued its work of advising on scientific questions, but left the job of implementing big projects to the IOC. Although some were critical of the fact that the commission was made up of national delegations, and thus susceptible to political uses, the IOC was to become the coordinator of the major international oceanographic activities in the years to come.

BIBLIOGRAPHY

Hamblin, J. D. 2005. *Oceanographers and the Cold War: disciples of marine science.* Seattle, Washington, University of Washington Press.

---

30   Initially, the 'S' in SCOR stood for 'Special'.

THE 'PERIPHERY PRINCIPLE'
CROSSING BORDERS
Contributing to the development of science in Latin America
Patrick Petitjean

THE meeting of the Panel of Experts in Latin America on the Development of Science, organized by UNESCO, opened on 6 September 1948, in Montevideo (Uruguay), pursuant to a decision adopted by the General Conference of UNESCO at its second session (Mexico City, Mexico, December 1947). It was the first of its kind and resulted from the application of the 'periphery principle' underlying the programmes of the Natural Sciences Department.

Some fifteen scientists from ten countries – including the Argentine Nobel Prize winner Bernardo Houssay and three Brazilians (Miguel Ozorio de Almeida, Enrique Rocha e Silva, and Joaquim Costa Ribeiro) – attended the UNESCO meeting, together with representatives of the International Labour Organization (ILO), the Rockefeller Foundation, the Organization of American States and the Smithsonian Institution. The participants stressed the relevance of discussions rooted in the reality of countries seeking scientific development.

Three basic lines of emphasis emerged from the meeting: the need for UNESCO to assist in the development of research on basic scientific issues in Latin American countries; the establishment of a 'full-time' employment system for researchers (who had previously been obliged to hold several jobs); and the development of scientific institutions and their coordination at the national level. The first attempt to establish a UNESCO science cooperation office for Latin America (at Rio de Janeiro and Manaus, Brazil) had failed because it had been confused with the International Hylean Amazon Institute, its only activity (with the same director). The decision was therefore taken to establish, at Montevideo, a new Field Science Cooperation Office, and its first director, appointed in 1949, was Angel Establier.[31] This Regional Office still exists today.

In Brazil, that meeting is viewed as having played a pivotal role in the establishment of the National Research Centre (CNPq, for which a bill was

---

31 Angel Establier: biochemist; previously, assistant to the director of the 'Sciences' section of the International Institute for Intellectual Cooperation; subsequently, the first liaison officer between UNESCO and ICSU (1947–48).

tabled in early 1949). The Brazilian Society for Scientific Progress (SBPC), established in 1948, drew on the UNESCO's Field Science Cooperation Office in Montevideo and adopted its initiatives. International scientific cooperation through UNESCO was the central theme of SBPC's second annual meeting (November 1950).

BIBLIOGRAPHY

Ciência e Cultura. 1948–1950. Various issues. SBPC (Brazilian Society for Scientific Progress).
LACDOS (Latin American Conference for the Development and Organization of Science). UNESCO Archives.
Motoyama, S. 1996. *O Almirante e o Novo Prometeu* [Admiral and the new Prometheus]. Sao Paulo, Brazil, EDUNESP.
——. 1985. A Genese do CNPq [Origins of the CNPq (National Council for Research)]. *Revista da Sociedade Brasileira de Historia da Ciencia*.

GOING GLOBAL
UNESCO Field Science Offices
Jürgen Hillig[32]

FROM its very beginning, science at UNESCO aspired to be something more than a think tank. It was understood that the pursuit of knowledge and the mission to better societies through scientific cooperation would require outreach that stretched to far-flung corners of the globe. Indeed as early as 1947 – well before the first education offices were established – the General Conference set up Field Science Cooperation Offices in the 'Middle East' (Cairo, Egypt, 1947), Latin America (Rio de Janeiro, Brazil, 1947; moved to Montevideo, Uruguay, in 1948), and 'East Asia' (Nanjing, China, 1947; moved briefly to Manila, the Philippines; and finally to Jakarta, Indonesia, in 1951, covering the South-East Asian region). An office for 'South Asia' (New Delhi, India) was set up in 1948. At that time, the main function of these offices was 'to maintain contact between the scientists and technologists in those parts of the world remote from the main centres of learning and research, and their colleagues at those centres'.[33]

---

[32] Jürgen Hillig: UNESCO Division of Science and Technology Policies, then Coordinator for Natural Sciences Sector, 1967–88; Director, Regional Office for Science and Technology, Jakarta, 1989–94; Director, then Assistant Director-General, Division of Decentralization and Field Relations, 1994–97.
[33] UNESCO Document C/3 1947.

These early Field Science Cooperation Offices were very small, consisting of only one or two professionals. Their resources were rather limited. But they were highly effective as information exchange centres and clearinghouses for the regions concerned. They also served liaison and representation functions. The fact that they had been set up so early in the life of the Organization, and the prominent place given to accounts of their work in official UNESCO documents, are indications of the importance assigned to these offices, irrespective of their size. They laid the ground for the numerous initiatives for regional scientific cooperation to be launched in the years to come. The first four offices, as well as those created later in Africa (Nairobi, Kenya, 1965) and Europe (Paris, 1972; later moved to Venice, Italy, in 1988) all had a regional or sub-regional vocation. Over the years, the heads of these offices came to act as de facto ambassadors of science to one or several countries in the region they were serving.

UNESCO's Science Cooperation Offices in 1952.

The Field Science Offices were eventually renamed, becoming 'UNESCO Regional Offices for Science and Technology' (ROSTs). As such, the scope of their activities – and the volume of financial and human resources they commanded – grew considerably. By the 1970s, the Regional Offices, in addition to their initial responsibilities for facilitating exchange of information between scientists, were participating in the planning, execution and evaluation of a broad range of scientific programmes. They advised Member States on science and technology policy. They became involved in the training of personnel through intensive courses, symposia and seminars. They participated in the preparation and follow-up of ministerial meetings and conferences of experts organized by UNESCO. Regional Offices oversaw the collection of scientific data and the production of detailed studies and research papers. Working closely with the United Nations Development Programme (UNDP), they provided support to Technical Assistance experts and helped supervise and coordinate operational projects.

However, while the Regional Offices' involvement in programme implementation expanded, their contribution to the planning and budgeting process at Headquarters was more limited. This can in part be explained by the fact that the international scientific programmes (mainly concerned with environmental sciences) that had taken shape during the 1960s and 1970s were all controlled by international rather than regional committees. These committees worked very closely with the respective science divisions at Headquarters. It is not surprising, then, that the Regional Offices were carrying out a relatively larger proportion of UNESCO's activities in the basic and engineering sciences, which were less structured and controlled by international bodies.

Prominent among these activities were the creation and subsequent operation of regional or sub-regional networks in fields such as chemistry, biology and engineering, especially in Asia, where Japan provided much valued support in the form of 'funds-in-trust' payments. The Regional Science Offices also played a major role in the preparation, holding and follow-up of the 'Regional Meetings of Ministers Responsible for the Application of Science and Technology to Development' (CAST) conferences. As a follow-up to the CASTASIA II conference in Manila in 1982, the Jakarta office was instrumental in setting up and administering the Science and Technology Policy Asian Network (STEPAN), which acted as a major channel for information exchange on science and technology policy issues of common interest.

Over time, the Regional Science Offices not only expanded the scope of their activities, they also absorbed an increasing amount of UNESCO's Regular Programme funds and staff resources. In 1986, for example, the Natural Sciences Sector's total staff allocation amounted to 342 posts, of which 125 were posts in the field. No precise budget figure is available for 1986, but an estimate of

Regular Programme funds allocated to the Regional Offices would probably have been in the 20–25 per cent range.

'Decentralization' became a major policy issue at UNESCO during the late 1980s and early 1990s and had a powerful impact on the Regional Offices. Additional funds were provided to field units, and their responsibilities were expanded even further. This meant that any UNESCO field unit would from now on be coping with *all* aspects of the Organization's programme, irrespective of the historical main focus of any given unit in science, education or culture. While that enlargement of responsibilities only affected the host country in which a Field Office was located, it was nevertheless the beginning of a new era: offices lost part of their specificity but gained broader responsibilities, including increased representational duties at country level. The emphasis on decentralization also led to a rapid increase in the overall number of UNESCO Field Offices, which grew from thirty-five in 1989 to fifty-two in 2005.

The world has grown much smaller since 1947, when the first UNESCO Field Science Cooperation Offices were established. Through the promotion of scientific cooperation, UNESCO has helped bridge the gaps in knowledge and expertise that existed between advanced nations and less developed countries. Much work remains to be done. But the Regional Science Offices are still going strong.

BIBLIOGRAPHY

UNESCO. 1950. *The Field Scientific Liaison Work of UNESCO*. Paris, UNESCO.
———. 2005. Report by the Director-General on the reform process, Part III (document 171EX/6, 6 March 2005). Paris, UNESCO.

MEANWHILE IN THE MOTHERLAND
Regional Office for Science and Technology in Africa (ROSTA)
Robert H. Maybury[34]

WHILE UNESCO established Field Science Offices in the Middle East, Asia and Latin America in the 1940s, it was not until 1965, with the

---

[34] Robert H. Maybury: UNESCO Natural Sciences Sector, 1963–72; UNESCO Regional Office for Science and Technology for Africa, Nairobi, Kenya (Deputy Director), 1973–80; UNESCO Headquarters (Managing Editor, *Impact of Science on Society*), 1980–83.

crumbling of colonialism, that such a bureau was inaugurated in Africa. Nairobi, Kenya, was selected as the location for the first Regional Office for Science and Technology in Africa (ROSTA). When reaching its operative size in the 1970s, ROSTA had a staff of five scientific members and one administrative officer. In addition, a programme officer financed under extra budgetary funds served as director of the regional Integrated Programme for Arid Lands. ROSTA's limited resources were complemented by the UN Economic Commission for Africa (UNECA) as well as assistance from such nations as the Federal Republic of Germany, the Netherlands and Norway. In its early days, the Nairobi office was responsible for scientific programmes in more than thirty countries in sub-Saharan Africa. A notable sign of UNESCO's commitment to ROSTA and its work on the African continent was the convening of the Organization's 1976 General Conference in Nairobi.

A framework for much of ROSTA's work emerged from the Conference of Ministers of African Member States Responsible for Application of Science and Technology to Development (CASTAFRICA), convened at Dakar, Senegal, in January 1974 by UNESCO in cooperation with UNECA and the Organization for African Unity. Indeed, no CASTAFRICA recommendation highlighted more clearly the opportunities for ROSTA's work than the one calling on governments to give absolute priority to the training of African scientists and technological personnel. As a follow-up to this CASTAFRICA recommendation, ROSTA hosted, in 1977, the founding conference of the Association of Faculties of Science of African Universities. Bringing together professors from roughly twenty different institutions, the conference was one of the earliest pan-African gatherings of university scientists.

Over the next several years, ROSTA cooperated with the United Nations Development Programme (UNDP) and the German Agency for Technical Assistance to initiate a network among universities in twenty-seven African countries. This became the African Network of Scientific and Technological Institutions (ANSTI), set up in 1980 with ROSTA as its secretariat. A report on ANSTI in the year 2000 portrayed its continuing role in human resource capacity-building in science and technology in Africa by providing exchange fellowships for university staff members, offering postgraduate fellowships, and convening consultative meetings and workshops.

# SCIENTIFIC NGOs AND UNESCO
A SPECIAL RELATIONSHIP
The early years of the UNESCO-ICSU partnership
Patrick Petitjean

THE International Association of Academies, established in October 1899, did not survive the First World War. The victorious countries established the International Research Council (IRC) in July 1919. Membership of the Central Powers nations became possible beginning in 1926. The IRC was renamed the International Council of Scientific Unions (ICSU) in July 1931. Whereas science was not a priority for the Organization of Intellectual Co-operation, linked to the League of Nations, ICSU gave a united voice to the eight scientific unions it encompassed and to the national constituencies of its forty-one member states. However, ICSU had only eight years to develop before the Second World War broke out, rendering the Council dormant.

With victory drawing near, the Allied countries and scientific communities started to discuss ways of reviving international scientific cooperation. In 1944, the US National Academy of Sciences sought out scientists and institutions in Allied countries to ask for their preferences. The result was the Cannon-Field report. For the majority of the scientists (most of them from the USA and the UK), the highest priority was the revival of ICSU. They wanted the unions to develop autonomously, free from the intervention of intergovernmental institutions. Joseph Needham, along with a minority of scientists, thought that ICSU alone was not equipped to meet the new challenges of scientific cooperation. They proposed an International Scientific Service, to be established and supported by Allied governments, as a complement to ICSU.

The context changed with the atomic bombs over Hiroshima and Nagasaki. The political decision was taken to include science in UNESCO in November 1945. A privileged partnership was soon established with ICSU. Needham's proposals to this end were accepted, in July 1946, by the UNESCO Preparatory Commission, as well as by the first post-war ICSU General Assembly. The final agreement was signed on 16 December 1946. Cooperation involved all aspects of UNESCO's scientific activities. It was based on mutual support. UNESCO helped ICSU and its unions to revive and develop, with the establishment of new unions and an increased participation by non-Western countries. In the early years, nearly one-third of the UNESCO science budget was earmarked to support ICSU. In return, ICSU participated in many UNESCO projects, offering its independent expertise, playing the de facto role of a 'scientific advisory council' for UNESCO.

The partnership was not always easy. The UNESCO Secretariat wished to have a hands-on operational role in many scientific endeavours; yet, some Member States wanted ICSU to be the executive body for all UNESCO activities in science. On the other hand, some ICSU scientists worried that UNESCO would bring political interference into the neutral realm of science. Nevertheless, a satisfactory balance was found, and the partnership turned out to be sustainable and very fruitful for both bodies.

A FAILED PARTNERSHIP
The WFSW and UNESCO in the late 1940s
Patrick Petitjean

UNESCO has, since its establishment, developed direct links with the scientific community through partnership with the International Council of Scientific Unions (ICSU, now the International Council for Science). The political climate of the Cold War, however, thwarted the establishment of a similar partnership with the World Federation of Scientific Workers (WFSW). Yet both UNESCO and the WFSW promoted the same ideals: the use of science for peace and the welfare of humanity, and scientists' social responsibility in that regard.

The WFSW was established primarily on the initiative of two associations of scientific workers, in the United Kingdom and in France, and involved smaller associations in the dominions and in the United States. The Russian associations refused to join the WFSW until 1952 (just as the USSR refused to join UNESCO before 1954). ICSU was established on a purely scientific basis, whereas the WFSW was defined as a 'science and society' movement: involving the social implications of science, its popularization, professional aspects, the social responsibility of researchers (including nuclear disarmament) and the promotion of unrestricted international cooperation.

UNESCO and the WFSW were founded in the same year (1946) with similar objectives, and it is not surprising that they were intended to be complementary by their founding fathers: Needham, Huxley and Auger at UNESCO; Frédéric

Joliot-Curie, John Desmond Bernal and James Gerald Crowther at the WFSW. UNESCO is an 'intergovernmental' body, and scientists participating in it are appointed by governments: it needs the support of the great mass of scientific workers through partnership with a body such as the WFSW, which is fully consistent with the concept of non-governmental organizations (NGOs) serving as UNESCO's supporting intermediaries.

Founded in London (UK) in July 1946, the WFSW was highly representative of the scientific communities in France and in the United Kingdom. Needham represented UNESCO at the founding meeting of the WFSW and drew up a partnership agreement modelled on the one binding UNESCO and ICSU, which covered funding (for travel, in particular), a liaison officer post to be paid by UNESCO, an office at UNESCO Headquarters in Paris, and other terms. From December 1946, the WFSW had an address and a temporary office at UNESCO for its French-speaking secretary. A grant-in-aid for the WFSW was included in the budget estimates.

But the ratification of the agreement by UNESCO was blocked in April 1947 by the delegation of the United States, concerned about the influence of French and British communists on the WFSW. Auger (who was then France's representative on the Executive Board of UNESCO) was the main supporter of the WFSW on the Board. The status of Observer was ultimately granted to the WFSW by UNESCO in July 1947 and was confirmed (against the objections of the American delegation) by the General Conference in Mexico City (Mexico) in November 1947. Such status meant that the WFSW could be invited officially to UNESCO initiatives, but without funding.

The WFSW thus participated in the round tables in Paris (October 1947) on the social implications of science and in UNESCO's General Conference in Mexico City (November–December 1947). Needham circumvented the lack of provision for funding by granting Crowther, Secretary-General of the WFSW, an official UNESCO mission (from December 1947 to April 1948) to the United States, in order to study the ways and means of UNESCO's participation in the Conference on the Conservation and Utilization of Natural Resources (organized by the UN Economic and Social Council, ECOSOC). Crowther was thus able, in addition to his work for UNESCO, to finance his participation in the Mexico City conference and to establish the WFSW in the United States.

The partnership continued in 1948: the WFSW's participation in the preparation of a 'science and society' journal in conjunction with UNESCO and ICSU, Frank Malina's participation in the WFSW's first General Assembly in Prague (Czechoslovakia) in November as UNESCO's representative, and Needham's participation in UNESCO's General Conference in Beirut (Lebanon) in November/December (for the British Government had refused to include Needham in its official delegation and so, only six months after leaving his post as Director of the Natural Sciences Section, Needham was obliged to join the WFSW contingent to take part in the UNESCO Conference).

But as the Cold War developed, the WFSW lost its Observer status in June 1950. The partnership was only revived fifteen years later, when the ethos of peaceful coexistence had thawed the frost of the Cold War.

SOCIAL ASPECTS OF SCIENCE
MEETINGS OF MINDS
UNESCO and the creation of the International Union of History of Science

Patrick Petitjean

FOR Joseph Needham, the history of science was a way of showing that 'there are few peoples or nations that have not contributed, even if to different degrees, to humanity's scientific heritage' (Cortesao, 1947). As the first Head of UNESCO's Natural Sciences Section, Needham wanted to break down the barriers of narrow nationalism and cultural bias. This, he believed, was part of 'UNESCO's task of enhancing international comprehension and understanding' (ibid.).

But Needham also wanted to break down the barriers of the human intellect. As an academic subject, the history of science had traditionally favoured history over science: all across the world, the history of science had been studied in the social sciences and humanities departments of universities. As an expert in both fields, Needham sought a dynamic synthesis of history and science. This was the impetus behind the creation of the International Union of History of Science (IUHS) in 1947.

The Union's precursor, the International Academy of the History of Science, was founded two decades earlier, in 1927. A rather elitist group, its members were mostly scholars focused on history. The Academy organized the International Congresses of the History of Science (ICHS), of which the second was the most famous. It took place in London in 1931, in the presence of a Soviet delegation headed by Nikolai Bukharin. This congress launched a social approach to the history of science, which greatly influenced British scientists, instigators of the movement to study the social relations of science. The young Joseph Needham participated actively in the second ICHS. It marked the beginning of his enduring commitment to the history of science.

Discussions in June 1946 on UNESCO's future scientific programme included the proposition for an 'institute of history of science' to enhance the already existing Academy. In December of that year, it was finally decided that a whole body should be created, one that would be attached to the International Council of Scientific Unions (ICSU). Armando Cortesao, member of the Academy, was recruited by UNESCO's Natural Sciences Section to set it up.

The International Union of History of Science became a reality in October 1947 when, at the fifth ICHS (in Lausanne, Switzerland), Cortesao presented the UNESCO project and the rationale behind the Organization's involvement in the field of study: scientific research must be linked to the history of science and vice versa. The IUHS automatically became a member of ICSU. If, within ICSU, there was some displeasure about the IUHS's membership, it was rapidly overcome. Cortesao worked with the IUHS for the first few months and, during 1947–48, UNESCO assumed a major part of its finances. Cortesao then joined the 'Philosophy and Civilizations' division to lead the *Scientific and Cultural History of Mankind* project. Jean Pelseener, also a member of the Academy, then joined UNESCO's Natural Science Section in order to edit a new journal, *Archives Internationales d'Histoire des Sciences*, again with financial support from the Organization. He, too, participated in the *Scientific and Cultural History of Mankind*.

The fact that the IUHS was a part of ICSU and not of the International Social Science Council was a clear break with scholarly tradition. In liaison with the Social Relations Commission of ICSU and UNESCO's Natural Sciences Section, the IUHS established, at the outset, a commission charged with the History of Social Relations of Science, chaired by the physicist Léon Rosenfeld. UNESCO requested a report from the commission on the 'Social Aspects of the History of Science', which Samuel Lilley finalized. The report's central theme was that contexts (social, economic, intellectual and political) facilitated or blocked scientific discoveries, without changing the overall direction that scientific progress takes.

Eventually, a 'P' was added to the Union's name, creating the International Union of History and Philosophy of Science (IUPHS). The original name may have changed but, after six decades, the Union's mission has remained the same.

BIBLIOGRAPHY

Cortesao, A. 1947 L'UNESCO, sa tâche et son but concernant les sciences et leur développement historique [UNESCO, its task and goal concerning the sciences and their historical development]. *Actes du Cinquième Congrès International d'Histoire des Sciences*. Lausanne, Switzerland, pp. 25–35.

Lilley, L. 1949. Social aspects of the history of science. *Archives Internationales d'Histoire des Sciences*, Vol. 2, pp. 376–443.

PART I: SETTING THE SCENE, 1945-1965

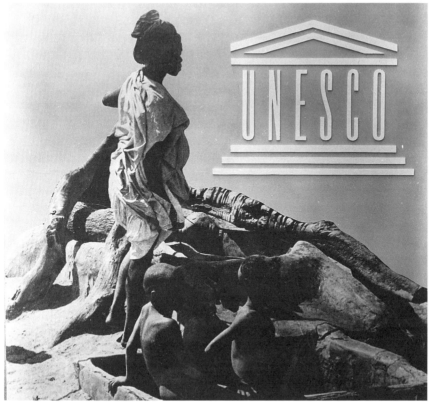

At UNESCO Headquarters, an exhibition on arid zones, 1953.

ON THE ROAD
UNESCO's travelling science exhibitions
Alain Gille[35]

The following is an extract from an article on UNESCO's travelling science exhibitions, by Alain Gille, which appeared in *MUSEUM* (Gille, 1954).

---

35 Alain Gille (1922–2005) joined UNESCO in 1949, when he was recruited to head its programmes concerning environmental protection and the popularization of science. It was within this mandate that he organized the travelling science exhibitions and also prepared a series of twenty-four publications providing an inventory of scientific equipment. Beginning in the 1960s, his attention focused on Africa's natural resources, first at Headquarters then as Director of the Regional Office for Science and Technology (ROSTA) in Nairobi, Kenya. He ended his career coordinating UNESCO's network of Regional Offices for Science and Technology.

UNESCO travelling science exhibition on geophysics. Alain Gille, in front of a Vanguard rocket, a copy of an American satellite and a part of Viking rocket, 21 October 1957. Alain Gille (1922–2005), an agronomist, joined UNESCO in 1949; he organized its travelling science exhibitions and prepared publications providing an inventory of scientific equipment.

AIMS, ORIGIN AND HISTORY OF THESE EXHIBITIONS

IN 1950, UNESCO's Department of Natural Sciences organized, as part of its programme for the dissemination of sciences, a travelling exhibition to acquaint the public in Latin America with the most important discoveries recently made in the field of physics and astronomy. In addition to explanatory panels, amply illustrated with drawings, diagrams and photographs, this exhibition included a large amount of apparatus that could be worked by visitors, as well as experiments for them to carry out themselves. It might, at first sight, have seemed a risky undertaking to send several tons of valuable, delicate equipment through countries where climatic conditions are unfavourable (tropical or equatorial countries) and transport is often difficult. But the experiment proved well worthwhile: between 1950 and 1952, the exhibition was on display in thirteen countries, and was visited by close to half a million persons, yet only very slight damage has been reported.

The public's sympathetic response was taken as an indication of the merit of this method for the dissemination of sciences, and UNESCO was accordingly encouraged to expand its work in this field. A second travelling science exhibition, on the theme 'Our Senses and the Knowledge of the World', was organized by UNESCO at Bangkok (Thailand) in November 1951, to coincide with the conference of National Commissions of its Member States in South-East Asia.

This second exhibition, consisting of some twenty panels with explanatory texts in three languages, fifty experiments and a large number of scientific instruments, is at present continuing its tour in the Far East, where it is proving as successful as the first. It has already been received by sixteen leading centres in eight different countries. A third exhibition was then prepared, in 1952, for UNESCO's Member States in the Near and Middle East. It is entitled 'New Materials' and deals with materials that have recently been placed at man's disposal by science and technology, i.e. mainly plastics and certain alloys. It has so far been shown in five countries.

Then quite recently, in July 1954, UNESCO opened in Paris a fourth science exhibition, entitled 'Man Measures the Universe'. This deals with the techniques and instruments used for the measurement of length and distance, from the smallest (intra-atomic) to the largest (intergalactic) distances now known. In the autumn it will leave for a tour of the European Member States of UNESCO.

BIBLIOGRAPHY

Gille, A. 1954. UNESCO's travelling science exhibitions. *MUSEUM*, Vol. VII, No. 4, p. 276.

THE ULTIMATE ODYSSEY
The birth of the *Scientific and Cultural History of Mankind* project
Patrick Petitjean

UNESCO'S founders were nothing if not ambitious. At the Organization's second General Conference (Mexico, November–December 1947), they decided to embark on a project whose very title is breathtaking in its scope: a *Scientific and Cultural History of Mankind*. Preparatory studies were launched to publish books for scientists as well as ordinary readers, 'allowing a better comprehension of scientific and cultural dimensions of the history of humankind and showing the mutual dependency of peoples and cultures, and their respective contribution to humankinds common patrimony'.

The main boosters of the project were Julian Huxley (UNESCO's first Director-General), Joseph Needham (first Head of UNESCO's Natural Sciences Section) and Lucien Febvre (French National Commission for the Organization). The project was initially nurtured by the Natural Sciences Section. It was entrusted to Armando Cortesao, who led the Department's History of Science division. Later, after the third General Conference (Beirut, Lebanon, December 1948), it became a joint undertaking with the Cultural Activities Department. Cortesao joined the 'Philosophy and Civilizations' division, but kept the secretariat of the project until 1952.

In his 1946 essay *UNESCO: Its Purpose and Its Philosophy*, Huxley had proposed that a central task for the Organization would be 'to help in constructing a history of the development of the human mind, notably in its highest cultural achievements' and claimed that 'the development of culture in the various regions of the Orient must receive equal attention to that paid to its Western growth'. His views were shared by Needham who, in October 1948, offered an outline of his study 'Science and Civilization in China' as part of the History. For Febvre, this project had to

> show that, since time immemorial, men have met peacefully with other men, that they have communicated by exchanges and the borrowing of one another's particular wealth, be it tools, technology, domesticated animals or improved plant specimens; that a network of peaceful relations has never ceased, through the ages, to cover a world that we want to see as permanently self-damaging; finally, that there are no insignificant peoples, no poor or destitute civilizations that have not had their glorious moments of invention, that have not contributed in one way or another to the building of our great and overconfident civilizations that, in fact, survive by borrowing (Febvre, 1949).

He believed that Westerners themselves demeaned their civilizations with colonialism and fascism: 'Entire parts of this civilization have been unsettled, demeaned by the Westerners themselves'.

Huxley, Needham and Febvre agreed that this History should not be encyclopaedic, nor mainly chronological. It was to be constructed around three basic hypotheses: exchanges of all kinds between civilizations are the driving force for the evolution of humankind; the exchanges of sciences and techniques among other cultural interactions are central to evolution; European civilization should not be considered the model, neither in the past nor in the future, for all civilizations.

Modern civilization – as Huxley, Needham and Febvre saw it – was a joint construction of various cultures. Their views shaped a report produced by a team of experts meeting in December 1949. Needham summarized the result in a letter to Cortesao (14 January 1950):

> After an opening part, introducing certain fundamental knowledge about Man and the world in which he finds himself, there would be a second part describing the series of chronologically successive *stages* in the progress of humanity in social organization and control over, and understanding of, Nature. The third part will be concerned with exchanges and *transmissions* in all branches of human knowledge, practice and experience; demonstrating the mutual indebtedness of all peoples, and bringing out the fact that there is no people or culture which has not contributed elements of essential value to the total human patrimony. The fourth part will outline the various patterns of the great cultures and civilizations, their particular world-outlooks which were characteristic of them, and which, though not transmitted in former times, are now fusing into the world-picture of universal man. The fifth concluding part would be of a synthetic character. In so far as the attainment of perfect historical objectivity might be considered to be impossible, the committee felt that emphasis might well be placed on the factors which have united mankind throughout history, rather than on those which have divided the various peoples.[36]

This orientation encountered opposition from traditional historians of science, especially in the United Kingdom. After the discussions of the British National Commission in early 1950, Febvre pointed out the source of difficulties: they were 'due to the obstinacy with which so many representatives of so-called "European" or "Western" civilization regard the latter – their own – as the only true civilization'.[37]

Nevertheless, the fifth session of the General Conference (Florence, Italy, June 1950) decided to undertake the publication of this History. UNESCO established an International Commission and an Editorial Committee, with a new group of scholars: Paulo de Berrêdo Carneiro, a Brazilian biologist; Ralph Turner, an American historian; Guy Métraux, a Swiss sociologist; and Charles Morazé, a French historian. They transformed the project, employing a more traditional, positivist and chronological approach. It became the history of the specific contributions of various civilizations to scientific and technological progress, rather than the history of mutual influences and exchanges.

---

36   Needham Papers, Cambridge University Library, Cambridge, UK.
37   Letter from the French National Commission, 24 March 1950, to UNESCO, in reaction to the project.

UNESCO launching the *Scientific and Cultural History of Mankind* project, February 1954. Left: Luther Evans, Director-General of UNESCO; centre: Paulo E. de Berrêdo Carneiro, President of the International Commission; right: C. Burckhardt.

The *Scientific and Cultural History of Mankind* was not published until the 1960s, the sixth and final volume appearing in 1969. UNESCO published a completely revised edition in the 1980s, and is currently working on a third edition. If the finished product did not quite conform to the vision of Huxley, Needham and Febvre, it remains, nonetheless, a work of remarkable scholarly ambition.

BIBLIOGRAPHY
Febvre, L. 1949. Report of the Beyrouth Conference to the French National Commission. *Notes et Études documentaires*, No. 1080, 26 February, pp. 9–13.

SPREADING THE NEWS
The natural sciences in the *UNESCO Courier*, 1947–1965
Lotta Nuotio[38]

COMMITTED to improving the world through international cooperation in the realms of education, science and culture, UNESCO, in the aftermath of the Second World War, embarked upon a long mission that would be composed

---
38 Lotta Nuotio, journalist at National Radio Broadcasting of Sweden, was a research assistant for *Sixty Years of Science at UNESCO 1945–2005*.

of a myriad of small but significant victories. The Organization recognized, early on, that it would have to keep the public informed of its progress. And so, in 1947, UNESCO began publishing the *Monitor*, which became, the following year, the *UNESCO Courier*, a monthly magazine aimed at a general readership. The natural sciences won their fair share of space in the pages of the *Courier*.[39] During the first years of the magazine's existence, science was presented as one of the paths to world unity.[40] Articles hailed scientific breakthroughs as a means of preventing future wars.

That point of view was still evident at the tenth anniversary of the Organization, when the *Courier* published a review of the work of natural sciences at UNESCO during its first decade.[41] Concrete examples given of the scientific work carried out by UNESCO[42] mainly concerned the Technical Assistance programme of the late 1940s and 1950s. The *Courier* reported on UNESCO's efforts at the rehabilitation of scientific education and research in the war-damaged countries of Europe and Asia and the establishment of a technology institute in India.[43] It also described UNESCO working for the unity of the scientific world community by, for example, coordinating scholarly publications or tackling the problems of comprehension of scientific results between different languages and cultures.[44]

Two special issues of the *Courier* concerned the social implications of modern science. The first, 'Science and Mankind', appeared in February 1959 and was focused on science's role in keeping the peace in the world. 'Science, Man and Society', in July/August 1961, discussed world trends in scientific research. The *Courier* told of UNESCO's collaboration with the Pugwash movement,[45] which stressed the responsibility of the scientist in world affairs.[46] It also reported on scientific efforts at wiping out disease, usually publishing articles on human health issues linked to a World Health Organization (WHO) special campaign or theme, particularly if UNESCO was somehow involved. This was the case when UNESCO's Executive Board proposed that the Organization initiate international action in the struggle against cancer in cooperation with WHO in 1954.[47]

---

39   This chapter reviews science articles in issues of the *UNESCO Monitor*, from August to October 1947, and the *UNESC Courier*, from February 1948 to December 1965.
40   April, May August 1948, Oct. 1950.
41   Nov./Dec. 1956.
42   May 1963, Oct. 1964, March 1965.
43   Aug. 1948, July/Aug. 1950.
44   July 1949, Feb. 1950, Jan. 1954, Sept. 1956.
45   A movement established in the mid-1950s – following an initiative of Bertrand Russell and Albert Einstein for disarmament – which brought together scientists from the two superpowers and the developing countries.
46   July/Aug., Sept., Dec. 1950, Jan. 1951, Feb. 1959, July/Aug. 1961, Nov. 1964, Oct. 1965.
47   April/May 1954, April 1956, May 1958, April 1960, April 1962.

There were also articles published on natural sciences that did not mention UNESCO or its programmes at all but scientific achievements in general.[48] By publishing all these articles, the *Courier* was participating in the general popularization of science that characterized the era. Still, the majority of the natural sciences articles published in the period from 1947 to 1965 at least mentioned projects or work done by UNESCO in whatever domain was under discussion. Sometimes staff members of UNESCO wrote to the magazine about their own experiences working in the field. The following themes were the most frequently discussed in the field of natural sciences in the pages of the *Courier*.

NATURE CONSERVATION

The Hylean Amazon project and the Conference on the Amazon, which led to the creation of the first international research institute under the auspices of a UN organization, were treated at length in the pages of the *Courier*. So were two major conferences on nature conservation, one in Fontainebleau, France, and the other at Lake Success, in New York State (USA).[49]

From the beginning of the 1950s, the magazine covered the evolution of the International Arid Zone Institute Council, publishing, for example, a special series entitled 'Men against the Desert'. There was also a sixty-eight-page special issue 'The Conquest of the Desert', published in July/August 1955. Coverage of work in the arid zones became a mainstay of the publication.[50]

There was a strong emphasis on the protection of natural resources. Another special issue, published in January 1958 and entitled 'Man against Nature' concentrated on the harm humankind had done to the natural world. The September 1961 special issue, 'Africa's Wildlife in Peril', focused on disappearing species.[51] Here, and in later articles, the *Courier* was able to highlight UNESCO's efforts at wildlife preservation in Africa.[52]

ENVIRONMENT: EXPLORING THE UNKNOWN

The *Courier* often plunged into the subject of oceanography. In May 1955, it published an issue devoted to 'exploring oceans with science'. The subject was covered more exhaustively after the first international Oceanographic

---

48   July/Aug., Dec. 1950, Dec. 1951, Feb., May, April, July 1952, Feb. 1954, Jan. 1955, July 1957, Feb. 1958, May 1961, Dec. 1962.
49   June, July, Oct., Nov. 1948, March, June, Sept., Oct. 1949.
50   April, May 1949, Jan., Feb. 1950, June 1951, July 1952, Aug./Sept. 1953, April/May, Aug./Sept. 1954, July/Aug. 1955, May 1956, March, June 1957, May 1959, Jan., March, Sept. 1960, Feb., Aug. 1961, Dec. 1962.
51   Jan., April 1958, Sept. 1961.
52   Nov., Dec. 1963, Oct. 1964, May 1965.

Congress in 1959, with a special double issue entitled 'The Ocean's Secrets – New Adventures in Sciences'. Published in July/August 1960, the issue stressed international cooperation and various forms of research in the domain, making special mention of UNESCO's participation in the activities. When the Intergovernmental Oceanographic Commission (IOC) was created, the *Courier* published articles about projects executed under IOC's auspices, such as the International Indian Ocean Expedition.[53]

In addition to oceanography, articles on water in general poured from the pages of the *Courier*. The International Hydrological Decade was highlighted in a July/August 1964 special issue, 'Water and Life' (the name of a worldwide programme of scientific research at the time). Sometimes water was treated as a health issue, as when WHO chose the theme 'Clean water means better health' for the World Health Day of 1955.[54] Another important environmental factor in human health was examined in the *Courier* when WHO had launched a campaign against atmospheric pollution in 1959.[55]

The International Geophysical Year (1957–1958) led to a series of *Courier* articles concerning the Earth sciences. An entire special issue (September 1957) was devoted to the Year, drawing attention to breakthroughs in research and, as usual, efforts in international cooperation. Another special issue (October 1963) concentrated on 'Probing the Interior of the Earth'. This was during the international Upper Mantle Project, which researched the inner depths of the planet. The work done in the field of warning systems for earthquakes also received attention, as UNESCO planned a 24-hour earthquake service at its Headquarters.[56]

THE ATOMIC AGE, SPACE AND COMPUTERS

Research on the atom was, naturally, the most explosive scientific issue of the era. The *Courier* reported on the topic after the machinery for planning an international laboratory and for organizing other forms of cooperation in nuclear research in Europe had been established at a UNESCO meeting in 1952. When the European Organization for Nuclear Research (CERN) was being set up, the *Courier* published a special issue on the subject in December 1953. 'The Promise of Atomic Power' discussed the many ways in which radioisotopes could be used for the 'good of mankind'. While the differences between peaceful and military nuclear research were examined, the *Courier*'s attitude towards the

---

53 Oct. 1953, July 1954, May, Nov. 1955, March 1956, May, June, Sept., Dec. 1959, Oct. 1960, June, July/Aug., Sept., Oct. 1962, Dec. 1965.
54 March/April 1955, Sept. 1962, July/August 1964.
55 March 1959.
56 Jan., Oct. 1955, Sept. 1957, April, Feb. 1959, Jan. 1962, Dec. 1965

Atomic Age was generally positive. There were, however, some articles published on radioactive waste and the risk of radioactive contamination of the air.[57]

The conquest of space began to get attention in the *Courier* at the beginning of the 1950s. The magazine eagerly recounted the 'countdown for space flight.'[58] Articles presented 'astronautics' as 'the new science of space' and also as a potential area of peaceful cooperation between nations. In 1963, the magazine covered a symposium that UNESCO helped to prepare on 'Man in Space'. Telling of the 'miraculous tools' needed for space exploration, the *Courier* described modern, room-sized calculating machines. In the same year, the magazine wrote, a little less breathlessly, about the first international computation centre, which had opened in Italy.[59]

MAKING SCIENCE MORE ACCESSIBLE

Another theme that appeared regularly in the *Courier* was the popularization of science. Some of the articles examined different ways of going about this, such as the cinema, comic strips and science fiction. The *Courier* also told readers about publications prepared by UNESCO to make science more accessible and about the Organization's travelling science exhibitions. There were more philosophical articles as well, which examined the question of the gulf between scientists and society. Recipients of UNESCO's annual Kalinga Prize for the Popularization of Science were usually interviewed or wrote about their work in the magazine. The *Courier* also published numerous articles about scientific education, including a piece about one of UNESCO's best-selling works, the 'Sourcebook for Science Teaching'.[60]

The important place that the *UNESCO Courier* allocated to science strengthened the notion that science and technology should and could be used to improve the welfare of humankind and, as such, make an essential contribution to peace. The magazine's readership was mainly UNESCO's partners: government administrations, national commissions, UNESCO Clubs, and so on. It could be said that the *Courier* exemplified the twofold objective of science at UNESCO: to demonstrate the social implications of scientific research and to share knowledge.

---

57   June 1952, Dec. 1953, Oct., Dec. 1954, June, Oct. 1955, April, July, Nov. 1957, July/Aug. 1959, July/Aug. 1960, July/Aug. 1963.
58   March, April, Nov. 1951, April 1960, June, Nov. 1961, Feb. 1962, Jan. 1963, Jan. 1964.
59   Feb. 1952, Jan. 1960.
60   April 1948, March, April, Aug., Dec. 1949, Jan., Feb., May, Sept., Nov. 1950, March, June 1951, April, July 1952, July 1953, Jan., March, Aug./Sept. 1954, July/Aug. 1955, Feb. 1956, Feb. 1958, April 1960, Feb., June, Nov. 1962, May 1964, Feb., March, July/Aug. 1965.

PART II
BASIC SCIENCES AND ENGINEERING

INTRODUCTION
AT THE HEART OF IT ALL
The basic and engineering sciences as the key to civilization
Vladimir Zharov[1]

THE inherent function of the basic and engineering sciences is to carry out a thorough inquiry, leading to new knowledge that results in an understanding of natural phenomena, provides the scientific basis for human activity, and gives rise to the educational, cultural and intellectual enrichment of humanity. This in turn leads to technological breakthroughs and offers unique opportunities to meet basic human needs, yield economic benefits and promote science-based sustainable development. Think of it: progress in medicine and biotechnologies, lasers, information and space technologies, the internet, materials sciences, and environmentally sound industrial and agricultural technologies – they all stem from the advances in, and the alliance between, basic and engineering sciences, as do many other fruits of science that society enjoys. The basic and engineering sciences constitute an integral element of the culture of the civilization. They form a cornerstone of education that provides scientific and technological knowledge and skills needed by every citizen in order to participate meaningfully in the emerging knowledge-based society.

Although the basic and engineering sciences have become an indispensable means for development, their benefits are still unevenly distributed, and many countries find themselves excluded from the endeavour to create – and, consequently, to profit from – scientific knowledge. This unequal distribution cannot but deepen the divide in science education, technology, agriculture, health care, informatics and, ultimately, between North and South.

Adequate national capacity in the basic and engineering sciences is a major prerequisite for harnessing science in the service of society. Efficient applied research, technology transfer, modern education, health care and industry call for a sound national basic science infrastructure and necessitate a commitment to strengthen basic sciences capacities through national efforts and international cooperation. However, there exists a lack of support for the basic sciences in

---
1   Vladimir Zharov: Director, UNESCO Division of Basic Sciences (1984–98).

many countries, including developed ones. Moreover, a strategy of investment in favour of applied research, which exclusively seeks immediate short-term returns, has an adverse long-term effect on national basic science and requires determined remedial action.

Hence, from the inception of UNESCO there have been significant motivations for the Organization to launch and sustain its action in the basic and engineering sciences as one of the principle elements of its 'S'. Within the United Nations (UN) system, UNESCO has a unique mandate for the basic sciences, one that seems particularly relevant for an Organization that encompasses under one roof education, science and culture. UNESCO's role in the development of fundamental programmes in scientific and technological research and training has often been stressed in UN documents and resolutions (as, for instance, in the World Plan of Action).

This section presents a historical review of the Organization's activity in the basic and engineering sciences, a fairly comprehensive collection of essays devoted to major branches of the programme activities. These essays have been prepared by selected scientists who were substantially involved, within and outside of the Organization, in the implementation and development of the programme. The review does not constitute a formal and complete analytical history of the basic and engineering sciences in UNESCO. Instead, the evolution of the programme is presented on the basis of information from, and the views of, its real actors, who originated from different continents and countries. Such a review can serve as a historical document, providing interesting personal insights. More than just an exhaustive recapitulation of facts, it takes into consideration the human background of an ambitious scientific endeavour.

The evolution of the programme in the basic and engineering sciences and of its position within the overall programme of the Organization has been a rather complicated story, some aspects of which need to be mentioned prior to consideration of constituent branches of the programme. For instance, in the 1984–1985 Programme and Budget of UNESCO (22 C/5), the basic and engineering sciences were represented, within the Major Programme VI, 'The sciences and their application to development', by three well-articulated programmes embracing nine goal-oriented sub-programmes. Twenty years later, the 2004–2005 Programme and Budget (32 C/5) contains only one programme consisting of one sub-programme within the Major Programme II, 'Natural Sciences'. Such a change should not be misunderstood. It indicates, mainly, a considerable shift in the logic UNESCO introduced in the presentation of its programmes. Over time, there have been many modifications in the logical breakdown of the programme on basic and engineering sciences. But the programme per se has remained steadfast in its goals, strategy and intrinsic structure.

PART II: BASIC SCIENCES AND ENGINEERING

Over sixty years, UNESCO activity in the basic and engineering sciences has been focused in four interdependent directions: general programmes for research and training, special assistance to developing countries, promotion of international and regional cooperation, and fostering science and technology education. The emphasis on one direction or another has varied in response to proposals by Member States. But two principal goals have always remained in sight: the building up of national capacities in science and technology, and the advancement of the basic and engineering sciences in areas of importance for development. The general programme for research and training sought to meet various demands from Member States of UNESCO, coordinating proposals from international and regional non-governmental organizations (NGOs) in many fields of science.

Think of the general programme as a tree with five major disciplinary branches: pure and applied mathematics, physics, chemistry, biology and engineering sciences. A sixth branch, representing science and technology education, grew in a cross-disciplinary shape, intertwining with the other branches. A common feature of the disciplinary branches was the emphasis on the training of researchers and university science teachers, mostly from developing countries, in the areas of priority for the implementation of national projects, technology transfer, and improvement of science and technology education. Although disciplinary branches had and still have a wide profile, each of them accentuated some selected fields of action more than others, in line with the developmental orientation of the programme.

In mathematics, attention was given to the training in applied mathematics, mechanics and computing science. The activity in physics addressed in particular solid state physics, laser physics and theoretical physics that underlie many advances in, and applications of, modern physical sciences. The chemistry branch concentrated on promoting the chemistry of natural products, leading to the proper appreciation of available natural resources, and to their efficient use in national development.

Since 1945 we have witnessed revolutionary advances in biological sciences that have already had, and will continue to have, a profound impact on the quality of life and sustainable development. In this context, UNESCO actions in the life sciences concentrated on those areas that are at the root of progress in modern biology, namely cell and molecular biology, microbiology and neurobiology. The training of specialists in these areas has become a key prerequisite for addressing national needs relating to basic research, biotechnologies, health care, food production, and environmental programmes.

There is a wide range of issues being dealt with in the engineering sciences programme, the fifth branch of the general programme tree. To cite a few: dissemination of technological innovations, technologies for rural development, exploration of alternative sources of energy, construction of low-cost housing in urban and rural areas, application of computers in engineering, university–industry cooperation, development of ethics and codes of professional practice, and a cross-cutting project on technology and poverty eradication.

While a broad spectrum of services was provided in the disciplinary branches of the programme, it was recognized that some areas of science and technology need to be given particular priority. Because of the exceptional role these areas play in economic development and/or the advancement of science and technology, they needed to be addressed within ad hoc international programmes. As a result, the programme on the basic and engineering sciences incorporated several sub-programmes that concentrated on key areas of science and technology: informatics, biotechnology and applied microbiology, renewable energy, and the human genome.

The sub-programme on informatics eventually gave birth to the Intergovernmental Informatics Programme (IIP), which became then an important constituent element of the programme of the Communication and Information Sector of UNESCO. The sub-programme on renewable energy was a cradle for the UN World Solar Programme 1996–2005, for which UNESCO is the leading executive agency. UNESCO action in the area of human genome research helped to ensure that developing countries would have access to information – as well as help in training specialists – in this new scientific frontier of medical and biological research. Moreover, this action spurred the establishment of UNESCO's International Committee on Bioethics, the elaboration of the Universal Declaration on the Human Genome and Human Rights, and further UNESCO activities in science ethics within the Sector of Social and Human Sciences. Society is now witnessing the considerable impact of developments in biotechnology and applied microbiology. It is noteworthy that UNESCO's strategy of concentration on certain key issues is also being pursued in the area of AIDS research, through cooperation with the World Foundation for AIDS Research and Prevention, and the UNAIDS programme of the United Nations.

When contemplating the entire basic and engineering sciences programme in the historical context, it is important to pay attention to the character and the scope of its accomplishments. According to a rough estimate, some 500,000 researchers and university teachers – the majority of them young scientists from developing countries – have received training within the programme during its first sixty years of operation. The training was given in the framework of fellowship programmes and at numerous short- and long-term training courses, workshops, and seminars organized in cooperation with highly competent non-governmental scientific organizations, centres of excellence and local universities. Up-to-date scientific information on research advances and innovations in university science teaching was disseminated in more than 2,000 UNESCO-supported international and regional conferences.

Many centres of excellence in the basic sciences have been established, developed or supported all over the world within the UNESCO programme. In the physical sciences, the list of world-renowned centres includes: the European Organization for Nuclear Research (CERN, Geneva, Switzerland), the Abdus Salam International Centre for Theoretical Physics (ICTP) and the International Centre for Science and High Technology (ICS) (both in Trieste, Italy), as well as the intergovernmental Latin American Centre of Physics (CLAF, Rio de Janeiro, Brazil) and the recently created Synchrotron-light for Experimental Science and Application in the Middle East (SESAME, Amman, Jordan). The International Centre for Pure and Applied Mathematics (Nice, France) concentrates its activity on the training of mathematicians from developing countries; and the Banach Mathematical Centre (Warsaw, Poland) is active in Eastern Europe. In chemistry, the International Centre for Chemical Studies (Ljubljana, Slovenia), and the International Centre for Membrane Science and Technology (Kensington, Australia) are successfully operating in south-eastern Europe, and Asia and the Pacific, respectively. In biological sciences, UNESCO took the lead in the establishment of the International Centre for Cell and Molecular Biology (ICCMB, Warsaw, Poland), and a number of the Biotechnology Education and Training Centres (BETCENs), such as those situated in China, Hungary, Mexico, the Palestinian Autonomous Territories, and South Africa. The Latin American Centre for Biology (Caracas, Venezuela) is also an important regional centre cooperating with the Organization. All these centres, particularly ICTP in Trieste, have made considerable contributions to human resources development and to building up national capacity in science.

A significant role in promoting North–South and East–West cooperation in science has always been played by international NGOs, such as the International Mathematical Union (IMU), the International Union for Pure and Applied Physics (IUPAP), the International Union of Pure and Applied Chemistry (IUPAC), the International Union for Biological Sciences (IUBS) and others. They were and still are close partners of UNESCO in providing services in the basic sciences sought by developing countries. The excellence in science and efficiency of NGOs motivated UNESCO to assist in the creation of new NGOs for promotion of international cooperation in the basic and engineering sciences. For instance, UNESCO played a key role in creating and promoting the activities of such NGOs as the International Cell Research Organization (ICRO), the International Organization for Chemical Sciences for Development (IOCD), the International Council of Engineering Sciences and Technology (ICET), and the World Federation of Technology Organizations (WFTO).

Promotion of the activities of NGOs and centres of excellence is an important prerequisite for strengthening national capacities in science through training and cooperative research programmes. Such action is particularly efficient if it is combined with a sustained effort to widen participation of many national institutions in international and regional cooperation within a developed system of networks. To this end, at the very early stage, UNESCO's programme in the basic sciences introduced and pursued the long-term strategy of creating networks and fostering their activity. The success of UNESCO's effort can be demonstrated by listing the regional and international networks established and developed by the Organization (see Box II.1.1).

These networks involve numerous national research institutions and local universities, which they integrate into international scientific infrastructures. The

---

BOX II.1.1: REGIONAL AND INTERNATIONAL NETWORKS ESTABLISHED AND DEVELOPED BY UNESCO

Mathematics, physics, chemistry
- Latin American Mathematics Network (RELAMA)
- Arab Physics Education Network (ARAPEN)
- Asian Physics Education Network (ASPEN)
- Latin American Physics Network (RELAFI)
- Latin American Astronomy Network (RELAA)
- International Network for Chemical Studies (INCS)
- Network for Instruments Development, Maintenance and Repair (NIDMAR)
- Natural Products Research Network for Eastern and Central Africa (NAPRECA)
- Asian Network for Analytical and Inorganic Chemistry (ANAIC)
- Asian Pacific Information Network on Medicinal and Aromatic Plants (APINMAP)
- South and Central Asian Medicinal and Aromatic Plants Network (SCAMAP)
- Regional Network for Chemistry of Natural Products in South-East Asia
- Mediterranean Network for Science and Technology of Polymer-based Materials
- Latin American Chemistry Network (RELACQ)

Biological sciences
- International Molecular and Cell Biology Network (MCBN)
- International Molecular and Cell Biology Network for Asia and Pacific (IMCBN)
- Microbiological Resources Centres (MIRCENs) Network
- International Biosciences Networks (IBN) encompassing regional networks in Africa (ABN), Arab States (AraBN), Asia and Latin America (RELAB)
- Latin American Network on Human Genome
- Regional Network for Microbiology in South-East Asia

Cross-disciplinary networks
- TWAS/UNESCO Network of Training and Research Centres of Excellence in the South.

World Conference on Science (WCS, Budapest, Hungary, 1999) proclaimed in its *Declaration on Science and the Use of Scientific Knowledge* that progress in science requires such cooperation as research networks, including South–South networking. The WCS thematic meeting on 'Science in Response to Basic Human Needs' also emphasized that 'networking is an important instrument to implement international cooperation and a most valuable proactive action to create local conditions for scientific research and, consequently, effectively avoid brain drain from developing countries'.

A comprehensive historical survey of the activity of all networks, NGOs and centres of excellence developed by UNESCO would go far beyond the framework of the present section and its chapters. These bodies should, however, be recognized for their important accomplishments.

A retrospective look at the programme in basic and engineering sciences shows that UNESCO made a considerable contribution to fostering international cooperation and strengthening national capacities in science. One of the major accomplishments of the Organization was the creation of the international infrastructures that put at the disposal of Member States a unique scientific community acting in the spirit of solidarity and cooperation. These infrastructures provided, and may still provide, vital services to Member States, so long as the services requested are not affected adversely by short-sighted strategies and the sometimes meagre resources made available.

There is, moreover, cause for optimism. The design of the programme stems from longstanding experience in responding to the needs of Member States, and from the vast pool of knowledge of efficient partners in cooperation. The governing bodies of UNESCO highlight the priority of the existing programme in the basic and engineering sciences and express overwhelming support for new initiatives in its development. In 2003, at the thirty-first session of the General Conference, the Organization decided to take measures to reinforce intergovernmental cooperation in strengthening national capacities in the basic sciences and science education through establishing an International Basic Sciences Programme (IBSP) focused on major specific actions involving a network of national, regional and international centres of excellence or benchmark centres in the basic sciences.

In 2004, the IBSP programme was launched and started its operations. Its principal goals are:

- the building of national capacities for basic research, training, science education and popularization of science through international and regional cooperation in development-oriented areas of national priority;

- the transfer and sharing of scientific information and excellence in science through North–South and South–South cooperation;
- the provision of scientific expertise for, and advice to, policy- and decision-makers, and increasing public awareness of science and the ethical issues that progress in science entails.

The IBSP is not intended to replace the existing programme in the basic sciences, which has a wide profile and yields considerable results. Instead, it seeks to provide a unifying tool to ensure concentrated actions involving a network of national, regional and international centres so as to maximize regional cooperation in the basic sciences. Relying on the services of existing centres, or newly created centres of excellence, IBSP's mission is to foster excellence in other national, regional and international institutions, involving them in cooperation with IBSP-associated centres. Furthermore, the IBSP constitutes a major UNESCO action to carry out the follow-up to the World Conference on Science in cooperation with the Academy of Sciences for the Developing World (TWAS), the International Council for Science (ICSU), and other partners such as the InterAcademy Panel (IAP).

There is a strong rationale for the coexistence of both the traditional activity in the basic sciences and the IBSP. The former provides a flexible means to respond to a variety of needs of Member States or proposals put forward by them. The latter serves as an instrument to concentrate efforts on selected major topics. It also provides a basic mechanism which allows Member States to have an ongoing direct impact on the planning and implementation of the programme through the Scientific Board – consisting of renowned scientists representing all regions and the major disciplines of basic science – that has been established to monitor the IBSP. The Scientific Board elected as its first chairman Herwig Schopper, a world-famous German physicist and former director of CERN, whose remarkable leadership in promoting cooperation in science is widely recognized by the international scientific community in industrialized and developing countries.

An important element of the IBSP strategy is that the new programme does not operate in isolation. Rather, it is a member of the family of international programmes of UNESCO in science (which includes the International Geoscience Correlation Programme [IGCP], the International Hydrological Programme [IHP], the Intergovernmental Oceanographic Commission [IOC], the Man and the Biosphere Programme [MAB] and the Management of Social Transformations Programme [MOST]). Hence, the Organization's programme in the basic sciences takes advantage of the type of intergovernmental mechanisms of cooperation practised in the above-mentioned programmes and coordinates the IBSP activity with them. Moreover, UNESCO was fortunate in creating, within its 'S', the symbiosis between the programme in the basic and engineering sciences and programmes addressing environmental issues and sustainable development. Such a symbiosis makes the science programme of the Organization particularly meaningful and balanced, and brings ample opportunities for further development.

The history of the basic and engineering sciences at UNESCO mirrors important developments in world science, offers instructive lessons, and reveals the pressing issues which must be addressed if the benefits of science are to be equitably shared by all nations. It is hoped that the chapters in this section will stimulate interest in the history of the programme and inspire future generations of scientists to continue UNESCO's mission of the pursuit of peace through knowledge and cooperation.

MATHEMATICS, PHYSICS AND CHEMISTRY
CALCULATED RISKS
Initiatives in pure and applied mathematics
Franck Dufour[2]

IN the wake of the devastation wrought by the Second World War, the international scientific community was eager to revive a spirit of peaceful cooperation. In 1951, the International Mathematical Union (IMU) was re-established and an interim committee set up to conduct the necessary negotiations with participating countries for the holding of the first General Assembly of the Union in 1952. That same year, the IMU was admitted to the International Council of Scientific Unions (ICSU, since renamed the International Council for Science) and became a major partner of UNESCO. As the only United Nations agency concerned with the discipline, UNESCO soon came to play a critical international role in the field of mathematics.

The necessity of developing research in mathematics arose from its broad scope of application in many growing fields such as physics, astronomy and computation. The emergence of the latter resulted in the founding, in 1952, of the International Computation Centre (ICC, in Rome, Italy) to conduct scientific research with a view to the improvement of calculating machines. The ICC also assisted in the training of specialists and scholars in the field, and provided an advisory and computation service to the scientific institutions of UNESCO's Member States. Progressively, the ICC steered its activities to the application of mathematics in informatics. It was later to become an independent International Bureau for informatics but was finally disbanded in 1987, since information technology was making considerable strides and being developed chiefly by the private sector.

Despite painful budget restrictions in the 1970s, UNESCO carried out major initiatives in mathematics. The Organization extended the activities of the International Centre for Theoretical Physics in the field of mathematics, to favour connections and exchanges between mathematics and physics. In 1972, the Banach Centre was established in Warsaw, Poland, under the auspices of UNESCO, to promote and stimulate international cooperation in mathematics,

---

2   Franck Dufour: UNESCO consultant in the basic sciences (2004–05).

especially between the East and West. In the 1970s and 1980s, Poland, with its geographic location, and cultural and mathematical traditions, served as a natural meeting place for East–West exchanges.

Recognizing the need for good mathematics training in the developing world, UNESCO, in 1962, created the Latin American Centre for Mathematics (CLAM) in Buenos Aires, Argentina. Seven years later, a regional pilot project in mathematics was set in motion in the Arab States. Following the recommendations of the nineteenth General Conference (1976), the International Centre for Pure and Applied Mathematics (ICPAM) was set up in Nice, France, in 1978. Its purpose was to increase manpower in mathematics and to provide assistance to the national institutions of the developing countries. At the same time, some projects were postponed owing to the lack of resources; but UNESCO maintained its assistance in the setting up and strengthening of regional scientific associations such as the South-East Asian Mathematical Society (SEAMS), created in 1972, and the African Mathematical Union, created in 1976.

The invention of the first microprocessors and the first microcomputers during the 1970s heralded a new era of information and communication that bloomed in the 1980s. From the early 1970s, UNESCO convened meetings of experts in order to submit its programmes in the field of computer sciences. Considering the central importance of mathematics and its applications with regard to informatics, relations between these fields were considerably expanded and reinforced in programme activities during the 1980s and 1990s. IMU and the International Centre for Pure and Applied Mathematics (ICPAM) were particularly active in organizing research workshops, courses and seminars with particular emphasis on the relations between mathematics and informatics. An internal evaluation of the training activities in mathematics (in cooperation with ICPAM) noted: 'Training received was considered by the participants as extremely useful for their work. Keeping up a correspondence with the former trainers is indicative of positive impact on their research and educational activities in their home institutions'. Before 1986, all ICPAM activities took place in France. But that year, ICPAM began to organize workshops and schools in developing countries, in cooperation with local institutions. The positive impact was immediate and strong. ICPAM went on to establish regional offices in Chile in 1992 and in China the following year.

On 6 May 1992, in Rio de Janeiro, Brazil, the IMU declared that 2000 would be the World Mathematical Year. In its 11 November 1997 plenary meeting, the UNESCO General Conference welcomed the initiative. That same year, UNESCO supported ICTP and the University of Wisconsin in the launching of the Programme for International Cooperation in Mathematics and its Applications (PICMA). This five-year project was designed to help the initiation and the enhancement of research and advanced study in developing countries. In addition to this initiative, the dissemination of mathematical information through African documentation centres and the establishment of the South-East Asian Mathematics Information Centre in Hong Kong (China) contributed to the strengthening of national and regional capacities for university mathematics education.

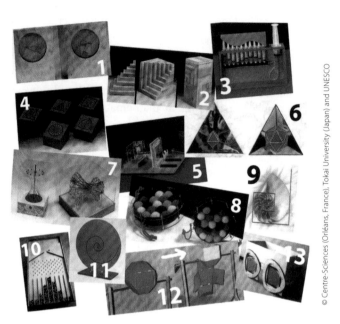

Following the World Mathematical Year 2000, a travelling exhibition entitled 'Experiencing Mathematics' was designed by prominent mathematicians from France and Japan, to show that mathematics is not only indispensable to daily life but can also be fun. This international exhibition, supported by UNESCO, began its world tour in 2004. Here, several models and experiments designed by the Research Institute of Educational Development, Tokai University, Japan.

During the 1990s, 'considering the central importance of mathematics and its applications in today's world with regard to science, technology, communications, economics and numerous other fields', and within the framework of the World Mathematical Year 2000 (WMY 2000), UNESCO supported activities to promote mathematics at all levels on a global scale. The celebration took place all around the world with a great number of events. All the International Mathematical Societies contributed to WMY 2000, by organizing special conferences, lectures for the general public, and other events. More than forty countries participated to the celebration of the year, with some – Argentina, Belgium, Croatia, the Czech Republic, Italy, Luxembourg, Monaco, Slovakia, Spain and Sweden – issuing special WMY 2000 stamps. It was a festive way to begin a new millennium, one in which the field of mathematics will only continue to grow in its centrality to our daily lives.

QUANTUM LEAPS FOR PEACE
Physics at UNESCO
Franck Dufour

THE end of the Second World War marked, of course, the disquieting dawn of the 'Atomic Age'. Particle physics became a prominent field. The international scientific community was acutely aware of the tremendous potential of discoveries in physics – and of the necessity for controlling their peaceful development.

Many actions carried out by UNESCO in physics during the 1950s and 1960s were aimed at improving scientific documentation and terminology. The International Advisory Committee for Documentation and Terminology in Pure and Applied Science (IACDT) and the Abstracting Board of the International Council for Science (ICSU) were the main bodies for implementing this initial step. They contributed significantly to standardizing terminology and compiling multilingual dictionaries in physics. UNESCO implemented its initial actions in physics through the financial support of existing programmes led by ICSU and other international scientific organizations. A substantial part of these funds were provided to the International Union of Pure and Applied Physics (IUPAP) for promoting research and training in the most promising fields of physics. Of particular interest was the growing field of quantum mechanics, which completely altered the fundamental precepts of physics. UNESCO's contribution to this field began by providing assistance to research in nuclear energy and by creating the European Organization for Nuclear Research (CERN) in 1954, pursuant to a resolution adopted by the General Conference in 1950. CERN was to build a large particle accelerator, on a site near Geneva, Switzerland, which would equip Europe with the most advanced facilities in fundamental nuclear research.

The same General Conference voted a resolution[3] for promoting the 'peaceful utilization of atomic energy'. The involvement of UNESCO in this field was extremely productive and resulted in 1957 in a major scientific conference convened by the Organization. The international Conference on Radioisotopes in Scientific Research highlighted the immense scope offered by radioactive elements, especially as tracers in a wide range of applications involving fundamental research, agriculture, medicine, industry and even archaeology. It was followed by a number of regional courses on the subject, and in 1957, the International Atomic Energy Agency (IAEA) was created to promote safe,

---

3   Resolution IV.1.2.2223.

secure and peaceful nuclear technologies. The collaboration between UNESCO, IAEA and the Italian Government resulted in the formation of the International Centre for Theoretical Physics (ICTP) of Trieste, Italy, in 1964. Its purpose was to train physics research leaders from developing countries at a genuine centre of excellence; it launched its first major programme in 1970.

In addition to these major implementations, and in order to fulfil its regional mission, the Division of Science Education, created in 1961, launched, two years later, a regional pilot project in physics in Latin America. This project started in São Paulo, Brazil, and extended its activities during the following years to several countries in the region. Meanwhile, eighteen countries signed the 1962 agreement for establishing the Latin American Physics Centre (CLAF). Holding its first General Assembly in 1966, CLAF became an intergovernmental organization that aimed at promoting the development of physics in Latin America, all the while maintaining a close cooperative relationship with UNESCO.

FROM BUDGET CUTS TO SUSTAINABLE DEVELOPMENT

Initiatives in the field of physics were deeply affected by budgetary cuts during the 1970s. However, many projects were eligible for special fund assistance from the United Nations Development Programme (UNDP), which made up partially for losses. Moreover, the implementation and the maintenance of existing programmes, and cooperation with non-governmental organizations, enabled UNESCO to sustain significant activities in physics. For instance, ICTP, which came into force in the early 1970s, continued to bear fruit, as the centre's activities enabled several hundred physicists, mostly from developing countries, to be trained each year. ICTP became a major forum for the international scientific community. In 1979, Abdus Salam, the founder and long-time director of ICTP, shared the Nobel Prize in Physics with Steven Weinberg and Sheldon Glashow, for the mathematical and conceptual unification of the electromagnetic and weak nuclear forces. The theory was subsequently confirmed by experiments carried out by the Italian physicist Carlo Rubbia at CERN, who was awarded the Nobel Prize in 1984.

In the late 1970s, UNESCO increased its activities in areas that were important for development, such as solid state physics, the physics of the oceans and the atmosphere, and the teaching of physics. The programmes in physics applied interdisciplinarity and intersectorality well before they became the official trends in UNESCO's policies. For instance, UNESCO contributed to many training courses and to the organization of conferences and symposia about medical applications of physics, such as medical imaging. ICTP and the IUPAP were the main partners for implementing projects in the field of physics for development during the 1980s. Training was provided in fundamental physics,

PART II: BASIC SCIENCES AND ENGINEERING

A UNESCO expert from the USSR instructing a class in spectroanalysis, at the Faculty of Sciences of Kabul, Afghanistan, 1964.

biophysics, medical applications of physics, nuclear physics, plasma physics, lasers, high energy and particle physics, solar energy and non-conventional energy sources, the physics of the atmosphere and oceans, geophysics, astrophysics, solid state physics and the physics of materials, and microprocessor applications in physics.

UNESCO pursed its regional mission by setting up the Asian Physics Education Network (AsPEN) in 1981, for providing state-of-the-art training for physicists in the Asian region. Five years later, it provided assistance for the creation of *Asia Physics News*, the first all-Asia bulletin for research and the teaching of physics. In Latin America, cooperation with CLAF was reinforced.

THE REVIVAL OF PHYSICS IN UNESCO'S PROGRAMMES

After a slight increase in the early 1980s, the budget dedicated to the Natural Sciences Sector decreased dramatically between 1985 and 1995. Although some planned activities were cancelled or postponed, a number of projects in the field of physics attracted support from extrabudgetary sources and budgetary reserve and were carried out. In 1993, the International Institute of Theoretical and

Applied Physics (IITAP) was founded to establish links between US universities and laboratories and educational institutions in the developing world; its funding cycle from Iowa State University (USA) and UNESCO ended in 2001. In 1994, the scientific programmes of the Organization in the field of physics were strengthened with the appointment of a Physics Action Council (PAC), responsible for advising UNESCO on the shape and implementation of physics programmes designed to promote the widest possible participation of the world's physicists in the international physics enterprise.

The World Conference on Science, held in Budapest (Hungary) in 1999, triggered a major phase of change in UNESCO's natural sciences programmes. Facing a global need for sustainable development, Member States were urged to foster capacity-building in science and to share technology with developing countries. In this respect, the most successful activity in the field of physics has been the development of the Synchrotron-light for Experimental Science and Applications in the Middle East (SESAME) in Jordan, inaugurated in January 2003. That same year, the thirty-second session of the General Conference of UNESCO adopted a resolution supporting the initiative of many organizations and physics societies to declare 2005 as the World Year of Physics. This initiative contributed significantly to promoting cooperation in research and training. People rarely speak anymore of the 'Atomic Age'. But the twenty-first century promises to hold ever-greater challenges, for scientists and for all of civilization, in the field of physics.

TAMING THE ATOMIC TIGER
The creation of the International Centre for Theoretical Physics
André M. Hamende[4]

ABDUS Salam was a man of vision. Born in Jhang, Pakistan, in 1926, educated at Panjab University (India), and at St John's College and Cavendish Laboratory, Cambridge (UK), Salam was professor of theoretical physics at the Imperial College in London and a member of the Atomic Energy Commission of Pakistan. His background gave him a special insight into the isolation faced by

---

[4] André-Marie Hamende has been associated with the International Centre for Theoretical Physics since 1964. At his retirement in 1990, he was Senior Administrative and Scientific Information Officer.

PART II: BASIC SCIENCES AND ENGINEERING

Professor Abdus Salam (1926–1996), founder of the International Centre for Theoretical Physics (ICTP).

physicists from developing countries who had been trained in advanced scientific institutions of the North and then chosen to return to their homelands. Salam also understood how that sense of isolation led to the dilemma of the 'brain drain': the migration of scientists from their home countries to more developed nations where they could find greater support for their work.

It was in September 1960, at a plenary meeting of the fourth regular session of the General Conference of the International Atomic Energy Agency (IAEA) in Vienna, Austria, that Abdus Salam made an inspired proposal. He called for the creation of an international institute for theoretical physics. The institute would host, at any given time, about fifty scientists: one-third from developing countries, one-third from Eastern Europe and one-third from industrialized nations. The majority of them would be short-term visitors doing research in theoretical nuclear physics, controlled thermonuclear fusion, nuclear reactor physics and elementary particle physics. Salam's vision would eventually result in a landmark of scientific cooperation: the International Centre for Theoretical Physics (ICTP) in Trieste, Italy.

Following a resolution of that 1960 General Conference, IAEA Director-General Sterling Cole convened a panel of eminent theoreticians and members of leading international research institutions, who were invited to express their opinion regarding the setting up of the new physics institute. Hilliard Roderick from UNESCO, a member of the panel, hinted that though his organization was not a funding agency, it would be prepared to collaborate, especially if the research fields covered would be extended to domains of interest to UNESCO.

Debates on the creation of the new institute continued at meetings of the General Conference and of the Board of Governors of the IAEA, then directed by Sigvard Eklund, where the most enthusiastic support came from the delegates of developing countries. At many of these sessions, A. Pérez Vitoria, representing UNESCO, conveyed the interest of the Organization and that of its Director-General, René Maheu, in joining the IAEA in running the proposed institution. Finally, the decision to establish ICTP was made by the Board of Governors in June 1963.

ICTP was inaugurated in October 1964. Abdus Salam was appointed as its director. Paolo Budinich, professor of theoretical physics at the University of Trieste and untiring promoter of the candidacy of his city as the seat of the Centre, was named deputy director. As the training function of the Centre had been strongly recommended at various phases of the debates in Vienna, an Advanced School for Theoretical Physics was created in Trieste upon agreement between the IAEA, UNESCO and the Istituto Nazionale di Fisica Nucleare (INFN), signed by the three parties in November 1964. The contribution provided by UNESCO was utilized to support research fellows jointly selected by the Organization and the IAEA. The bulk of the Centre's annual budget came from the Italian Government (US$278,000). The other contributors were the IAEA (US$55,000) and UNESCO (US$110,000, spread over five years).

During the first four years of its existence, the Centre was located in downtown Trieste, as a temporary seat. ICTP moved to its permanent location near the Miramare Park in 1968. Early research and training was focused on theoretical physics of elementary particles, of the nucleus and of plasma. The academic year 1965/66 featured a ten-month workshop on plasma and fusion physics with the participation of outstanding experts from the United States, Western Europe and the Soviet Union. Five extended meetings and seminars, some lasting from two to three months, were held during the same period. Eight Nobel laureates took part in a three-week symposium on contemporary physics, and nine of the other lecturers would go on to receive the Nobel Prize themselves in the years which followed.

Abdus Salam's vision of ICTP as an intellectual gathering place for scientists from developing countries quickly came to fruition. During those first four years,

270 scientists from about 30 developing countries[5] took part in the activities of the Centre, in one capacity or another. By 1970, the total annual number of scientific visitors was 580, of whom 220 were nationals from 35 developing countries. From that point on, participation grew steadily. In 1980, out of 1,500 visitors, 615 came from a total of 72 developing nations. A decade later, in 1990, there were 4,000 scientific guests, 2,300 hailing from 92 developing countries. The high quality of the activities carried out at ICTP was unanimously acknowledged in the international scientific community.

As the agreement for an annual contribution from UNESCO for the operation of the Centre was to expire in 1968, negotiations for new arrangements with the IAEA started in 1967, taking the experience gained since 1964 into account. In its final version, the agreement stated that the scientific activities of the Centre would constitute a joint programme carried out by both organizations. It was decided that the members of the ICTP Scientific Council would be jointly selected and appointed by the directors-general of UNESCO and the IAEA but that the director of the Centre and the professional staff would be members of the Agency. Regular consultations would take place regarding the scientific activities and the preparation of annual budgets. The financial contribution of both organizations would be US$150,000 each per calendar year. The agreement was concluded for a period ending on 31 December 1974. It was signed by the two directors-general in July 1969.

UNESCO brought in its wake a financial contribution for several years from the United Nations Development Programme (UNDP). Other notable financial support to ICTP programmes came from the Ford Foundation (from 1967 to 1973), and from the Swedish International Development Authority (Sida, beginning in 1969). Sida's assistance has continued under the auspices of the Swedish Agency for Research Cooperation. Enhanced funding allowed for the expansion of the Centre's programmes into scientific areas that were not included in its initial scheme. There was a strengthening of the activity in solid state physics and the introduction of mathematics. Physics of the oceans, of the atmosphere, of the Earth, of non-conventional energy, of atoms, molecules and lasers, as well as informatics, were all gradually introduced in the curriculum.

As the prestige of ICTP grew in the 1980s, the Italian Government increased its annual financial support to the Centre, and this led to further expansion. With this enlargement, the centre of gravity of the disciplines covered at ICTP shifted from fields of relevance to nuclear energy, to others pertaining to the sphere of

---

5   From Africa: Ghana, Morocco, Nigeria, South Africa, Sudan, Tunisia, United Arab Republic; from Asia: Ceylon, China, India, Iran (Islamic Republic of), Iraq, Israel, Jordan, Lebanon, Republic of Korea, Singapore, Syrian Arab Republic, Turkey, Vietnam; from Latin America: Argentina, Brazil, Chile, Jamaica, Mexico, Peru, Venezuela, Uruguay.

interest of UNESCO. As a result of these developments, both parent organizations of ICTP decided that the administrative and managerial responsibility of the management of the Centre ought to be transferred from the IAEA to UNESCO. This change was implemented in 1996 under the mandate of Hans Blix and Federico Mayor as respective directors-general of the two organizations.

Decade after decade, Abdus Salam's vision thrived. In all, some eighty Nobel laureates have lectured at ICTP. More than 60,000 scientists from 150 countries have taken part in its activities. The Centre has continued to be a magnet for scientists from developing countries and a crucial meeting place for thinkers from around the world. It has embodied the spirit of international scientific cooperation, the free exchange of ideas across cultures. In 1979, Abdus Salam himself was honoured with the Nobel Prize in Physics. He often said that 'scientific thought is the common heritage of mankind'. After his death in 1996, his institutional brainchild was given a new name. It is now known as the Abdus Salam International Centre for Theoretical Physics.

WARM RELATIONS AFTER THE COLD WAR
The Physics Action Council, 1993–1999
Irving A. Lerch[6]

IN the early 1990s, physics – one of the most global of scientific enterprises – began to look to UNESCO as an instrument to unify and coordinate the international outreach of the global physics communities. Several developments fuelled this interest: the signing of a new tripartite agreement between UNESCO, the International Atomic Energy Agency and the Government of Italy assigning governance responsibility to UNESCO for the International Centre for Theoretical Physics (ICIP); the sponsorship by UNESCO of the European Organization for Nuclear Research (CERN), the rise of developmental programmes in Latin America, Africa and Asia; and the growing realization that a new framework was needed to coordinate research, education and intellectual exchange worldwide. Most important, physics had become a tool for intellectual, cultural and economic development, in furtherance of UNESCO's fundamental goals.

On 24–25 June 1993, Federico Mayor, Director-General of UNESCO, and Adnan Badran, Assistant Director-General for the Natural Sciences Sector,

---

6   Irving A. Lerch: Member of Physics Action Council of UNESCO (1993–99).

convened a consultative meeting in Paris entitled 'UNESCO and the International Physics Community: An Agenda for Scientific Cooperation'. Invited participants included distinguished physicists from Russia, the American, French, and German Physical Societies, the Association of Asia-Pacific Physical Societies, the European Physical Society, the Commission of the European Communities, ICTP, the International Union of Pure and Applied Physics (IUPAP), the Organisation for Economic Co-operation and Development (OECD) and other organizations. The organizers of the conference were Siegbert Raither, head of mathematics and physics programmes in the Division of Basic and Engineering Sciences, and Vladimir Zharov, the director of the Division.

At the conclusion of the meeting, the participants submitted a recommendation to Director-General Mayor recognizing UNESCO's growing importance to international science and recommending that priority be given to physics in developing countries, sustaining excellence of the physical sciences in east-central Europe and the emerging states of the former Soviet Union, and promoting mega-projects in the physical sciences. A statement of general principles emphasizing the important roles of learned societies and partnerships between the public and private sectors recommended that a Physics Action Council (PAC) be convened to superintend the goals of UNESCO.

The Director-General appointed ten senior physicists to the council and charged them with promoting international cooperation and collaboration and to provide counsel and guidance to UNESCO and its management. Donald Langenberg, president of the American Physical Society and chancellor of the University of Maryland System (USA) was made Chair. The other appointees were Carlos Aguirre (Bolivia), F. K. A. Allotey (Ghana), Sivaramakrishna Chandrasekhar (India), Yang Guo-Zhen (China), Michiji Konuma (Japan), Norbert Kroó (Hungary), Irving Lerch (United States), Yuri Novozhilov (Russia) and Herwig Schopper (Germany). Siegbert Raither was assigned as secretary.

The PAC convened an organizing meeting in April 1994 and developed a work plan based on three working groups: Large Physics Facilities (Schopper, Chair; Aguirre and Kroo), Communications Networks for Science (Lerch, Chair; Novozhilov and Langenberg), and University Physics Education (Konuma, Chair; Yang, Chandrasekhar and Novozhilov). The PAC remained an active component of the Organization's programmes throughout the period of Mayor's tenure as Director-General (1994–99).

Almost immediately, the Council's working groups began an aggressive programme of international outreach, organizing meetings and workshops on telecommunications, access to large international research facilities and new approaches to invigorate physics education. The physics facilities working group held meetings in Belgium, Cuba, France and Japan and focused on projects in developing countries, such as: the Pierre Auger giant air shower project in

Latin America, small accelerators in the Caribbean, and the SESAME project in Jordan. Telecommunications workshops were convened in China, Ghana, Japan, the Philippines, Russia and Ukraine – supported with resources provided by UNESCO, NATO, the National Science Foundation (USA), and other funding organizations. Training for network administrators, programmers and technicians designed to promote internet access was the focus of these efforts. The role of physics education in capacity-building led the education working group to survey successful programmes and to examine ways of finding exceptional talent at the M.Sc. and Ph.D. levels in developing countries. In all cases, these efforts were directed to complementing and strengthening the programmes of the Natural Sciences Sector and ICTP.

Today, the success of the PAC's activities may be seen in the UNESCO-supported SESAME project, the programmes of the Communication and Information Sector of the Organization, and in the increased awareness of the crucial role that science education plays in developing the intellectual capacity of a nation.

OPENING SESAME
A landmark of scientific cooperation in the Middle East
Clarissa Formosa Gauci[7]

THE SESAME (Synchrotron-light for Experimental Science and Applications in the Middle East) Centre was established under the auspices of UNESCO, along the model of CERN (the European Organization for Nuclear Research). SESAME is intended to provide a first-class, fully competitive source for synchrotron light, which has a multitude of research and development applications, in many areas: for instance, material research, nanotechnology, biology, environmental problems, medical applications and archaeology, to name but a few. Like UNESCO, it has also the goal of seeking the promotion of peace through science.

At the turn of the twenty-first century, there were about sixty synchrotron-light sources in the world (including those in Brazil, France, Germany, Russia,

---

7   Clarissa Formosa Gauci: Assistant Programme Specialist (since 1981), UNESCO Division of Basic and Engineering Sciences.

Thailand, the UK and the USA). The use of synchrotron radiation is considered an important means of promoting many modern technologies, as well as of fostering interdisciplinary activities. Yet no such facility existed in the Middle East, although a need had been recognized by eminent scientists, such as Nobel laureate Abdus Salam, more than twenty years earlier.

It was in 1997, during a workshop organized by the CERN-based Middle East Scientific Cooperation (MESC) group headed by the world-renowned Italian physicist Sergio Fubini, that a concrete proposal to set up an international synchrotron source in the Middle East was first put forward. At the time, Germany had just decided to decommission its facility, BESSY I, since a newer one was being built in Berlin. When initially constructed, this BESSY I synchrotron-light source was valued at US$60 million. At the request of Sergio Fubini and Herwig Schopper, former director-general of CERN, the German Government agreed to donate its components to the proposed centre for the creation of a state-of-the-art facility.

The plan was brought to the attention of Federico Mayor, then Director-General of UNESCO, who, in July 1999, called a meeting at the Organization's Headquarters of delegates from the Middle East and other regions. The outcome of the meeting was the launching of the project and the setting up of an International Interim Council (under the chairmanship of Herwig Schopper) so that the necessary measures could be taken to prepare the establishment and operation of such a centre. UNESCO agreed to provide direct help for the creation of the Center. Delegates at the World Conference on Science (WCS) – convened in Budapest, Hungary, in June–July 1999 by UNESCO and ICSU – welcomed the proposal to establish the SESAME Centre.

A detailed technical study and proposal for the SESAME synchrotron-light source was prepared in October 1999. It confirmed that using the essential components of the BESSY I machine would be substantially cheaper than building a totally new machine.

Jordan – and, more specifically, the Al-Balqa' Applied University in Allan – was elected to host the Centre by the International Interim Council. Allan is situated 30 km from Amman and 30 km from the King Hussein/Allenby Bridge, which crosses the Jordan River. The total area for the facility is 6,200 m². The basis for choosing this location was the guarantee that all scientists of the world would have free access to the Centre and the commitment by the Government of Jordan to provide the land, and the existing buildings on the land, as well as up to US$6 million for the construction of the building to house the project.

In January 2000, Koïchiro Matsuura, Director-General of UNESCO, informed the German Federal Ministry of Education and Research that he was ready to take the necessary steps for the setting up of SESAME. This prompted arrangements to be made for the dismantling, packing and temporary storage in

Berlin of the BESSY I machine. In June 2002, the component parts of BESSY I were shipped to Jordan.

In May 2002, the 164th session of UNESCO's Executive Board approved the establishment of the SESAME Centre under the auspices of UNESCO, and the Director-General of UNESCO, invited Member States to become Members or Observers of the Centre. Six countries rose to the call, and in April 2004 SESAME formally came into existence as a fully fledged intergovernmental organization cooperating with UNESCO, the depository organization of the Centre.

In July 2004, the Council of SESAME (the Centre's statutory governing body) ratified various governing instruments such as the Centre's Rules of Procedure, Financial Rules, and Staff Rules and Regulations, and it signed a Seat Agreement with its host country, giving the Centre privileges similar to those which CERN obtains from its host states. In April 2005, a Royal Decree was issued in Jordan for the approval of the decision of the Jordanian Cabinet ratifying the Seat Agreement. UNESCO helped SESAME in the drafting of all these governing instruments.

In July 2004, a tripartite cooperative agreement was signed between CERN, Jordan and SESAME. Proposals for other cooperative arrangements have been made to SESAME.

The SESAME Council is advised by four committees: the Beamlines Committee for the conceptual design of some of the phase I beamlines; the Scientific Committee for the planning of the overall scientific management of the programme; the Technical Committee for the design and upgrading of the SESAME machine; and the Training Committee for the training of personnel and users.

In November 2004, the Technical Advisory Committee of SESAME endorsed the overall concept for the upgrading of the accelerator system, taking into account the request of the potential users formulated by the Beamlines and Scientific Advisory Committees of SESAME. In December 2004, the Council of SESAME approved the final design of the SESAME machine for a final energy of 2.5 GeV. This implied that the period of the conceptual design was terminated and the engineering design phase started.

The six Phase I beamlines were identified by the Beamlines and Scientific Advisory Committees, based on discussions with future users; their layout was considered and the type of experiments to be carried out with them examined. Five scientific directions have been identified for SESAME: physical science, biological and medical sciences, environmental sciences, industrial applications, and archaeology.

SESAME also covered much ground in the training of scientists and technicians from the region. In fact, it launched a training programme virtually immediately after the International Interim Council was set up. Initially, emphasis

PART II: BASIC SCIENCES AND ENGINEERING

A brochure on the Synchrotron-light for Experimental Science and Applications in the Middle East (SESAME), explaining its scientific background, history, activities and administrative organs (UNESCO, 2005, 12 pp.).

was on the training of accelerator experts. This training, which received support from the International Atomic Energy Agency (IAEA) and synchrotron radiation facilities in Europe and the United States, was satisfactorily completed, and in 2004 emphasis was moved to training people who will design, operate and carry out maintenance of the SESAME beamlines and to training potential users. In line with this, it is placing young scientists and technicians in world-class synchrotron-light source centres so that they may receive training in beamlines. These centres have frequently granted fellowships to young scientists and technicians. SESAME is also organizing road shows and workshops in the region for the attention of users, and at the Third Users Meeting (Turkey, October 2004) some 100 persons took part. The Fourth Users Meeting was held in Jordan in December 2005. SESAME plans to expand its training programme thanks to extrabudgetary resources it expects to receive.

Because of the sensitive and very specific political situation in the region and the need to promote international cooperation there, UNESCO will continue to be closely associated with SESAME and will offer assistance as required. It will also continue to encourage other countries to join the Centre, whether as Members or Observers, and will help them comply with the necessary conditions for this.

UNESCO will also help develop networking with synchrotron radiation centres in other countries. It will also associate SESAME with the Organization's newly created International Basic Sciences Programme (IBSP). Promoting the use of the SESAME Centre for the implementation of UNESCO's programme in physics and allied areas of the basic sciences, the Organization will help to spread excellence in the region by conducting training activities using the facilities available at the Centre. As stipulated in the Statutes of SESAME presented to the 164th session of the Executive Board, UNESCO will remain the depository for the Statutes of SESAME, and a representative of the Director-General of UNESCO will be on the Council of SESAME.

The construction of the building that is to host the Centre started in August 2003 and is scheduled to be completed by January 2007. It is being fully financed by the Jordanian Authorities. In addition, Jordan is also financing a technical building and an independent electrical power line for SESAME, thereby ensuring that the site will have a fully independent supply of energy.

In September 2005, membership stood as follows: Bahrain, Egypt, Israel, Jordan, Pakistan, Palestinian Authority, and Turkey. On 7 July 2005, the Iranian Parliament ratified the membership of Iran (Islamic Republic of). Observers were Germany, Greece, Italy, Kuwait, the Russian Federation, Sweden, the United Kingdom and the United States. Other countries are expected to join either as Members or Observers. It is hoped that SESAME will be a model of international scientific cooperation in the region.

FINDING THE RIGHT CHEMISTRY
The International Union of Pure and Applied Chemistry (IUPAC) and UNESCO
Mohammed Shamsul Alam[8]

UNESCO is an acronym that is familiar to people everywhere. One of its most steadfast partners is less famous among the general public; but for chemists around the globe, IUPAC (the International Union of Pure and Applied Chemistry) is a household name. Founded in 1919, the Union has succeeded in

---

8   Mohammed Shamsul Alam: UNESCO (Since 1982), Head of UNESCO Offices in New Delhi and Tehran; UNESCO representative to Iran (Islamic Republic of); Senior Programme Specialist for Chemistry, Division of Basic and Engineering Sciences.

promoting worldwide communications between chemists in academia, industry and altruistic organizations. Since the founding of UNESCO in 1945, IUPAC has supported the Organization in its ambitious goal to promote peace and progress through international scientific cooperation. One of the most successful aspects of this partnership has been the range of efforts to assist chemists in developing countries.

With funding from the United Nations Development Programme (UNDP), UNESCO in the 1960s began efforts to strengthen the science faculties of universities in newly emerging countries. The Natural Sciences Sector, through its Division of Basic and Engineering Sciences, had the principal responsibility in this endeavour. It was up to staff members at UNESCO's five Regional Offices for Science and Technology to ascertain the needs of Member States and to coordinate the day-by-day administration of the various projects. Direct aid was provided in the form of expatriate teaching staff and scientific equipment.

Progress in university chemical education moved very quickly – helped undoubtedly by the involvement of IUPAC right from the planning stage. Efforts included a joint UNESCO-IUPAC project on the production of low-cost, locally produced equipment; a programme for university–industry cooperation, and another focused on the training of technicians. While there was also a programme for bringing students from developing countries to more advanced nations to pursue their education, the emphasis was always on improving teaching and research in local environments.

In 1974, planning began for two new programmes in chemistry: one was focused on research and training in natural products chemistry; the other related to laboratory curriculum development at the university level. The former began almost immediately as the Regional Network for the Chemistry of Natural Products in South-East Asia. The latter originated as a series of laboratory workshops, in different geographical areas, devoted to the development of a series of experiments which could be used in the early phases of an undergraduate chemistry course. Natural products chemistry eventually spread to other areas, particularly South and Central Asia, and to other associated fields, such as medicinal and aromatic plants. Meanwhile, programmes on environmental chemistry were launched and, after a long planning period, programmes in analytical and organic chemistry became operational. In 1981, UNESCO created the International Organization for Chemical Sciences in Development (IOCD), which was designed as a mechanism through which chemists in both developing and industrial countries could collaborate in improving and strengthening the chemical sciences in the poorer countries.

In 1975, a special project in university chemical education began with a laboratory-based workshop at the Seoul National University (Republic of

Korea), attended by teachers from countries in South-East Asia. Participants carried out experiments on specimens collected from the region prior to the workshop and then wrote up the results in the best manner for presentation to students. The experiments were subsequently rechecked and then published as *A Sourcebook of Chemical Experiments, Volume 1*. A second workshop was held at the University of Jordan in 1976, which followed much the same pattern, with university teachers from seven Arab States. This workshop resulted in the *Sourcebook of Chemical Experiments, Volume 2*. In 1977, a workshop for Latin American countries was held in Mexico. Here the Asian approach was adapted to Latin American needs. The result was the *Manual de Experimentos Quimicos, Tome 3*, which served as background material for a series of experiments in Latin American countries. The next working group met in Lomé, Togo, in 1977, and prepared a series of experiments (in English and French), which have been circulated to universities in African countries for student testing. The first laboratory-based workshop in France was organized at l'Université des Sciences et Techniques du Languedoc, Montpellier, in association with Société Chimique de France; *Manuel d'expériences de Chimie, Vol. 5* was the result.

All these workshops also produced sets of recommendations, which laid down priorities for future activities. These included the need to find alternative sources for laboratory teaching equipment in countries that could not hope to import it. The search for new sources of laboratory equipment resulted in a programme in designing, developing and producing laboratory teaching equipment locally, coordinated by IUPAC and directed by the University of Delhi (India). It expanded into other Asian and African nations, and eventually equipment for undergraduate laboratories was being produced in many locations in developing countries.

Developing countries have emphasized the need for their universities to help in their socio-economic development, and UNESCO endeavours in science and technology have reflected this through programmes in university–industry cooperation. The start of such a programme for chemistry began with an International Symposium in Toronto, Canada, in 1978, attended by participants from seventy-eight countries, followed by a series of Regional Symposia. Important elements of the university–industry project in chemistry include: national committees with representatives from university, industry and government; student placement in industrial jobs on an inter-country basis; training courses for young graduates wishing to set up small-scale industrial companies; and research and development services, including the monitoring and control of pollution.

Under the UNESCO fellowship programme, awards are given for scientists from developing countries to continue their studies overseas, both at the

PART II: BASIC SCIENCES AND ENGINEERING

Postgraduate student, the first female research scholar of the UNESCO-assisted Indian Institute of Technology, working on solid catalysts, Kharagpur, India, 1955.

UNESCO pilot project to teach university teachers in Asia how to use modern chemistry equipment, Bangkok, Thailand, 1965.

pre-doctoral and postdoctoral levels. At the same time, UNESCO has sponsored postgraduate, postdoctoral courses of one year at selected universities. Examples are courses in Physical Chemistry at the Catholic University of Leuven in Belgium; Natural Products Chemistry at the University of Uppsala in Sweden; Physical Chemistry at the Charles University, Prague, Czechoslovakia; and an introductory course in research techniques at the University of New South Wales in Australia.

When it comes to the creation of regional networks, UNESCO has always favoured an approach that relies on existing local facilities, institutions and resources. This approach offers an efficient way of using the very limited funding available, provided mainly by the Member States themselves (most of which are developing countries), supplemented by contributions from UNESCO's Regular Programme budget. The most successful of these UNESCO Regional Networks is in the South-East Asia region, where the chemistry of natural products is being studied. The region extends from Malaysia to Fiji, from the Republic of Korea to New Zealand – an area with a real potential for development and exploration of natural products. The network consists of ten Member States, and the programme is directed by a coordinating Board (one chemist from each participating Member State) and monitored by Chulalongkorn University, Bangkok, Thailand.

UNESCO established another network in Africa, where the sponsored activities relate to environmental chemistry training courses and use the various analytical chemical techniques associated with environmental monitoring. In particular, annual training courses on pesticide analysis have been held in both East and West Africa. A joint UNESCO-FACS (Federation of Asian Chemical Societies) Asian network for analytical and inorganic chemistry includes workshops and training courses in field-oriented analytical chemistry and in management. This network also took over responsibility for instrument technician training in Asia.

UNESCO also organizes or supports research seminars and symposia in developing countries. One of the more prominent series is the Asian Symposia on Medicinal Plants and Spices, held every four years. As with all UNESCO scientific cooperation programmes, the basic aim is to bring together scientists from across the region. It is hoped that this will encourage the free flow of information on research findings throughout the region, to promote mechanisms for working out priorities and strategies, and to foster research cooperation between individuals and institutions, thus minimizing costs and maximizing participation of scientists in the region.

The approach used by UNESCO to achieve these objectives involves offers of financial assistance to aid in the exchange of scientists and to promote the holding of scientific meetings. The chemistry meetings sponsored by UNESCO mainly take the form of workshops – roughly twenty-five people meeting for about one week. These meetings are organized in areas of interest to the countries in the region to highlight current trends in fundamental or applied research. This has allowed the setting up of cooperative research carried out by a network of those institutions in the region working in similar or related fields.

Another critical endeavour in the field of chemistry has been the Trace Element Institute, established in 1996 at Lyon, France, under the auspices of the UNESCO Natural Sciences Sector, Division of Basic and Engineering Sciences. So far, sixteen satellite centres of the Trace Element Institute have been established all over the world, and the multidisciplinary nature of the centres has enhanced international scientific collaboration.

UNESCO has also been instrumental in the Microscience Experiments Projects. In twenty-two African countries, this project was developed in accordance with the Gaddafi International Foundation for Charity Associations and, in some other countries, under the auspices of the Islamic Educational Scientific and Cultural Organization. The Microscience Experiments Project has provided an innovative methodology for practical work in science teaching that is safe, affordable, and adaptable to various situations in developing countries. Member States have recognized the potential of the project for strengthening science and technical education.

Finally, the donation programme, which has existed on a large scale since 1997, has allowed for scientific books, journals, chemicals and small-scale equipment to be provided to poorly resourced universities in developing countries. These donations have invariably proven to be very popular. And they represent more than charity. A simple but powerful ideal has always been at the heart of the UNESCO-IUPAC endeavour in implementing chemistry programmes in developing countries. It is the ideal of helping people help themselves.

INFORMATICS, INFORMATION AND COMMUNICATION TECHNOLOGIES
MAKING EVERYTHING COMPUTE
ICTs at the crossroads of applied mathematics, physics and chemistry
Rene Paul Cluzel[9]

INFORMATION and communication technologies (ICTs) have long been considered a revolutionary factor in industries, services, research, teaching and cultural activities. However, discrepancies in access to ICTs abound, fuelling the digital divide between North and South, rich and poor.

Based on research and development and then on industrial needs, the information technology explosion was mainly concentrated in the industrially and economically advanced countries. In the 1990s, the costs of developing information technology research were estimated at about US$50 billion, the same amount as the annual investment in new software. About 95 per cent of all information technology was situated in the industrialized countries and represented between 3 and 5 per cent of the GNP of these countries, whereas in developing countries it represented less than 1 per cent of their GNP. During this same period, spending on information technologies in industrialized countries represented about 90 per cent of total costs in North America, about 40 per cent in Europe, and between 20 and 25 per cent in Japan and South-East Asian industrialized countries. The rest of the world covered less than 20 per cent of the costs.

The risks of increasing the inequality between developed and developing countries were obvious, both in the production of information technology and in its usages and applications. This danger was discussed at the Intergovernmental Conference on Strategies and Policies for Informatics (SPIN) in Torremolinos, Spain, in 1978. The conference brought together experts and politicians

---
9   Rene Paul Cluzel: UNESCO Programme Specialist in Informatics; administered and coordinated projects on application of ICTs for capacity-building and education, Natural Sciences Sector (1987–89), Communication and Information Sector (since 1989).

Three books published by UNESCO (1998–2000) on information and communication technologies (ICTs).

responsible for information technology policy worldwide and set up a framework for UNESCO to define its programme of cooperation, specifically dedicated to what was not yet called ICTs.

Based on the results and recommendations of the SPIN conference, discussions took place on how UNESCO could act in this field in conformity with its mandate. The result was a recommendation by the twenty-second session of its General Conference in 1983 to establish the Intergovernmental Informatics Programme (IIP). An interim committee, created in 1984, drew up IIP's mandate, priorities, work methods and means of finance. This led to a recommendation to launch IIP, adopted in 1985 at the twenty-third session of the General Conference.

The creation of the IIP corresponded with a strategic change in international cooperation concerning information technologies, moving from a centralized to a decentralized model. In 1971, the Intergovernmental Bureau for Informatics (IBI) had been set up in Rome, Italy, under the patronage of UNESCO (where its statutes were kept), as a centralized computing centre for developing countries, with substantial support from France, Italy and Spain. Later, thanks to technological progress, information technologies were, even if not equitably, diffused worldwide, thus obviating the need for a central computing centre. This led to a decentralized network structure based on IIP focal points; and donors progressively transferred their support to IIP.

IIP's major objectives were complementary. On the one hand, they sought to help the least-equipped countries to use and manage information technologies for their own development, through the training of specialists, reinforcement of infrastructures, and research and applications. On the other, they sought to analyse the changes brought about by information technologies in society. IIP, as part of UNESCO, aimed at linking reflection and concrete action via operational projects.

The management of the programme was assumed by a thirty-three-member intergovernmental committee, elected and periodically renewed by UNESCO's General Conference and by a bureau composed of members, from various regions of the world, elected for their expertise and experience in the various aspects of information technologies. The programme was carried out through institutes designated as national focal points, which in turn built up a global expertise network. The committee's working methods were based on project propositions, originating with Member States, which focused on several priority areas: training of users and specialists; software development, both methods and applications; networking; research and development; and regional and national policies and strategies. In twelve years, 170 projects were financed for about US$10.7 million. Member States were the main source of financing.

In the 1990s, technology, content and services converged. During this period, important rapprochements and mergers took place between principal producers of computer hardware, services, telecommunications, and text and audio-visual software possibilities, creating large industrial and economic groups, which covered simultaneously all aspects of ICTs. Furthermore it was during this period that the internet developed with an unheard-of rapidity, giving rise to numerous new enterprises, even if many were short-lived.

UNESCO had another intergovernmental information programme, the General Information Programme (PGI), which centred on contents from libraries, archives, and science and technical information systems. Within the context of the emergence of the 'information society', it was natural that UNESCO reflect on how to bring together the two programmes, IIP and PGI. The distinction between technology, on the one hand, and content, on the other, had lost its relevance. Technological progress meant that it was no longer necessary to be highly qualified in ICTs to be able to use them efficiently.

The IIP and PGI intergovernmental committees examined a possible fusion between the two bodies, which led to a recommendation at the 160th session of the Executive Board to replace them by the Information for All Programme (IFAP) on 1 January 2001. It is hoped that as ICTs continue to evolve, IFAP will play an important role in bridging the digital divide.

ORGANIZING INFORMATION
The origins and development of UNISIST
Jacques Tocatlian[10]

THE United Nations Information System for Science and Technology (UNISIST) grew out of a concern in the international scientific community – expressed through the International Council for Science (ICSU) – that the uncoordinated development of incompatible information systems and services in the 1960s was jeopardizing the international exchange of scientific and technical information (STI). A joint study to address this problem was conducted by ICSU and UNESCO, and the results were submitted to an intergovernmental conference in 1971, known as the UNISIST I Conference, which, in turn, gave shape to the UNISIST Intergovernmental Programme. The programme was concerned with improving access to STI and designed to provide a conceptual framework for the establishment of national, regional and international STI systems. In that respect, UNESCO became unique within the UN system by treating 'information', through UNISIST, as a subject in and of itself. UNISIST provided guidelines, methodology, norms and standards, and assistance to Member States, for the development of national and regional information systems and services.

However, UNESCO had at the time another programme in the Communication Sector addressing information issues in the library, documentation and archives communities. The overlap between this programme and UNISIST was such that, in order to avoid risks of duplication, competition, and conflicting advice to Member States, the General Conference of UNESCO, in 1976, combined the two, creating the General Information Programme (PGI). The integration of issues related to library, documentation and archives services with those related to transfer of STI proved smoother and easier than expected. The PGI was first placed in UNESCO's Bureau of Studies and Programming (BEP) but was later moved to a newly created sector on Communication, Information and Informatics (CII).

The emphasis on science in the original UNISIST Programme was thought by some critics in developing countries to indicate a primary concern with the 'elite'. They believed that some of the information requirements and needs of the most deprived international partners might be ignored. While

---

10 Jacques Tocatlian joined the Natural Sciences Sector of UNESCO in 1969 in the division of Scientific Documentation and Information (DIS). In 1979 he became the Director of the General Information Programme (PGI) and later the Director of Information Programmes and Services (IPS).

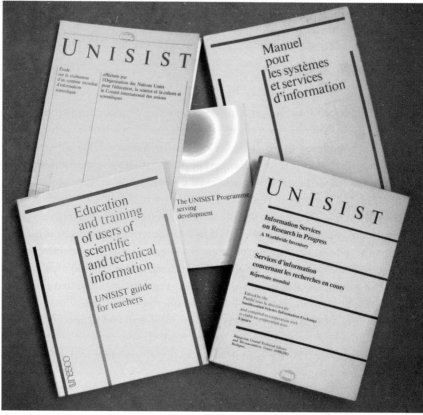

Manuals and bulletins published by UNESCO as part of the UNISIST programme.

pure science was particularly important for the industrialized countries, the developing nations expressed a greater need for applied science, technological know-how, and solutions to social and economic problems. Information users were thought to range from economic planners to 'grass-roots' workers in local communities.

In 1979, the Intergovernmental Conference on Scientific and Technological Information for Development (known as the UNISIST II Conference) evaluated the work achieved up to that point and gave new orientation to the programme. A positive result was that several developing countries created and strengthened their infrastructures in STI and established national information policies.

THE WAY IT WAS
UNESCO and informatics, a memoir
Sidney Passman[11]

A MODEST historical review may remind readers that when it comes to the computerization of information and its applications, UNESCO was present at the creation. In 1950 or thereabouts, informed people believed the world might need only a handful of the then-gigantic computers. A United Nations committee was set up to think about research institutes of the future. It was chaired by UNESCO's first Head of Natural Sciences, Joseph Needham, who recommended an International Computation Center (ICC), which was later founded, with help from UNESCO, as an intergovernmental organization. It eventually mutated into the Intergovernmental Bureau for Informatics (IBI), headquartered in Rome, Italy.

It was deemed essential to nurture the non-governmental organization (NGO) community to master this technical area. To this end, UNESCO sponsored the Paris World Conference on Information Processing in 1959. This led in turn to the formation of the International Federation for Information Processing (IFIP), still the principal world society for data processing. It goes without saying that professionals from the United States played a key role in this effort.

By the time I joined UNESCO in 1973, computers had found their niche. They were already being used for manifold applications in the industrialized world, but there was a growing need to consider coherent policies for their successful adaptation in development. During this process, the term 'informatics' was coined, from the French *informatique*, to cover these various issues.

John E. Fobes, then UNESCO's Deputy Director-General, immediately saw the implications. As a master of the complexities of the UN system, he was a superb guide. With his help, I was able to work with the United Nations Development Programme (UNDP), the IBI, and the UN Office of Science and Technology in organizing a UN-wide computer programme for development. We wanted government policies to support these efforts, and thus we called for national and international informatics policies. With IBI, we organized the First World Conference on Strategies and Policies for Informatics (SPIN-1978, Torremolinos, Spain), for which I served as secretary-general.

---

11   Sidney Passman: Director, UNESCO Division of Scientific Research and Higher Education (1973–81).

I have come to realize over time that this effort, including the now-standard regional preparatory meetings, played an important role in adapting computer developments in science, education, communications and commerce to the needs of dozens of countries. I confirmed this later, as a consultant for the US Information Agency in India and the US Agency for International Development in Tunisia on a project to assist in computerization and institutional development. India is now a leader in the field, and Tunisia has gone on to significant progress, hosting the second round of the World Summit on the Information Society in 2005.

UNESCO recognized early on that the private sector would play an essential role in computerization. My contacts with IBM, the giant in the field, led to its agreeing to assist UNESCO in training personnel, supplying computers to scientific and educational centres, and making their worldwide application centres available to facilitate development applications. The memo of understanding between IBM and UNESCO was, I believe, a pioneering step in such partnerships. Today there are many such partnerships, including the latest with Microsoft and Hewlett Packard. There are also extrabudgetary programmes with national and UN sponsors, which supplement the very limited funds available under UNESCO's Regular Programme. Over the years these have enabled the Organization to assist in generating capacity for worldwide efforts in informatics development.

With financial support from the UNDP, the Organization set up computer facilities for education and research at a number of universities, including at Bucharest, Romania, where it was eminently successful and helped produce several generations of computer specialists. UNESCO also established postgraduate training courses in a number of countries, which accepted students from abroad. The computer sciences curriculum was also improved as a result of a programme established with IFIP. The programme of library information profited from database software developed by innovative members of UNESCO's own computer centre. Today, UNESCO maintains a software portal, as part of its excellent internet presence. The Organization later established the Intergovernmental Informatics Programme (IIP) to support developments as a supplement to its regular programme; IBI, having run out of momentum, was dissolved in the late 1980s.

During this time, parallel projects were being carried out at UNESCO in the fields of scientific information and library information, in close cooperation with the professional community, culminating in the UNISIST conference, which led to the establishment of the UNISIST Intergovernmental Programme. It eventually became clear that these subjects were heavily concerned with computerization; and ultimately both intergovernmental programmes were merged into the present Intergovernmental Programme for Information for All (IFAP). IFAP provides a forum for international policy discussions, and guidelines for action, concerning the preservation of information, and the universal access to it, as well as the participation of all in the emerging global information society, and the ethical, legal and societal consequences of developments in information and communication technologies (ICTs).

UNESCO played a central role at the first World Summit on the Information Society (WSIS), the International Telecommunications Union meeting in Geneva, Switzerland, in December 2003. The meeting was part of the global interest in using ICTs for improving the welfare of humankind. Such issues as the digital divide, access to scientific information, freedom of information, preservation of digital archives, government policies for information processing, and information and development, are now 'on the front burner' of the United Nations and its specialized agencies (as well as the World Bank). For WSIS, UNESCO organized various important symposia, including one for the scientific community, convened with CERN (the European Organization for Nuclear Research), the Geneva-based high-energy physics centre which was also the initiator of the World Wide Web. For WSIS, UNESCO's documentation, seminars, and consultants – as well as Director-General Matsuura's personal dedication – all succeeded in bringing that conference into the mainstream of support for the knowledge society and the free flow of information.

In all, admirers of UNESCO's work may take pride in its half century of accomplishments in informatics. The Organization was not the only important body to recognize the importance of these issues. But it was most definitely one of the first.

BIOLOGICAL SCIENCES
TO BE OR NOT TO BE
UNESCO's biological and microbiological programmes
Franck Dufour, Julia Hasler, Lucy Hoareau[12]

'**OMNIS** cellula a cellula', the biologist Rudolph Virchow stated in 1855. 'All cells arise from cells.' It is an idea that expresses how biology is, in a sense, the most basic of the basic sciences. Biology is the very science of existence. And it has been at the core of UNESCO's science activities since the Organization was created sixty years ago.

1945–1960: EMERGING PRIORITIES IN BIOLOGY

The International Council for Science (ICSU) was the main beneficiary of UNESCO financial support for the purpose of implementing programmes in biology through scientific organizations such as the International Union of Biological Sciences. UNESCO and the World Health Organization (WHO) created a Council for the Coordination of International Congresses of Medical Sciences, which in 1949 became the Council for International Organizations of Medical Sciences (CIOMS), the second UNESCO partner in carrying out programmes in the biological sciences. Many international training courses were organized, and meetings of experts were convened to build up future programmes with a special focus on interdisciplinary brain research, cell biology, biochemistry, microbiology and botany. In light of the multitude of challenges facing the scientific community, and considering the limited resources available to UNESCO, the Organization defined itself as a catalyst to prompt the creation of international and regional research institutes through the coordination of activities implemented by existing organizations. The first Field Science Cooperation Offices, initiated by Joseph Needham soon after the creation of the Organization, were aimed at providing assistance for researchers working in

---

12   Julie Hasler: Programme Specialist for the life sciences, UNESCO Division of Basic and Engineering Sciences (since October 2003).
Lucy Hoareau: Programme Specialist for the life sciences and biotechnology, UNESCO Division of Basic and Engineering Sciences (since April 1997).

regions remote from major scientific centres. These offices were located in Latin America (Montevideo, Uruguay), South Asia (New Delhi, India), South-East Asia (Jakarta, Indonesia), and the Middle East (Cairo, Egypt). They became the major means for implementing UNESCO's biological programmes and were the launching bases for training courses and international scientific meetings within biological programmes.

BOX II.11.2: NEUROBIOLOGY IN UNESCO: THE INTERNATIONAL BRAIN RESEARCH ORGANIZATION (IBRO)

During the 1950s, the growing knowledge of and interest in neurobiology prompted UNESCO, in a close collaborative effort with the Council for International Organizations for Medical Sciences (CIOMS), to hold many consultations with scientists from interdisciplinary fields related to the brain. In 1960, the International Brain Research Organization (IBRO) was founded, under the aegis of UNESCO, intending to offer an international forum to encourage scientific discussion and training of neuroscientists.

In collaboration with UNESCO and the International Council for Science (ICSU), IBRO fostered worldwide collaborations in the community of neuroscientists through the organization of international meetings and training courses. During the 1970s, molecular approaches of neuroscience were encouraged through the collaboration of IBRO with the International Cell Research Organization (ICRO). In 1983, considering the dynamic development of large multinational neuroscience societies working to promote neuroscience through the world, IBRO reoriented its structure and mission to focus on the training and education of students and scientists in regions with special needs. Ten years later, IBRO was admitted by ICSU as a member of the category of Scientific Unions.

Today, IBRO is an independent non-governmental organization that aims at: (1) developing, supporting, coordinating and promoting scientific research in all fields concerning the brain; (2) promoting international collaboration and interchange of scientific information on brain research throughout the world; and (3) providing for and assisting in education and the dissemination of information relating to brain research by all available means. It implements its action on the basis of competitive applications and by sponsoring symposia, workshops, fellowships, travel grants and the organization of meetings. To ensure its regional role, many programmes are administered by the Regional Committees in Asia/Pacific, Africa, Eastern and Central Europe, Latin America, USA/Canada and Western Europe. The School Programme is the perfect illustration of the crucial role played by the Regional Committees. The purpose of this programme (launched in 1999) is to provide support for organizing neuroscience schools that will be accessible to students from less developed countries.

More recently, a modern website has been established, in order to provide information to members and to the public. Online applications for fellowships and travel grants are available. Other innovations include a 'Brain Awareness' portal, and an international registry of neuroscience training programmes. The *IBRO Reporter*, a monthly email newsletter, is sent to all members registered in the extensive Membership Directory.

> **BOX II.11.3: ICRO-UNESCO COOPERATION IN CELL BIOLOGY** *Georges Cohen*[1]
>
> The International Cell Research Organization (ICRO), a non-governmental organization specifically designed to assist UNESCO in the implementation of its cell biology programme, was founded in 1962 under the initiative of Adam Kepes. ICRO's activities have concentrated on the organization of advanced experimental training courses in various fields of cell biology and biotechnology for young scientists from both developing and developed countries. Over the last forty-three years, ICRO has organized altogether 456 such courses in 80 countries, attended by more than 12,000 students from all over the world. Several of these training courses have been organized in collaboration with UNESCO Microbial Resources Centres (MIRCENs), the Biotechnology Action Committee (BAC), the Molecular and Cell Biology Network (MCBN) and the Human Genome Programme of UNESCO, as well as various ICSU bodies, the European Molecular Biology Organization (EMBO), International Centre for Genetic Engineering and Biotechnology (ICGEB), Federation of European Biochemical Societies (FEBS) and Academia de Ciencias de América Latina (ACAL). Since 1988, ICRO is a scientific associate of ICSU.
>
> The present programme of ICRO courses falls into five areas of Cell Biology: Molecular Structure and Function, Animal Cell Biology, Plant Cell Biology, Applied Microbiology, and Instrumentation and Information. Each course includes both lectures and laboratory work, and is led by a selection of invited foreign teachers, as well as a local teaching staff. Some of the courses are regional, recruiting students from, for example, Africa, Latin America, the Arab States or South-East Asia; some are international, selecting participants from all over the world. The teachers are invariably chosen among worldwide top-level scientists in their respective fields.

1961–1980: IMPLEMENTATION OF MAJOR PROGRAMMES

The multitude of consultations and analyses performed during the 1950s shaped the course of UNESCO's actions in biology for the following decades. In 1960, the International Brain Research Organization (IBRO) was founded in a collaborative effort between UNESCO and CIOMS, providing an international forum to encourage scientific discussion and the training of neuroscientists. In 1962, UNESCO launched the International Cell Research Organization (ICRO), in order to promote research and knowledge of cell biology, one of the emerging priorities in biology. The subsequent collaboration of UNESCO with IBRO and ICRO is the most representative example of the contribution of the Organization to the field of biology during this period, through its interactions with non-governmental organizations (NGOs). During the 1970s, collaboration between IBRO and ICRO was strongly encouraged with a view to stimulating emergent interdisciplinary fields (see Box II.11.2 and Box II.11.3).

Microbiology was another field in biology that attracted UNESCO's support. In order to foster communication and understanding of applied science needs in developed and developing countries, the Global Impact of Applied Microbiology (GIAM) was initiated with the help of UNESCO, the United Nations Environment

PART II: BASIC SCIENCES AND ENGINEERING

> Experience has shown that ICRO courses held over the past forty-three years have served as an important basis for bringing together senior and junior scientists from different parts of the world, and that these courses have in fact constituted a most successful mechanism for stimulating international contacts between scientists in cell biology across linguistic, cultural and ideological boundaries. In this context, it is especially important to emphasize the role that these courses have played and continue to play in establishing contacts between students and teachers from developing and developed countries. To respond to the needs and requests of the Member States, the ICRO training programme has been intensified in Africa, particularly in South Africa.
>
> During all these years, ICRO has organized these courses with a partial financial support from UNESCO. At present, this support covers between 15 and 40 per cent of the total expenses involved in organizing the various courses, including travel and subsistence allowance of students and teachers, and varying in extent from one course to another within the range indicated.
>
> Although diminishing in recent years, UNESCO's contribution plays an essential role as 'seed money', opening the possibility for obtaining further support from other international organizations, as well as from local sources. ICRO is therefore most grateful to UNESCO and its Member States for this contribution that makes it possible to continue its task, which was initiated by UNESCO forty-three years ago.
>
> ICRO looks forward to a continued friendly and fruitful collaboration with UNESCO.
>
> ---
> 1. Georges Cohen: ICRO Executive Secretary, since 1987.

Programme (UNEP) and ICRO in 1963, and this special conference programme was pursued over the following decades. In 1965, in collaboration with the International Union of Microbiological Societies (IUMS) and ICRO, UNESCO launched a research programme in the field of microbiology in accordance with the resolutions of the thirteenth General Conference. In 1970, UNESCO and ICRO created the World Federation for Culture Collections (WFCC) for the listing and conservation of microbial strains of importance for medicine, agriculture and industry. With regard to its goal of fostering scientific capacity in developing countries, in response to their regional needs, UNESCO launched a worldwide network of Microbial Resources Centres (MIRCENs) in 1975 (see Box II.11.4).

The programmes launched during this period benefited from UNESCO's goal of encouraging the establishment of regional networks for fundamental sciences, and the request from the General Conference in 1963 to give a higher priority to sciences, to equal the priority then given to education. To this end, the Cairo and New Delhi offices became Regional Offices for Science and Technology, and similar offices were subsequently set up in Montevideo, Jakarta, and Nairobi (Kenya). Numerous fellowships, symposia, training courses and publications in the field of biology were supported and funded by UNESCO through the actions of these Regional Offices.

Microbial culture collection, laboratory of the network of Microbial Resource Centres (MIRCENs), Tehran, Islamic Republic of Iran, 2004.

1980 TO THE PRESENT: SCIENTIFIC BOOM AND PREPARATION FOR THE NEW MILLENNIUM

UNESCO's support and activities in the area of applied microbiology anticipated the revolution in biology that occurred in the late 1970s and 1980s and positioned the Organization to respond to the emergence of recombinant DNA technologies, biotechnology, and the use of genetic engineering to solve the problems of both old and new diseases. Several new programmes were created, including the Biotechnology Action Council Programme (BAC) in 1990, formed to promote biotechnologies in developing countries.

Molecular and cell biology also benefited from great technological advances during this period, and UNESCO disseminated knowledge to developing countries by creating the Molecular and Cell Biology Network in 1990, which became an independent NGO in 2002. It also helped to create the World Foundation for AIDS Research and Prevention, with which it is still promoting scientific research in the prevention of HIV/AIDS transmission. Finally, UNESCO participated in one of the most challenging scientific issues of the 1990s, the Human Genome Project, by adopting the *Universal Declaration on the Human Genome and Human Rights* at the twenty-ninth session of the General Conference in 1997.

BOX II.11.4: THE GLOBAL NETWORK OF MICROBIAL RESOURCES CENTRES (MIRCENS) *Rita Colwell*[1]

Recognizing the high potential of applied microbiology and biotechnology to produce a great number of substances and compounds essential to human life and welfare, UNESCO launched the global network of Microbial Resources Centres (MIRCENs) in 1975. Its original mission was to preserve microbial gene pools and to make them available to developing countries for further development in medicine, agriculture and industry. Today, MIRCENs has five main objectives:

- to provide a global infrastructure incorporating national, regional and interregional cooperating laboratories geared to the management, distribution and utilization of the microbial gene pools;
- to reinforce the conservation of micro-organisms – with emphasis on Rhizobium gene pools – in developing countries, with an agrarian base;
- to foster the development of new inexpensive technologies native to specific regions;
- to promote the economic and environmental applications of microbiology; and
- to serve as focal centres in the network for the training of human resources.

To achieve these objectives, a large number of MIRCENs (thirty-four in 2005) have been established throughout the developing world and collaborate through four thematic networks:

- the Biological Nitrogen-Fixation MIRCENs (five institutes);
- the Culture Collection MIRCENs (seven institutes);
- the Biotech MIRCENs (fourteen institutes);
- the Aquaculture and Marine Biotech MIRCENs (eight institutes).

The international scientific cooperation between research institutes is fostered through international cooperation sustained by the involvement of governments, UNESCO National Commissions, UN agencies and programmes (FAO, WHO, UNIDO, UNU, UNDP and UNEP), and the international scientific community (ICRO, IUMS, IOBB, WFCC, SCOPE, AABNF). The dynamism of these networks rests on support to programmes such as fellowships, grants, lectures, courses, publication of the *World Journal of Microbiology and Biotechnology*, facilitated access to scientific publications, and supplying of light laboratory equipment and reagents.

Overall, and despite a small budgetary input, the MIRCENs network has established a sustainable and dynamic framework to promote and significantly develop the knowledge and application of modern microbiology and biotechnologies in developing countries.

---
1. Rita Colwell: Director, MIRCENs; UNESCO Panellist for the MIRCENs and BETCENs programmes, 1980–98.

BOX II.11.5: THE BIOTECHNOLOGY ACTION COUNCIL (BAC) *Indra Vasil*[1]

Significant and rapid advances were being made during the 1980s in the new field of biotechnology. They had generated considerable discussion and debate about the potential of this powerful new genetic tool to address problems of food security, human health and the environment, especially in the developing countries, which suffered from large and rapidly increasing populations, chronic food shortages and malnutrition, poor health, and profound environmental problems. Yet most of the biotechnology research and development (R&D) at the time was taking place in industrially advanced countries/regions, such as the United States, Japan and Western Europe. There was genuine concern on the part of many in the scientific community and international organizations that the countries which could benefit most from the emerging technology were being deprived of this opportunity because they lacked scientific infrastructure and manpower.

Consequently, several leading scientists took the initiative to express their concerns to Federico Mayor, the newly elected Director-General of UNESCO. He responded by inviting an international panel of distinguished scientists to meet in Paris in 1989, to discuss the problem and recommend appropriate action. After considerable discussion about support for the human genome project, HIV/AIDS, biotechnology, and so on – and keeping in mind the limited financial resources available to UNESCO – the panel recommended the creation of the Biotechnology Action Council (BAC) to promote plant molecular biology and biotechnology, and aquatic biotechnology, in developing countries.

BAC was formally constituted in 1990 and chaired by Indra K. Vasil (1990–2001). It was charged with promoting the development and strengthening of national and regional capabilities in biotechnology, by providing opportunities for advanced education and training, and the efficient and rapid exchange of information. BAC immediately embarked on a number of activities targeted at young scientists, with the goal of contributing to the building of a sound scientific and technological base and manpower in the developing

BOX II.11.6: THE UNESCO GLOBAL NETWORK FOR MOLECULAR AND CELL BIOLOGY (MCBN) *Angelo Azzi*[1]

The phenomenal progress of molecular and cell biology in developed countries during the 1980s, and its growing potential for solving contemporary problems of humanity, inevitably led UNESCO to promote this scientific field in developing countries. The UNESCO Global Network for Molecular and Cell Biology (MCBN) was launched in 1990, and continued as a UNESCO programme until 2002, when it was restructured as a non-governmental organization (NGO), which is still supported as part of the Basic Science programmes.

MCBN – now a non-profit association and NGO – aims to provide opportunities to solve local and regional problems in developing and restructuring countries through the use of molecular and cell biological approaches. In order to achieve these goals, MCBN promotes global West–East and North–South international cooperation designed to countries, leading eventually to their independence and self-sufficiency in biotechnology and the ability to solve their own problems. These included Short-Term Fellowships to developing country scientists for advanced training in the laboratories of their choice anywhere in the world, free distribution of state-of-the art laboratory manuals in plant tissue culture and molecular biology, organization of intensive training courses in plant molecular biology and biotechnology taught by distinguished teams of international scientists, and the award of biotechnology professorships to eminent scientists for extended lecture tours or resident work in the developing countries.

In 1995, BAC activities were further supplemented and strengthened with the establishment of Biotechnology Education and Training Centres (BETCENs) in Qingdao (China), Godollo (Hungary), Pretoria (South Africa), Irapuato (Mexico) and Bethlehem (Palestinian Autonomous Territories). The BETCENs created new and valuable opportunities for closer regional and interregional cooperation by offering longer-term fellowships, and organizing workshops and training courses that were tailored to regional needs and depended on regional talent.

During the past fifteen years, several thousand young women and men have directly benefited from education and training opportunities provided by BAC. At least fifty developing countries now have active biotechnology R&D programmes, leading to commercialization of biotechnology products in China, Egypt, India, the Philippines, South Africa and elsewhere. UNESCO was one of the first international organizations to invest in plant biotechnology in developing countries. It can be justifiably proud of its pioneering role in helping to build scientific infrastructure and manpower, leading to the production of more and better food, reducing the use of harmful agrochemicals, improving the environment, and contributing to job creation and economic development.

1. Indra Vasil: Founding Chairman of Biotechnology Action Council, 1990–2001.

stimulate and facilitate research and training in various area of molecular and cell biology. Collaboration and scientific exchange are enhanced through support provided for the participation of scientists from developing and restructuring countries, in high-level international symposia, conferences and collaborative scientific activities. In addition to an active fellowship programme, the Network implements other actions including lectures, scientific advice, and establishment of molecular and cell biology centres of excellence.

The establishment of the International Institute of Molecular and Cell Biology (IIMCB) in Warsaw, Poland, is a perfect illustration of the dynamism of MCBN. IIMCB was founded in the late 1980s and developed in a global collaborative effort of MCBN, UNESCO, Polish scientists and the Polish Government. Beginning its research activities in 1999, it has grown and matured very rapidly, becoming one of the most modern, competitive and productive Polish institutes in its field.

1. Angelo Azzi is Chairman of the UNESCO Global Network for Molecular and Cell Biology.

BOX II.11.7: HUMAN GENOME PROJECT Santiago Grisolía[1]

Following the First Workshop on International Cooperation for the Human Genome Project – held in Valencia (Spain) in October 1988, with representatives and support of UNESCO – an advisory group of scientists was assembled to consider the role UNESCO might play in the advancement of the Human Genome Project. A first meeting was held in Paris in February 1989. The participants supported the Director-General's initiative and agreed that UNESCO could help facilitate international cooperation, particularly among developing countries and between developing and developed nations. To consider plans in greater detail, a second consultative meeting was held in Moscow (USSR) in parallel with the Human Genome Organization (HUGO) the following June. The conclusions from the second UNESCO advisory meeting were summarized and presented at the twenty-fifth session of UNESCO's General Conference.

A programme on the Human Genome for 1990–91 was subsequently approved by the participants at the General Conference, and UNESCO immediately confirmed its active involvement in the COGENE/UNESCO/ICSU/EEC/FEBS/IUB[2] 'Symposium on Human Genome Research: Strategies and Priorities', held at its Headquarters in January 1990. Federico Mayor then set up a Scientific Coordinating Committee to help plan and implement the programme as proposed in the Moscow recommendations. Santiago Grisolía was invited to assume the chairmanship of the Committee.

The Committee was created in 1989 and comprised thirteen scientists: Jorge Allende, Giorgio Bernardi, C. Cantor, R. Cook-Deegan, C. Coutelle, K. Dellagi, S. Grisolía (Chairman), K. Matsubara, V. McKusick, A. Mirzabekov, O. Ole-MoiYoi and S. Panyim. Its main activities were a large number of workshops held in the developing countries and large South–North symposia in Caxambu (Brazil), Beijing (China), New Delhi (India), Guadalajara (Mexico) and Namibia. During these years, the Scientific Coordinating Committee developed an extensive Short-Term Fellowship programme, which benefited some 200 scientists from over fifty countries in development (Africa, Mediterranean and Arab States, South and Central Asia, South-East Asia, Europe and Latin America). These scientists were trained in genetics in the best laboratories of Europe and America.

The author wishes to acknowledge the support of the personnel of the Natural Sciences Sector of UNESCO – especially Vladimir Zharov and Svetlana Matsui – and also that of the personnel of UNESCO in all the countries where we acted. In 1991, because of the ethical and social considerations of the Genome Project, the Director-General organized an International Bioethics Committee, presided by Mrs Noëlle Lenoir, which has continued very successfully under the direction of George Kutukdjian.

---
1. Santiago Grisolía: Chairman, UNESCO Scientific Coordinating Committee (SCC) for the Human Genome Programme, 1990–2000.
2. Committee for Genetic Experimentation (COGENE) of the International Council for Science (ICSU); European Economic Community (EEC); Federation of the European Biochemical Societies (FEBS); International Union of Biochemistry (IUB).

UNESCO also established, with ICSU, International Biosciences Networks (IBN) in Latin America, Asia, Africa and the Arab Region. These networks of centres of excellence, in collaboration with the Regional Offices, have contributed to the decentralization of UNESCO activities, a policy that took effect in the early 1990s. Moreover, in 1991, the General Conference voted for the creation of UNESCO Chairs that endeavour to promote twinning and networking, and to provide support for advanced postgraduate studies and research in developing countries. A multitude of Chairs have been created in the field of life sciences and biotechnologies. In 2003, UNESCO launched the International Basic Sciences Programme IBSP, which should contribute significantly to the Organization's future achievements in the realm of life sciences.

MAKING CONNECTIONS
International Biosciences Networks (IBN)
Julia Marton-Lefèvre[13]

INSPIRED by the success of the UN Development Programme/UNESCO Regional Programme for Postgraduate Training in Biological Sciences (in operation in South America since 1975, under the leadership of Jorge Allende), several scientists active in the International Council for Science (ICSU) decided to establish a worldwide network of biosciences, to be known as the International Biosciences Networks (IBN). The IBN, set up as a partnership between ICSU and UNESCO, came into being in 1980, when its first Steering Committee meeting was held in Paris.

The Steering Committee, under the chairmanship of Richard Darwin Keynes (Cambridge, UK), agreed on the terms of reference for the IBN, which was created to:

open up and exploit areas of biological research of particular promise and relevance to the needs of developing countries by:

- helping to provide education and training in the basic biological sciences to scientists from developing countries;
- disseminating in developing countries a knowledge of and expertise in appropriate findings in biology;

---

[13] Julia Marton-Lefèvre was Deputy then Executive Director of ICSU (1978–97). Currently, Rector of the (UN-affiliated) University for Peace.

- facilitating an interchange of knowledge between biologists worldwide;
- creating a favourable climate for basic biological research in the developing countries;
- acting as a source of advice on priorities in biological research in developing countries; and
- utilizing biological resources endogenous to developing countries.[14]

The IBN partnership was a unique one for those days, as seven of the Biological Sciences Unions of ICSU (in the fields of biology, biochemistry, biophysics, immunology, nutrition, pharmacology and physiology) decided to join forces with UNESCO's Basic Sciences Division to provide assistance to developing countries, based on the potential of the biological sciences. A network in Asia, built around a similar theme, had already been set up by the ICSU Committee on Science and Technology in Developing Countries (COSTED); this Asian Network of Biological Sciences, established in 1978, was brought under the umbrella of the IBN.

The IBN Steering Committee therefore decided to concentrate on Africa, and to this end convened a Symposium on the State of Biology in Africa, which was held in Ghana in April 1981. The outcome of the excellent symposium – attended by around 100 eminent scientists and science policy experts – was a clearer understanding of Africa's existing capacities and scientific needs, and the commitment to establish a single, continent-wide African Biosciences Network (ABN). The symposium pointed to a number of common problems, which the ABN was to address, including the serious gaps in access to scientific information, the inadequacy of teaching materials and lack of training programmes.

With the establishment of the ABN, the next area of the world receiving priority attention by the IBN Steering Committee was the Arab Region. An Arab Regional Biosciences Network was established a few years later, with an opening symposium held in Jordan. Between 1979 and 1993, a large number of scientific meetings took place under the IBN umbrella in Africa, the Arab Region, Asia and Latin America. These meetings were often scheduled alongside the meetings of the IBN Steering Committee, thus allowing the committee to remain in close touch with the evolution of the Network.

In order to streamline its activities in the developing world, ICSU decided, in 1993, to merge the IBN and COSTED. The resulting body, COSTED-IBN

---

14   Terms of Reference of IBN adopted at 1980 IBN Steering Committee meeting.

BOX II.12.1: INTERNATIONAL BIOSCIENCES NETWORKS (IBN) IN LATIN AMERICA: RELAB (LATIN AMERICAN NETWORK OF BIOLOGICAL SCIENCES)   *Jorge Allende*[1]

In April 1973, I had a long talk with Gabriel Valdés, United Nations Development Programme (UNDP) director for Latin America, regarding ways to stimulate scientific collaboration in Latin America, and he asked me to explore the possibility of obtaining support from three governments for a regional project in postgraduate training in biological sciences. As UNDP would work with UNESCO as the executive agency in the project, Mr Valdés and I visited Paris to meet with staff from the Natural Sciences Sector.

Following trips to several countries, the Regional Project was endorsed by five governments – Colombia, Ecuador, Bolivia, Peru and Chile – and the first meeting of the Regional Executive Committee was held in September 1975. Soon other countries asked to join: Venezuela (1976), Argentina (1977), Brazil (1979), Uruguay and Paraguay (1982). The Regional UNDP/UNESCO Project was renewed three times and lasted ten years (1975–85), investing US$3 million in fellowships, collaborative research projects, training courses and workshops.

In 1981, when the International Council for Science (ICSU) and UNESCO decided to organize regional networks to promote biology in developing regions, the UNDP/UNESCO Regional Programme on Postgraduate Training in Biological Sciences was taken as the Latin American Network, and it was decided to start similar efforts in other regions. The four new networks (Africa, Arab Region, Asia, and Latin America) were collectively known as the International Biosciences Networks (IBN), whose biggest success was the approval of a UNDP/UNESCO project to support activities of the African Biosciences Network (ABN), for approximately US$2 million for five years.

In 1985, when UNDP funds stopped, the Regional Executive Council (REC) of the Latin America Project decided – with UNESCO and ICSU – to continue activities under the scheme of the Latin American Network of Biological Sciences (RELAB). Costa Rica, Cuba, Honduras, Mexico and Panama have joined since then. In 1993, the IBN was merged with COSTED (ICSU Committee on Science and Technology for Development), leading to the disappearance of the other Regional Biosciences Networks, with the exception of RELAB. In the same year, the REC decided to establish the RELAB Corporation, whose sole purpose is to finance initiatives approved by the REC.

In 1994, ICSU and UNESCO used the RELAB model to organize networks in physics, astronomy, chemistry and mathematics, using mainly the regional organizations of national scientific societies. This network of networks received the enthusiastic support of the Director-General of UNESCO and of the Natural Sciences Sector, especially V. Zharov. In recent years, financial support from both ICSU and UNESCO to the Latin American Scientific Networks has been reduced. Today, the most active networks are the Mathematical Union of Latin America and the Caribbean (UMALCA) and RELAB.

---

1. Jorge Allende: President, RELAB Corporation; Expert, UNDP/UNESCO Regional Programme on Postgraduate Training in Biological Sciences; Member, UNESCO Scientific Board for International Basic Sciences Programme.

(with the distinguished Indian scientist and former president of ICSU, M. G. K. Menon, as president), continued to be co-sponsored by UNESCO.

The first joint meeting was held in Ghana in April 1994, where it was decided that COSTED-IBN would work along the lines of the IBN, by allowing the regional networks to set their own priorities and modalities for action. It was decided to continue to merge their regional operations – with offices located in Accra and Dakar (for Africa), Amman (the Arab Region), Madras (Asia) and Santiago de Chile (Latin America). The Central Secretariat was located in Madras (India), in a building that had been made available to ICSU for COSTED by the Indian Government.

In 2002, ICSU – satisfied with the progress it had made since the 1960s in linking scientists from all parts of the world – decided to dissolve COSTED-IBN and set up a Policy Committee on Developing Counties, as well as regional ICSU offices in all of the major regions where the IBN had established roots.

ENGINEERING SCIENCES
THE SPIRIT OF INNOVATION
Engineering and technology at UNESCO
Tony Marjoram[15]

> 'Those who forget the past are condemned to repeat it.'
>
> —George Santayana

> 'It is important to remember the vital contribution of engineering and technology to development. We need to encourage international commitments to promote engineering and technology to lasting development around the world.'
>
> —Koïchiro Matsuura, Director-General of UNESCO, 2000

INTRODUCTORY OVERVIEW

ENGINEERING and history are closely connected. Indeed, the broader history of civilization is largely the history of engineering and engineering applications: the Stone Age, Bronze Age, Iron Age and Information Age all relate to engineering and shaping our interaction with the world. Human beings are partly defined as tool users, and it is this use of tools, and innovation, that accounts for so much of the direction and pace of change of history. The Pyramids, Borobudur Temple, the civilizations linked to metal smelting at Zimbabwe and water engineering at Angkor, the medieval cathedrals, and the Industrial Revolution are all testament to the engineering skills of past generations. Contemporary engineers deserve credit for the fact that engineering and technology is so extensive and reliable that it is hardly noticed or appreciated – except when something goes wrong – and for the use of engineering to conserve our heritage. Engineering plays a vital role in social and economic development as part of an increasingly close continuum of activity with science, although in many fields practical activity preceded scientific understanding. We had steam engines before thermodynamics, and rocket 'science' is more about engineering than science.

---

15   Tony Marjoram: Programme Specialist in Engineering Science, Technology and Informatics, Regional Office for Science and Technology, South-East Asia, Jakarta, 1993–97; then. Senior Programme Specialist, Division Engineering Sciences and Technology, (now Division of Basic and Engineering Sciences), 1997 onwards.

Researching the history of the engineering and technological sciences at UNESCO – reading through archival material, including past Medium-Term Strategy (C/4) and Programme and Budget (C/5) documents, staff lists and old telephone directories – it is interesting to note the similarities and resonances between the programme priorities in engineering and technology today and those of the 1960s, 1970s and the intervening years. It is also interesting to note the importance of engineering and technology in those earlier years – when engineering was the biggest activity in the Natural Sciences Sector, in terms of personnel and budget, before the rise of the environmental sciences (although one notes that a programme on environmental engineering began in the late 1970s).

There has also been long-term interest in renewable energy, beginning with an international congress in 1973. The Division of Applied Social Sciences was part of the Department of Application of Sciences to Development up to 1965, and there has been close cooperation with the social sciences in the field of 'science and society', with the journal *Impact of Science on Society* being published from 1967 to 1992. Other recurrent themes are: the reform of engineering education, the need for greater interdisciplinarity and intersectoral cooperation, women (and gender issues) in engineering, innovation, and the development of endogenous technologies. All these are as important today as they were thirty years ago. Indeed, it is interesting to note that programme activities appear to have been more interdisciplinary twenty years ago than they are today.

Apart from these similarities, there are of course differences between programme activities over the last forty years, and also differences in definition and context over time and in different places. The difficulties of defining 'engineering' and 'engineering science' – not to mention 'engineer', 'technologist' and 'technician' – is illustrated by the discussions over the Bologna Accord in 1999 regarding the harmonization of graduate and postgraduate education in Europe by 2010. In Germany, for example, there are over forty definitions of an engineer.

The context of 'development' has also changed, although development specialists continue generally to overlook the role of engineering and technology in development at all levels. At the macroeconomic level and at the grass roots – small, affordable technologies can make a tremendous difference in people's lives, as well as in poverty reduction. Most development specialists have a background in economics, and continue to view the world in terms of the three classical factors of production (capital, labour and natural resources) where knowledge – in the form of engineering, science and technology – is not easily accommodated. This is unfortunate, given the obvious importance of engineering, science and technology in development (a prominent example of this is the Industrial Revolution). This importance is recognized by economists such as Schumpeter and Freeman, in their work on the role of knowledge and innovation in economic change. We in the twenty-first century now live in 'knowledge societies'.

The context of UNESCO has also changed from the early days, when engineering was the main activity area in the Natural Sciences Sector, largely supported by United Nations Development Programme (UNDP) special funding. The decline of such funding led to the decline of engineering, and of the Sector as a whole, in terms of personnel and budget. UNESCO faced a crisis from the mid-1980s, with the decline of UN funding, the withdrawal of the United States and the United Kingdom (in 1984), and a consequent budget cut of 25 per cent.

There have also been changes in governance, management and administration at UNESCO – relating to the 'three organs' of the Organization (General Conference, Executive Board and Secretariat) – which have contributed and continue to contribute to the decline of engineering. Richard Hoggart (former Assistant Director-General of Education) has commented that UNESCO is 'over-governed and under-regulated' – although former staff members will be interested to hear that regulation has caught up dramatically in the past few years, with the introduction of the SISTER and FABS computer systems for programme and financial management, ushering in the administrative advent of the new millennium.

Engineering was part of UNESCO from the beginning. It was the intention of the founders of UNESCO that the 'S' refer to science and technology, and that this include the applied sciences, technological sciences and engineering. The engineering and technological sciences have always played a significant role in the Natural Sciences Sector at UNESCO. Indeed, UNESCO was established in a conference that took place in London (in November 1945) at the Institution of Civil Engineers, the oldest engineering institution in the world. This fact reflects the stark realization of, and emphasis on, the importance of science, engineering and technology in the Second World War – during which many new fields and applications were developed in such areas as materials, aeronautics, systems analysis and project management – and the success of the Marshall Plan in rebuilding capacity and infrastructure after the war. This is mirrored in the support by other UN agencies for programme activities at UNESCO of the basic, applied and engineering sciences and technology, before the development of operational activities by UNDP in the mid-1980s.

ENGINEERING PROGRAMME AND ACTIVITIES

The engineering programme at UNESCO, as the main programme in the Natural Sciences Sector until the 1980s, has been active in a diverse range of initiatives. These include the implementation of multimillion-dollar projects supported by UN special funds, project development and fundraising, support to international professional organizations and non-governmental organizations (NGOs), conferences, training and seminars, information and publications, consultancy and advisory activities, and programme activity areas, including engineering education and energy. The primary focus of the engineering programme, until the late 1980s, was on core areas of engineering education (what would now be called human and institutional capacity-building), where the emphasis turned increasingly towards renewable energy. The focus on core areas of engineering education and capacity-building is presently returning with the new millennium, albeit with much less human and financial resources. Much of this activity was conducted in close cooperation with the five main science Field Offices, which were established in the early days to facilitate implementation of projects supported by the UNDP special funds. With the decline of funds in the 1990s, this field network also declined, with fewer specialists in engineering in the field and at Headquarters.

The field of energy was increasingly a focus of the engineering programme in the late 1970s and 1980s. Energy activity at UNESCO effectively began in the early 1970s, with the International Congress on the 'Sun in the Service of Mankind', held in Paris in 1973, organized by UNESCO, with the World Meteorological Organization, the World Health Organization, and the International Solar Energy Society (ISES), at which time the International Solar Energy Commission was also created. In the late 1980s and 1990s, this interest on renewable energy continued with the creation of the World Solar Programme (WSP, 1996–2005) and the associated World Solar Commission (WSC, which clearly borrowed from the earlier activity of ISES).

WSP/WSC activity accounted for over US$4 million of UNESCO funds, with over US$1 million alone from UNESCO and other donors supporting WSP/WSC activity in Zimbabwe, where the World Solar Summit was held in 1996, which in turn led to the creation of the World Solar Programme and World Solar Commission. Declining funding in the late 1980s and 1990s gave rise to increasing budgetary anxiety and what some consider an overly ambitious programme.

PROJECT ACTIVITIES WITH UNDP SPECIAL FUNDS

From the early 1960s until the late 1980s, the engineering programme, the largest of the three activity areas of the Natural Sciences Sector, peaked – with over ten staff at Headquarters, another ten in the five main Regional Field Offices that were developed over this period, and a biennial budget of up to US$30 million. A diverse range of activities and initiatives were implemented, including the establishment and support of engineering departments at universities, research centres, standards institutions, and the like in numerous countries. Most of this activity is what we would now call human and institutional capacity-building. So it is interesting to reflect on the current emphasis on technical capacity-building, and on the lessons we may learn from the past.

PROJECT DEVELOPMENT AND FUNDRAISING

Programme staff have long been active in the development of new project proposals, in the earlier days primarily for UNDP funding. More recent project development activity includes the DaimlerChrysler–UNESCO Mondialogo Engineering Award, one of the three pillars of the UNESCO partnership with DaimlerChrysler to promote intercultural dialogue, in this case between young engineers and the preparation of project proposals to address poverty reduction, sustainable development, and the Millennium Development Goals. Among other proposals is the *Encyclopedia of Life Support Systems* (EOLSS) project, which took off in 2002, based initially in the Engineering Division and subsequently transferred to the Science Analysis and Policy Division.

Proposals that did not take off include a low-orbit satellite project designed to promote education in Africa, using Russian military rockets to launch the satellites (an idea borrowed from Volunteers in Technical Assistance [VITA] in the USA, and which VITA continued to develop, with continued lack of success, leading to its near collapse in 2001), and a proposal for the World Technological University.

NETWORKING, INTERNATIONAL PROFESSIONAL ORGANIZATIONS AND NGOs

The engineering programme has been continuously active in the development and support of networking, international organizations, and NGOs in engineering. It helped create the Union Internationale des Associations et Organismes Techniques (UATI) in 1951, and the World Federation of Engineering Organizations (WFEO/FMOI) – the main 'umbrella' organization for national and regional engineering institutions and associations – in 1968, and the International Council for Engineering and Technology (ICET, jointly with WFEO and UATI) in 1999. The African Network of Scientific and Technical Institutions (ANSTI) project, started in 1979, had a budget of around US$5 million in its first phase and continues today. Network support activity continues: UNESCO hosted a meeting to establish Engineers Without Borders International in May 2005.

CONFERENCES AND SYMPOSIA

The organization and support of conferences and symposia constitutes another long-term activity of the engineering programme. Most recently the programme was involved in organizing and supporting the first World Engineers' Convention (WEC) in Hannover, Germany, in 2000, followed by WEC2004 (in Shanghai,

UNESCO and the International Council for Engineering and Technology (ICET) – which groups some 10 million engineers – signing a framework agreement aiming to reinforce cooperation of engineers' organizations in favour of environmental protection and sustainable development. Left and right: members of ICET Executive Board; centre: Federico Mayor (biologist), Director-General of UNESCO, 1999.

China) and is currently involved in the organization of WEC2008 in Brazil (in cooperation with WFEO). From the early days, the engineering programme was involved in the International Conference on Trends in Engineering Education, held in Paris in 1968, and the World Congress on Engineering Education (in cooperation with WFEO) held first in India (1988), with the seventh Congress planned for Budapest in 2006.

UNESCO was involved with the organization of the First International Congress for Engineering Deans and Industry Leaders (in Ohio, USA,1989) and continued the series by hosting them in 1991, 1993, 1995 and 1996. It was involved in the UN Conference on Science and Technology for Development in 1979, the World Conference on Science in 1999 (neither of which appears to have been conspicuously successful), and the CAST (Conference on the Application of Science and Technology to Development) series of conferences and related activities in Asia (CASTASIA), Latin America (CASTALA) and Africa (CASTAFRICA), held in the late 1960s and 1970s, as well as the second series CASTASIA-II and CASTAFRICA-II in the 1980s. In the field of energy, UNESCO was closely involved in the UN Conference on New and Renewable Sources of Energy, held in Nairobi, Kenya, in 1981, the first major international forum on energy alternatives.

TRAINING, WORKSHOPS AND SEMINARS

The engineering programme was particularly active in the organization and presentation of training and seminars from the 1960s to the 1980s, with UNDP special funds, although this activity has inevitably declined since those golden years. In the field of energy, the former Section of Energy Development and Coordination focused on training and seminars in cooperation with other international agencies and NGOs. These included seminars on: 'Non-Technical Obstacles to New and Renewable Energies', held at the Rockefeller Foundation's Bellagio (Italy) centre in 1981; 'New Technologies in Coal Utilization' in Essen (Germany) in 1980; 'Solar Energy' in Hangzhou (China) in 1982; and 'Fuel Cells: Trends in Research and Applications' at Ravello (Italy) in 1985. More recent activities include workshops on 'Technology and Poverty Reduction' and 'Technology, Enterprise Development and Poverty Reduction' in Ghana and the Republic of Tanzania in 2003, and an 'International Focus: Engineering and Technology for Poverty Eradication' held in Washington DC in 2004 (the first UNESCO-sponsored meeting to take place in the United States since it rejoined the Organization).

INFORMATION AND PUBLICATIONS

The production of information and publications is a vital part of capacity-building and the engineering programme continues to contribute in this domain. Activities included the development of the UN Information System for Science and Technology (UNISIST) programme, based at UNESCO. Earlier publications include the first edition of the *International Directory of New and Renewable Energy Information Sources and Research Centres* in 1982, the UNESCO Energy Engineering Series with John Wiley beginning in the early 1990s (some titles have since been republished). More recent publications include *Small is Working: Technology for Poverty Reduction* and *Rays of Hope: Renewable Energy in the Pacific* as video/booklets and CD-ROM, *Solar Photovoltaic Project Development* and *Solar Photovoltaic Systems: Technical Training Manual*, as UNESCO toolkits of learning and teaching materials, all in conjunction with UNESCO Publishing.

OTHER ACTIVITIES

Other programme activities that have continued since the establishment of engineering in UNESCO include consultancy and advisory services. This includes, most recently, participation in the UN Millennium Project Task Force 10 (TF10) on Science, Technology and Innovation, and contribution to the

Glass blowing at the UNESCO-assisted Faculty of Technology of Tehran, Islamic Republic of Iran, 1958.

UNESCO-UNDP project for the training of engineers in mechanics, electricity and chemical processing, National University of Engineering, Lima, Peru, 1967.

TF10 report 'Innovation: Applying Knowledge in Development'. Pilot projects have also been supported, most notably relating to energy. It is interesting to note that the UNESCO Chairs programme was developed with particular input from Geoff Holister, director of the engineering programme from 1984 to 1990.

Interest at UNESCO was also developing in the early 1990s regarding university–industry cooperation and innovation. The University-Industry-Science Partnership (UNISPAR) programme was created in the engineering programme in 1993. The UNISPAR activity included an innovative International Fund for the Technological Development of Africa (IFTDA) – established with an investment of US$1 million, the interest from which (around US$60,000 per year) was used to fund activities from the UNESCO office in Nairobi. After supporting the development of many small-scale innovations, the IFTDA project was closed by the Director-General, Federico Mayor, in 1994, not long after his second-term re-appointment in 1993.

THE RISE AND DECLINE OF ENGINEERING: POLICY CONSIDERATIONS AND FUTURE PROSPECTS

Engineering at UNESCO rose in the early years to be the largest of the three initial and continuing theme areas of the Natural Sciences Sector (the other two being the basic sciences and the environmental and ecological sciences). Over the last fifty years, the engineering programme has had around 100 professional and support staff, a Regular Programme budget of over US$50 million and extrabudgetary funding of over US$200 million (mainly UNDP special funds in the mid-1960s to the early 1990s). Engineering at UNESCO began to decline (in terms of staff and budget, in real terms) in the 1990s – reflecting the decline of the Sector, and indeed of the Organization as a whole, over this period.

This decline was due to various external and internal factors. The 1980s marked a general decline in overseas aid. The withdrawal of the United States and the United Kingdom (in 1984) precipitated a funding crisis. And the fall of the Berlin Wall (in 1989) lead to the end of the Cold War. UNDP special funds began to decline from the late 1980s, with the establishment and development of its Operations Division.

There were also various internal factors relating to the three 'organs' of UNESCO – the General Conference, Executive Board and Secretariat. Due to its technical background, the Natural Sciences Sector is perhaps the least-well understood in the Secretariat; and engineering, for various reasons, is less well understood than science. Engineering is distinct from science; and, with a declining budget, science priorities have tended to predominate. Engineering and technology have also declined, owing to limited incorporation of contemporary issues and priorities in science and technology policy. This situation is reflected in the Executive Board (where there are very few scientists and engineers) and, to a lesser extent, in the General Conference (where education interests predominate).

Other internal factors include the choice of programme priorities based on personal interaction and lobbying – rather than a strategic approach, based on broader policy issues and a more democratic determination of needs and priorities. This was compounded – in the late 1980s and 1990s – by limited programme achievements in an area that was peripheral to core engineering issues, and by the recruitment of non-engineering personnel. This contributed to the further decline of engineering and its administrative merger into the Basic and Engineering Sciences Division in 2002 – with obvious potential consequences for the future of engineering at UNESCO.

The programmes at UNESCO with the most secure budgets and effective lobbying are those that are linked to international and intergovernmental programmes – such as the Man and the Biosphere Programme (created in 1971), the International Hydrological Programme (1975), and the Intergovernmental Oceanographic Commission (1960). There is clearly an advantage for programmes to have such a background, as evidenced by the new International Basic Sciences Programme (which may have further implications for engineering).

People were also a factor in this decline, and the engineering programme has suffered by having staff without professional backgrounds, contacts, credibility or interest in engineering (not to mention a cast of colourful characters – one of whom lived in his office for a while – and the wider governance, management and administrative challenges of work in UNESCO). This situation has led to constraints in terms of programme leadership, management and priority-setting. While the engineering programme successfully focused on core engineering issues up to the 1980s, the diversification and almost complete allocation of human and budgetary resources onto renewable energy in the 1990s – coupled with limited programme achievements – can be seen as a misguided policy of 'picking winners', leading to the serious decline of engineering and eventual merger into the Basic and Engineering Science Division.

This left the engineering programme with limited human and budgetary capacity to address core issues of engineering. In the people context, it is also worth noting that the science and engineering programmes have until recently paid limited attention to women and gender issues in science and engineering. It appears that there have only been two women professional staff members in engineering: one of them left the Organization, while the other moved to a Field Office.

The struggle to continue engineering activities at UNESCO rallies into the new millennium, with significant external support. Notable in this respect was the return of the United States to UNESCO in 2003, and strong support of the

US engineering community and the US Mission to UNESCO. This support was reflected in a Draft Decision for the development of 'Cross-Sectoral Activities in Technical Capacity-Building', presented by the USA and twenty-three other countries to the Executive Board in April 2005. This Draft Decision (for which twenty-five speakers expressed their support) and the proposal (accepted with unanimous support and acclaim – the greatest support for any draft decision or resolution in recent years) refers particularly to the need to strengthen engineering in UNESCO, to focus on capacity-building in engineering and engineering applications for poverty reduction and sustainable development around the world (with a progress report due for the Executive Board in spring 2006). It is hoped, therefore, that this decision to develop 'Cross-Sectoral Activities in Technical Capacity-Building' will help resurrect and strengthen engineering in UNESCO and around the world, in conjunction with the development of international programme activity.

UNESCO has a unique mandate and mission in the natural sciences, including engineering and technology, to assist Member States (especially developing countries) in capacity-building for poverty eradication and sustainable development. This it can do, as it has in the past, in connection with other areas of UNESCO activity in the sciences, the social and human sciences, education, communication, information and culture, in such fields as technological innovation, engineering ethics and codes of practice, information and communication technologies (ICTs), and the 'knowledge society' and 'engineered' parts of our cultural heritage. UNESCO looks forward to the development of cooperation in this activity.

APPENDIXES: PERSONAL REFLECTIONS ON ENGINEERING

These reflections by former colleagues working in the engineering sciences and technology programme are intended as personal commentaries, designed to put some flesh on the bones of history. Attempts were made to contact other former directors and colleagues, unfortunately without success.

APPENDIX 1

James McDivitt[16]

AT THE DIVISION OF TECHNOLOGICAL RESEARCH AND STUDIES, 1965–1967

I JOINED UNESCO in 1965, at which time Alexei Matveyev was Assistant Director-General for Natural Sciences. It is my recollection that the Natural Sciences Sector had recently been split into two departments. One, the Department for the Advancement of Science, included Education, Information and perhaps the Basic Sciences. Even at that time Michel Batisse was a key player; he had been instrumental in setting up the Intergovernmental Oceanographic Commission (IOC) and the water programme and was then working on the International Geological Correlation Programme (IGCP) and Man and the Biosphere (MAB) programme, which came along a bit later. The other department was the Department of Application of Sciences to Development, which included Engineering; as I recall, a man from Stanford Research Institute (Dale Krause) was head of Application.

Much of our work at that time was running UN Development Programme (UNDP) projects, and I don't recall much, if any, Regular Programme activity in engineering. I was one of about a dozen programme officers dealing with engineering projects, and, of these, one may have worked on Regular Programme. I worked on Asia with an Englishman named James (Jimmie) Swarbrick, who had been head of the New Delhi (India) office. Other names were Eneberg

---

16   James McDivitt: UNESCO Natural Sciences Sector, Programme Specialist (1965–67), Director (1972–82; Director, Jakarta Office, 1968–78), 'Director, Division of Technical Research and Higher Education (1978–82).

and Papa Blanco, who dealt with Latin America; and there were always some Russians, including Evstafiev and Evteev and others. As an example of what we did: I handled four projects – one to develop a Fine Instruments Centre in Seoul (the Republic of Korea), a Polytechnic in what is now Bangladesh but was then East Pakistan, the College of Engineering at Lahore (Pakistan), and a Technical Research Centre at Bhopal/Bangalore (India).

These were all existing projects that I took over from someone else, so the details had already been agreed on and the documents approved and signed by the governments, UNDP and UNESCO. My job was to search the roster and submit candidates for the various expert posts, to arrange for the fellowships, and to monitor and follow up on the equipment orders. There were divisions within UNESCO that did the detailed work in each of these areas (the Personnel Division, the Equipment Division and, I believe, a Fellowship Division), and we were expected to work closely with each of these and coordinate all our activities. Essentially we were the link between HQ and the field. I recall that one of the great challenges was getting letters through the system, as each letter had to be signed off on (visaed) by each Division in the house that might have any interest in it. Thus a normal letter would require three or four visas, and anything important might require ten – so things did not move fast. Every letter that went out had to pass through Science Administration and Mr Guintoli – who was a master bureaucrat, in a positive way, as he handled a very difficult job efficiently and with good spirit.

One of our other jobs was to develop new projects for UNDP funding. This involved visiting the countries, or meeting with delegates who were attending meetings and the General Conference to identify and promote possible projects. Many times the requests were originated by the Resident Representative office in the countries and were assigned to the appropriate agency. Thus we would get any Engineering Education projects, but in some areas, such as the Earth sciences, we would be in competition with the UN Technical Cooperation unit in New York.

Another line of programme development was to help set up international unions and structures to promote engineering. Here we were following the lead of the Basic Sciences, and again Batisse was very involved in this. In the Basic Sciences, they had ICSU (International Council for Scientific Unions, now International Council for Science) and all of its Unions. We used Regular Programme money to help set up the Union of International Technical Associations (UATI) and the World Federation of Engineering Organisations (WFEO).

AS DIRECTOR OF THE UNESCO OFFICE IN JAKARTA, 1968–1978

After two and a half years at Headquarters, I was transferred to Jakarta (Indonesia) as director of the Regional Office, where I was able to see the programme from a much different point of view. There, our emphasis was much more on Regular Programme activities that often had a regional dimension. We were the Regional Office for Science and Technology (the name changed quite often, but that was a reasonably accurate version), and this gave equal weight to the science and engineering dimensions. At the same time, we were serving the whole Natural Sciences Sector, and there was much more Regular Programme money in science than in technology. However, because of my personal working link with the engineering side of the Sector, I probably gave relatively more attention to this than to other programmes.

One project in Indonesia, the restoration of the Buddhist temple of Borobudur (Java, Indonesia), might warrant special attention here. This was a major activity costing many millions of donated dollars and lasting more that ten years. Although the project officially fell within the Culture Sector, the actual restoration was mainly an engineering exercise involving the disassembling, cleaning, treating chemically, and reassembling of tens of thousands of carved stones that had stood on the site for well over 1,000 years. The same would be true of many of the other large cultural projects that UNESCO has implemented over the years, as for example with Abu Simbel (two temples built by the Egyptian king Ramses II, near the border of Egypt with Sudan – relocated due to flooding with the construction of the Aswan high dam). These activities are carried out because of their cultural importance and implications, but there can be no question that the actual restoration projects involve, and are dependent on, engineering.

AS DIRECTOR OF DIVISION OF TECHNOLOGICAL RESEARCH AND HIGHER EDUCATION, HQ, 1978-1982

One of the early projects of the Engineering Division was to prepare a directory of engineering education programmes throughout the world, with emphasis on programmes in developing countries. There are a number of such reviews in the developed countries – for example, the American Association of Engineering Education does a detailed annual review for the United States, and there are similar programmes in Canada and many of the developed countries. With the advent of the computer, a project was developed in the Division to create a database with information on all the engineering programmes; and significant progress was made on this, although, for budgetary reasons, it has been put on hold. After I retired, I worked on this with Fumin Zhang, and we actually put the bit of it that dealt with Canada on the Division website.

PART II: BASIC SCIENCES AND ENGINEERING

APPENDIX 2
Charles Gottschalk[17]

THE UN system-wide effort to establish a network for all science and technology (S&T) information – at the core of which was UNESCO's UNISIST (UN Information System for Science and Technology) programme – was peaking in 1979. I was then seconded by the United States Government to a senior officer post in PGI (General Information Programme, where UNISIST had been housed since its inception in the early 1970s) to provide expertise gained over the prior twenty years with major US Government S&T agencies involved in designing computerized information networks that led to today's internet. Moreover, I had served five years with the International Atomic Energy Agency (IAEA) in Vienna (Austria), as one of the founders of the International Nuclear Information System (INIS), the first computerized international system of its kind.

Shortly after my arrival, the Director-General (Amadou-Mahtar M'Bow) called on PGI for project proposals related to scientific information under the UNISIST umbrella and in support of UNESCO's contribution to the 1979 UN Conference on S&T for Development (UNCSTD). My proposal to conduct a detailed study on the need for, and feasibility of, an international information system to focus on new and renewable energy sources was selected for special funding of US$100,000 in late 1979, and for submission to the UNESCO General Conference in Belgrade (Yugoslavia) in October 1980 (at that time, General Conferences were held around the world). Two major events favoured my proposal: the energy crisis of 1973, and the UN Conference on New and Renewable Sources of Energy (UNCNRSE), held in Nairobi (Kenya) in 1981. The latter provided the first major international forum for a reasoned discussion of energy alternatives, with its emphasis on the need for a global transition to a mix of energy sources. UNESCO was ideally positioned – by dint of its multidisciplinary mission – to foster the many less tractable issues that affect the diffusion of any development innovation, including: information, education and training, social and cultural factors, institutional structures, and environmental impact, particularly for the developing world.

A variety of sound administrative decisions led to the transfer of my post, budget and the study project (on its completion in late 1980), from PGI to

---

[17] Charles Gottschalk: UNESCO, Chief of Section, Section of Energy Information Systems (1979–88).

the Natural Sciences Sector's TER (Division of Technological and Engineering Research, headed by Jim McDivitt). There, it became the Energy Information Section (which grew to a staff of five by 1986) and functioned in close cooperation with the Energy R&D Section (headed by Ted Beresowski), until about 1988, when I retired. However, I continued to serve the Division (which had become EST, Engineering Sciences and Technology) as a consultant in various energy and engineering capacities until 1996, working from an office in the Temporary Building behind the Bonvin Building (which had housed EST on the third floor since the 1960s.) In 2002, when EST and its sister division BSC (Basic Sciences) were merged into one division (BES) for Basic and Engineering Sciences, EST became a Section of BES, and was re-dubbed the Engineering Sciences and Technology Section.

The major recommendation from the New and Renewable Energy Information study of 1979–80 was to assess the capacity of global efforts to meet the demand for new and renewable energy information and to create a cohesive international computer network linking information providers and users. The programme initiated to accomplish this was funded with US$1.5 million over 1981–83, although the projected (modest) activities of UNESCO's programme alone called for an estimated US$18 million over the same period. Major accomplishments over the initial three-year period were:

- The establishment of a computerized and centralized database of ultimately online information and data on global new and renewable energies.
- The publication and worldwide distribution in 1982 of the first edition of the *International Directory of New and Renewable Energy Information Sources and Research Centres*, generated from the aforementioned database. A second edition of 660 pages was published in 1986, which contained some 3,600 entries, representing 156 countries. Information collection and input preparation continued well into the early 1990s, in order to underpin the foundations of the networking effort initiated by UNESCO.
- The launching of Regional Pilot Projects (RPPs) in East Africa, Latin America, and the Arab States (Asia and the Pacific was created in 1985); these RPPs engaged in activities involving information generation and collection, information analysis and packaging, dissemination, networking, and training.

In the early 1990s, EST launched a distance learning publishing activity, called the UNESCO Energy Engineering Series (Energy Engineering Learning

Package), under a contract with the UK publisher John Wiley. It was targeted to train undergraduate and postgraduate students with a particular interest in new and renewable energies, and to promote information – sensitizing specialists and the general public to the possible uses of renewable energy sources, with particular regard to environmental concerns and the requirements of sustainable development. Some twelve texts were published in this successful series, which ended in the late 1990s (some titles were later updated and republished, and a related Renewable Energies Series continued from this was produced, first by Akio Suzuki, then Tony Marjoram).

As funding for EST waned over the course of three successive short-lived EST directorships from the mid-1980s to the mid-1990s, novel programmes were thought up in the hope of attracting extrabudgetary funds and improving UNESCO's reputation in the field of technology. This strategy of desperation resulted in the creation of EST's 'World Solar Programme 1996–2005' (WSP) in 1996, which promptly annexed all extant energy activities of the Division. As of this writing (February 2005), the jury's verdict on the termination of an overly ambitious and politically incorrect WSP, sadly lacking in accomplishments, has not as yet been rendered.

APPENDIX 3

Edward Beresowski[18]

A FEW HIGHLIGHTS OF PAST SUCCESSFUL UNESCO ENDEAVOURS MANAGED BY THE TRAINING AND ENGINEERING RESEARCH (TER) DIVISION

BELIEVE the Division got some early distinction from a 'summit' meeting to plan its activities, organized by James McDivitt in early 1979 in northern Spain. Its activity probably peaked circa 1980–82, when McDivitt set up four branches. Under the impetus of the preparations for the United Nations System's Conference on New and Renewable Energies in 1981, most of the action was in two energy-area branches, with Chuck Gottschalk's energy information activity and my energy development and training activity.

---

18 Edward Beresowski: UNESCO, Chief of Energy Development and Coordination Section (1979–86).

The regular budget of the Division was quite modest (on the order of US$1 million per year), so my branch's activities focused on training and seminars in cooperation with international agencies such as the European Communities Agency in Brussels (Belgium), the Council of Europe in Strasbourg (France) and the International Labour Organization in Geneva (Switzerland), as well as non-governmental groups such as the International Solar Energy Society (chaired at the time by Bill Charters from Australia). We pooled activities and tried to activate work groups and organize seminars and networks – primarily in areas such as energy planning, and solar, wind and biomass energy training.

I believe the most successful activity during my tenure was the management of an extrabudgetary energy training project for Brazilians, from about 1981 to 1986. These projects were adequately funded by the United Nations Development Programme (UNDP) and Brazil at a level of a few million dollars over the five-year period. We organized training seminars on various energy topics in Brazil and other countries; we sent energy specialists from various countries to tour and lecture in Brazil; and we organized visits for selected Brazilian energy specialists to tour in energy laboratories and enterprises in their areas of expertise, throughout various developed countries. Some several hundred Brazilian industry, laboratory and university energy specialists improved their effectiveness in energy planning and in conventional and renewable energy activities.

Aside from the above, I believe that the most significant activities with some useful impact were:

- A week-long seminar organized at the Rockefeller Foundation's Bellagio (Italy) centre, circa 1981, on 'Non-Technical Obstacles to New and Renewable Energies', with participation from the European Commission – the results of which were contributed to the United Nations System Conference on New and Renewable Energy in Nairobi (Kenya).
- A seminar in Essen, Germany, on 'New Technologies in Coal Utilization' (such as fluidized bed), circa 1980. At the time, its results and report were of great interest and very well received by countries such as Poland, the USSR and China.
- A two-week seminar on solar energy (photovoltaic and solar heat), organized in Hangzhou (China), circa 1982, for specialists from various Chinese laboratories, with some teacher contributions from the French Solar Energy Agency and the European Commission.
- A 1985 pioneering seminar in Ravello (Italy), on 'Fuel Cells: Trends in Research and Applications'. At the time there was almost no attention – outside the USA and Japan – to this

technology, and undoubtedly this was the first international meeting on the subject. It led to our sponsorship of case studies in several countries just before I left UNESCO. Since then, this technology has become of significant interest and is considered promising for several applications.

APPENDIX 4
Benjamin Ntim[19]

SOME REFLECTIONS ON THE DIVISION OF ENGINEERING SCIENCES

I JOINED UNESCO in 1980, at a time when the Engineering Division, officially known as the Division of Higher Education and Technological Research, was one of the three pillars of the Natural Sciences Sector (the other two being Basic Sciences and Environmental Sciences). The Engineering Division, with four sections – dealing with Engineering Education, Engineering Research, and (two sections on) Renewable Energy Development and Information – was by far the most prominent among the three, in terms of its budgetary allocation. This allocation had two components, the regular and the extrabudgetary, with the latter component mainly sourced from UNDP.

The Engineering Division, by this period, was executing several projects on the improvement of curricula in engineering schools and institutions, as well as on methods for increasing efficiency in the engineering profession in many developing countries scattered across Latin America, Africa and Asia. The magnitude of extrabudgetary projects executed by the Division was usually above the US$1 million ceiling. This is in contrast to the projects executed by the Division in the late 1980s and 1990s – when the ceiling on projects was of the order of a few hundred thousand dollars. The ANSTI project (started in 1979 for Africa), for example, had around US$5 million in its first phase.

Soon after this era – around 1987 – the extrabudgetary sources for engineering activities began to dwindle, and as the scientific community began to put much emphasis on environmental issues, engineering at UNESCO suffered a loss in prominence. This period also coincided with the withdrawal of the USA (1984)

---

19   Benjamin Ntim: UNESCO, Programme Specialist, Engineering Sciences and Technology (1980–2002).

and the UK (1985); both have since returned (in 2003 and 1997, respectively). It was also around this period when the UNDP formed an Operations Division to execute its projects at its Headquarters in New York. This series of events lead to the pruning of the Division from four sections to three, and even some voluntary staff retirements. Indeed, there was talk of amalgamating the two Divisions of Engineering and Basic Sciences, though this did not materialize until after 1998.

DIRECTORS AT THE ENGINEERING DIVISION

During the sixteen years I spent at the Division, its name changed from the Division of Higher Education and Technological Research, to the Division of Engineering Sciences. Three directors served the Division under the first name, while one served under the second name. The first two directors – James McDivitt (1978–82) and F. Papa Blanco (1982–84) – between them served five years and made much impact through innovative programmes that they developed with national and regional professional institutions. They were followed by Geoff Holister (1984–90) and Boris Berkovski (1990–2000), who principally used personal connections within the Organization and the Permanent Delegations in Paris to develop equally interesting programmes on UNESCO Chairs and Renewable Energy. Whether this tactic is of legitimate worth is a matter for debate.

HEADQUARTERS VERSUS FIELD SERVICE

Prior to the 1990s, and before decentralization became a recognized policy, the vast majority of the staff spent their life at Headquarters. Indeed, during this period, it was regarded as a punishment to be sent to the Field Offices. I recall that a colleague in the Division became ill three times on the three occasions that he was proposed to be assigned to a Field Office. Indeed, his continued refusal eventually led to his separation from the Organization. After spending sixteen years at Headquarters and six in the Field, I can safely say that it is immensely satisfying to serve in the Field – both from the point of view of job satisfaction and the close varied human interaction that one enjoys there.

FUEL FOR THOUGHT
UNESCO initiatives on renewable energy sources
Osman Benchikh[20]

INTRODUCTION

MANKIND'S use of energy has been inextricably intertwined with its history. From lighting a fire to nuclear fission – the production of energy through the ages has followed the development of scientific thought (and the available forms of energy) and often had a direct influence on the structure of society. In many respects, the production of energy has been one of civilization's greatest constraints.

Demand for energy continues to grow even though governments adopt vigorous policies to conserve it. Energy problems are today so acute at the international level that it is no longer possible to satisfy the world's constantly growing needs by continuing to exploit, as before, too limited a range of resources. This growth of energy demand must be increasingly satisfied by diversified energy resources, including sustainable and renewable sources.

Over the past decades, the expectations for renewable energies have not been fulfilled. Research did not receive the necessary support, and its findings were often viewed solely in immediate economic terms. Furthermore, the countries that most needed to exploit these energies did not have the capital required for the initial investments, nor often the infrastructures to profit from the achievements made in renewable energy technologies.

The United Nations Conference on Environment and Development, held in 1992 in Rio de Janeiro, Brazil (known as the 'Earth Summit'), highlighted the important role that renewable energy sources and technologies can play in helping to address the twin challenges of development and environmental protection. Developing and industrialized countries alike stand to benefit a great deal from the exploitation of plentiful indigenous energy resources, thereby lessening their dependency on expensive, environmentally damaging forms of energy.

More recently, the World Summit on Sustainable Development (Johannesburg, South Africa, August 2002) addressed emerging and critical issues for the future, reflecting a global shift in emphasis since 1992, from environmental concerns to the more holistic approach of sustainable development, focusing on the interrelationships of environment, society and economy. Within this framework,

---

20   Osman Benchikh: Programme Specialist responsible for energy and renewable energies (since 1998), UNESCO Division of Basic and Engineering Sciences.

the Summit also set up a process to promote the use of sustainable and renewable sources of energy to improve the living conditions of those without access to conventional sources of energy. Almost one-third of people living in rural areas, mainly in developing countries, have no access to electricity. Another third is severely undersupplied, relying mainly on renewable energy sources to improve living conditions. Today, the world community is aware of the role and importance of these energies, especially for sustainable development and improved living conditions for the rural poor population. However, to meet the challenging task of energizing the non-energized, specific efforts have to be undertaken to adopt and make use of renewable energy technologies.

The main pillars of a global strategy for better management of energy are the development of human resources and the adaptation of technologies development to local needs. Other elemental aspects include: austerity in the consumption of energy, scientific rigour in considering energy alternatives, the promotion of research on new more 'environmentally friendly' energy sources, and higher investment in Earth security. Energy constitutes one of the keys to the social and economic development of all nations. Together with education, it is the fundamental element in the development of societies, and the world of tomorrow will be shaped irreversibly by the particular energy technologies used.

The importance attached to increasing the use of renewable sources of energy arises from concern over five issues. *First*, the most urgent concern is for the more than 2 billion people who currently have no access to modern energy services (Geller, 2003). Most live in rural areas, where they rely on non-commercial energy sources, such as biomass, firewood, charcoal and animal waste. Indoor burning of these materials seriously affects the health of women and children in particular. The objective of the Millennium Development Goal of halving, by 2015, the proportion of the world's population whose income is less than one dollar per day will depend on providing this population with access to modern energy services for their needs, as well as for income generation. For widely dispersed and low-density rural populations, decentralized energy technologies based on renewable sources provide a viable alternative to expensive grid extensions. Decentralized renewable energy systems constitute a valuable component of efforts at poverty eradication.

Second, there is the urgent need to reduce the emission of greenhouse gases, which are primarily associated with fossil-fuel extraction and use. Of primary importance are carbon dioxide emissions from the combustion of fossil fuels, and the release of methane during the extraction of natural gas, oil or coal, and the transport of natural gas. The third assessment report of the Intergovernmental Panel on Climate Change (IPCC), entitled *Climate Change 2001: the Scientific Basis*, attributes a significant portion of the increase in average global temperature observed over the past fifty years to increasing atmospheric concentrations of greenhouse gases resulting from human activities (United Nations, 2001). The thinning of Arctic ice, the retreat of the glaciers, and the elevation of sea levels that has been reported constitutes predictable manifestations of global warming.

Third, there is a serious concern about localized pollutants emanating from fossil-fuel use, including sulphur and nitrogen oxides, carbon monoxide and suspended particle matter. Variously, these gases contribute to the depletion of the stratospheric ozone layer and acid precipitation, as well as to increased incidents of ill health and death. Replacing fossil fuel with renewable sources of energy greatly reduces, and often completely eliminates, the emission of greenhouse gases and localized pollutants associated with fossil-fuel combustion.

The *fourth* major issue remains the need for more energy from all sources, including renewable sources, particularly for developing nations, which are faced with a rising energy demand, associated with the twin pressures of a significant per-capita economic development and increase in population. *Fifth*, the persistent threat of the ultimate depletion of economically recoverable quantities of fossil fuels – especially oil, but also natural gas and eventually coal – cannot be dismissed. It raises the issue of the need to begin phasing in an energy supply mix using renewable sources, which can ultimately replace depletable fossil fuels.

As a result of the concern over these five issues, the international community has gradually come to recognize the importance of developing and utilizing renewable sources of energy. Since the World Solar Programme 1996–2005[21] was launched as an instrument at the service of the international community for the promotion and enhanced utilization of environment-friendly, renewable sources of energy, this recognition has, in particular, led to a major emphasis on renewable energy for achieving the objectives of sustainable development.

HISTORICAL BACKGROUND

UNESCO involvement with energy problems and their scientific and technological applications dates back to the first years of the Organization's history. In particular, UNESCO was one of the initiators of international cooperation towards long-range scientific and technological solutions to energy problems. At the beginning of the 1950s, the Organization pioneered the promotion of research in non-conventional and renewable sources of energy (particularly solar and wind energy) when the large-scale International Arid Zone Programme was established, in which the energy problems confronting arid regions were considered. It was the first example of international cooperation in the field of

---

21 The term solar is used herein in a generic sense and refers to all forms of renewable energy including the solar thermal, solar photovoltaic, biomass, wind, mini-hydro, tidal, ocean and geothermal forms.

Solar oven at the UNESCO-assisted Solar Energy Laboratory of Dakar University, Senegal, 1962.

new and renewable sources of energy, and it culminated in the International Symposium on Solar and Wind Energy of the Arid Zone, convened in New Delhi, India, in 1954. The Arid Zone Programme was terminated in 1962 but created a solid international scientific basis for later UNESCO initiatives.

In the early 1970s, the world community began to suffer from shortages of energy as a result of a crisis in oil production and distribution as well as the inefficient use of available energy resources. Shortages and the rising cost of energy contributed considerably to the slow-down of the process of social and economic advancement of many developing countries in the past decades. This is why scientific problems connected with developing new methods for the use of energy, as well as the search for new sources, has become once again a preoccupation of the United Nations (UN), its specialized agencies, and various international intergovernmental and non-governmental organizations.

In 1973, the Natural Sciences Sector of UNESCO convened the world congress 'The Sun in the Service of Mankind', which took an optimistic and enthusiastic view of the perspectives for renewable energy sources. At its seventeenth session in 1974, the General Conference of UNESCO approved a resolution on 'fostering international cooperation in selected breakthrough fields which will permit the development of pollution-free sources of energy', giving impetus to UNESCO's energy programme. Since then, new and renewable sources of energy have played a major role in UNESCO's activities.

PART II: BASIC SCIENCES AND ENGINEERING

UNESCO'S CONTRIBUTION TO THE 'EARTH SUMMIT' FOLLOW-UP

Twenty years later – and as a follow up to the UN Conference on Environment and Development, held in 1992 in Rio de Janeiro, Brazil (the adopted 'Agenda 21' contains a chapter on climate change with special attention to renewable energies) – the Engineering and Technology Division of the Natural Sciences Sector took the initiative of organizing, in 1993, a high-level Expert Meeting – recycling the title 'The Sun in the Service of Mankind' – in order to look into the potential of all forms of renewable energies (solar, wind, biomass, geothermal, tidal, ocean, etc.) and their applications. Experts from fifty-three countries prepared sixty-six in-depth critical assessments of solar energy and related fields, including 'Solar Energy and Health', 'Solar Energy: A Strategy in Support of Environment and Development', and 'Financing Solar Energy Development'. These reports were debated in thirty-five specialized round tables.

This meeting suggested undertaking a three-year campaign – the World Solar Summit Process – leading to the organization of a World Solar Summit at the highest political level, that is, heads of state or government. This recommendation was based on the fact that one of the main reasons renewable energies were not sufficiently exploited was the lack of political will on the part of the top decision-makers, partly due to insufficient information on the potential of these energies. The World Solar Summit Process (1994–96) was a concrete response to the recommendations of the Earth Summit concerning energy and the decisions of the twenty-seventh and twenty-eighth sessions of the UNESCO General Conference.

In following the meeting's recommendations, UNESCO initiated a process leading to the holding in 1996 of a World Solar Summit of heads of states and governments, aiming at the definition of a World Solar Programme 1996–2005 and fostering a global mobilization towards use and applications of renewable energy. UNESCO has in fact conceived and initiated this programme as an instrument at the service of the world community, for the promotion of renewable energies and the dissemination of the relevant technologies.

The coordination by UNESCO of a 'World Solar Summit Process' (WSSP) was included in the Organization's Programme and Budget for 1994–1995. During the first year (1994) of this preparatory process leading to the Summit, the Executive Board, at its 145th session, approved the establishment of a 'World Solar Commission' to advise the Organization and the participants in the WSSP on measures for reinforcing global and regional cooperation in the promotion of renewable sources of energy.

Recognizing that the supply of skilled manpower is a critical element in the transfer of technology, and that relatively little is known about training needs and potential in the field of new and renewable sources of energy, UNESCO also launched the Global Renewable Energy Education and Training (GREET) Programme, based on a request made by Member States and consultations made in the different regions. The GREET Programme aims at raising awareness, capacity-building, provision of related policy advice, and development of competent human resources, geared to improve the use, maintenance and management of renewable energy systems. It also constitutes a framework that favours the transfer of technological know-how and sharing of experiences.

The demand is steadily growing for trained workers possessing basic scientific and technical knowledge related to rapid economic development, with its vital dependence on energy supplies. UNESCO efforts (focused on renewable energy sources) confirm the need for education in fundamental disciplines and the additional requirement for specialized training of personnel for different levels of responsibility. The majority of the present specialists in renewable energy have had basic scientific or technical training without specific emphasis on new energies, and have acquired their knowledge by on-the-job experience. This category of specialists will exist for many years to come, and their training will need to be encouraged and augmented by an expanded and coordinated project within the GREET programme.

The World Solar Summit was held in September 1996, in Harare, Zimbabwe. A total of 104 countries were officially represented, along with twelve organizations of the UN system, and ten regional intergovernmental organizations (including the European Commission and the Organization of African Unity). Representatives of private industry and non-governmental organizations (NGOs) also participated. The Summit was preceded in Harare by the first session of the World Solar Commission (composed then of sixteen heads of state and government) which reviewed and endorsed the two basic documents prepared for the Summit: the draft 'Harare Declaration on Renewable Energy and Sustainable Development', and the draft outline of the World Solar Programme 1996–2005. The Harare Declaration calls 'on all nations to join in the development and implementation of the World Solar Programme 1996–2005', invites 'the World Solar Commission to continue to provide high-level leadership and guidance' in this respect, and 'invites UNESCO to continue to play a leading role in the development of the World Solar Programme'.

The main aim of the World Solar Programme (WSP) 1996–2005 is to encourage the wider use of all forms of renewable energy to alleviate poverty and promote sustainable development. The WSP has the purpose of sensitizing governments, intergovernmental organizations, NGOs, financial institutions and academia on the need to support the development and utilization of renewable energy for sustainable development. The master document of the WSP, resulting from the regional consultations and the proposals made by the countries, outlined an important programme of national priority projects, regional projects, and projects with universal value, ranging from education and training, information, policy and planning, to issues such as rural electrification, and water supply and treatment using renewable energies. Resolutions of the General Assembly of the United Nations in the period 1998–2004 further endorsed the aims of the WSP within the framework of sustainable development and in promoting the special interests of UNESCO in education and training activities in the field of renewable energy through the GREET Programme.

During 1996–2004, the different approved Regular Programmes and Budgets of UNESCO all included provisions for the World Solar Programme 1996–2005. The twenty-ninth session of the General Conference adopted a specific resolution co-sponsored by nearly seventy Member States, inviting the Director-General to 'undertake further consultations ... with a view to ... transforming the WSP into an interdisciplinary endeavour'. In general, UNESCO's contribution to the implementation of the World Solar Programme involved the support and promotion of a selected few national projects that served as a showcase at the national and/or regional level. The Organization provided assistance to Member States, especially developing countries, in defining their national energy policies and strategies, as well as securing financing from funding sources. UNESCO played mainly a catalytic role by launching new concepts and initiatives and offering to the world community a framework for promoting clean and sustainable sources of energy.

UNESCO's input was primarily aimed at drawing the attention of a number of governments to the need to increase the use and application of renewable energy sources and to set up the structures necessary for the development and dissemination of the relevant technology. The Organization also made a major contribution in presenting the concept of a 'Solar Village' to the international community. The goal of stand-alone power for local consumption through the wider acceptance of the Solar Village concept could prove to be the ultimate solution to some of the energy needs of the majority of the world's population living in poverty. Renewable energy systems can be used to meet a range of needs, including low-power applications, water pumping, vaccine refrigeration, radio and television.

THE WORLD SOLAR PROGRAMME IN UNITED NATIONS GENERAL ASSEMBLY RESOLUTIONS

Over the past decade, the WSP has been introduced on the agenda of the debates of the UN General Assembly. The General Assembly has adopted several resolutions endorsing the World Solar Programme 1996–2005 as a contribution to the achievement of the goals of sustainable development and calling for further action to ensure that the programme is fully integrated into the mainstream of the efforts of the UN system towards attaining the objective of sustainable development. The UN General Assembly also invited the international community to support – as appropriate, including by providing financial resources – the efforts of developing countries to move towards sustainable patterns of energy production and consumption.

The General Assembly called upon all relevant funding institutions and bilateral and multilateral donors, as well as NGOs, to support the efforts being made for the development of the renewable energy sector in developing countries on the basis of environment-friendly, renewable sources of energy of demonstrated viability, and to assist in the attainment of the levels of investment necessary to expand energy supplies beyond urban areas.

In expressing its appreciation of the continued efforts of the Secretary-General in bringing the WSP to the attention of relevant sources of funding and financial assistance, the General Assembly encouraged him to continue his efforts to promote the mobilization of adequate technical assistance and funding, and to enhance the effectiveness and the full utilization of existing international funds for the implementation of national and regional high-priority projects in the area of renewable sources of energy.

PRESENT SITUATION AND PROSPECTS FOR THE FUTURE

Many countries have hailed UNESCO's support and called for its further involvement in implementing renewable energy activities, including the World Solar Programme. At a more global level, concern over five issues has led to a determination to increase the use of new and renewable sources of energy: poverty elimination, climate change, localized pollution, increased energy demand, and eventual fossil-fuel depletion. The Plan of Implementation of the WSSD calls for substantially increasing, with a sense of urgency, the global share of energy from renewable sources. The United Nations and its relevant member organizations, other international organizations, national governments, private corporations and NGOs are currently engaged in promoting new and renewable sources of energy. The partnerships for sustainable development resulting from the WSSD process have contributed significantly to the increased interest in renewable sources of energy; and the prospects for accelerating the development, dissemination, utilization and commercialization of new and renewable sources of energy have become markedly more optimistic in the past two years.

UNESCO's long experience in the field of renewable energy, all the initiatives and programmes launched by the Organization over the past decades, constitute a major contribution to sustainable development. UNESCO's achievements represent a key element of any overall energy strategy that could be agreed upon at the international level in order to promote a balanced and reinforced approach to economic, social and environmental aspects of sustainable energy development. Countries are increasingly aware of the fact that the wider-scale use of renewable energies can have a crucial impact on two major issues confronting them: environmental protection and social development. This understanding alone should make the issue of renewable energies a top priority for responsible leaders.

BIBLIOGRAPHY

Geller, H. 2003. *Energy Revolution: policies for a sustainable future.* Washington DC, Island Press, 256 pp.

United Nations. 2001. Report of the Secretary-General: Protection of the Atmosphere (E/CN.17/2001/2).

SCIENCE EDUCATION
LIVE AND LEARN
The early days of science education at UNESCO
Albert Baez[22]

UNESCO, PARIS, 1961

IN 1961, my family and I moved to Paris, so that I could become involved in the new activities in science education there. To my pleasant surprise, I found two individuals who had been instrumental in sending me to Baghdad almost ten years earlier: Mme Thérèse Grivet and Malcolm Adiseshiah, who had now become Deputy Director-General of UNESCO. There were two other individuals, at UNESCO, who were influential in the direction that my Division took. One was Hilliard Roderick, who was a Deputy Director of Sciences, and Victor Kovda, who was Director of the Natural Sciences Department. In those days, science education was in the Natural Sciences Department. In fact, it was because of that, that I was attracted to come to UNESCO.

I was working on the assumptions that Zacharias had utilized at PSSC, namely that scientists should be in charge of the science education process. I felt that I could put into practice many of the things that I had learned with PSSC and avoid some of the mistakes that I thought we had made there. Before coming to Paris, I went to visit the heads of the other projects in the other sciences which had by now sprung up: the chemistry project called 'Chemstudy', the biology project – there were two projects instead of one there – and the mathematics project that had begun. In other words, curriculum reform in the sciences and mathematics had begun to flourish in the United States, and I thought I'd better take advantage of what had been done there, and take these ideas with me to utilize as the basis of whatever activities I would develop at UNESCO.

---

22 The first Director of UNESCO's Science Education Department (1961–67), the American physicist Albert Baez taught at the University of Redlands and Stanford University (both in California, USA) before coming to UNESCO. Baez's experience with the Physical Science Study Committee (PSSC) – whose goal was to improve physics education in US secondary schools – convinced him that science-education reform should be in the hands of scientists, rather than educators, since scientists know the content. The following are extracts from 'The Early Days of Science Education in UNESCO' (available at: http://www.unesco.org/education/pdf/BAEZ.PDF).

PART II: BASIC SCIENCES AND ENGINEERING

PHYSICS, CHEMISTRY, BIOLOGY AND MATHEMATICS AT UNESCO

I was given the title of Director of the Division of Science Teaching at UNESCO, and it was really my task to develop a program; I didn't have yet much of a staff, although Mme Grivet, the old pioneer, had been assigned to work with me. I came to the conclusion that, because of limitations of budget, we had better have some criteria for limiting what we would do. The limitations that came to mind were to consider what was fundamental and basic. My experience in the US programs led me to think that physics, chemistry and biology were the basic sciences and that mathematics was, as Eric Temple Bell had once said, the 'queen and handmaiden of science', so I figured that the Division of Science Teaching should have sections, with activities in physics, chemistry, biology and mathematics. These would be four basic areas in which we would do some work.

They were, fortunately, the areas in which work had been done in the United States, and subsequently in England and many other countries. So there was a backlog of material and man- and woman-power out there which could be tapped as we were generating a programme for UNESCO. Because of my own background, it seemed clear that I should take responsibility for developing the physics activities, and so I began seeking help in developing chemistry and biology activities. Once again, because of the prejudices and biases that I had picked up from the curriculum reform activities in the United States, I decided and later got the support of my director in this choice, namely to concentrate on secondary school science. Mme Grivet, who had sent me to Baghdad to work at the university level, continued to work in my Division with the focus on university-level science teaching activities. Because of my own training, I decided that I would take responsibility for physics and that I would immediately be on the lookout for a chemist and a biologist.

At about that time, it was rumoured that a fairly large sum had been set aside for new methods, techniques and approaches for teaching at UNESCO. No mention of science. But when I learned that money had been set aside for new activities, new types of activities and teaching, I immediately thought that the experience that I had had with PSSC, and subsequently in my connections with other leaders of the curriculum reform movement in the United States, could be useful. I thought that it was my task to find the best that had been done or was being done around the world and somehow utilize the talents of the people that had been involved to help us at UNESCO in developing a strong UNESCO program.

Although the wording of the proposed new activities within UNESCO included the concept of improvements in teaching, it didn't specify what those improvements might be, nor what areas they would deal with. I saw the opportunity of getting assistance for my Division by latching onto some of the funds that were being set aside for improvements in teaching. So I sat down and wrote the draft of a proposal for a pilot project on new methods, techniques and approaches in the teaching of physics in Latin America. I had chosen the subject area, namely physics, the geographical area, namely Latin America because I spoke Spanish, and what I had in mind was a substantial program in which scientists and science professors from the region would help us generate a program that would develop new materials, new techniques for the teaching of that particular subject. I was invited to state my proposals before the authorities of UNESCO at the time; so the Director-General, René Maheu was present, Malcolm Adiseshiah was present, and of course, all the other relevant people including the director of the Department of Education and the director of the Department of Science. I expounded my ideas for a project in Latin America that would help improve the teaching of physics in Latin American countries. The funds that had been set aside were substantial by the standards of those days; I think it was US$140,000 that had been set aside for a special project in new approaches, methods and techniques. When I had finished my presentation, the Director-General asked for questions and comments. Mr Adiseshiah rose and said: 'The funds that have been set aside for a new project are in the Education Department', and essentially he was saying that Baez was in the Science Department. The way I remember it, Mr Maheu said: 'Yes, but I think that Baez knows what he is doing, and I propose we allow him to use this money', or words to that effect.

That was the beginning of a very strong support for me and for the Division of Science Teaching, but at the same time it generated problems because I had alienated the members of the Department of Education, because they felt that this new project rightly belonged to them. So the compromises that had to be made were of this nature: that if someone in the Education Department had the knowledge and the know-how that was needed in the running of our proposed project, that we would call upon the UNESCO staff member who had those responsibilities. For example, in the audio-visual field, there was someone in the Education Department and I was supposed to consult with him very closely on whatever activities we developed along those lines. There was a new area that involved programmed learning – programmed instruction was something very new – and since we were talking about new activities, it was clear that we would have to utilize the person who was in charge of programmed learning in the Education Department. Fortunately I knew most of these people in the Education Department, and I had developed fairly good personal relationships with many of them and, therefore, they did not take my proposals as a threat and we actually found ways to collaborate. The point is that we got the green light to start something new and very important from the point of view of education: new approaches, methods and techniques in the teaching of science at UNESCO.

THE PHYSICS PILOT PROJECT IN LATIN AMERICA

I immediately began my search of the people who might run the project. We needed a director and an assistant director. When I asked this question in several different countries, one name came up more than once, and that was Nahum Joel of Chile. So I made up my mind that I had to go to Chile to meet this man, which I did. That must have been around 1962, and when I did meet Nahum Joel, I think there was a mutual attraction between us which was to last many, many years. In any case, I discussed the matter with Joel, who said he didn't know whether he could be relieved of his duties at the university, but that the idea was interesting to him. At the same time, we had been adding members to the staff of our Division, and before too long we had a chemist, who was Robert Maybury from the United States, and a biologist, a Mr Wroblewski, from Poland. So it looked as if, within the House, I would be in charge of physics, chemistry would be in the hands of Maybury, and the biology program would be in the hands of Wroblewski.

Finally, Bergvall, a Swede, was chosen as director of the project, and Nahum Joel was chosen as assistant director, and before long we were planning what the actual content, what the actual approach would be. We decided that it should last a year, that we should invite participants from all the Latin American countries, and that we should emphasize the creation of text materials. In one year we knew we could not do a whole textbook in physics, so we chose an area of physics, namely the physics of light, which again showed my preferences because I thought I knew something about that. We decided to produce films and programmed instruction materials, and that we would produce, perhaps, some television materials.

THE USE OF FILMS IN SCIENCE TEACHING

Our charge was to find what was new as regards methods, techniques and approaches in teaching, and certainly, one of the things that was new at that time was the use of films that fitted a special new projector which the Technicolor company had put out. The films were of short duration, about four minutes; they were in continuous loops, meaning that when you projected a film, when it came to the end, it was ready to start at the beginning again. The teacher could easily stop the projector, and start it again, by pressing a button, so it had some of the features which are now utilized in video. The notion that they should be short had already been explored in many different parts of the world. There had been recognition of the fact that long films really did not belong in the classroom. We decided that in the physics pilot project we would produce a set of film loops, and that each participant in the pilot project would be given a projector of this type to take back home for utilizing the film.

UNESCO-UNDP project to train scientific research personnel, science teachers and engineers (here in electronics) at the University of Brasilia, Brazil, 1967.

UNESCO-UNICEF project to train 1,500 primary-school teachers, over a two-year period, in modern methods of natural sciences teaching, Domasi, Malawi, 1967.

WHY WAS BRAZIL CHOSEN?

At a meeting in UNESCO House in Paris, among a group of educators from many parts of the world, was a woman from Brazil, Maria Julieta Ormastroni. She came to my office and said: 'This project has to be in Brazil.' And I said: 'Well, here's what we'll need', so I outlined what we needed in terms of space and other facilities, and she said: 'I can get all of those for you in Brazil.' In fact, she wanted me to come to Brazil to visit so that I could see that they did indeed have the facilities that we needed. I told her that I didn't have money in my budget to make that trip, and she said: 'I'll find the money.' So, indeed, they bought a ticket for me, they invited me to come to Brazil, and they showed me indeed that they had facilities in São Paulo. It actually happened to be in the medical school, but there was an enthusiast, Dr Isaias Raw, who had himself started some curriculum reform projects, quite independently of UNESCO. They said: 'Whatever you produce in Spanish can be produced in Portuguese almost simultaneously, and we have the television station if you want,' and they did indeed make everything available that we thought was necessary. So we decided to go to Brazil.

THE CONSTRUCTION OF LABORATORY KITS

The construction of inexpensive laboratory materials in the form of kits, which the teacher could utilize, was one of the important aspects of our work. It was clear to everyone that one of the weaknesses of science education, not only in Latin America but around the world, was that students did not have access to laboratory materials. We decided to invent materials that would be of low cost, and that could be repaired if things went wrong. They would be boxed separately so that any given topic would be covered by a kit which is in a box, so that you could put all the materials back into that box at the end of the laboratory hour. A lot of ingenuity went into the invention of the experiments and of the materials that were needed to do those experiments, and so a set of eight kits were invented, manufactured, produced during the pilot project. All of these materials were interlinked with one another, so that the lab material supplemented the text as did the films. We ended up with about eight kits, twelve short films and about four books, in programmed instruction form, which were teacher's manuals for the use of the films. They also produced one long 16 mm film the title of it was 'Light, is it a particle or a wave?'

Because of the success of the physics pilot project, we were able to find funding in subsequent biennia for a pilot project on chemistry in Asia, a pilot project on biology in Africa, and then one on mathematics in the Arab States. The actual development of these projects turned out to be somewhat different from the approach we used in Latin America, but they were adapted to the needs of the region by the people who were chosen to run them.

WORKING KNOWLEDGE
Symbiosis of programmes in science teaching, environmental education, and technical and vocational education
Saif R. Samady[23]

UNESCO'S work in science teaching began shortly after the Organization was established. After the Second World War, many schools in Europe were in great need of science equipment. To meet this need, the Organization sponsored the publication of a small volume entitled *Suggestions for Science Teachers in Devastated Countries*. The book was further developed to include a wide range of guidelines for simple equipment and science experiments and published in 1956 as the *UNESCO Source Book for Science Teaching*. Over the years, the *Source Book* was revised several times and translated into more than twenty-five languages. During five decades, the Organization promoted worldwide exchange of information and innovations in science education and assisted many Member States, especially the developing countries, in setting up science teacher training programmes, curriculum development centres, and projects for design and development of science equipment.

Working closely with the International Council for Science's (ICSU) Education Commissions, the Organization supported major conferences on education in basic sciences and mathematics. Based on the studies commissioned by UNESCO, the series of *New Trends* in the teaching of basic sciences was published. The first volumes of this series appeared in 1967–68, and the series continued until the early 1990s. During this period, five volumes were published on mathematics, five on biology, six on chemistry, four on physics, and six on integrated science teaching.

Through advisory services and execution of operational projects funded by extrabudgetary resources, the Organization has cooperated with a large number of Member States during these five decades in national teacher training and curriculum development programmes and institutions. Training science teachers has always been a priority in UNESCO's education programme. A variety of activities was carried out to support and improve teacher training for science education in primary and secondary schools. International and

---

23 Saif R. Samady: former Director, Division of Science, Technical and Environmental, Education Sector, UNESCO.

PART II: BASIC SCIENCES AND ENGINEERING

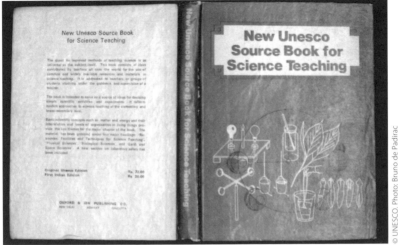

The *New UNESCO Source Book for Science Teaching* (original UNESCO edition issued in 1973) was a UNESCO 'best-seller', with many reprints and translations. It is an update of the *UNESCO Source Book for Science Teaching*, produced in 1956, which was itself an expansion of a booklet entitled *Suggestions for Science Teachers in Devastated Countries*, published by UNESCO in 1948. The various versions of this book have been used worldwide for teacher training, for over three decades. Here the Indian edition, published in 1978 (Oxford & IBH Publishing Co., New Delhi, India).

regional meetings and training workshops were organized; teachers' guides and source books were produced; a series of pilot and experimental projects on teacher training methodology were conducted. UNESCO was involved in the establishment of the International Council of Associations for Science Education in 1973 and supported national science teachers associations in their efforts for improvements and innovations in science and technology education.

In 1981, UNESCO convened in Paris an International Congress on Science and Technology Education and National Development. The programme and activities of the Organization in science and technology education during the 1980s were largely inspired by the recommendations of the Congress. An International Network for Information in Science and Technology Education (INISTE) was established in 1985. A series of pilot projects relating science and technology education to development needs were conducted; and a new publication on innovations in science education was launched. Attention was given to the importance of teaching technology in general education, out-of-school scientific activities, and access to science education (especially for girls). A regional meeting on 'Science for All' was organized (Bangkok, Thailand, 1983). The Organization supported a series of activities on 'Science and Society' and cooperated with the ICSU Committee on the Teaching of Science in the organization of an international meeting on this theme (Bangalore, India, 1984), which produced teaching materials on science and society issues.

UNESCO-UNDP project to establish a Faculty of Engineering: a UNESCO electronics expert from Egypt (United Arab Republic) demonstrating characteristics of semiconductors, at the University of Lagos, Nigeria, 1968.

To improve the content and methods of science and technology education, the Organization supported worldwide reform and innovations and conducted a number of pilot and experimental projects. The first UNESCO pilot projects in the 1960s focused on curriculum development and modernization of the content of science subjects in secondary schools; prototype teaching materials in basic science disciplines (chemistry, physics and biology) and mathematics were prepared. In the 1970s, attention was given to promotion of innovation in primary and integrated science teaching and out-of-school activities for young people, including science olympiads, science clubs, and so on. In the 1980s, the issue of relevance, technology education, cost-effective training, and application of science and technology education to community development (including nutrition and health education) were the main concerns of the pilot projects and experimental activities. In cooperation with institutions in fifty Member States in all regions, ten pilot projects were conducted during the decade. Institutions from several countries participated and cooperated among themselves in the design and execution of each pilot project. The salient results of these endeavours were disseminated through technical documents and the publication on innovations in science and technology education.

In the 1990s, special attention was given to the promotion of scientific and technological literacy for all. Based on studies and experiences in different regions, the trends, prospects and constraints of scientific and technological literacy were published in Volume IV of *Innovations in Science and Technology Education* (1992). A worldwide survey in science, mathematics and technology education in primary and secondary school was prepared and disseminated. UNESCO – in cooperation with the International Council of Associations for Science Education (ICASE) and several other partners – organized an International Forum (Paris, 1993), which adopted a Declaration and an 'International Project 2000+ on Science and Technology Literacy for All'. This project focused on networking and supporting innovations in science education, development of training materials, linking science and environmental education, technology education in schools, and promotion of access of girls and women to scientific and technological studies. However, during this period, the resources of the Organization for science education were significantly reduced.

Capacity-building for curriculum renewal and teacher training were supported, through subregional workshops and advisory services. A broadened approach to science and technology education, emphasizing societal issues (notably in relation to the environment, renewable energy resources, health and nutrition) was encouraged. The scope of the widely distributed environmental education newsletter *Connect* was expanded to include science and technology education. A special project on 'Scientific, technical and vocational education of girls in Africa' was conducted (1996–2001). Twenty-one African Member States participated in this project, which involved national surveys, awareness programmes, and training of national education decision-makers through two major regional meetings in 2000. In view of the overall success of the special project, its expansion to other regions was envisaged in future UNESCO programmes.

ENVIRONMENTAL EDUCATION

International concern about the environment and the urgent need for environmental awareness and education were expressed in the UN Conference on the Human Environment (Stockholm, Sweden, 1972). UNESCO, in cooperation with the United Nations Environment Programme (UNEP), launched in 1975 the International Environmental Education Programme (IEEP), which for over twenty years contributed to promoting environmental awareness worldwide. Member States were assisted in the development of programmes and projects to incorporate the environmental dimension in formal and non-formal education, including teacher training, technical and vocational education, general university education and adult education. A series of high-level regional meetings for national experts and officials of ministries of education was organized to exchange views on the role and strategies of developing environmental education in the education systems.

The Intergovernmental Conference on Environmental Education (Tbilisi, USSR, 1977), organized by UNESCO in cooperation with UNEP, was an important forum in providing conceptual guidance in the field of environmental education. The conference adopted a Declaration and Recommendations on the role, objectives and guiding principles of environmental education and strategies for the development of environmental education. It was recommended: that the environment should be considered in its totality (including natural, built, and social components); that environmental education be interdisciplinary, use problem-solving, and be community based; and that it cover all age-groups, both inside and outside the school system, in the context of lifelong education.

As a follow-up to the recommendations of the Tbilisi Conference, IEEP started a series of activities for promotion and development of environmental education. In the 1980s, over 100 pilot and research projects were conducted; 50 prototype curricula and modules for formal and non-formal education, teacher training and vocational education were developed; and about 200 national, subregional and international training seminars and workshops were supported or organized. The quarterly newsletter *Connect* was produced in several languages and distributed to thousands of institutions and specialists around the world. The evaluation of IEEP by external consultants in late 1980s confirmed that these activities had a significant impact on the promotion of environmental education in Member States.

Ten years after the Tbilisi Conference, UNESCO and UNEP organized the International Congress on Environmental Education and Training (Moscow, USSR, 1987). The Congress took stock of worldwide progress in environmental education, reconfirmed the principles and guidelines, and adopted an 'International Strategy for action in the field of environmental education and training for the 1990s', covering the following themes: access to information; research and experimentation; educational programmes and teaching materials; training of personnel; technical and vocational education; educating and informing the public; general university education; specialist training; and international and regional cooperation.

In the 1990s, UNESCO's programme in environmental education reflected the worldwide concern for the environment and sustainable development. The United Nations Conference on Environment and Development (Rio de Janeiro, Brazil, 1992) agreed on the 'Rio Declaration on Environment and Development', which set out twenty-seven principles supporting sustainable development. Also agreed was a plan of action, 'Agenda 21', which included a chapter (Chapter 36) on education, for which UNESCO was designated the Task Manager by the Commission on Sustainable Development. The Organization developed a programme and activities based on Chapter 36, which states: 'Education is critical for achieving environmental and ethical awareness, values and attitudes, skills and behaviour consistent with sustainable development and for effective public participation in decision-making. Both formal and non-formal education are indispensable to sustainable development'.

In 1994, a transdisciplinary project was launched that included environmental education along with population education and information for development. This project was a follow-up to the Rio conference. It involved elaborating the concepts and messages for a sustainable future, and developing national policies and action plans, working together with other agencies of the United Nations (UN). UNESCO's action was largely decentralized to Field Offices to nurture a transdisciplinary approach, with appropriate linkages with other relevant development activities and the promotion of training at the community level. Technical and financial support were given to Member States in all regions to reorient curricula and teacher education, and develop demonstration projects on education for sustainability. A number of teaching and learning packages in environmental education and population education were produced. A multimedia programme on learning and teaching for a sustainable future – composed of twenty-five modules, on the internet and in CD-ROM format – was developed for the use of teachers in formal and non-formal settings. An interregional workshop on 'Reorienting environmental education to sustainable development' (Athens, Greece, 1995) and an international conference on 'Education and public awareness for sustainability' (Thessaloniki, Greece, 1997) were organized.

In the 1990s, educational programmes and projects for the environment were promoted within the broader concept of sustainable development. This meant that the existing education policies and programmes should focus on building concepts, skills and commitments for sustainable development. It is recognized that basic education, appropriate technical and vocational education, higher education and teacher education are vital ingredients of capacity-building for a sustainable future. Since 1975, UNESCO, together with UNEP and other partners, has made a significant contribution to worldwide environmental awareness and education. However, promotion of environmental quality and sustainable development remains a long-term challenge for humanity, but one in which education has an important and continuing role to play. Recognizing the essential role of education to sustainability, the UN General Assembly declared the period 2005–2014 as the UN Decade of Education for Sustainable Development. UNESCO has been designated as 'lead agency' for international coordination of the work of the Decade.

TECHNICAL AND VOCATIONAL EDUCATION

As early as the 1950s, Member States were seeking UNESCO's cooperation in the establishment of training institutions for technicians. Later, the programme was concentrated on the training of technical teachers. These efforts contributed to the initial development of technical and vocational education in many countries. However, it soon became clear that technical and vocational education had to be considered from the point of view of its relations with the overall education system and with the world of work, in the context of the social and economic evolution of societies. The UNESCO Recommendation concerning Technical and Vocational Education (adopted by the General Conference in 1962, and revised in 1974) provided guidance on policy and basic principles in Technical and Vocational Education. In the 1960s and 1970s, the Organization assisted in the formulation and execution of a considerable number of national projects in technical and vocational education, financed by extrabudgetary resources. In the 1980s, more than seventy operational projects, with a budget of about US$70 million – representing nearly half of all education projects – were concerned with technical and vocational education and training.

The Organization's programme and activities in technical and vocational education can be grouped around three themes: normative action, innovations in content and training methodology, and support to the development of national infrastructures. The UNESCO Recommendation was the first international normative instrument in technical and vocational education. In view of the growing importance of this field of education and training, and the need for increased awareness among national authorities for legislation, the General Conference at its twenty-fifth session (1989) adopted a Convention on Technical and Vocational Education. The Convention deals with the following aspects: definition of technical and vocational education, comprehensive development for young people and adults, standards in the context of lifelong education, adapting technical and vocational education to scientific change, training of teachers, and international cooperation.

During the 1980s, the Organization conducted a number of activities for improvement of the quality of technical and vocational education. Over thirty regional and subregional seminars and workshops, as well as several pilot projects, were conducted to promote innovations in the content and methodology of technical and vocational education. A graphic communication course, consisting of a teacher's guide and eighteen learning modules, was developed to facilitate the introduction of technology in general education. A number of units of international technical illustrations were prepared to facilitate the development of learning materials in national languages. Guidelines were prepared to develop modular approaches to adapting vocational education programmes to evolving needs of the labour market. Case studies and a newsletter on innovative practices in technical and vocational education were prepared and disseminated to Member States.

Prototype curricula and modules for technical teacher training programmes were developed. A prototype curriculum for computer technicians was prepared and distributed in several countries. Pilot projects on cooperation between technical schools and industry, technical and vocational education for rural development, and access of women to technical and vocational education were carried out in cooperation with national institutions in all regions of the world. In the 1980s, a number of international training workshops were organized (in cooperation with the International Labour Organization Training Centre in Turin, Italy) for key personnel from developing countries, on the planning and management of technical and vocational education, curriculum development, new technologies, and so on.

The first important international meeting organized by UNESCO in this field was the International Congress on the Development and Improvement of Technical and Vocational Education (Berlin, Germany, 1987). The Congress reviewed major trends: the contribution of technical and vocational education to social progress and human resources development, the increased participation of women in technical and vocational education, and the implications of rapid scientific and technological progress for the field.

Based on the recommendation of the Congress concerning the setting up of an international centre in this field of education, after carrying out a feasibility study, UNESCO (in cooperation with Germany) launched in 1992 the International Project on Technical and Vocational Education (UNEVOC). Throughout the 1990s, the Project carried out the following actions: promotion of the international exchange of ideas and studies on policy issues, support for national research and development capabilities, and facilitation of access to data bases and documentation. The Project developed guidelines and references for national policy reform and legislation, adaptation of technical and vocational education to technological changes and the world of work, and the introduction of new concepts and innovative practices.

UNEVOC supported the training of key personnel in technical and vocational education and organized a number of international and regional meetings, which included: a European symposium on technical and vocational education and training in countries in a period of transition towards market economies, seminars for key personnel on cooperation between educational institutions and enterprises, a conference of experts on promotion of equal access of girls and women in technical and vocational education, and a workshop on development of rural vocational education. After an evaluation in 2000, the UNEVOC Project was transformed into an International Centre (in Bonn, Germany), which acts as an international hub for a global network of institutions active in the area of technical and vocational education and training. The Centre addresses such key issues as the promotion of innovative practices, providing support to systems, improving access, and assuring quality. It has functional links with the network centres and experts from over 100 countries. A regional clearing house and electronic communication network for Asia and the Pacific has also been developed.

In order to reflect on strategies for the first decade of the twenty-first century, taking into consideration globalization and rapid changes in technology, as well as the need for sustainable development, the Second International Congress on Technical and Vocational Education was organized in Seoul (Republic of Korea) in 1999. The Congress made a number of recommendations for national and international action, including the preparation, by UNESCO, of a long-term international programme for development of technical and vocational education. In order to take into account the emerging challenges of the twenty-first century, an updated version of the Revised Recommendation concerning Technical and Vocational Education was prepared and adopted by the General Conference in 2001.

The growing importance of technical and vocational education for social and economic development in Member States during the 1960s and 1970s was reflected in the education and training programme of the Organization. UNESCO – in cooperation with the International Labour Organization (ILO), the United Nations Development Programme (UNDP), the World Bank and other funding sources – assisted many developing countries in all regions in the establishment of vocational education programmes, training institutions and schools. In the 1980s and 1990s, greater attention was given to the quality and efficiency of technical and vocational education, taking into account rapid changes in technology and the needs of the world of work. To facilitate a systematic exchange of information and cooperation among national institutions, UNESCO established regional networks and the International Centre for Technical and Vocational Education and Training. Cooperation between UNESCO and the ILO was reinforced. On the basis of a memorandum signed by the two organizations in 1954, UNESCO's concern was centred on technical and vocational education in the formal education system, while the ILO focused on training for employment in non-formal settings and in enterprises. However, education and training were rapidly becoming inseparable, as the notion of a job for life was being replaced by the necessity of lifelong learning.

BIBLIOGRAPHY

UNESCO. 1981. Report of the International Congress on Science and Technology Education and National Development. Paris, UNESCO.
——. 1986–2003. Innovations in Science and Technology Education, Vols I–VIII. Paris, UNESCO.
——. 1987. UNESCO-UNEP International Congress on Environmental Education and Training (Moscow, USSR, August 17–21, 1987). Paris, UNESCO.
——. 1988. Report of the International Congress on the development and improvement of technical and vocational education (Berlin, Germany, 1987). Paris, UNESCO.
——. 1989. Convention on Technical and Vocational Education, Paris, 10 November 1989.

PART II: BASIC SCIENCES AND ENGINEERING

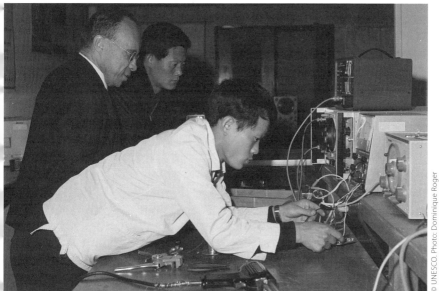
UNESCO-UNDP project to develop the training programme at the Fine Instruments Centre in Seoul: experimental work in an electronics laboratory, Republic of Korea, 1968.

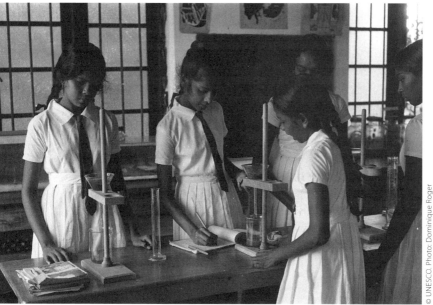
Laboratory equipment designed by the UNESCO-assisted Asian Regional Institute for School Building Research, Colombo, Ceylon (Sri Lanka), 1969.

———. 1993. *Final Report: International Forum on Scientific and Technological Literacy for All.* Paris, UNESCO.
———. 1997. *Fifty Years for Education.* Book and CD-ROM. Paris, UNESCO.
———. 1999. *Lifelong Learning and Training: a bridge to the future.* Final Report of UNESCO's Second International Congress on Technical and Vocational Education (Seoul, Republic of Korea, 26–30 April 1999). Paris, UNESCO, 134 pp.
———. 2002. Revised Recommendation concerning Technical and Vocational Education. Records of the General Conference, Paris, thirty-first session, 15 October to 3 November 2001, Vol. 1. Paris, UNESCO.
———. Reports of the Director-General on the activities of the Organization (C/3), 1994–1995, 1996–1997, 1998–1999, 2000–2001.
———. Consultations with Field Offices of Regional projects to be launched in 2005.
UNESCO-UNEP. 1978. *The Tbilisi Declaration.* Final Report of the Intergovernmental Conference on Environmental Education. Organized by UNESCO in cooperation with UNEP (Tbilisi, USSR, 14–26 October 1977). Paris, UNESCO.
United Nations. 1992. Declaration and Plan of Action (Agenda 21) of the United Nations Conference on Environment and Development (Rio de Janeiro, Brazil, 1992).

PART III
ENVIRONMENTAL SCIENCES

PART III: ENVIRONMENTAL SCIENCES

INTRODUCTION
EARTH MATTERS
The environment and sustainable development focus in UNESCO's science programmes
Gisbert Glaser[1]

TECHNOLOGICAL progress with little regard for the environment. Exponential population growth. Poor management of natural resources. Unbridled consumption of ecosystem goods and of non-renewable resources, including fossil fuels. Various forms of pollution and uncontrolled urbanization. Taken together, these problems have formed the most daunting challenge that humanity has ever faced. Human history from the Stone Age to the beginning of the twentieth century was characterized by a limited and slowly increasing impact on nature, affecting ecosystems locally and leaving vast areas in their natural state. Within the twentieth century, and much more profoundly during the sixty years since the creation of UNESCO, a global environment and development crisis has emerged. Imperfect understanding of the changing natural environment and systems that make possible the maintenance of life on Earth – and provide the natural resources basis for human development – contributed to the woefully insufficient response to the crisis. The pace of the negative changes occurring in our life support systems has dramatically increased since the end of the Second World War. The scientific challenge is no longer limited to a better understanding of nature alone, but has shifted during the last decades towards the challenge of understanding the ever-closer interaction between natural and human systems.

The seeds of UNESCO's programmes in the environmental sciences were planted in 1946 by the Organization's first Director-General, Sir Julian Huxley, a zoologist. However, the steady growth of UNESCO's environmental programmes up to the first years of the 1980s is directly linked to the parallel development of a collective international awareness of what are now commonly called global environmental problems, and the increasing national and international efforts to deal with these problems. After 1970, when the scientific proof of many

---
[1] Gisbert Glaser joined UNESCO in 1971 as Programme Specialist (Division of Natural Resources Research); member of the new Division of Ecological Sciences in 1974, main responsibility in MAB; appointed Deputy Coordinator for Environmental Programmes in 1990 and Coordinator Environmental Programmes in 1993. Assistant Director-General for Natural Sciences a.i., March 2000 to February 2001.

environmental problems had become evident, thanks to the work of environmental scientists, the Organization's environmental sciences programmes and activities became the largest element in the Natural Sciences Sector. In budgetary terms, environmental programmes – which in the 1971–1972 biennium amounted to roughly US$4 million (including staff costs) – received over US$16 million from UNESCO's regular budget for the 1979–1980 biennium. This represented 35 per cent and 50 per cent, respectively, of the overall budget of the Sector. At the same time, important additional extrabudgetary resources became available after 1970, often more than doubling the total funds available. The environmental sciences programmes as a whole have remained the largest programme element in the Sector ever since. In the Draft Programme and Budget for 2006–2007, it is proposed that the environmental sciences programmes will receive US$17.8 million from the regular budget (without staff costs) which represents 71 per cent of the total programme funds proposed for the Sector.

The term 'environmental sciences' refers to all scientific areas and disciplines having as their research object a part of the natural environment. With the exception of the atmosphere – which at the international level is the domain of the World Meteorological Organization (WMO) – the environmental sciences programmes in UNESCO cover all major component parts of our natural environment: terrestrial and coastal ecosystems, freshwater, oceans, and the Earth crust. Consequently, four key chapters in Part III of this book deal with the activities in scientific disciplines focused on these four components: ecological sciences, water sciences, Earth sciences (Earth crust), and oceanography. UNESCO's activities in these four scientific areas have in common that they have been structured, for most of the sixty-year period (albeit each for a different duration), around four international scientific cooperation programmes, which have their own institutional identity and coordination mechanisms. Legally speaking, three of these 'undertakings' are integral programme parts of UNESCO. They are: the Man and the Biosphere Programme (MAB), focusing on terrestrial and coastal ecosystems; the International Hydrological Programme (IHP); and the International Geoscience Programme (IGCP, formerly the International Geological Correlation Programme). The latter has a unique feature, as it is a cooperative undertaking of UNESCO and the International Union of Geological Sciences (IUGS). The fourth undertaking, the Intergovernmental Oceanographic Commission (IOC), has a different legal status. It represents a semi-autonomous intergovernmental body, with its own intergovernmental General Assembly. UNESCO's governing bodies have, however, the final say on policy and budgetary matters. A general feature of all four undertakings is that UNESCO provides their respective international secretariats.

These four chapters are preceded by a chapter on the 'early years', 1945–1965, which provides detailed information on the evolution of UNESCO's still-fledgling activities in the environmental sciences in the 1940s and early 1950s. Moreover, it dwells on the crucial decade 1955–1965, when the environmental sciences activities started to emerge as a hallmark of UNESCO's scientific programmes as a whole, and when the first medium-term programmes were launched and given special separate institutional structures.

The origin of UNESCO's activities related to the environment was characterized by the desire to enhance international scientific cooperation in a number of natural sciences disciplines – such as hydrology, geology, oceanography, and biology/ecology – having a geographical and environmental scope, and thus lending themselves very much to such cooperation. After all, nature recognizes no frontiers between states. At the same time, from the beginning these scientific activities had a declared second objective: to promote science for the benefit of society at both the national and international level. More precisely, the activities were designed to lay the foundation for more applied environmental sciences, those in support of a more rational use of natural resources and conservation of nature. To meet these goals, these scientific efforts early on moved away from a single discipline approach, first to become multidisciplinary (combining several natural sciences disciplines) and later interdisciplinary (combining natural and social sciences disciplines). Interdisciplinary scientific undertakings are essential for a better understanding of the interactions of human societies – and human systems – with natural systems. Interdisciplinary knowledge is required for applying integrated approaches towards meeting development goals, while at the same time maintaining our natural life support systems, and improving the quality of the human environment. In this vein, UNESCO has played a major role in the paradigm shift from environmental sciences to science for sustainable development.

The world's coastal zones and small islands at the interface of marine and terrestrial systems face particularly conflicting development and environment goals. Moreover, they are zones of great vulnerability. It is thus no accident that coastal zones and small islands became the focus of enhanced efforts by UNESCO to promote multi- and interdisciplinary activities. Two chapters provide a history of these efforts, which started in the context of the Marine Science Division (which existed from 1972 to 1989) and gained greater momentum in 1996 with the launching of the Coastal Regions and Small Islands Platform.

Small island communities and local fishermen are among the holders of a particular type of knowledge, commonly referred to as 'traditional knowledge'. The short chapter on the intersectoral Project 'Local and Indigenous Knowledge Systems (LINKS) for Sustainable Development', begun in 2002, tells the story of this recent initiative, which in its geographical scope is, however, not limited to coastal zones and small islands.

The best-known UNESCO undertaking across all fields of the Organization's work (at least in the developed countries) is the World Heritage programme, developed around the implementation of the World Heritage Convention. Currently, there are 788 sites inscribed on the World Heritage List, located in 134 countries. Of these, 154 are natural World Heritage sites, inscribed on the basis of specific 'natural criteria', which have been developed within UNESCO's environmental sciences programmes. Another 23 properties, called 'mixed sites', have been listed under both sets of criteria, cultural and natural. For twenty years (1972–1992), the Division of Ecological Sciences in the Natural Sciences Sector has acted as the Secretariat for the natural part of the World Heritage programme. Consequently, an account of the history of the World Heritage Convention has been included, with particular attention to the involvement of UNESCO's environmental sciences programmes in the work of the Convention concerning the Protection of the World Cultural and Natural Heritage.

The last chapter in this section, focusing on UNESCO's environmental sciences programmes, is of a somewhat different nature than the others. Each of these other chapters presents the history of the activities/programmes covering one of the major scientific fields or topical areas which make up UNESCO's environmental sciences undertakings. The chapter 'Acting Together' concentrates on the history of how, together, these UNESCO undertakings – and UNESCO as a whole – have been involved in the development of an international environmental agenda, including international policy agreements at world conferences in Stockholm, Sweden (1972), Rio de Janeiro, Brazil (1992) and Johannesburg, South Africa (2002). In this context, it dwells on UNESCO's contribution and reaction to a critical paradigm shift: the movement from dealing with environmental protection as a sectoral issue to today's holistic sustainable development approach. This aspect of the history of science in UNESCO is intrinsically linked to the history of 'coordination of environmental programmes' within the Sector, the Organization as a whole, and the entire United Nations system. UNESCO's environmental activities are not limited to the Natural Sciences Sector. They include important contributions from the other areas of the Organization's competence: social sciences, the humanities, education, culture and communication. Thus, its unique institutional setting gives UNESCO a distinct comparative advantage in the international arena: to bring to bear well-coordinated intersectoral efforts, with contributions from all its programme sectors, aimed at helping countries to resolve interlinked economic, social, and environmental problems that make current development paths unsustainable.

A word on the relationships of the environmental sciences programmes to the two other programme poles in the Natural Sciences Sector: the basic and engineering sciences, and science and technology policies. Some of the basic sciences carry out important environment-related research on their own: for example, research on toxicity in chemistry and research on physical aspects of the climate system in physics. Capacity-building in the basic sciences is essential also for the environmental sciences. Environmental research cannot be undertaken without mathematics, chemistry, physics and biology. The engineering sciences are crucial partners in addressing environment and development problems by working to produce technologies that are less polluting, more energy efficient and less harmful to the environment in general. Sound national science and technology policies must include strengthening of environmental sciences and ensuring that the national scientific and technological enterprise is geared towards national development goals, embedded in overall sustainable development policies. A few successful examples of developing synergies between the environmental sciences programmes and the two other programme poles are referred to in this section; other examples are provided in the two sections of the book dealing with the basic and engineering sciences, and science and technology policies, respectively.

A specificity of UNESCO's international scientific cooperation programmes is that they have been embedded in an intergovernmental structure. Governments and groups of scientists work together to define particular needs in knowledge and in scientific capacity-building. This process is intended to ensure that scientific activities undertaken within the framework of these programmes have a high degree of policy relevance. At the same time, the scientific quality of these programmes depends first and foremost on the quality of the many scientists involved in the planning and implementation of the activities at national, regional and global levels.

At the global level, all of UNESCO's environmental sciences programmes have worked in close partnership with the International Council for Science (ICSU; the name was changed from the International Council of Scientific Unions in 1998, but the acronym has been maintained) and, depending on the scientific area, with relevant international scientific unions, part of the ICSU family. A piece of the history of this cooperation is presented in all chapters of this section of the book. Moreover, a general historical overview of UNESCO-ICSU cooperation is provided in Part V. The involvement of ICSU and its scientific unions, as well as several international scientific associations, has contributed significantly to the scientific credibility of UNESCO's programmes in the environmental sciences.

Cooperation with other United Nations organizations, including with the secretariats of major international environmental conventions, represents another common feature of all of these programmes. While information on the history of this cooperation has been included as part and parcel of the different chapters on the history of the individual programmes, the last chapter provides more detailed information on the particularly close cooperation between UNESCO and the

United Nations Environment Programme (UNEP) during the first decade after the creation of the latter organization. This chapter also dwells on some aspects of the history of UNESCO's contribution to UNEP-led UN system-wide coordination efforts in the area of the environment, and to coordination efforts by the United Nations Department of Social and Economic Affairs in the area of sustainable development.

UNESCO programmes have realized great achievements in the field of human and institutional capacity-building, particularly in developing countries. Yet it is true also, for the sciences related to environment and sustainable development, that the North–South knowledge divide has been widening during the last decades, rather than narrowing. However, this in no way diminishes the immense value of the efforts made by the UNESCO programmes, and the success in training and retraining tens of thousands of scientists, and in strengthening a large number of scientific institutions on all continents of the developing world.

Generally speaking, the history of UNESCO's environmental sciences programmes is undoubtedly a story of success, with impressive achievements to show by each individual programme. Yet, there have also been numerous shortcomings. Examples can be found in several chapters of this section. However, during the last fifty years – not counting the fledgling activities during the first decade of UNESCO's existence – the Organization's programmes have shaped international science in a broad scope of environmental areas. They have done this together with complementary activities undertaken by, or under the aegis of, ICSU, in cooperation with other UN system partners and other international scientific organizations.

After five decades of successful international programmes, we have accumulated a remarkable body of knowledge. At the same time, we have also been able to define better what we do not know yet. These gaps in knowledge remain huge, representing a major challenge to present and future generations of scientists. The other huge tasks that must be addressed in the future concern the building-up of relevant human and institutional scientific capacities at national and regional levels, particularly in the developing world, as well as establishing much-improved links between science and the public domain. In the decades ahead, UNESCO will face a major responsibility: to effectively fulfil its role as a global intergovernmental leader in science and education for sustainable development.

PART III: ENVIRONMENTAL SCIENCES

NATURE TO THE FORE
The early years of UNESCO's environmental programme, 1945–1965
Malcolm Hadley[2]

FIRST STEPS

THE Earth's environment and its natural resources did not figure explicitly in the initial set of actions in the natural sciences field, proposed by the Preparatory Commission. Rather, these actions were couched in more generalized cross-cutting terms: speeding up the work of scientific rehabilitation after the ravages of the Second World War, throwing a network of regional science cooperation offices round the world, supporting and extending the scientific unions and their work, organizing and operating an international science service system, cooperating with the work of the United Nations Organization and its specialized agencies, informing the people of all countries on the international implications of scientific discoveries, and originating new forms of scientific cooperation.

But in starting to shape and implement activities in these various domains, environmental and natural resource issues quickly came to prominence – not surprisingly perhaps, given the professional interests and personal philosophies of the Organization's founding Director-General, Julian Huxley, and its first Head of Natural Sciences, Joseph Needham. Huxley was a biologist and evolutionary humanist, visionary of the world. Needham, a morphologist and biochemist, bridge-builder, twentieth-century Renaissance man, integrative generalist, polymath, historiographer and historian of Chinese science and technology, political activist in socialism and social responsibility, philosopher in linking science and religion (Goldsmith, 1995).

Thus, in UNESCO's 1947 report to the United Nations Economic and Social Council on scientific research laboratories and observatories under United Nations auspices (UNESCO, 1947), among the issues addressed was scientific research in climatic zones where comparatively little work had been carried out – the polar (Arctic, Antarctic) and the arid and the humid tropical (equatorial). Specific sections in the UNESCO report took up the issues of centres for

---

[2] Malcolm Hadley: Member of staff of UNESCO's Division of Ecological Sciences from the early 1970s to 2001.

study of tropical life and resources, international ornithological observatories, international oceanographic institutes and 'stockrooms, type-collections and standards'. As described later in this chapter, the issue of research capacities in arid and humid tropical zones and in oceanography remained a continuing feature of the Organization's environmental work in the 1950s and thereafter.

CONSERVATION

Among UNESCO's early activities was taking part in a process for establishing an international organization for the protection of nature. The idea of such a body had been discussed on various occasions in the first half of the twentieth century (Nicholson, 1972), including US President Theodore Roosevelt's proposal in 1909 to convene an international conservation conference (plans for that conference were abortive). Subsequently, at a meeting of the International Union of Biological Sciences in Poland in 1928, plans were discussed for an international office for the protection of nature, mainly for documentation purposes. In 1934, this office at Brussels (Belgium) was afforded legal recognition. But war was looming. After being hurriedly transferred to Amsterdam (the Netherlands) in 1940, the office was compelled to cease functioning.

After the Second World War, the Swiss League for Nature took the lead in reviving the ideas for an international union for nature protection. A preparatory meeting was convened in Brunnen, Switzerland, in mid-1947 to discuss options. According to one internal memo of the time,[3] that meeting was a somewhat agitated one, with some of the delegates insinuating that UNESCO was trying to be imperialistic in the field of nature conservation, alleging also that it was weighed down by a cumbersome administrative machinery and that its Museum Section was intending to call a (parallel) meeting on nature protection. But the general feeling among the seventy-odd delegates present was evidently that UNESCO's particular contribution was in assuring government support for the new organization; and the Brunnen Conference duly drew up and adopted unanimously a draft constitution establishing a Provisional International Union for the Protection of Nature on a governmental level. This draft constitution was submitted to UNESCO for onward transmission to governments.

An outstanding issue was the status of the new body. The governments of France and the United States had indicated informally that they were 'not favourably disposed towards the formation of new international inter-governmental

---

3   UNESCO memo dated 15 July 1947, from Eleen Sam to Joseph Needham (Head of Natural Sciences): Report on the Brunnen Conference for the International Protection of Nature – 28 June to 3 July 1947.

PART III: ENVIRONMENTAL SCIENCES

organizations.'[4] In view of reservations such as these, the Director-General suggested (on 5 April 1948) that the formation of an international institute, not governmental in character, should be considered. With Julian Huxley energetically embracing the project (Nicholson, 1972, p. 225), UNESCO sent a revised draft constitution to governments, interested organizations and UN specialized agencies, asking for comments. And from 30 September to 5 October 1948, UNESCO joined forces with the French Government and the Swiss League of Nature in convening a conference at Fontainebleau (France), which gave birth to the International Union for the Protection of Nature and Natural Resources (IUPN), since renamed IUCN (International Union for the Conservation of Nature and Natural Resources, now known as the World Conservation Union).[5]

In contrast to the constitutions of other international unions of a strictly scientific character, that of IUPN provided not only for membership by non-governmental national bodies but also for inclusion of international organizations whether intergovernmental or unofficial, and of public services and governments themselves. The initial signatures thus included eighteen governments, but not until 1963 did actual governmental membership surpass that figure. Jean-Paul Harroy was appointed secretary of the new Union, with an office in the Natural History Museum at Brussels, provided by the generosity of the Government of Belgium.

Over the five decades that have elapsed since the 1948 Fontainbleau Conference, IUCN has grown to become the world's leading non-governmental organization (NGO) devoted to the conservation of nature and the world's largest environmental knowledge network,[6] and has also carried out many activities in close cooperation with UNESCO.[7]

One early cooperative undertaking of UNESCO and IUPN was an international technical conference for the protection of nature, held in August 1949 at Lake Success in New York State (USA), in parallel with the United Nations Conference on the Conservation and Use of Natural Resources, jointly organized by UNESCO, the Food and Agriculture Organization of the UN

---

4   UNESCO memo dated 8 April 1948, from Eleen Sam to Deputy Director-General: Nature Protection: Resume of Brunnen Conference Proposals and Suggested Plan of Action for UNESCO, April–September, 1948.
5   A description of IUCN's founding is included in an account of its first fifty years by Martin Holdgate, IUCN's Director-General from 1988 to 1994 (Holdgate, 1999).
6   The World Conservation Union brings together 81 states, 114 government agencies, over 800 NGOs, and some 10,000 scientists and experts from 181 countries in a unique worldwide partnership. IUCN's mission is to influence, encourage and assist societies throughout the world to conserve the integrity and diversity of nature and to ensure that any use of natural resources is equitable and ecologically sustainable.
7   Among the most remarkable of these cooperative ventures have been the development of the biosphere reserve concept and IUCN's promotional and advisory functions for the natural part of the World Heritage Convention. More recently, at the Third IUCN World Conservation Congress in Bangkok in November 2004, IUCN and UNESCO announced a partnership to develop indicators to assess the progress in education for sustainable development.

(FAO), the World Health Organization (WHO) and the International Labour Organization (ILO) (Conil-Lacoste, 1994, p. 43).

The technical conference for the protection of nature brought together 140 delegates[8] from thirty-two countries and eleven international organizations, who adopted twenty-three resolutions covering such subjects as education, human ecology, the ecological impact of major engineering works, the monitoring of pesticides and the conservation of food resources. Among the UNESCO inputs to the discussions at Lake Success was a 113-page survey on education for the conservation of natural resources and their better utilization, prepared by UNESCO agricultural engineer Alain Gille (Gille, 1949).[9]

During the next two decades, conservation issues figured in UNESCO's programmes related to arid lands, the humid tropics and natural resources, described in the pages that follow. Specific activities included the setting up in 1959 of the Charles Darwin Foundation for the Galapagos Islands (Dorst, 1970; Kramer, 1973).[10] Previously (as early as 1957), several missions had been organized under the aegis of UNESCO with the help of IUCN to study the means of saving the natural and unique heritage which the flora and fauna of the Galapagos Islands represent. The impetus for the Foundation came from a group of nature conservationists including Julian Huxley, Roger Heim and Victor Van Straelen (who became the first president of the Foundation). The Foundation later set up a research station, whose director was provided by UNESCO for several years. UNESCO was also instrumental in the setting up of several other bodies linked to the conservation of nature and study of its flora and fauna, including the Organization for Flora Neotropica in 1964 (Maguire, 1973).

Another type of activity concerned field missions to advise on conservation issues in particular localities, countries and regions, often organized in cooperation with other international organizations such as FAO, IUCN and the World Wildlife Fund (WWF). Recommendations and findings were published in a series of reports submitted by UNESCO to the governments concerned, with digests[11] prepared for more accessible sources, such as the *UNESCO Courier* and the quarterly magazine *Nature and Resources*.

---

8   Delegates at the Technical Conference for the Protection of Nature at Lake Success in 1949 included many leading figures in the conservation and scientific community, including Enrique Beltrán, Charles Bernard, Harold Coolidge, Frank Fraser Darling, Ray Fosberg, Jean-Paul Harroy, Roger Heim, Ernst Mayr, Théodore Monod, Fairfield Osborn, Dillon Ripley and Solly Zuckerman.
9   Alain Gille (1922–2005) spent most of his professional career with the Natural Sciences Sector of UNESCO. He was a lifelong environmental activist, who took part (though in failing health) in the first informal meeting on this history of the Natural Sciences in UNESCO, on 11 May 2004.
10 See also http://www.darwinfoundation.org/.
11 Examples of potted or semi-popular accounts of advisory missions on conservation included: Huxley, 1961; Brown, 1966; and Worthington, 1971.

An early example was a three-month advisory mission of the founding Director-General, Julian Huxley, to East Africa in 1956, which was among the activities that led UNESCO to be among the advocates of an African Convention on the Conservation of Nature and Natural Resources.[12] Later missions included those on wildlife conservation in Ethiopia (1963), and on ecology and conservation in Jamaica (1970). In some cases, field missions were combined with technical symposia of various kinds, such as that on the rational use and conservation of natural resources in Madagascar in October 1970. And in the second half of the 1960s, several well-known conservationists were members of the Natural Sciences Sector staff, including Gerardo Budowski and Kai Curry Lindahl.[13]

HYLEAN AMAZON PROJECT

One of the earliest proposed projects in UNESCO's fledgling programme was that of creating an International Institute of the Hylean Amazon ('Hylea' being the name used by the early-nineteenth-century naturalists and explorers Humboldt and Bonpland for the vast area of equatorial America). The idea of setting up such an institute was initially suggested by the Brazilian delegation (headed by Paulo Carneiro) at the May 1946 Preparatory Commission of the Natural Sciences Sub-Commission of the UNESCO General Conference.

Subsequently, the proposed institute was selected by the first General Conference in December 1946 and put among the four all-UNESCO major initiatives in April 1947. At first the proposal had limited scope: international support to the Museu Goeldi in Belém, Brazil – to save its precious botanical, zoological and ethnological collections – and to develop natural sciences in the Amazon region through the cooperation of all the countries of this geographical zone (Petitjean and Domingues, 2000; Domingues and Petitjean, 2001). The project was soon enlarged by UNESCO to an international institute for natural sciences in the Amazon region, the institute being perceived by some of the protagonists as a prefiguration of other tropical institutes elsewhere.

A scientific and diplomatic conference took place in Belém in August 1947 to establish the International Institute of the Hylean Amazon (IIHA). The meeting transformed the project: the economic development of the Amazon

---

12   The main impetus to begin work on this Convention came later, in 1964–65, when the Organization of African Unions requested IUCN to assist in the preparation of a draft text for presentation to the Council of Ministers. The Convention was subsequently adopted by the OAU at its fifth ordinary session in Algiers (Algeria) in September 1968, one of the first major events in the history of the continental body.
13   Gerardo Budowski was programme specialist in ecology and conservation in the Division of Natural Resources Research from 1967 to 1970, when he became Director-General of IUCN. Swedish conservationist Kai Curry Lindahl served as regional ecologist in the UNESCO Field Science Office in Nairobi (Kenya).

Two scientists centrally involved in the aborted plans to set up an International Institute of the Hylean Amazon. Left: tropical botanist E. J. H. Corner, who served as UNESCO's Principal Science Cooperation Officer for Latin America from 1947 to 1949; right: Brazilian chemist Paulo E. Berrêdo Carneiro, who proposed the setting up of the Institute at the General Conference of May 1946, and who was a prominent member of the Executive Board of UNESCO.

region became a major focus of the plans for the IIHA, though the idea of a basic research programme remained as well.

The Belém conference took place when UNESCO was preparing the second session of its General Conference in Mexico and was under strong pressure by the United States and the United Kingdom to reduce its programmes and expenses. Huxley, Needham and Carneiro feared a complete withdrawal of UNESCO from the IIHA project (Domingues and Petitjean, 2004). The lack of practical commitments in the final Belém conclusions was a handicap. Finally, a compromise was devised in Mexico City. The priority for the IIHA was lowered, and a new meeting was called to establish the juridical basis for it, with the Director-General being instructed to take steps to bring the institute into being in 1948.

A conference for the creation of the IIHA – called jointly by the governments of Peru and Brazil and UNESCO – took place in Iquitos (Peru) from 30 April to 10 May 1948. The conference agreed upon a convention establishing the institute, which would come into force upon final acceptance by the founding nations. The conference recommended that Manaus, Brazil, be established as the seat of the IIHA, because of its geographical situation, and that centres for study be developed in Belém to cover the Lower Amazon, in Iquitos for the Upper Amazon, in San Fernando de Atabapo (Venezuela) for the northern region, in Riberalta (Bolivia) for the southern region, and in Archidona (Ecuador) and Sibundoy (Colombia).

As visualized in the UNESCO submission to the United Nations enquiry on international research laboratories, the first objective might be 'the creation, in Hylea itself, of an international museum to house collections of plants, animals, minerals and rocks, as well as ethnographical documentation collected from the Amazon Basin. No inventory has yet been made of Hylea's natural wealth, most of which for this reason remains unknown' (UNESCO, 1947). The next task would be to establish a number of research laboratories, both in pure systematics and in plant and animal chemistry, geophysics, physiology, microbiology, and so on. The International Institute of the Hylean Amazon could immediately tackle the various food problems in these tropical and subtropical regions. A food office, with as many centres of study and research as possible, would then enable a scientific treatment of this vital problem, of concern to millions of people inhabiting damp and hot climates and living under conditions of chronic scarcity.

The UNESCO report to the United Nations laid out the aims of these laboratories:

- Primarily the aims will be to promote and guide research by visiting scientists. Many of them will be outstanding specialists who come to do a special piece of work that has been carefully planned and prepared for in advance.
- For persons less interested in special research and for those specialists who want to learn about things other than their own narrow fields, a few courses in tropical biology will have to be organized. These will be very suitable also for local residents interested in biology, and may well succeed in raising the level and activities of the natural history amateurs of the region.
- For visitors without formal training, it would seem to be desirable to develop short courses (a few hours every morning). Many of the increasing numbers of tourists would also appreciate such courses, and arrangements might be made with the large steamship companies to include them in their programmes. The latter group should of course be kept separate from the scientific visitors as this 'extension work' must not interfere with the real task of the institute, the promotion of research on the flora and fauna of the region (UNESCO, 1947).

To the modern reader, these quaint objectives smack of the self-serving interests of scientists from outside the region. They are a reminder, first, of how recent environmental transformation and hence the concern for conservation of the Amazon forest has been. There is no mention of conservation as an issue throughout the report. Remarkable also, scant attention was paid to formal training and institution-building within the region.

Whatever one might think in hindsight, the proposal got off to a good start. Science writer Ritchie Calder wrote an article (Calder, 1948) referring to the Hylean Amazon project in the February 1948 issue of *Discovery* magazine (see Box III.2.1). The renowned tropical botanist E. J. H. Corner[14] became executive-secretary of the Institute.

A budget of US$300,000 was set for the first year of the Institute's operations. A financial protocol was also signed by Bolivia, Brazil, Colombia, Ecuador, France, the Netherlands, Peru and Venezuela, which determined the scale of national contributions towards financing the Institute.

BOX III.2.1: SCIENCE IS UNESCO'S STRONG POINT

'Without question, the section which proved itself the most efficient both in execution of this 1947 programme and in its proposals for 1948 was the science section. Some dubiety was expressed (and removed) about the Hylean Amazon project. This was initiated at the first UNESCO Conference in Paris from an offer by Brazil to make available an institute, as an international centre for the study of the problems of the Amazon area. That embraces a third of the continent of South America and involves seven sovereign states and three European dependencies – the Guianas. It is largely unexplored and there is little systematic knowledge about its natural characteristics.

There is the obvious wealth of timber and convertible vegetation ... Some of us thought it exciting. Or it could be made so and, heaven knows, UNESCO needs something colorful to catch the public imagination ... it is not just a Pan-Hylean project but an international one to establish a lien on the development of this area not merely for the governments concerned but for the people of the world.'

—**Ritchie Calder**, from 'Science is UNESCO's strong point' (Calder, 1948).

Among the initial activities of the IIHA was a 'team survey' of a limited geographical area of the Hylean Amazon – the Rio Huallaga in Peru – with the aim of guiding the planning of future explorations to be undertaken in the Hylean Amazon on behalf of the IIHA. The survey was headed by C. Bolivar, with component reports including those on botanical research (by R. Ferreyra), geography (by E. B. Doran) and ethnology (by A. Bultron).

---

14 For an account of Corner's work with UNESCO, see Ashton and Hadley, 2001.

> '... that office which may yet be the highest in the secular world; I mean the Director-General of UNESCO'.
>
> Tropical botanist E. J. H. Corner, 1981, p. ix.

But little more was accomplished, owing to the failure of the participatory governments to agree to a convention of cooperation. At the close of 1948, UNESCO concluded its role of bringing the IIHA into being. Cooperation was continued with the Institute's Interim Commission (under the presidency of Dona Heliaso Torres of the National Museum in Rio de Janeiro), but the development of the Institute floundered over the next few years, as the Convention failed to achieve ratification.

Although some countries did ratify the Convention (including Colombia, Ecuador and France), the key ratification of Brazil was not forthcoming; and the leading Brazilian proponents of the Institute were accused of compromising the national security of the country, by permitting an international institute to be set up in the 'natural defence zone' of Amazonia.

Though the idea of the IIHA did not come to fruition, some of the thinking and plans of the Institute did contribute to the subsequent setting up by Brazil of the Instituto Nacional de Pesquisas da Amazônia (INPA) in Manaus, which continues to provide a major focus for research in the Amazonian region. The IIHA was perhaps an idea whose time had not come, at least in terms of the ways in which the institute was being perceived during that period.

ARID ZONE RESEARCH PROGRAMME

Arid lands have been another long-term concern in UNESCO's science programmes.[15] As mentioned above, in 1947 the Organization recommended to the United Nations that an international institute be set up for arid zone studies. The following year, during the third session of the General Conference held in Beirut (Lebanon), the Indian delegation put forward a resolution, later adopted, demanding action on this recommendation. A committee of experts would decide if such an institute was needed.

This committee was convened in Paris in December 1949. The idea of an institute was not maintained. Rather, the committee suggested the setting up of an 'arid zone research council', which saw the light of day as an Advisory Committee on Arid Zone Research, which met for its first session in April 1951 in Algiers (Algeria). This was the origin of UNESCO's Arid Zone Research Programme.

---

15   For semi-popular accounts of UNESCO's early work on arid lands, see: Swarbrick, 1955; White, 1961; Behrman, 1964; Behrman, 1972a; Batisse, 1994.

Distribution of Arid Zones

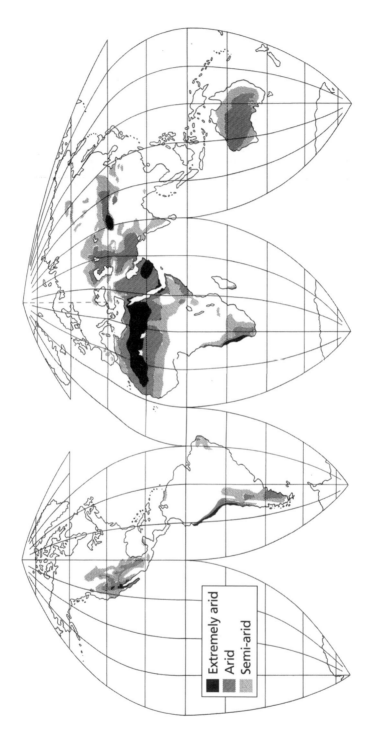

Early map of the world's arid zones. The map of the distribution of arid homoclimate zones in 'orange peel' projection was prepared for UNESCO in 1951 by geographer Peveril Meigs. Besides serving as the logo of the Arid Zone Programme, the map (and more detailed regional components) was widely used for information and teaching purposes for more than two decades, until new maps were prepared in 1977. Image: © UNESCO

PART III: ENVIRONMENTAL SCIENCES

Leading scientists from different countries and disciplines served on the advisory committee, under whose guidance a series of innovative activities was carried out for a modest outlay for more than a decade. The first task was to draw up a complex and detailed map showing the world's arid zones and their degree of aridity. The map was based on an innovative projection of the globe that resembles the four quarters of an orange. It became the programme's emblem, and appears on the cover of a collection of thirty volumes – the Arid Zone Research Series – the contents of many of which remain a store of valid information to this day. These publications report the results of a sequence of regional and international symposia and associated reviews of research, aimed at taking stock of existing knowledge on the ecology and natural resources of arid zones. An important feature was to focus activities in a given year on a particular problem area: hydrology (1951–52); plant ecology (1952–53); sources of power available in the arid zone, especially wind power and solar energy (1953–54); human and animal ecology (1954–55); and arid zone climatology (1955–56) (Baker, 1986).

In 1957, UNESCO upgraded what had hitherto been one project among many to the rank of Major Project on Arid Zones. The new status helped to raise the profile of activities that were already underway, as well as spurring multi- and interdisciplinary research in the field. A significant capacity-building component was developed, with the training of hundreds of specialists and the creation and reinforcement of national centres to promote the development of arid regions. This capacity-building work generally involved technical and financial support from UNESCO's Regular Programme or from the Expanded Programme of Technical Assistance, or through the Special Fund component of the United Nations Development Programme (UNDP), or through a combination of these sources.

Institutions that were created or strengthened in this way under the auspices of UNESCO[16] included the Desert Institute in Cairo (Egypt), the Central Arid Zone Research Institute in Jodphur (India), the Arid Zone Research Institute of Iraq at Abu Ghraib (later to become the Institute for Applied Research on Natural Resources), Israel's Negev Desert Institute in Beersheba, the Mexico City Institute of Applied Sciences, the Geophysical Research Institute of Quetta (Pakistan), the Tunis Arid Zone Research Centre (Tunisia) and the Ankara Botanical Institute (Turkey).

Among the active groups of researchers associated with the Arid Zone Research Programme were members of the American Association for the Advancement of Science (AAAS) and its Committee on Arid Lands. Among the activities organized by this group – with the support of UNESCO and

---

16   Information culled from an introductory article on 'UNESCO's natural resources research programme' in the inaugural issue of the magazine *Nature and Resources* (Vol. 1, Nos 1–2, 1965). Although unattributed, the article was most probably written by Michel Batisse.

> ## BOX III.2.2: 'ONE OF THE VERY BEST THINGS UNESCO EVER DID'
>
> 'My relationship with UNESCO started in 1956 and then developed quite strongly... They started a major programme in arid zones, which internationally meant not only deserts but also subhumid (seasonally dry) environments such as in northern Australia. Just after the war there had been an attitude that somehow science would "make the deserts bloom" as important places for rural production and for people to live in. Australians already knew that idea was wrong; we knew how fragile the arid zones were. But UNESCO's programme, more than any other single activity, really put the idea to bed.
>
> Through a consciousness-raising exercise of symposia and reviews of various areas of research related to the use of arid zones, the programme drew in many scientific people from developing countries and put them in contact with outstanding researchers from developed countries ... and did a whole lot of bridge-building. It was one of the very best things UNESCO ever did.'
>
> —Ralph Slatyer,[1] in an interview with the Australian Academy of Science (Slatyer, 1993).
>
> 1   Ralph Slatyer: Ecophysiologist, Chair of the International Coordinating Council for MAB (1977–81), Australian Ambassador to UNESCO (1978-81), and first Chief Scientist and Adviser to the Australian Prime Minister on Science and Technology (1989–92).

institutions such as the National Science Foundation and the Rockefeller Foundation – was an amalgam of International Arid Lands Meetings held in New Mexico (USA) in April–May 1955 (White, 1956). Some years later, the AAAS and UNESCO again joined forces in organizing another international conference in New Mexico, on the theme of Arid Lands in a Changing World. Participants from twenty-three nations took part in an assessment of experiences with arid environments (Dregne, 1970). Drawing together those experiences, University of Chicago geographer Gilbert White – one of those associated with the Arid Zone Research Programme from the very beginning – put forward seven unresolved issues that impinged (and indeed continue to impinge) on the 'long-run aims of enhancing social growth without undue destruction of natural resources' (White, 1970, in: Dregne, 1970):

- What is the role of irrigation?
- Should new development expand or intensify?
- Should the single big project or diverse smaller measures be stressed?
- What is the role of amenities?
- Should basic relationships or problem-solving take precedence?
- How can world linkages be achieved?
- Can integrated study and action be achieved?

PART III: ENVIRONMENTAL SCIENCES

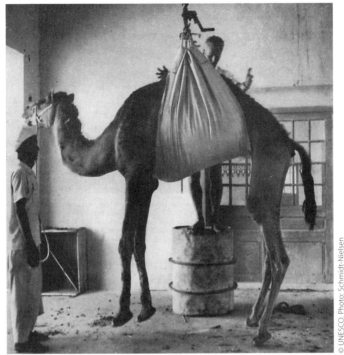

One of the emblematic images of the Arid Zone Research Programme in the 1950s: a dromedary being weighed at Beni-Abbès Research Station in Algeria, as part of metabolic studies by Knut and Bodil Schmidt-Nielsen on the adaptations of desert fauna to arid conditions.

The Major Project on Arid Lands formally came to an end in 1964. But its legacy continued in many types of follow-up activity, including educational and learning materials of various kinds.[17] One specific follow-up was in the field of international cooperation on water resources, with the launch of the International Hydrological Decade (1965–1974) and the subsequent International Hydrological Programme (IHP). Work on the scientific basis for the sustainable management of marginal lands continued through such field studies as that in Tunisia on the use of saline water for irrigation purposes. At the international level, the Man and the Biosphere (MAB) Programme has provided a framework for cross-disciplinary work on the ecology and natural resources of arid and semi-arid lands. And within the broader international community, many of the lessons of UNESCO's early work on arid lands found expression in the debates and outcomes of the United Nations Conference on Desertification held in Nairobi in 1977.

---

17   One example was a reader for 'advanced geography classes or for the educated public, giving a general perspective of arid zone geography' (Hills, 1966).

> **BOX III.2.3: ARID ZONE PROGRAMME: A POTTED BALANCE SHEET**
>
> "In the end, and after approximately a fifteen-year period, the UNESCO programme had neither shrunk the deserts nor stopped erosion, which then more than ever before threatened the world. But it had contributed to clarifying problems in arid lands and their economic, ecological and social repercussions. It had stimulated interest in questions previously neglected such as groundwater and salinity. It had opened the way towards an interdisciplinary approach to developing lands. It had served as the loom for weaving a lasting worldwide network of human contacts and dependable interchanges. It had acted as a catalyser for a multitude of national and local initiatives whose list would undoubtedly be impressive. It had led to fruitful cooperation not duplicated by other organizations. It left a precious heritage of achieving success in finely tuning and synthesizing a number of topics on which the scientific community had worked very hard. Not insignificant in this regard is the fact that the 600 authors from 42 countries who participated in writing the some 7,600 pages of the 30 volumes of the Arid Zone Research series never asked for and never received a cent for their work."
>
> —**Michel Batisse**, from *The UNESCO Water Adventure: From Desert to Water* (Batisse, 2005).

HUMID TROPICS RESEARCH PROGRAMME

In the light of the enthusiasm generated by the launching in 1951 of the Arid Zone Research Programme, the eighth session of the General Conference of UNESCO, held in Montevideo (Uruguay) in 1954, authorized the Director-General to promote the coordination of research on scientific problems relating *inter alia* to the humid tropical zone and to promote international or regional measures to expand such research.[18] By virtue of this authorization, the Director-General convened a preparatory meeting of specialists in humid tropics research in Kandy, Sri Lanka, in March 1956. The preparatory meeting recommended, among other things, the creation of an International Advisory Committee for Humid Tropics Research. The Director-General accepted this recommendation, and the Advisory Committee of nine members came into being at the end of 1956.

The purpose of this committee was to consider matters pertaining to research in the fundamental aspects of the natural sciences, of interest in the humid tropical zones, and to stimulate the carrying on of such research in areas considered at the time as important for long-term action.

Organization of symposia on well-defined aspects of humid tropics research was one of the means adopted in realizing the objectives of the Humid Tropics

---

18 This digest of the Humid Tropics Research Programme is based on Hadley (1999), whose paper includes a tabular chronology and some forty bibliographic references on UNESCO-related research in the humid tropics.

BOX III.2.4: HUMID TROPICS PROGRAMME – PRIORITY THEMES:

- tropical herbaria and key zoological collections
- the termite problem
- the preparation of regional floras
- the problem of laterization
- the scientific problems related to the delta areas of the humid tropics
- the principles and methodology of integrated research in the humid tropics
- the advance of the savannah into the tropical forest
- the chemistry and biology of tropical soils

Research Programme. The first such symposium was held in Kandy immediately prior to the preparatory meeting, on the topic 'Study of Tropical Vegetation'.

It is fascinating to leaf through the proceedings of the 1956 Kandy Symposium, and to reflect on the contributions that the participants at that symposium of fifty years ago made during their professional careers to science, and to studies on tropical vegetation in particular.[19] The Kandy symposium saw the presentation and discussion of several ideas and concepts which have contributed to the development of understanding of humid tropical vegetation. One example is the seminal paper by C. G. G. J. van Steenis on 'Rejuvenation as a factor for judging the status of vegetation types: the biological nomad theory', including the distinction between the stationary species (or dryads) of the climax rain forest and the temporary species (or nomads), which 'cannot rejuvenate in the rain forest and can maintain themselves only at its borders'. A specific output of the field trip associated with the symposium was the discovery and naming of a new species of subcanopy tree in the rain forest (*Cullenia rosayroana* Kostermans, Bombacaeae), following discussions in the field between A. J. G. H. Kostermans and R. A. de Rosayro (I. A. U. N. Gunatilleke, personal communication).

Following the Kandy symposium of March 1956, the Humid Tropics Research Programme entailed a series of somewhat similar symposia and seminars, including those organized in Tjiawi (Indonesia, 1958), Abidjan (Côte d'Ivoire, 1959), Goroka (Papua New Guinea, 1960), New Delhi (India, 1960), Kuching (Malaysia, 1963), Dhaka (Bangladesh, 1964). In addition to the proceedings volumes, outputs of the programme included a directory of zoology (and entomological) specimen collections of tropical institutions

---

19 Participants included (in the order of proceedings): Puri, Dilmy, Kostermans, de Rosayro, Wyatt-Smith, Richards, Bedard, Fosberg, Chatterjee, van Stennis, Misra, Purseglove, Bharucha, MacFadden, Holmes, Mangenot, Taylor, Stewart, Janaki, Hedayetullah, Browne, Wood, Webb, Senaratne, Schmid and Anderson.

(1962), a manual for tropical herbaria (1965) prepared under the auspices of the UNESCO Visiting Committee on Tropical Herbaria, and an ethnobotanical guide for anthropological research in Malayo-Oceania (1967).[20] Looking back at the various publications generated within the programme, once again one is struck by the participation of scientists whose work has marked the pages of scientific understanding of the humid tropical regions.[21]

The Humid Tropics Research Programme came to an end in the mid-1960s. It was followed by several other UNESCO initiatives related to the environment and its natural resources, which have continued to include work in humid tropical regions, as described in subsequent chapters within this 'Environment' section.

CARTOGRAPHY AND SCIENTIFIC MAPS

The importance of developing cooperation and coordination across the United Nations system was stressed by the General Conference at its second session in late 1947, in commending to the UN Secretary-General the urgency of adequate planning in the basic field of cartographic science. Since that time, cooperation among UN organizations and the international non-governmental scientific community has been a leitmotiv in long-term work on cartography, including the development of internationally recognized norms and protocols, and the preparation of regional and global maps in such fields as climate, geology, soils and vegetation.

Although these small-scale maps cannot replace large-scale maps produced on a national level, they nevertheless constitute a valuable source of information for broad planning and study of resources-development activities. They are also useful working tools for secondary and university courses. Some examples of UNESCO's work on scientific mapping follow.

CLIMATIC MAPS

Asia, North and Central America, South America, and Europe are four regions for which regional climatic atlases have been prepared as joint initiatives of UNESCO and the World Meteorological Organization (WMO), in cooperation with specialist bodies such as Cartographia (Budapest, Hungary) and Goscomgidromet (Moscow, Russia).

---

20   A list of publications of the Humid Tropics Research Programme is given in Hadley, 1999.
21   In addition to many of those who attended the Kandy symposium, participants in the Tjiawi seminar included Ashton and Whyte; in the Goroka seminar: Corner and Smitinand; and the Kuching seminar: Brünig, Greig Smith, Kartawinata and Whitmore.

OCEANIC MAPS

As described in a later chapter, work on ocean mapping within the Intergovernmental Oceanographic Commission dates back to the late 1960s, and includes improved bathymetric charts of the world ocean, as well as regional initiatives such as the preparation of the Geological and Geophysical Atlas of the Indian Ocean.

GEOLOGICAL MAPS AND RELATED CARTOGRAPHY

The series of geological maps includes the Geological Map of the World at a scale of 1:25,000,000 (four sheets) and the Geological World Atlas (twenty-two sheets), as well as Regional Geological Maps of Europe and the Mediterranean Regions and of Africa. Related cartography includes: an International Hydrogeological Map of Europe, Metallogenic Maps of Europe and of South and South-East Asia, International Mineral Deposits Map of Africa, Tectonic Maps of Africa and of the Carpathian-Balkan system, a Metamorphic Map of Europe, and Quaternary Maps of North-West Africa and of Europe. Accompanying or complementing many of the maps are guidance and explanatory materials of various kinds (such as a guide to the preparation of engineering geological maps).

SOIL MAPS

Among UNESCO's best-known projects is that which started in 1961, at the instigation of Victor Kodva, a Russian soil specialist and Director of the Department of Natural Sciences. This was a joint initiative with FAO to prepare a 1:5,000,000 scale soil map of the world. Under this project, which lasted for seventeen years, all the world's soils were plotted on eighteen detailed maps, in accordance with a standard terminology and system of classification.

The project had to contend with three major difficulties. In the first place, the significant differences between the systems of representation used by the American, French and Russian schools of soil science had to be reconciled, which proved difficult. Second, it was necessary to gather the requisite field data, which were lacking in many countries; and specialists from around the world (in particular from FAO and among the correspondents of the International Society of Soil Science) had to be mobilized for the purpose. Third, the project had to be spread over an adequate length of time, and continuity had to be ensured, notably as far as its budget was concerned; and this was only possible because UNESCO and FAO each in turn took over responsibility for the project (Conil-Lacoste, 1994, p. 107; see also: Dudal and Batisse, 1978).

VEGETATION MAPS

Vegetation is a domain where UNESCO's activities did not translate into methodologically consistent mapping products, in part because of fundamental differences between specialists within the botanical-science community. A Standing Committee on Classification and Mapping of Vegetation on a World Basis was set up in the early 1960s, under the leadership of Heinz Ellenberg, professor of botany at the University of Göttingen. A classification scheme was developed[22] and a programme of mapping at a scale of 1:5,000,000 initiated, with maps published for the Mediterranean Basin (jointly with FAO) and for South America. It was intended that all maps in the series should use a uniform legend and colour scheme. But because of the complexity of the subject matter and the diversity of approaches, this objective was not achieved. Hence the *UNESCO/ AETFAT/UNSO Vegetation Map of Africa*[23] differs in some important respects from those for the Mediterranean Basin and Latin America.

FURTHER DEVELOPMENTS

The major thrust of UNESCO's early cartographic work was on the preparation of large-scale maps relating to the physical environment and particular categories of natural resources. One follow-up to that work was to promote the preparation of maps and atlases of natural resources at the national level (Salishchev, 1967, 1972). The field of integrated cartography is another area that has been explored, in cooperation with the International Geographical Union and its working group on Cartography of the Environment and its Dynamics. Methodological and field studies were carried out to explore the role and usefulness of integrated maps in interdisciplinary research, planning and integrated land-use management (Journeaux, 1987).

Another feature of more recent work has concerned the use of modern information and communication technologies in producing maps geared to the needs of particular user groups, and involving specific partnership arrangements with space agencies and other remote sensing facilities and with bilateral aid agencies. One such initiative with the European Space Agency relates to the use of space technologies in protected area management, and specifically in support of the World Heritage Convention and sites inscribed on the World Heritage List. Among the contributing projects is one to prepare accurate maps derived from satellite images for World Heritage sites and other protected areas in central Africa that host populations of endangered mountain gorillas.[24]

---

22   Published as *International Classification and Mapping of Vegetation* (UNESCO, 1973).
23   The *UNESCO/AETFAT/UNSO Vegetation Map of Africa* was compiled as a joint initiative of UNESCO, the Vegetation Map Committee of the Association pour l'Etude Taxonomique de la Flore de l'Afrique Tropicale (AETFAT) and the United Nations Sudano-Sahelian Office (UNSO). Accompanying the map is a descriptive memoir by Oxford University botanist Frank White (White, 1983). With its twenty-two chapters describing the vegetation of the main floristic regions, a bibliography of more than 2,400 items, and an index of nearly 3,500 plant species, the memoir is an example of meticulous scholarship that remains a basic reference work several decades after its compilation and publication.
24   The 'Build Environment for Gorilla' (BEGo) project is a joint initiative of the European Space Agency (ESA) and UNESCO (through its World Heritage Centre and the Remote Sensing Unit in the

PART III: ENVIRONMENTAL SCIENCES

HYDROLOGY AND FRESHWATER RESOURCES

Issues related to freshwater resources figured prominently in UNESCO's work in arid zones in the 1950s. Hydrology was the first problem area addressed (in 1951–52) in the Arid Zone Research Programme, with arid zone hydrology featuring in the first two volumes in the Arid Zone Research Series. Several subsequent volumes in the series were concerned with more finely focused aspects, such as the use of saline waters (1954), recent progress in arid zone hydrology (1959), salinity problems in the arid zone (1961), plant–water relationships in arid and semi-arid conditions (1960), evaporation reduction (1965) and physical principles of water percolation and seepage (1968).

Capacity-building included a five-year research and training project on the use of saline water for irrigation in Tunisia, launched in 1962 with the support of the United Nations Special Fund (subsequently, the UNDP). And at about the same time, during 1961–64, the international community of hydrologists set in train a process that was to lead to the launching by UNESCO in 1965 of the International Hydrological Decade. In the late 1960s, support was provided to work on biological dimensions of freshwater resources, through ICSU's Committee for Research on Water (COWAR) and the Productivity of Freshwater Communities section of the International Biological Programme (Rzóska, 1970; Saunders, 1972). Among the aims was to highlight the role of biological phenomena in issues related to water quality[25] – including the control of water weeds and the role of detritus in aquatic ecosystems.

MARINE SCIENCES

If Aristotle can be considered the first oceanographer, and the Polynesians among the first true observers of the marine world (navigating the Pacific without chart, compass or sextant), it was in the seventeenth century that the first burgeoning of marine science took place, following the great age of discovery and navigation.[26] The sea was one of the early concerns of the Royal Society when it was founded in 1660. Scientists started to seek explanations for the observations of seamen,

---

Division of Ecological and Earth Sciences), in partnership with nature conservation bodies in central Africa and donor organizations. Among the products is a 1:50,000 map of the Virunga Massif (www.gorillamap.org).
25 The Scientific Coordinator of IBP's Freshwater Productivity Section, Julian Rzóska, commented somewhat pithily: 'Water in its manifold aspects is the concern of the International Hydrological Decade, but water in nature can never exist without biological phenomena. Hydrologists may regard this as a nuisance, but they have become more and more aware of the need to consult their biological colleagues concerned with waters on problems with which they themselves cannot cope. The cooperation between the different fields of fundamental and applied sciences is recognized nowadays as very important for the proper management of resources, aquatic and terrestrial' (Rzóska, 1969).
26 These paragraphs are based on 'The marine environment', in Behrman, 1972b.

and Newton solved the problems of the tides. In the nineteenth century, when the sailing ship reached the height of its development, there was a quick pay-off on knowledge of wind and current systems. Marine geology got its start with the laying of the first transatlantic cable, and the *Challenger* brought back thousands of samples of marine life from the depths. The period following the First World War saw such achievements as the *Discovery* expeditions to the Antarctic and the *Meteor* studies of the Atlantic. But oceanography was very much the preserve of rich countries or rich individuals.

Following the Second World War, as UNESCO's first programmes were taking shape, scientific interest in the sea arose anew. The war had left a surplus of ships that could be had cheaply, and scientists were not slow in proposing uses for them, as witness the Organization's reply in 1947 to the enquiry on the possibility of United Nations research laboratories. J. A. Fleming, the president of the International Council for Science (ICSU), wanted the UN to operate 'floating laboratories' to carry out work in oceanography, volcanology, geomagnetism and seismic studies of the Earth's crust under the sea. Harald Sverdrup, director of the Scripps Institution of Oceanography in the United States, proposed a technical body to work in physical oceanography. UNESCO also reported a proposal made by the Indian delegation to the UNESCO Sub-Commission on Science in 1946, on the need to establish an oceanographic station in South India or Ceylon.

UNESCO kept dipping into the pool of ideas that had been put forward in replies to the United Nations enquiry. In 1950, the General Conference, meeting in Montevideo, adopted a resolution that led to the establishment five years later of an International Advisory Committee on Marine Sciences (IACOMS), whose members included some of the world's leading oceanographers.[27] Among the matters addressed were fellowships in the marine sciences, a travelling marine sciences expedition, and contributing to events to mark the founding of the Oceanographic Museum in Monaco. The committee also facilitated contributions of major oceanographic institutions to observations at sea, within the International Geophysical Year (IGY), which took place under the aegis of ICSU during an eighteen-month period in 1957–1958.

But perhaps the most important action of the Advisory Committee was to call for an intergovernmental oceanographic conference. Initially the idea was to focus on the old theme of an international research vessel, a floating laboratory, a sort of marine equivalent to the cyclotron that was being operated by CERN (the

---

27   Among the members of IACOMS were Len Zenkevitch (the 'grand old man' of marine biology in the USSR), G. E. R. Deacon (head of the National Institute of Oceanography in the UK), Roger Revelle (director of the Scripps Institution of Oceanography at the time), Anton Bruun (leader of the *Galathea* expedition, the first to find life in the deepest ocean), and Marc Eyries (a pioneer in tidal measurements in the open sea).

European Organization for Nuclear Research, also helped by UNESCO during its formative stages). However, many oceanographers had other ideas.[28]

With utmost discretion, the name of the meeting was changed to the Intergovernmental Conference on Oceanographic Research, thereby enlarging options and possibilities of action. The conference took place in Copenhagen (Denmark) from 11 to 16 July 1960, with delegates in attendance from thirty-five countries. And it was there that the Intergovernmental Oceanographic Commission (IOC) was founded.

GEOLOGICAL SCIENCES AND NATURAL HAZARDS

While the early programmes on arid zones and the humid tropics had included a number of component studies related to the Earth sciences (in fields such as laterites), it was only in the 1960s that more concerted action was taken with regard to geology and related sciences such as geomorphology, geochemistry and geophysics. In close liaison with the International Union of Geological Sciences (IUGS) and its various branches, activities aimed 'to promote the general advancement of knowledge and at the same time to provide the scientific and institutional foundation for the rational evaluation and use of natural resources'.

A first group of activities (mentioned above) concerned the standardization or harmonization of nomenclatures, terminologies and cartographic methods, and the synthesis of existing knowledge, leading to the establishment of international and intercontinental correlations and comparisons, and to the preparation and publication of small-scale geological, tectonic and metallogenic maps.

Research and the exchange of scientific information regarding certain key problems in the geological sciences was a second major focus. Activities included the organization and servicing of expert missions to carry out stratigraphic correlation of geological formations on a regional basis. Field studies and associated symposia were supported in such fields as the correlation of the granites of West Africa and comparison with other regions (e.g. north-eastern South America), the origin of stratiform ore deposits, and the study of geochemical prospecting methods. International symposia organized in cooperation with international scientific associations included that on the geology and genesis of Precambrian iron/manganese deposits, continental drift formations and ore deposits (Kiev, Ukraine, USSR, August 1970).

---

28 Dan Behrman comments that many oceanographers did not want to put all their budgets into one ship: 'Outside observers, too, were bemused by the possibility of a ship being run by a UN agency. "Hard a starboard", cries the captain, when he spots breakers ahead. "Point of order", replies the helm...' (Behrman, 1972b, p. 51).

As in other environmental sciences fields, special efforts were made to help developing countries to create or strengthen research and teaching institutions in the Earth sciences. The Technical Assistance and Special Fund programmes of the period provided support to such institutions as the Bandung Institute of Geology and Mining in Indonesia. Regional training courses were held in fields such as applied geology, applied geomorphology and geochemical prospecting methods. A network of extended postgraduate courses took shape based on training for specialists from developing countries at specialized institutions in developed countries, such as courses on petroleum geology at Bucharest (Romania), applied geology at Vienna (Austria), and geothermal energy at Fukuoka (Japan).

It was also in the early 1960s that work on natural hazards started to take shape, with a focus on earthquakes and volcanic eruptions.[29] As an intergovernmental organization, UNESCO is concerned less with promoting scientific research for its own sake (this, at the international level, is the task of the international scientific unions and associations), than with the application of scientific and technological knowledge to the improvement of methods of risk assessment, warning systems and protection measures. Efforts therefore focused on a three-pronged approach: first, seismic and volcanic zoning, and the estimation in statistical terms of the risks to life and property within these zones; second, the development of improved monitoring systems and of methods of forecasting seismic or volcanic activity; and third, the refinement and wider application of earthquake-resistant design and construction. In pursuing these objectives, the Organization's main functions have been to provide opportunities and means for international consultation and cooperation – by convening conferences, symposia, and meetings of working groups on particular topics – and to assist developing countries in carrying out research and training in the appropriate scientific and technical disciplines.

A major component was the sending of reconnaissance missions of experts to the sites of destructive earthquakes immediately after their occurrence.[30] A first earthquake reconnaissance mission to Iran in 1962 was followed by another nineteen missions in the period 1962–76, such as those in Latin America after earthquakes in El Salvador (1965), Peru (1966, 1970, 1974), Venezuela (1967) and Chile (1972). Early experiences were brought together in presentations to

---

29 This discussion is largely based on a three-page article on 'Current UNESCO programmes in seismology and volcanology' in *Nature and Resources*, Vol. 11, No. 3, 1975. In turn, this unsigned review was based on a presentation to a meeting on Seismic and Volcanic Risk (San José, Costa Rica, July 1975) by E. M. Fournier d'Albe, who was the main driving force within the Secretariat for the Organization's early work on natural hazards in the 1960s and 1970s.

30 Reviews of UNESCO's early work on natural hazards are given in: 'When nature runs wild' in Behrman, 1972c (pp. 62–77); and in: Ambraseys, 1976; Roberts, 1979.

PART III: ENVIRONMENTAL SCIENCES

several conferences on the assessment and mitigation of earthquake risk, such as those intergovernmental conferences held in Paris in 1965 and February 1976.

The findings of the various missions sent out by UNESCO added to our knowledge, not only of earthquakes and their effects, but also of ways of mitigating such disasters through the use of local building materials and methods of construction. Missions contributed significantly to demonstrating the importance of tectonics and faulting in the assessment of earthquake risk, to the mapping of major tectonic features, and in retrieving valuable information about historical earthquakes. In addition, on a number of occasions, earthquake missions were followed by UNESCO-sponsored seminars and courses, which heightened awareness of the need for disaster prevention studies, particularly among policy-makers.

Although the related field of volcanology has, from a scientific point of view, much in common with seismology (as witness the many scientists who have worked in both fields), from the point of view of disaster prevention, the two are very different. First, the problem of predicting the occurrence and intensity of volcanic eruptions, though by no means an easy one, appears to be more tractable than that of predicting earthquakes. Second, the zones at risk are relatively small and well defined. And lastly, since the only protection against violent eruptions lies in timely flight from the danger zones, the main problem here is that of accurate warning and rapid evacuation in cases of emergency.

For this reason, the early programme on volcanology focused on the improvement of methods of surveillance and prediction of volcanic activity. A review of research on this subject prepared in 1971 was followed by a series of regional seminars held in Guadaloupe (1974), the Philippines (1975) and southern Europe (1976).

SOIL SCIENCES

In a somewhat analogous way to the Earth sciences, soil-science issues received some but relatively limited attention in the first decade and a half of UNESCO's activities, as part of activities in the arid zone and humid tropics (e.g. work on the chemistry and biology of tropical soils).

In the late 1950s, the profile of soils in UNESCO's science programmes underwent a major transformation, under the impetus of one of the Organization's best-known science projects – the FAO/UNESCO Soil Map of the World (UNESCO, 1974). From the mid-1960s until the early 1970s, soil sciences were to have a distinctive and explicitly recognized place in UNESCO's programmes, under the aegis of the Advisory Committee on Natural Resources Research (see below). Perhaps coincidentally, the first Chair of this Committee was the French pedologist Georges Aubert. Activities in this period included a major international symposium on methods of study in soil biology (with the International Biological Programme). Scientific missions were organized, such as for surveying, collecting and comparing the soil fauna in various parts of the tropics (Congo, Australia, New Guinea, Central and South America). Another project involved surveying and mapping soils along a 2,000-km coastal desert strip in Peru.

Soil-survey work also featured prominently in an FAO-UNESCO-WMO programme on agricultural bioclimatology. Its accomplishments included the development and field testing of a protocol for agroclimatological studies and field studies in such regions as the Near East, the Sahel region of West Africa, the highlands of eastern Africa, and the Andean plateaux.[31]

In the early 1970s, scientific work involving soils became an integral part of the various intergovernmental environmental programmes. This has included studies on soil erosion and sedimentation, tropical soil biology and fertility, soil survey and land evaluation, and soil regeneration with erosion products and other waste. It has also included joint activities with such bodies as the International Soil Reference and Information Centre (ISRIC) and the International Society of Soil Science (ISSS) (now the International Union of Soil Sciences, IUSS) and its World Congress of Soil Sciences series.

NATURAL RESOURCES RESEARCH PROGRAMME

As described above, since UNESCO's earliest days, a good part of the Organization's science programme was concerned with selected problems relating to the natural environment and to natural resources, whether under the primary heading of nature conservation or particular biogeographical zones, such as arid lands and the humid tropics.

As these various projects developed,[32] three facts became increasingly clear: some activities undertaken by a given programme concerned an entire continent or the entire world (e.g. saline soil problems); on the other hand, there were others which ought to be undertaken that did not exactly fit in with any of the existing programmes (e.g. geological mapping); finally, there was a great similarity of methods and problems between such apparently different programmes as those concerning the arid zone and the humid tropical zone (e.g. agroclimatological studies).

For these reasons, in 1960, all these activities were regrouped within a single division known as the Division of Studies and Research relating to Natural

---

31   For glimpses into the interagency agricultural bioclimatology programme, see: Cochemé, 1966; UNESCO, 1968; Brown and Cochemé, 1970.
32   Paragraph cited *in extenso* from an unsigned article on 'UNESCO's Natural Resources Research programme' in the first issue of UNESCO's quarterly *Nature and Resources* (Vol. 1, Nos.1/2, June 1965, pp. 1–5).

Resources, with Michel Batisse as director. As the programme took shape in the first part of the 1960s, the principal activities concentrated on surface and subsurface hydrology, geology and geomorphology, soil science, and the ecological sciences and conservation. In terms of budget resources, the Regular Programme budget was nearly US$800,000 for 1965–1966 (excluding staff costs and the Technical Assistance and Special Fund programmes).

In the planning and implementation of these activities, the Secretariat was guided by an Advisory Committee on Natural Resources Research, a group of fifteen internationally renowned specialists, which met for the first time in UNESCO House in Paris in September 1965.[33] At its biennial meetings, the Advisory Committee considered issues related to the overall approach and balance of activities, as well as the structure and contents of the proposed programme. Among points emphasized at various sessions were the importance of the concept of integrated surveys of land and water resources, and proposals for international scientific cooperation on the rational utilization of the resources of the biosphere, which were to take shape around the international 'Biosphere Conference' of September 1968 and the subsequent launching of the Man and the Biosphere (MAB) Programme.

Some of the substantive results of the Natural Resources Research Programme were published in two series of books launched by UNESCO in the 1960s. The Natural Resources Research series was started in 1963; twenty-one volumes were produced over the following two decades. Some volumes comprised commissioned reviews of research in particular technical fields or biogeographic regions, such as: the natural resources of the African continent or humid tropical Asia, research on laterites and on salt-affected soils, computer handling of geographical data, and state of knowledge reports on tropical forest ecosystems and tropical grazing land ecosystems. Descriptive memoirs or explanatory notes included those accompanying the Geological Map of Africa, and the Vegetation Maps of Africa and South America. Symposium proceedings addressed such topics as: the functioning of terrestrial ecosystems at the primary production level, aerial surveys and integrated studies, agroclimatological studies, West African granites, use and conservation of the biosphere, soils and tropical weathering, and conservation, science and society. Other titles included a bibliography of African hydrology and case studies on desertification.

The second series, on 'Ecology and Conservation', comprised just six volumes, treating: the ecology of the subarctic regions, methods of study in soil ecology, the origin of *Homo sapiens*, productivity of forest ecosystems, plant responses to climatic factors, and vegetation classification.

---

33 Subsequent meetings were held in June 1967 (Paris), June 1969 (Paris) and August 1971 (Canberra, Australia).

BOX III.2.5: NATURE AND RESOURCES (1965-1999)

The first issue of the quarterly magazine *Nature and Resources* was produced in 1965. With the end of the Organization's Arid Zone Programme and publication of the final issue of the *Arid Zone Newsletter* in December 1964, the new periodical was designed to cover the different aspects of the Division of Natural Resources Research in the fields of hydrology, geology, soil sciences, ecology and conservation of nature. Particular emphasis was given in the first few years to the International Hydrological Decade (which started in early 1965), with the magazine including the subtitle 'Bulletin of the International Hydrological Decade'.

Each issue carried from three to six feature articles. Some articles were anonymous, and mainly reported official UNESCO texts such as the reports of the Advisory Committee on Natural Resources Research or of major UNESCO events (such as the 1968 Biosphere Conference). But the majority were signed and comprised accounts of the substantive findings and approaches of projects and activities contributing to the Organization's environmental sciences programmes. From the beginning, each 28–36 page issue included short sections of news items and abstracts of publications.

Beginning in December 1970, the banner became a double one of 'Bulletin of the Man and the Biosphere Programme/Bulletin of the International Hydrological Decade', with 'Bulletin of the International Geological Correlation Programme' being added in 1973. During this period, the magazine was published by UNESCO in English, French and Spanish, with Arabic, Chinese and Russian editions published through arrangements with collaborating institutions in Cairo, Beijing and Moscow.

In 1990, the magazine took on a new look, with a more modern layout and the use of colour. Editorially, an important change was the focus of each issue on a particular topic, comprising five or six signed articles reviewing research on resource management for sustainable development. The theme of each issue continued to be closely linked to the Organization's environmental sciences programmes. Issues addressed such topics as the coastal environment, water quality and availability, ocean and coastal research, natural disasters, tropical forests, biotechnology, managing our common resources, sustainable development for all, environmental awareness building, different faces of World Heritage.

In the 1990s, inputs from the Natural Sciences Sector focused on substantive, editorial and design matters. Little or no attention was given to the many other aspects of producing a viable publication, including the key issue (in the second half of the 1990s) of electronic publication. Also largely ignored was the need to rebuild and mobilize a broad constituency of support for the magazine within UNESCO and its intergovernmental environmental research programmes, and within the international community of scientists, educators and policy-makers interested in environmental and resource management issues.

The director and staff of the division responsible for the magazine became increasingly concerned that with budgets and resources shrinking, three staff members were involved for a fair amount of their time on a publication covering a broad agenda of environmental issues. Efforts to generate additional support for the magazine were not successful. In late 1999, the decision was taken that pressure on scarce human and financial resources could no longer be borne. Publication was halted with the last issue of 1999.

PART III: ENVIRONMENTAL SCIENCES

© UNESCO. Photo: I. Fabbri

The changing face of *Nature and Resources*. Covers of the quarterly magazine (from top left), in 1965, 1974, 1984, 1992 and 1999. In 1976, the cover design was adopted by the French Post Office for stamps issued in connection with the 'Année de la Qualité de la Vie' (Quality of Life Year).

LINKS WITH THE NGO COMMUNITY

From its very early days, UNESCO's work in the natural sciences field has been carried out in close cooperation with the non-governmental scientific community. Indeed, as described elsewhere in this volume, rebuilding and reinforcing the non-governmental scientific associations was one of the early priorities of the Organization after the Second World War, and those links have continued throughout the six decades under review. In some cases, cooperation has taken the form of joint activities. More frequently, it has taken the form of mutual support to projects and programmes launched and piloted by one organization or the other.

An early example in the geosciences field was the International Geophysical Year (IGY), which took place in 1957–1958, under the aegis of ICSU. UNESCO provided financial support to the Special Committee for the IGY for the organization of the first plenary meeting in Brussels (Belgium) in 1952, and provided a total of US$110,000 for the programme over the next six years (Baker, 1986, p. 12). But it was not only financial support. UNESCO also helped to bring the IGY programme and the issues that it was addressing to the attention of the general public. A travelling exhibition dedicated to geophysics was organized, which toured many regions of the world in 1957–58. The IGY and its research programmes and findings were featured in several issues of the *UNESCO Courier* magazine.[34] In addition, UNESCO assisted scientists in developing countries to participate in the scientific programme of the IGY by providing financial support, equipment and training.

The IGY had a great impact on geoscientists but also on the broader scientific and lay communities. It led to the setting up by ICSU of a number of Scientific Committees, such as those on Oceanic Research (SCOR), Antarctic Research (SCAR) and Space Research (COSPAR). Cooperation has continued to this day between these Special Committees and UNESCO's own programmes in such fields as oceanography, Earth sciences, space activities and remote sensing. The Upper Mantle Programme was another initiative of ICSU to which UNESCO contributed, through support to such projects as a study of the East African Rift System. And this cooperation in the geosciences continued in 1966–68, in the planning – by UNESCO and ICSU's International Union of Geological Sciences (IUGS) – of the International Geological Correlation Programme (IGCP, later renamed the International Geoscience Programme).

Among those inspired by the IGY were biologists interested in the biological basis of productivity and human welfare, who formulated plans for

---

34 For example, the entire September 1957 issue of the *UNESCO Courier* was devoted to the IGY, with an article on 'Operation Globe: History's greatest science research project', which ends: 'For the first time, the peoples of the earth have joined to study their common and fundamental scientific problems together.'

the International Biological Programme (IBP), formally launched by the tenth General Assembly of ICSU in Vienna in November 1963. The first General Assembly for the IBP took place in UNESCO House in July 1964. Direct contracts between the Scientific Committee for the IBP (SCIBP) and UNESCO for projects of mutual interest and for joint meetings began in 1965, reached a maximum of US$56,000 and amounted to almost 16 per cent of the total budget of the Programme (Baker, 1986, p. 11).

From the mid-1960s, close links were developed between the IBP and UNESCO's natural resources programmes, and the experience gained in developing the IBP helped in the planning and convening by UNESCO of the 1968 Biosphere Conference and the subsequent development of the Man and the Biosphere (MAB) Programme. As noted by ICSU's long-serving (1965–89) executive secretary, F. W. G. (Mike) Baker, of the eighty-odd scientists who made presentations to UNESCO's 1981 conference, 'Ecology in Action: Establishing a Scientific Basis for Land Management', held to mark the tenth anniversary of the MAB Programme, about half had also been active in the IBP (ibid.).

If the IGY and IBP are examples of ICSU activities supported over a sustained period by UNESCO, many of the science-based activities of UNESCO have likewise benefited from the advice and cooperation of the international scientific community, as attested to in many chapters of this volume. Though perhaps less frequent, there have also been joint ICSU–UNESCO activities planned and carried out on a partnership basis, mostly since the 1970s. These have included shared responsibility for long-term programmes and projects of scientific cooperation, in fields such as: geological correlation, biosciences, scientific hydrology, tropical soil biology and fertility, savannah ecology, and biological diversity.

Though ICSU and its specialized unions and committees was the principal NGO contact of UNESCO in the environmental sciences over the two decades following the Second World War, other NGOs also figured prominently in the Organization's work. Indeed, among UNESCO's actions was helping in the establishment of international organizations and scientific associations of various kinds. Examples included the International Union for the Protection of Nature (1948, now IUCN, the World Conservation Union), the Charles Darwin Foundation for the Galápagos Islands (1959), and the Organization for Flora Neotropica (1964).

TOWARDS INTERGOVERNMENTAL PROGRAMMES, UNDERTAKINGS AND FRAMEWORKS

Over the first fifteen or twenty years of its work on natural resources and the environment, one of UNESCO's priorities was to help the international scientific community to get back on its feet after the ravages of the Second World War.

Actions supporting the re-establishment of international cooperation within the scientific community had included grant-in-aid support to ICSU, and working with partners for the creation of new institutions such as IUCN. Programmes such as those on the arid zones and the humid tropics had helped bring together specialists from different regions 'to share findings and problems, ignoring the usual boundaries of discipline and nation' (White, 1970, p. 486).

From the late 1950s on, it became time for governments to join in the formulation and implementation of research programmes directed towards more specific development needs. In the year 1960 alone, eighteen countries – all but one from Africa – joined UNESCO, and throughout the 1960s many newly independent and developing countries became responsible for the use of their natural resources. Additionally, to borrow the words of the Chair of the US President's Council on Environmental Quality, Russell W. Peterson, 'at this stage of our history, the involvement of governments in all aspects of environmental research and activity' was widely considered to be 'absolutely essential' (Peterson, 1974).[35]

Considerations such as these led to many new initiatives at the intergovernmental level. Within UNESCO, they included the establishment of the Intergovernmental Oceanographic Commission (1960), the launching of the International Hydrological Decade (1965), the organization of the intergovernmental meeting of experts on the natural resources of the biosphere (1968) and the founding of the Man and the Biosphere Programme (1970), the creation of the World Heritage Convention (1972), and the start of the International Geological Correlation Programme (1974). And within the broader international community, this involvement of governments in the whole environmental debate was assured by the Stockholm Conference on the Human Environment in 1972, and the subsequent setting up of the United Nations Environment Programme.

BIBLIOGRAPHY

Ambraseys, N. 1976. Field studies of earthquakes. *Nature and Resources*, Vol. 12, No. 2, pp. 12–16.
Ashton, P. S. and Hadley, M. 2001. Corner's contribution to ecology and conservation. In: L. G. Saw, L. S. L. Chua, K. C. Khoo (eds), *Taxonomy: The Cornerstone of Biodiversity. Proceedings of the Fourth International Flora Malesiana Symposium, 1998*, pp. 5–10. Kepong, Malaysia, Forest Research Institute Malaysia.
Baker, F. W .G. 1986. *ICSU-UNESCO: Forty Years of Cooperation*. Paris, ICSU Secretariat. http://unesdoc.unesco.org/images/0008/000812/081277eo.pdf
Batisse, M. 1994. A long look at the world's arid lands. *UNESCO Courier*, January 1994, pp. 34–39.
——. 2005. *The UNESCO Water Adventure: From desert to water*. Paris, UNESCO, p. 77.

---

35 Peterson adapted his article from an address given at Raleigh, North Carolina (USA), on 23 September 1974, to delegates attending the third session of the International Coordinating Council of the MAB Programme.

Behrman, D. 1964. Taming an 'Act of God'. In: *Web of Progress: UNESCO at Work in Science and Technology*. Paris, UNESCO, pp. 51–63
——. 1972a. On the environmental extremes. Chapter 3 in: *In Partnership with Nature: UNESCO and the Environment*. Paris, UNESCO, pp. 31–46.
——. 1972b. The marine environment. Chapter 4 in: *In Partnership with Nature: UNESCO and the Environment*. Paris, UNESCO, pp. 47–61
——. 1972c. When nature runs wild. Chapter 5 in: *In Partnership with Nature: UNESCO and the Environment*. Paris, UNESCO, pp. 62–77
Brown, L. H. 1966. Wildlife conservation in Ethiopia. *Nature and Resources*, Vol. 2, No. 1, pp. 5–9.
Brown, L. H. and Cochemé, J. 1970. Agrometeorology survey of the highlands of eastern Africa. *Nature and Resources*, Vol. 6, No. 3, pp. 2–10.
Calder, R. 1948. Science is UNESCO's strong point. *Discovery* magazine. February 1948.
Cochemé, J. 1966. FAO/UNESCO/WMO agroclimatology survey of a semi-arid area south of the Sahara. *Nature and Resources*, Vol. 2, No. 4, pp. 1–10.
Conil-Lacoste, M. 1994. *The Story of a Grand Design, UNESCO 1946–1993: People, events, and achievements*. Paris, UNESCO.
Corner, E. J. H. 1981. *The Marquis: A Tale of Syonan-to*. Singapore, Heinemann Asia.
Domingues, H. M. B. and Petitjean, P. 2001. A UNESCO, o Instituto Internacional de Hiléia Amazônica e a antropologia no final dos anos 40. In: P. Falhauber, P. M. de Toledo (eds), *Conhecimento e Fronteira: Historia da Ciência na Amazônia* [Knowledge and Frontier: History of science in the Amazon region]. Belem, Museu Paraense Emilio Goeldi.
——. 2004. International science, Brazil and diplomacy in UNESCO (1946–50). *Science, Technology & Society,* Vol. 9, No. 1, pp. 29–50.
Dorst, J. 1970. The Charles Darwin Foundation for the Galapagos Islands. *Nature and Resources*, Vol. 6, No. 3, pp. 11–14.
Dregne, H. A. (ed.). 1970. *Arid Lands in Transition*. Publication No. 90. Washington DC, American Association for the Advancement of Science.
Dudal, R. and Batisse, M. 1978. The Soil Map of the World. *Nature and Resources*. Vol. 14, No. 1, pp. 2–6.
Gille, A. 1949. *Education for the Conservation of Natural Resources and Their Better Utilization*. Paris, UNESCO.
Goldsmith, M. 1995. *Joseph Needham: Twentieth-Century Renaissance Man*. Paris, UNESCO.
Hadley, M.1999. Humid tropics research and UNESCO: an introduction and overview. In: *Proceedings of the Regional Seminar on Forests of the Humid Tropics of South and South-East Asia (Kandy, Sri Lanka, 19–22 March 1996)*, pp. xxv-xxxv. Colombo, Natural Resources, Energy and Science Authority of Sri Lanka and MAB National Committee.
Hills, E. S. (ed.). 1966. *Arid Lands: A Geographical Appraisal*. London, Methuen & Co., and Paris, UNESCO.
Holdgate, M. 1999. *The Green Web: a union for world conservation*. London, Earthscan Publications.
Huxley, J. 1961. Wild life as a world asset. *UNESCO Courier*, September 1961, pp. 15-16.
Journeaux, A. (ed.). 1987. *Integrated Environmental Cartography: A Tool for Research and Land-Use Planning*. MAB Technical Notes 16. Paris, UNESCO.
Kramer, P. 1973. Wildlife conservation in the Galapágos Islands (Ecuador). *Nature and Resources*, Vol. 9, No. 4, pp. 3–10.

Maguire, B. 1973. The Organization for 'Flora Neotropica'. *Nature and Resources*, Vol. 9, No. 3, pp. 18–21.

Nicholson, M. 1972. *The Environmental Revolution: A Guide for the New Masters of the World.* Chapter 9, Towards worldwide action. Harmondsworth, UK, Penguin.

Peterson, R. W. 1974. Towards a bill of ecological rights. *Nature and Resources*, Vol. 10, No. 4, pp. 2–5.

Petitjean, P. and Domingues, H. M. B. 2000. A redescoberta da Amazônia num projeto da UNESCO [Rediscovering the Amazon Region with a UNESCO project]. *Estudios Históricos*, Vol. 14, No. 2, pp. 265–292.

Roberts, J. L. 1979. Reflections on environmental hazard. *Nature and Resources*, Vol. 15, No. 4, pp. 24–30.

Rzóska, J. 1969. Hydrology and ecology: control of aquatic vegetation. *Nature and Resources*, Vol. 5, No. 2, pp. 16–17.

——. 1970. Productivity problems of fresh waters. *Nature and Resources*, Vol. 6, No. 3, pp. 15–17.

Salishchev, K.A. 1967. The scientific basis of Soviet integrated atlases. *Nature and Resources*, Vol. 3, No. 4, pp. 13–16.

——. 1972. National atlases of natural resources. *Nature and Resources*, Vol. 8, No. 4, pp. 20–27.

Saunders, G. 1972. Detritus in aquatic ecosystems. *Nature and Resources*, Vol. 8, No. 4, pp. 17–19.

Slatyer, R. 1993. Interview with Australian Academy of Science ('Interviews with Australian scientists' programme), conducted by Max Blythe, of the Medical Sciences Video-archive of the Royal College of Physicians and Oxford Brookes University in the United Kingdom. http://www.science.org.au/scientists/rs.htm

Swarbrick, J. 1955. One-quarter of the Earth is arid. *UNESCO Courier*, Nos 8–9, August/September 1955, pp. 4–5.

UNESCO. 1947. Report on the Question of United Nations Research Laboratories and Observatories. Report submitted to ECOSOC (20 February 1947). In: United Nations, 1948, *The Question of Establishing United Nations Research Laboratories*. Lake Success, New York, United Nations, pp. 37–110.

——. 1968. *Agroclimatological Methods: Proceeedings of the Reading Symposium/Méthodes agroclimatologiques: Actes du colloque de Reading.* Natural Resources Research Series 7. Paris, UNESCO.

——. 1973. *International Classification and Mapping of Vegetation.* Ecology and Conservation Series 6. Paris, UNESCO.

——. 1974. *FAO/UNESCO Soil Map of the World*, 1:5,000,000, 10 vols. Paris, UNESCO.

White, F. 1983. *Vegetation of Africa: a descriptive memoir to accompany the UNESCO/AETFAT/UNSO Vegetation Map of Africa.* Natural Resources Research Series 20. Paris, UNESCO.

White, G. (ed.). 1956. *The Future of Arid Lands. Papers and Recommendations from the International Arid Lands Meetings.* Publication No. 43. Washington DC, American Association for the Advancement of Science.

——. 1961. *Science and the Future of Arid Lands.* Paris, UNESCO.

——. 1970. Unresolved issues. In: Dregne, 1970, pp. 481–91.

Worthington, E. B. 1971. Ecology and conservation: Jamaica. *Nature and Resources*, Vol. 7, No. 1, pp. 2–9.

PART III: ENVIRONMENTAL SCIENCES

THE ESSENCE OF LIFE
UNESCO initiatives in the water sciences
Sorin Dumitrescu[36]

BEFORE 1961, the governing bodies of UNESCO did not recognize water as a specific field of interest to the Organization and its Member States. Yet, by their very nature, water resources constitute a subject requiring crucial international cooperation. This cooperation fully corresponds with the purposes and functions of UNESCO as outlined in its Constitution. Over the years, the planning and management of water resources have formed a critical problem of economic and social development, a *sine qua non* condition for combating poverty and sustaining decent living standards for people throughout the world. Thus it was inevitable that water-related activities began to appear in UNESCO's programme in the early 1950s; and, from about 1965 onward, these grew steadily to constitute in recent years a top priority area.

ARID LANDS AND WATER (1948-1964)

In 1948, the General Conference adopted at its third session a resolution aiming at the creation of an International Institute of the Arid Zone. A committee of experts was set up to study the programme and structure of such an institute. Meeting at the end of 1949, the Committee considered that the conditions were not propitious for the creation of the Institute. The experts proposed instead to set up an International Council for the problems relating to the resources of arid and semi-arid regions. The General Conference, at its fifth session in 1950, endorsed the proposal and decided to set up an interim International Arid Zone Research Council for promoting research and development. The Council met in November 1950. Its recommendations led to the establishment of the Consultative Committee on Arid Zone Research, which was meant to advise the Director-General on the preparation and implementation of projects in that area. The Consultative Committee remained in existence until 1964. Under its guidance, what became known as the UNESCO Programme on the Arid Zone took shape.

The members of the Consultative Committee represented various disciplines, including hydrology and water development. Within the framework of the

---

36   Sorin Dumitrescu: Chairman, International Hydrological Decade Coordinating Council (1967–69); joined UNESCO in 1969, as Director of the Office of Hydrology (subsequently Division of Water Sciences); Deputy Assistant Director-General for Science, focal point environmental activities (1985–88), Assistant Director-General from July 1988 to January 1989.

Programme, experts were requested to prepare state-of-the-art reports, which were then presented at international symposia. The reports and the procedure of symposia were published by UNESCO under the heading Arid Zone Research (in total, thirty publications).

The first publication reported on arid areas in eight geographic zones of the world. The second contained the papers presented at the Symposium on Arid Zone Hydrology (Ankara, Turkey, 1951). In 1959, a third publication concerned progress in Arid Zone Hydrology. Others in the series included the Use of Saline Waters (1954), Salinity Problems in the Arid Zone (1961), Plant–Water Relationship in Arid and Semi-Arid Conditions (1960), Evaporation Reduction (1965), and Physical Principles of Water Percolation and Seepage (1968).

The state-of-the-art reports and the papers presented at symposia drew attention to the prominent role of water resources research and management in the overall development of arid regions. As never before, international experts and local specialists were able to share their knowledge on subjects such as groundwater and water salinity. In addition, water-related consultant missions were sent to developing countries, and fellowships (usually of six months) were granted for the training of hydrologists. Thus, from 1951 to 1958, twenty-three fellowships were granted to specialists from developing countries in the area of hydrology, hydrogeology and hydraulics.

The success of the Arid Zone Programme led the General Conference, at its ninth session in 1956, to decide that one of the three principal concentration areas of UNESCO's 'Major Projects' be devoted to scientific research on arid lands. For this Project, assistance was provided for the establishment and functioning of national research centres, of which some (e.g. the Negev Desert Institute at Beersheba, Israel; and the Geophysical Research Institute at Quetta, Pakistan) included water-related activities. In 1962, under a UNESCO special fund project, the Research Centre for the Utilization of Saline Waters for Irrigation (CRUESI) was created in Tunisia. Following UNESCO's recommendations, national committees were created for coordination of activities related to the Project. A General Symposium on arid zone problems was organized in Paris in 1960 to take stock of the progress achieved under the Major Project in various fields. Among the papers presented at the Symposium, were reports on: surface water hydrology and sedimentation processes, hydrology, hydrometeorology and climatology, and desalting. Jointly with the Food and Agriculture Organization of the United Nations (FAO), an International Manual on irrigation and drainage of arid lands in relation with salinity problems was prepared.

The Arid Zone Programme constituted a milestone in the development of UNESCO programmes in the fields of natural resources. The water-related activities described above can, in fact, be considered as a 'sub-programme on water', although not officially defined as such, within the framework of an interdisciplinary programme that prefigured the subsequent intergovernmental programmes in hydrology and ecology. Retrospectively, one could consider that the emphasis on water was quite natural in a programme dealing with arid lands. However, at that time, in the absence of any major international venture in the field of water resources, what was achieved is truly remarkable.

THE INTERNATIONAL HYDROLOGICAL DECADE

PREPARATORY WORK, 1961-1964

In implementing its programme on the arid zone, UNESCO had established a fruitful cooperation with other UN organizations – in particular FAO and the World Meteorological Organization (WMO) – and with non-governmental organizations such as the International Association of Scientific Hydrology (IASH). UNESCO provided assistance for the organization of an IASH symposium on groundwater resources in the arid zone (Athens, Greece, 1961). On the occasion of the symposium, Raymond L. Nace, a hydrologist at the United States Geological Service, proposed the launching of a research and educational programme in hydrology and suggested that UNESCO take the

Studying plants' resistance to salt water and reaction to heat, radiation and drought (1964). One of the early field projects on water resources was assisting the Tunisian authorities in setting up the Research Centre for the Utilization of Saline Waters for Irrigation (CRUESI). UNESCO's role included the deployment of seven experts, study grants, and the purchase of equipment for laboratories and experimental basins (1964-67).

lead in its development. He mentioned that such a programme might have a ten-year duration.

The proposal for an International Hydrological Decade (IHD) grew out of the recognition: (1) that certain highly significant hydrological problems and phenomena can be studied effectively only on an international scale, and (2) that foreseeable crises in human affairs due to water problems could be met or forestalled only by applying extraordinary stimulation to hydrologic studies and the development of adequate hydrological services (Nace, 1964).

The proposal was endorsed by the Executive Committee of the International Association of Hydrological Sciences (IAHS). Soon afterward, the US delegate to the sixtieth session of UNESCO's Executive Board (November 1961) relayed the proposal. After examining the matter, the Executive Board adopted Decision 11.1 (IV), to the effect that:

> Considering that hydrology; the science of the waters of the Earth, occupies an area in the world of science that can make a great beneficial impact on society;
> Considering that the occurrence and distribution of water in any country is a consequence of the circulation of water in the whole planet and therefore that water, among all the world's resources, occupies a position especially adaptable to international cooperation;
> Recommends (i) that the Acting Director-General include provisions in the Proposed Programme and Budget for 1963–1964 to convene an intergovernmental conference to explore the ways in which Member States might cooperate in developing coordinated international research and training programmes in the field of scientific hydrology ...

This decision constitutes the first statement by the governing bodies of UNESCO regarding the importance of international cooperation in the field of water resources and the inclusion of this domain in UNESCO programmes and budgets.

The intergovernmental conference was scheduled to take place in spring 1964, to be preceded by a preparatory meeting of experts in 1963. In 1962, the Secretariat conducted intensive consultations with national authorities, UN organizations and NGOs (in particular the International Council for Sience [ICSU] and affiliated bodies), with a view to ensuring their participation in the programme.

The Executive Committee of WMO, at its meeting in May 1962, expressed interest in participating in a hydrological decade. The Advisory Committee on Arid Zone Research, at its session in Tashkent (Uzbekistan, USSR) in August

1962, expressed strong support for the proposed programme. It stressed both the scientific and practical objectives and recommended that UNESCO devote particular attention to the education and training of hydrologists, something that was deficient throughout the world. The UNESCO Secretariat started the preparation of a draft programme for the proposed decade with the assistance of three consultants: R. L. Nace (USA, rightly considered the founder of the IHD), V. I. Korzun (USSR), and J. Rodier (France) (see Box III.3.1).

IMPLEMENTATION OF THE INTERNATIONAL HYDROLOGICAL DECADE (1965-74)

The Decade started, officially, in January 1965. However, by that date, activities were already under way. Numerous countries had established National Committees for the International Hydrological Decade, which were planning their contribution to the programme. International organizations that had pledged their participation in the IHD were making their own plans. Thus, the first IHD symposium on hydrological networks, organized jointly by WMO and IASH with the support of UNESCO (to be held in Quebec), was in an advanced stage of preparation.

The IHD programme followed the recommendations adopted at the preparatory meetings in 1963 and 1964, with small amendments introduced by the IHD Coordination Council at its successive meetings. The programme included a definition of hydrology: 'Hydrology is the science which deals with waters of the Earth, their occurrence, circulation and distribution on the planet, their physical and chemical properties and their interactions with the environment, including their responses to human activity' (US Committee for Scientific Hydrology, 1962).[37]

Inclusion of this definition in the IHD programme document was justified by the fact that, at the beginning of the Decade, hydrology was still a young science and there was no unanimous acceptance of the term. Thus, in the countries of Central Europe, 'hydrology' meant the study of groundwater (known now as geohydrology or hydrogeology), while surface-water hydrology was designated as 'hydrography'. In this situation, one of the priority tasks of the IHD was to bring about a standardization of terminology. UNESCO and WMO undertook, with the assistance of a panel of experts, the preparation of an International Glossary of Hydrology in four languages, which was published in 1974 and particularly welcomed by the hydrological community all over the world. This work was complemented by the publication in 1978 of an International Glossary of Hydrogeology, prepared by another IHD panel of experts with the assistance of the International Association of Hydrogeologists (IAH).

---

37 This definition reproduces, with minor changes, the one formulated by the Committee for Scientific Hydrology of the US Federal Council of Science and Technology (1962).

BOX III.3.1: AN EFFECTIVE TROIKA OF CONSULTANTS

Extracts from *The UNESCO Water Adventure: From desert to water*, by Michel Batisse (2005)

The next step was to prepare the documents that would help obtain the General Conference's approval to hold the two meetings in 1963 and 1964, as recommended by the Executive Board. But here, the conventional procedure was to take place in two phases. In order to have the programme examined by the most qualified advisors possible, the Board requested, as an experiment, that commissions of experts be convened a week before the General Conference in order to prepare the General Conference decisions. The hydrological programme was one of four topics chosen for this exercise. The documents to be prepared were going to be scrutinized by a group of experienced hydrologists and not just by the usual delegates. Careful attention thus had to be paid to the drafting of these documents.

Very fortunately, and as had been agreed, Nace accepted to come to Paris for over a month to help. But in light of my experience with the Major Project and in keeping with the spirit of our discussions in Moscow, I felt it was indispensable to call in a Soviet expert to help prepare the documents, to accentuate the accord of the two 'superpowers' of the day. Kodva thought so too and called in someone from the executive management group of the Hydrometeorological Department, who was none other than Korsun. With a view to further smoothing rough edges and diversifying the points of view, I also asked Jean Rodier, head of the Hydrological Department of the French Office of Overseas Scientific and Technical Research (ORSTOM), to complete the team.

During their weeks of working together, this trio of consultants were to display a remarkable spirit of cooperation and a great desire to achieve results. This said, everything would appear to conspire to separate them: nationality, mother tongue, and more importantly experience. Nace was a hydrogeologist and was only familiar with the United States. Korsun only dealt with surface water and only knew the Soviet Union. But fortunately Rodier was familiar with Africa and the problems of developing countries. He was convivial and brought a touch of lightness to the interminable discussions that often took place in my office, my only contribution being simply to keep the discussion from getting bogged down in too much technical detail. On his part, Korsun had a great deal of common sense and despite his rather austere manner, was not devoid of humour, however subtle, which sometimes nonplussed Nace a little ...

The work of the three consultants primarily consisted of drafting a strategic document to convince the commission of experts that was to meet at the beginning of November of the merits of an international programme, without attempting to precisely define either its content – a task for intergovernmental meetings – or the mechanisms for its implementation, which is always an extremely delicate question. A second rather short text was also needed to give an idea of the topics that could be included in the proposals the Secretariat would have to write up for the preparatory conference of specialists planned for February or March 1963. A list of the major hydrological questions was consequently drawn up; it included the following nine topics: hydrological networks, discharges of rivers, changes in riverbeds and sedimentation, precipitation, evaporation, groundwater, water balance of basins, hydrological forecasting and man's impact on continental waters.

The Preparatory Meeting 'on the long-term programme of research in scientific hydrology' took place in May 1963. Forty-eight Member States and a number of governmental and non-governmental international organizations were represented. The Meeting recognized the need for an international programme in hydrology; the overall objective:

> to accelerate the study of the regimen and resources of waters with the view to the rational management in the interest of mankind, to promulgate the need of hydrological research and education in all countries and to improve their ability to evaluate and utilize their resources in the best possible way. That is, the programme will focus on science but will give strong consideration to utilitarian factors.

The Meeting considered that: 'To be fully effective and to have the desirable coverage, the proposed long-term programme should be spread over a minimum period of ten years, from 1965 onwards, and this is why the programme should be named the "International Hydrological Decade"'. Subsequently, this name and its acronym IHD were used in all references to the programme.

One could note that in 1963 the concept of a 'decade' for implementing a thematic action plan was rather unusual. Subsequently, the concept was frequently used by the UN system. In the field of water, two other decades were launched over the years.

The Preparatory Meeting adopted an outline of the scientific activities to be carried out during the IHD. It also emphasized the paramount importance of education and training in hydrology. In this respect, in his opening speech, the Director-General, René Maheu, had stressed that one cannot 'do' hydrology without hydrologists and without training facilities.

On the basis of the recommendations formulated by the Preparatory Meeting and of additional documents worked out by the Secretariat, the Intergovernmental meeting of experts, which took place in April 1964, adopted the final programme of the IHD. The Intergovernmental meeting also identified the three basic elements of the mechanism needed for the implementation of the Decade's programme:

- the establishment, in each participating State, of a National Committee for the IHD in charge of planning and implementing of all national activities within the Decade's framework;
- the creation at intergovernmental level of an IHD Coordinating Council composed of representatives of twenty-one Member States elected by the General Conference (in 1970 the number of members of the Council was raised to thirty);
- the cooperation with international governmental and non-governmental organizations in the execution of the various IHD activities.

At its thirteenth session, in November 1964, the General Conference approved these recommendations and declared open the IHD 'as a worldwide enterprise of scientific cooperation among nations'. Thus, less than twenty years after the creation of UNESCO, hydrology (and water, more generally) was definitively recognized as an important programme, laced squarely among the Organization's fields of competence (Batisse, 2005).[1]

---

1   One has to mention the outstanding role – in the whole preparatory process that led to this result – of our late lamented colleague Michel Batisse. He was, from 1956, the coordinator of the Major Project on Arid Lands, and then, from 1961, head and subsequently director of the Natural Resources Division, which had under its responsibility (until 1969) hydrology.

The research themes of the IHD comprised, among others, the following:

Water balances
The adequate estimation of the freshwater resources of the Earth, as of the various components of the world water balance (as well as continental and local balances), was considered a main objective of the Decade. It should be noted that previous estimates of these components, in particular concerning river flow, showed differences of as much as 50 per cent (of the lowest figures). The IHD Council established a working group on world water balance to prepare a guide (published in 1974) on balance computations. The USSR National Committee issued, as an IHD publication (also in 1974), an authoritative monograph on the subject. This was subsequently translated into English and published by UNESCO in 1978, under the title *World Water Balance and Water Resources of the Earth*.

Network planning and design
No worldwide study of water resources and hydrological processes would be possible without adequate observation data. In the early 1960s, national networks of hydrological observation and measurement stations were often deficient (especially in developing countries), with poor geographical coverage and unreliable measurement techniques. The improvement of data acquisition systems was, therefore, considered a priority. Special attention was given to hydrological elements such as evaporation, groundwater, sediment transport, and water quality, for which data were particularly scarce. Participating countries were invited to select a number of stations, representative for their hydrological conditions, to be designated as Decade stations, for which they would pledge adequate operation and supply of reliable data.

Representative and experimental basins
'Representative basin' and 'experimental basin' are concepts which existed prior to the launching of the Decade. The basic idea is to carry on intensive observations and investigations within relatively small catchment areas, in order to obtain information applicable to a broader region. However, until the early 1960s, only a few countries used the approach. The IHD programme recognized these concepts as an important tool for supplementing the limited hydrometric data available and for developing the study of hydrological processes. During the Decade, the number of representative and experimental basins (R&EB) increased throughout the world to reach, at the end of the ten-year period, a reported figure of about 800. An international guide on research in, and practical applications of, R&EB was prepared and published in 1970. Later on, the R&EB were adopted as a key link in the investigation scheme of the hydrological regimes.

Snow and ice
Glaciers and permanent snow cover represent 69 per cent of the world's total freshwater resources. It was therefore fully justified for the IHD to include snow and ice investigations as a major section of its programme. Three main projects were carried out: the global inventory of perennial and annual snow and ice masses; a worldwide study of glacier variations (this project being of particular relevance to the study of climate changes); and estimation of water-, ice-, and heat balance at selected representative glacier basins. The coordination of these activities was entrusted to the IASH International Commission on Snow and Ice (ICSI).

Education and training
Education and training constituted a top priority area of activities under the IHD, based on the principle 'no hydrology without hydrologists'. Since in most countries no specialized education in hydrology was available, the IHD Council – through its Working Group on Education – undertook an evaluation of the situation and identified the most urgent needs. Methodological guides were prepared and issued on: *The Teaching of Hydrology*; *Teaching Aids*; *Curricula and Syllabi*; and *Textbooks in Hydrology*. To cope with the lack of adequately trained hydrologists, it was decided to put emphasis on training at the postgraduate level. UNESCO provided financial assistance as seed money for the establishment and functioning of ten postgraduate courses in hydrology in various host countries. More than 1,500 specialists – mostly from developing countries – were trained in such courses. In addition, training courses for hydrology technicians were organized in Africa.

UNESCO'S PARTNERS IN THE IHD
The International Hydrological Decade was conceived as a joint venture of several UN organizations and NGOs concerned with hydrology. Within the UN system, UNESCO's closest partner was the World Meteorological Organization (WMO). WMO was associated with all planning and implementation phases of the Decade, and it assumed the lead role in activities related to hydrological networks' methods of measurement and hydrological forecasting. In 1973, a working agreement was signed, delineating the responsibilities of UNESCO and WMO in the field of hydrology, and providing mechanisms for cooperation, including the convening of joint international conferences.

Close cooperation was also established with FAO (concerning, in particular, the human influence on hydrological processes), the World Health Organization (for aspects related to water quality), the International Atomic nergy Agency (IAEA, concerning the application of nuclear techniques in hydrology), the UN Secretariat, the UN Economic Commissions and, later on, the United Nations Environment Programme.

A number of regional intergovernmental organizations also participated in certain programme activities.[38] Among NGOs, the main partners were ICSU and affiliated associations, as well as the various organizations affiliated with the Union of International Technical Associations (UITA). Special mention must be made of the International Association of Scientific Hydrology (IASH, renamed in 1972 as the International Association of Hydrological Sciences, IAHS). As noted above, the idea of a hydrological decade was first promoted in an IASH symposium. The secretary-general of IASH, Léon Tison, was a strong supporter of the IHD and participated in the preparatory process as well as in the implementation phase. IASH/IAHS contributed to most IHD research projects. Another association affiliated with ICSU, the International Association of Hydrogeologists (IAH), participated in all activities related to groundwater.

ICSU established in 1964 a Committee on Water Research (COWAR) to provide scientific advice to the IHD Council and to promote an integrated approach to water-related issues among its various scientific associations. COWAR later became a joint body of ICSU and UATI, but its real contribution did not equal the direct inputs from the individual associations cited above.

MAIN ACHIEVEMENTS OF THE IHD

The major achievement of the IHD was to create awareness at the national and international levels about water resources, and to mobilize efforts for the implementation of a worldwide research and educational programme in this field. Hydrology, which was relatively unknown in many countries – either as a science or as a profession – received universal recognition.

One hundred and eight Member States officially participated in the programme and established IHD National Committees or focal points. The efficiency of National Committees greatly varied from one country to another, but their mere existence was an event in itself, confirming that hydrology had begun to be considered an activity of national significance. During the Decade, in many countries the development of measurement networks was markedly enhanced and the quantity and reliability of hydrological data substantially increased.

Cooperation on scientific and practical aspects among the participating countries, both East–West and North–South, was fostered and contributed to the fruitful exchange of experience and transfer of knowledge. Scientists from around the world were able to meet in forty international symposia, an unprecedented figure not only for hydrology but for other disciplines.

---

38   Notably the Arab Center for the Studies of Arid Zones and Dry Lands (ACSAD,) the Arab League Educational, Cultural and Scientific Organization (ALECSO), the International Commission for the Hydrology of the Rhine Basin (CHR) and the Comité Regional de Recursos Hidráulicos (CRRH).

Hydrology was recognized as a distinctive geoscience and, due to the development of its various branches, its domain became defined as 'hydrological sciences'. Scientific knowledge and research have since been developed in practically all subject areas of hydrology: study of water balances, research on hydrological processes in experimental basins, flood and low flow studies, sediment transport, groundwater, and so on.

The many hydrologists trained by the IHD-sponsored postgraduate courses helped create and consolidate national hydrological services in developing countries and, for the author, it was a rewarding opportunity to frequently meet such former trainees at various technical and executive levels in their respective countries. For the UNESCO Secretariat, the IHD constituted a valuable experience – particularly with regard to the implementation mechanism – for launching other intergovernmental programmes.

One could state that no other 'decade' or 'year' launched subsequently within the UN system so influenced a specific area of activity and led to so many practical results. In 1964, R. L. Nace expressed the hope that the Decade 'may also usher in the New Age of Hydrology'. His wish was largely fulfilled.

The Regular Programme activities under the IHD were complemented by the execution of water-related projects under extrabudgetary funding (mainly UNDP). Such projects mainly concerned: the assessment of water resources in various geographic areas (Lake Chad basin; the Canary Islands; the upper basin of Paraguay River, including the Pantanal area; the groundwater resources of Northern Sahara); capacity-building in the field of water research and education (National Centre for Hydraulics, Ezeiza, Argentina; Centre for Applied Hydrology, Porto Alegre, Brazil; National School of Engineering, Bamako, Mali; the Hydraulic Research Station, Wad Medani, Sudan, etc.); and technician training courses for countries in north-east Africa. Most of these projects had a significant impact on the technical and economic development of the recipient countries.

THE INTERNATIONAL HYDROLOGICAL PROGRAMME (1975 TO THE PRESENT)

FROM SCIENTIFIC HYDROLOGY TO WATER RESOURCES ASSESSMENT AND MANAGEMENT

The need for continuing the efforts undertaken within the IHD through a long-term international programme in the field of hydrology was recognized by the International Conference on the Practical and Scientific Results of the IHD and on International Cooperation in Hydrology, held in December 1969 (known as the Mid-Decade Conference). Its recommendations were approved by the General Conference of UNESCO at its seventeenth session (1970).

The IHD Coordinating Council and the Director-General were requested to submit proposals to be considered by the General Conference at its next session. The idea was strongly opposed by the Assistant Director-General for Natural Sciences, A. Buzzati-Traverso, who believed that all UNESCO's activities in the field of water should be transferred to WMO. Amazingly, Michel Batisse himself was not in favour of a continued intergovernmental programme (he later conceded that this was a mistake [Batisse, 2005]). However, the Director-General, René Maheu, followed the advice given by the director of the newly created Office of Hydrology (the author had been appointed to that post in November 1979) and submitted the recommendations of the IHD Council to UNESCO's governing bodies.

Why was it necessary to capitalize on the work done by the IHD and develop an open-ended long-term international programme on hydrology and water resources? Freshwater resources, like oceanography or ecology, constitute a subject area where international cooperation is needed on a permanent basis. From the outset, it was obvious that:

> in ten years no substantial share of the world's scientific or practical water problems will be solved ... Fortunately, it is expected that after 1974 essential cooperation will continue under its own impetus and inertial guidance. Hydrologists, like other scientists, look to the time when international cooperation will be the order of the day rather than a special event (Nace, 1964, p. 415).

In 1972, the General Conference (through its Resolution 18C/2.32) instructed the Director-General to:

> take appropriate measures for the implementation, beginning in 1975, in full cooperation with those other organizations of the United Nations system whose competence relates to hydrology, of a long-term programme of intergovernmental cooperation in hydrology, called the International Hydrological Programme (which is intended to contribute) to the general development of hydrological activities throughout the world.

Two years later, in September 1974, another Intergovernmental Conference (the 'End-of-Decade Conference') took place, jointly convened by UNESCO and WMO, which adopted an outline plan for implementation of the International Hydrological Programme (IHP) for 1975–80. Subsequently, the General Conference at its nineteenth session approved this plan and elected the thirty members of the new IHP Intergovernmental Council.

Some people believed that the new long-term programme was a mere continuation of the Hydrological Decade. To a certain extent this was true. The IHP inherited from the IHD a number of research and educational themes. It also inherited the mechanism that proved so successful during the Decade: the Intergovernmental Council, the fruitful partnership with various international organizations and the National Committees ensuring the participation of a great number of countries in the Programme. But over the thirty years of its existence, the IHP has had its own dynamics: some earlier themes were becoming less interesting for international cooperation, and new themes emerged as vitally important problems at national and international levels. While the qualifier 'hydrological' remained in the title of the Programme, a shift was gradually made from the traditional subjects of scientific hydrology to the practical aspects of water resources assessment and management.

Already in 1974, in his opening address at the End-of-Decade Conference, Deputy Director-General John Fobes pointed out that the new Programme 'should respond to the needs of tomorrow and should contain therefore some major innovations ... taking into account the major water-related problems facing the world'. He added:

> I believe strongly that your plan should reflect, even more than the plan which is before you, what may be called a UNESCO approach to the problems of the contemporary world. I mean by this an approach which tries to integrate the contributions from the environmental sciences, the social sciences, culture, education and the communication media in the service of human progress – which today means survival, social justice and quality of life.

These words were prophetic, but it took twenty-five years to fully achieve that objective.

It was agreed that the IHP would be planned in successive phases. Under the plan for the first phase – IHP-I (1975–80) – a number of research projects initiated during the IHD were continued and developed (water balances, computation of average, maximum and minimum river flow under various natural conditions, representative and experimental basins, etc.). Among the new lines of emphasis, one should mention projects concerning the following topics:

- hydrological and ecological effects of human activities, and their assessment;
- hydrological and ecological aspects of water pollution; and
- urban hydrology, including socio-economic aspects.

IHP-II was designated for a shorter duration (1981–83), in order to enable, in the following years, integration of the successive phases of the IHP in the Medium-Term Plans of UNESCO. New research subjects included:

- the preparation of reports on the status of knowledge on the hydrology of humid tropical zones and on the hydrology of arid and semi-arid regions;
- the definition of hydro-environmental indices to be used in the assessment of the environmental impacts of water projects;
- rational use and management of water resources in coastal areas; and
- methodologies to ensure optimal allocation of water resources among users.

Parallel to the normal IHP activities, in accordance with the decision of the General Conference at its twenty-first session concerning the launching of 'major regional projects', three such projects were started in the field of water. The purpose of these projects (conducted in Latin America and the Caribbean, the Arab States and Africa) was to seek the most appropriate ways of developing and conserving water resources in order to meet the economic and social needs of rural communities, placing emphasis on low-cost technologies. The projects produced interesting results, with immediate practical application, but the expected extrabudgetary financial contributions did not materialize.

IHP-III (1984–1989)

The general title of this programme phase was Hydrology and the Scientific Bases for the Rational Management of Water Resources for Economic and Social Development. It included eighteen themes, grouped under four major sections:

- Section I: hydrological processes and parameters for water projects;
- Section II: influence of man on the hydrological cycle;
- Section III: rational water resources – assessment and management;
- Section IV: education and training, public information and scientific information systems.

Theme 4 ('Hydrology of particular regions and land areas') deserves special mention. It stressed the need to take into account specific features of the hydrological regime under different climatic, geomorphologic and geological conditions. It marked the first time that international research addressed various aspects concerning the hydrology of flatlands or the hydrology of small islands. One of the remarkable outputs of Theme 4 was the preparation and publication of a monograph on *Comparative Hydrology* (Falkenmark and Chapman, 1989). Theme 10 ('Methodologies for integrated planning and management of water resources') and Theme 11 ('Systems management for redaction of negative side-effects of water resources developments') were representative of the new emphasis placed on the practical aspects of water management related to socio-economic development.

IHP-IV (1990-1995)

IHP-IV comprised three sub-programmes:

- H: Hydrological Research in a Changing Environment
- M: Management of Water Resources for Sustainable Development
- E: Education and Training and the Transfer of Knowledge and Information

The content of the programme was strongly influenced by the international debate on global change and sustainable development which followed the publication of the Brundtland Report and the preparation of the IPCC report on Climate Change. Special attention was given to the humid tropics and to arid and semi-arid zones. The need for a holistic approach was also emphasized by undertaking activities with policy-makers and planners within the framework of integrated water resources management.

IHP-V (1996-2001)

Eight themes were retained for this IHP phase, under the general heading 'Hydrology and Water Resources Development in a Vulnerable Environment':

- Theme 1. Global Hydrological and Geochemical Processes;
- Theme 2: Ecohydrological Processes in the Surficial Environment;
- Theme 3: Groundwater Resources at Risk;
- Theme 4: Strategies for Water Resources Management in Emergency and Conflict Situations;
- Theme 5: Integrated Water Resources Management in Arid and Semi-Arid Zones;
- Theme 6: Humid Tropics and Water Management;
- Theme 7: Integrated Urban Water Management;
- Theme 8: Transfer of Knowledge Information and Technology.

During the whole period under review, the IHP became increasingly intertwined with other activities carried out in the field of water by organizations of the UN system. UNESCO contributed to the preparation of the first United Nations Conference on Water held at Mar del Plata, Argentina, in March 1977. This conference, attended by representatives of 116 governments and of many international organizations, had a paramount importance in the international recognition of water as a key factor in socio-economic development. The Mar del Plata Action Plan recommended that countries should 'ensure that national water policy is conceived and carried out within the framework of an interdisciplinary national economic, social and environmental development policy'. The Action Plan identified the IHP as one of the international efforts contributing to a better assessment of water resources as a prerequisite of rational water management.

Ten years later, the World Commission on Environment and Development, appointed in 1983 by the UN General Assembly, published a report, *Our Common Future*. Its publication constituted a milestone in the adoption, at the global level, of the concept of sustainable development. Surprisingly, in spite of the fact that no human and economic development can be sustained without adequate water supply, freshwater issues were almost ignored in the report. This is a direct consequence of the fact that none of the twenty-three members of the Commission was familiar with water resources development and management.

In 1992, fifteen years after Mar del Plata, water resources were again brought to the forefront of the global debate on sustainable development. In January of that year, UNESCO was one of the twenty bodies and agencies of the UN system which organized (in Dublin, Ireland) the International Conference on Water and the Environment (ICWE). The Dublin Statement enunciates four basic principles:

1. Freshwater is a finite and vulnerable resource, essential to sustain life, development and the environment.
2. Water development and management should be based on a participatory approach, involving users, planners and policy-makers at all levels.
3. Women play a central part in the provision, management and safeguarding of water.
4. Water has an economic value in all its competing uses and should be recognized as an economic good.

The ICWE was designed as an input to the UN Conference on Environment and Development ('The Earth Summit') held in Rio de Janeiro, Brazil, in June 1992. The 'Agenda 21' adopted by the conference mentions, *inter alia*, that 'The holistic management of freshwater ... and the integration of sectoral water plans and programmes within the framework of national economic and social policy, are of paramount importance for action in the 1990s and beyond'. The above-mentioned principles are fully in line with the basic philosophy of the IHP.

The IHP and water-related activities during 1975–99 suffered from a lack of adequate funding. Despite the diversification of its activities, the relative importance of water sciences in the Natural Sciences Sector decreased over the last years of the century, in terms of both programme budget and personnel. The author of this chapter raised the problem of the water-resources programme's insufficient funding with the two directors-general of that period and insisted on the growing importance of water in the world and on the unique role that UNESCO could play in that area, by virtue of its multi-faceted competences. Both directors-general agreed, in principle, to the need for increasing the staff resources and budgetary allocations for the programme and on the usefulness of an intersectoral approach. Regrettably, they eventually followed the poor advice received from the successive assistant directors-general for Natural Sciences and the Secretariat unit responsible for planning; thus, the situation remained the same. This advice originated from the false assumption that 'environment' can be reduced to 'ecology', and from the lack of understanding of the global issues related to water and their relevance to UNESCO's mandate. The mistake appears even more striking when one recalls that, during the 1980s, replies to two questionnaires sent to Member States on the priorities they attached to the various programmes of the Organization clearly showed that Programme X.3 (Water Resources) enjoyed the highest rating.

WATER BECOMES A TOP PRIORITY (2000 ONWARDS)

This situation changed radically with the inception of Mr Koïchiro Matsuura's mandate as Director-General. He repeatedly emphasized the global importance of water and the role that UNESCO should play – at the crossroads of natural sciences, education, social sciences, ethics and culture – in order to efficiently contribute to the solution of water problems (see Box III.3.2). The Director-General took resolute measures to reinforce the budgetary allocation for Programme X.2 – Water and Associated Ecosystems. Out of the total budget for Major Programme X, 'Natural Sciences', this programme represented 7.6 per cent in the 30 C/5[39] (Approved), 17 per cent in the 31 C/5 (Approved) and 33 per cent in the 32 C/5 (Approved). If one takes into account only the programme activities (without staff costs), the respective percentages became 45 per cent in the 32 C/5 and 48 per cent in the Draft 33 C/5.

---

39 UNESCO's biennial Programme and Budget documents are referred to as C/5.

The priority attached by UNESCO in recent years to its water-related programme coincides with an unprecedented interest shown by the international community in water-related problems. Consider the several major international conferences on water held since 2000, such as the Second World Water Forum and the Ministerial Conference on Water Security in the Twenty-first Century (the Hague, March 2000), the International Conference on Freshwater and the ensuing Ministerial Declaration (Bonn, Germany, 2001), and the Third World Water Forum (Kyoto, Japan, March 2003). At the World Summit on Sustainable Development ('Rio+10', Johannesburg, South Africa, August/September 2002), the relevance of water issues for human health, environmental protection, and development strategies was duly acknowledged. One should recall that the United Nations Millennium Declaration stressed the need 'to stop the unsustainable exploitation of water resources, by developing water management strategies at the regional, national and local levels, which promote both equitable access and adequate supplies'. UNESCO was one of the lead agencies in the International Year of Freshwater (2003) and is contributing to the International Decade for Action 'Water for Life' (2005–2015), both of which were proclaimed by the United Nations General Assembly.

The first pillar of UNESCO's programme in the field of water remains the International Hydrological Programme. The fifth phase of the IHP was completed in 2001. Following a decision of the Intergovernmental Council, an external evaluation was carried out in 2002–03 by a team of five independent scientists. The evaluation report concludes that:

> taken as a whole, IHP-V appears to have achieved its objectives to a considerable degree. However, in more detail, its achievements seem to be unevenly distributed across the themes and regions. It stimulated the hydrological sciences, particularly through Theme 3, Groundwater resources at risk, and through the FRIEND project of Theme 1. It was praised for the transfer of knowledge motivated by Theme 8, for its publications and website, with the possible exception of Africa (UNESCO, 2003).

IHP-V continued to be a broadly based programme, which extended beyond the traditional fringes of hydrology into a number of allied areas. It was a phase of the IHP that attracted considerable support from the hydrological community. Like previous phases, IHP-V was a collaborative programme with 90 per cent of the effort and resources being committed nationally. Hence, the evaluation team stressed the key role of the IHP national committees and the need to strengthen them.

At present, the sixth phase of the IHP (2002–07) is being executed. IHP-VI is structured along five major themes, including a number of focal areas, as

BOX III.3.2: THE DIRECTOR-GENERAL ON THE PLACE OF WATER IN UNESCO'S PROGRAMMES

Extracts from the address by Mr Koïchiro Matsuura at the information meeting of Permanent Delegations on the occasion of the launching of the International Year of Freshwater, 12 December 2002.

The problem of water … comes at the crossroads between economic, ecological, social and – let us not forget – cultural issues. The interactions between and among these different areas need to be better understood if we are to have the proper scientific bases for action.…

In all cases, there is a need to develop and to promote new concepts of water resources management. Here again, what is needed is a comprehensive change of attitude about resources and the environment, where the long-term consequences are taken into account too. There can indeed be no doubt that if we are to master the water challenges ahead, we will need to have full recourse to our senses of human solidarity, compassion and ethical standards. Without this, we are jeopardizing the quality of life, or even the survival, of future generations.

From all these viewpoints, UNESCO's role is fundamental. Apart from the support for prospective studies in the area of water, for research in hydrology, the oceans and climate, and in the other disciplines concerned, as well as for innovation, the Organization can federate on the international level the drive for education, information and training in the area of water. Since education is a vital issue that lies at the very heart of the process of sustainable development, then education has to address all issues of water and place itself at the service of water. And education and capacity-building programmes must take fully into account – as UNESCO has long been advocating – the cultural values and culture-based traditions and techniques that are the precondition for successful water management.

[It is in this that] … lies UNESCO's strength and comparative advantage within the United Nations system, and this fact is being recognized widely. It is certainly true that the other specialized agencies have a clear mandate on specific callings bearing a great relationship to water: WHO in water and health, FAO in water and food, Habitat in water and cities, WMO in water and climate, and so forth. But it is equally true that nowhere more so than in UNESCO is the subject matter integrated in a holistic and transdisciplinary manner, ranging from science to ethics to consensus-building, and those principles and practices distilled into human capacity-building.…

In my inaugural address when I assumed my post as Director-General, one of the great world issues I recognized was the global importance of water and the role UNESCO could and should play in [this area] …

In the light of the serious nature of the water challenges in connection with UNESCO's mandate, I proposed to the Executive Board and the General Conference to set 'water resources and supporting ecosystems' as the principal priority for the Natural Sciences Sector, not only for the 2002–2003 biennium, but for the full Medium-Term period of 2002–2007.

indicated in Box III.3.3. Box III.3.3 indicates the shift in the lines of emphasis of the IHP, mentioned above. While Theme 5 – relating to education and training – remains a priority, the most innovative part of the programme is Theme 4, focusing on the inter-relationship between human welfare and water resources. The theme seeks to provide guidelines by which water can be managed in an equitable, sustainable, and ethical manner.

In the field of water-related ethics, the IHP has established a fruitful cooperation with the World Commission on the Ethics of Scientific Knowledge and Technologies (COMEST), which has established a sub-commission on the Ethics of Fresh Water.

> The social context of ethical questions concerning water tends to revolve around notions of water as a common good: water and its connection to human dignity and basic needs for life; water as a facilitator of well-being of people; rights and responsibility towards water access; water and social justice; and the wealth generating and development roles of water infrastructure ... There is no life without water, and those to whom it is denied are denied life. Water for all and meeting minimum basic needs are vitally tied to the principle of human dignity ...The principle of participation means that individuals, especially the poor, must not be shut out from participating in those institutions which are necessary for human fulfilment (Priscoli et al., 2004).

Under Focal Area 4.3, it is worthwhile mentioning the project 'From Potential Conflict to Cooperation Potential' (PCCP). It seems that this is the first case within the Natural Sciences Sector (with the possible exception of the IOC) when such a theme, with major political implications, has been included in the current programme. The purpose of the project is to foster cooperation among the various stakeholders in the management of shared water resources (there are more than 260 river basins falling under this category), in order to mitigate the risk of potential conflicts turning into real ones. The PCCP programme attempts to demonstrate that a situation with clear potential for conflict can be transformed into a situation in which the potential for cooperation can emerge. Many countries have shown interest in PCCP. An international conference on this subject was held in Delft (the Netherlands) in November 2002, and more than thirty publications have been issued. This project is being implemented in cooperation with Green Cross International.

Under IHP-VI, there are in addition two cross-cutting programme components, FRIEND and HELP as outlined below.

PART III: ENVIRONMENTAL SCIENCES

FRIEND

The overall objective of the FRIEND (Flow Regimes from International Experimental and Network Data) project is to ease the problems of water resources assessment and management through applied research targeted at regionally identified problems. The project is an international collaborative study intended to develop – through the mutual exchange of data – knowledge and techniques at a regional level and a better understanding of hydrological variability and similarity across time and space. The advanced knowledge of hydrological processes and flow regimes gained through FRIEND helps to improve methods applicable in water resources planning and management. FRIEND also provides support to

BOX III.3.3: THE STRUCTURE OF IHP-VI

Theme 1: Global changes and water resources
- Focal Area 1.1: Global estimation of resources – water supply and water quality
- Focal Area 1.2: Global estimation of water withdrawals and consumption
- Focal Area 1.3: Integrated assessment of water resources in the context of global land-based activities and climate change

Theme 2: Integrated watershed and aquifer dynamics
- Focal Area 2.1: Extreme events in land and water resources management
- Focal Area 2.2: International river basins and aquifers
- Focal Area 2.3: Endorheic basins
- Focal Area 2.4: Methodologies for integrated river basin management

Theme 3: Land habitat hydrology
- Focal Area 3.1: Drylands
- Focal Area 3.2: Wetlands
- Focal Area 3.3: Mountains
- Focal Area 3.4: Small islands and coastal zones
- Focal Area 3.5: Urban areas and rural settlements

Theme 4: Water and society
- Focal Area 4.1: Water, civilization and ethics
- Focal Area 4.2: Value of water
- Focal Area 4.3: Water conflicts – prevention and resolution
- Focal Area 4.4: Human security in water-related disasters and degrading environments
- Focal Area 4.5: Public awareness raising on water interactions

Theme 5: Water education and capacity-building
- Focal Area 5.1: Teaching – techniques and material development
- Focal Area 5.2: Continuing education and training for selected target groups
- Focal Area 5.3: Crossing the digital divide
- Focal Area 5.4: Institutional development and networking for W.E.T.

researchers and operational staff of hydrological services in developing countries, thereby contributing to their capacity to assess and manage their own national water resources. It thus contributes to the goal of providing a reliable supply of fresh clean water to the world's poor. A major achievement of regional FRIEND projects has been the establishment of regional databases and the sharing of data between participating countries.

HELP

HELP (Hydrology for the Environment, Life and Policy) is an initiative establishing a global network of basins to improve the links between hydrology and the needs of society. The vital importance of water in sustaining human and environmental health has been widely recognized by numerous national and international forums. However, no worldwide programme has addressed key water resources management issues in the field and integrated them with policy and management. HELP is designed to change this by creating a new approach to integrated basin management. By 2005, more than fifty catchments were established as demonstration or operational projects in about forty countries around the world.

New initiatives under IHP-VI include:

- The International Sedimentation Initiative (ISI). Erosion and sedimentation processes affect the management of catchments, river systems and reservoirs. It was estimated that within the next few decades more than half of the global storage capacity might be lost due to sedimentation. Under this initiative, an Advisory Group has been set up to provide guidance to all related activities of the IHP.
- The establishment of an International Groundwater Resources Assessment Centre (IGRAC) under the auspices of UNESCO and WMO. The Centre aims to tackle the general lack of information and awareness about the status of groundwater resources. It will act as a catalyst for stimulating national and regional efforts in the monitoring and assessment of aquifer systems. Arrangements are underway for establishing the Centre in Utrecht (the Netherlands).
- A project on International Shared Aquifer Resources Management (ISARM), aiming to compile a world inventory of transboundary aquifers and to develop wise management practices, is being implemented by UNESCO, IAH, FAO and the United Nations Economic Commission for Europe (UN/ECE).

Hydrological measurements for a flood event at Tougou in the Nakambé HELP basin, Burkina Faso.

- The establishment of an umbrella network: the Global Observatory of Units for Teaching, Training and Ethics of Water (GOUTTE of Water). This initiative is meant to foster intellectual cooperation among academic, scientific and professional water institutions, and to promote environmental ethics as a basic concept of education, training and public awareness.
- The launching of the Global Network of Water and Development Information for Arid Lands (G-WADI) in response to the urgent need for increased regional and international cooperation contributing to sustainable development of arid and semi-arid zones. Its primary aim is to build an effective community through integration of existing materials from networks, centres, organizations and individuals.

- The International Flood and Landslides Initiative. This interagency[40] initiative was launched by the Director-General at the World Conference on Disaster Reduction (Kobe, Japan, January 2005). The headquarters of the new project will be based at the Centre for the Water Hazard and Risk Management, hosted by the Public Works Research Institute in Tsukuba (Japan). The project is set to integrate the scientific, operational, educational and public awareness aspects of flood management, and aims at increasing disaster preparedness.
- The preparation of a comprehensive series of publications on world water history. The series intends to provide the public, academics and policy-makers with an overview of the role of water throughout the history of humankind. UNESCO supported the creation of the Water History Association, with which the IHP cooperates in this area.

Regional cooperation was enhanced, and several Member States assumed responsibility for the hosting of international and regional water-related centres under the auspices of UNESCO. Ten such centres are now functioning:

- International Resources and Training Center on Erosion and Sedimentation (IRTCES), Beijing, China
- Regional Humid Tropics Hydrology and Water Resources Center for South-East Asia and the Pacific (WRCSEAP), Kuala Lumpur, Malaysia
- Regional Water Center for the Humid Tropics of Latin America and the Caribbean (CATHALAC), Panama City, Panama
- Regional Center on Urban Water Management (RCUWM), Tehran, Iran (Islamic Republic of)
- Regional Center for Training and Water Studies of Arid and Semi-Arid Zones (RCTWS), Egypt
- International Center on Qanats and Historic Hydraulic Structures (ICQHHS), Yazd, Islamic Republic of Iran
- International Center for Water Hazard and Risk Management (ICHARM), Tsukuba, Japan
- Center for Water Law, Policy and Science, Dundee, United Kingdom

---

40   Cooperating in the initiative are: UNESCO, WMO, United Nations University (UNU), the United Nations Inter-Agency Secretariat of the International Strategy for Disaster Reduction (UN-ISDR) and IAHS.

- Regional Center for Arid and Semi-Arid Zones of Latin America and the Caribbean (CAZALAC), La Serena, Chile
- European Regional Center for Ecohydrology, Lódz, Poland

The number of these centres is larger than in any other UNESCO programme area, another indication of the priority attached by Member States to water. The second pillar of the water programme is UNESCO's participation in the World Water Assessment Programme (WWAP), created as a United Nations system-wide effort in response to the decision of the UN Commission on Sustainable Development. WWAP (which involves twenty-three UN agencies and programmes) works to develop the tools and skills needed to better understand the basic processes, management practices and policies for improving the quality and supply of global freshwater resources. Its Secretariat is provided by UNESCO, based on the extrabudgetary support from the Government of Japan. Several other countries have joined as donors. The programme's mission is to:

- assess the state of the world's freshwater resources and ecosystems;
- identify critical issues and problems;
- measure progress towards achieving sustainable use of water resources;
- help countries develop their own assessment capacity;
- improve water resources planning, policy and management; and
- document lessons learned, and publish a *World Water Development Report* at regular intervals.

The first *World Water Development Report* was published by UNESCO, on behalf of the UN system, in March 2003, in conjunction with the Third Water Forum in Kyoto. The second Report is under preparation, with the title 'Water, a Shared Responsibility'. It will be launched at the fourth World Water Forum in Mexico City (March 2006). The Forum will be organized in conjunction with the World Water Day (22 March), to be guided by the theme 'Water and Culture', under the leadership of UNESCO.

The third pillar, reflecting the absolute priority given in the water programme to education and training, is the UNESCO–IHE Institute for Water Education (see Box III.3.4). This is a unique arrangement under which an institute is established as part of UNESCO ('Category I' of UNESCO institutes and centres) and *entirely* financed from extrabudgetary sources. The Institute is now the world's largest graduate facility offering training in hydraulic engineering, sanitary engineering, urban-water infrastructure, hydroinformatics, water

> **BOX III.3.4: UNESCO-IHE INSTITUTE FOR WATER EDUCATION**
>
> UNESCO-IHE continues the work that was started in 1957 when IHE, based in Delft (the Netherlands), first offered a postgraduate diploma course in hydraulic engineering to practicing professionals from developing countries. Over the years, IHE has developed into an international education institute providing a host of postgraduate courses and tailor-made training programmes in the fields of water, environment and infrastructure; conducting applied research; implementing institutional capacity-building and human resource development programmes; participating in policy development; and offering advisory services worldwide.
>
> The Institute has gradually expanded its academic base to include disciplines such as sociology, economics, and environmental and management sciences. Its range of activities has broadened, accordingly, from identifying solutions to engineering problems, to designing holistic and integrated approaches in the development and management of water and environmental resources, and urban infrastructure systems. The Institute's services now also comprise integrated water resources management, effective service delivery and institutional reform, all of which aim to enhance full stakeholder involvement, equity, accountability and efficiency in water sector development and management.
>
> In November 2001, UNESCO's 31st General Conference decided to make IHE an integral part of the Organization. By March 2003, the necessary treaties and agreements between the IHE Delft Foundation, UNESCO, and the Netherlands Government were signed, allowing for the entry into operation of the new UNESCO-IHE Institute for Water Education. UNESCO-IHE is governed by a thirteen-member Governing Board appointed by the Director-General, and is managed by a director and deputy director, both of whom are UNESCO staff members. The IHE Delft Foundation provides all staff and facilities to UNESCO-IHE.

resources management, hydrology, aquatic ecosystems, and other subjects. About 250 master's degrees are awarded each year. Most trainees come from developing countries and receive fellowships. The Institute serves also as a converging point for all other IHP academic networks (courses, Chairs and centres). During the 2004–2005 biennium, it was expected that UNESCO-IHE would benefit from extrabudgetary resources amounting to US$50 million, an amount six times greater than the whole budget for the regular programme on water.

PROSPECTS FOR THE FUTURE

In 2005, the programme on water is firmly anchored in the overall programme of the Organization. It has become a top priority and is a good example of intersectoral cooperation around a vital environmental and socioe-conomic problem. By reinforcing the resources allocated to this programme, UNESCO was able to assume a leading role within the UN system in the field of water.

The growing importance and urgency of water issues throughout the world is a guarantee for the pursuit of this trend. While contributing to a broad international cooperation involving many partners, UNESCO should consolidate its own water-related niche, so as to reflect the complexity of the Organization's mandate in natural and social sciences, culture and communication, and the field of human rights. Following the pattern of Education for All, the water-related programme should also promote Water for All.

BIBLIOGRAPHY

Batisse, M. 2005. *The UNESCO Water Adventure: from desert to water.* Paris, UNESCO, p. 155.

Committee for Scientific Hydrology of the US Federal Council of Science and Technology. 1962. *Transactions AGU* [American Geophysical Union], Vol. 43, No. 4 (December), p. 493.

Falkenmark, M. and Chapman, T. (eds). 1989. *Comparative Hydrology: an ecological approach to land and water resources.* Paris, UNESCO.

Nace, R. L. 1964. The International Hydrological Decade. *Transactions AGU* [American Geophysical Union], Vol. 45, No. 3 (September).

Priscoli, J. D., Dooge, J., Llamas, R. 2004. Overview. In: *Water and Ethics, IHP and COMEST.* Paris, UNESCO.

UNESCO. 2003. *Comprehensive Evaluation Report for the Fifth Phase of the International Hydrological Programme: 'Hydrology and Water Resources Development in a Vulnerable Environment'* (December 2003). Paris, UNESCO. http://www.unesco.org/water/ihp/bureau/36th/eval.pdf

A PRACTICAL ECOLOGY
The Man and the Biosphere (MAB) Programme
Malcolm Hadley

DURING the 1960s, the clearly visible stresses of population, natural resources and the environment became matters of public and political concern. Rachel Carson's *Silent Spring* (1962) served as a clarion call to what was happening to our living world and was one of the catalysts of a whole series of initiatives designed to mobilize resources and attention. The United Nations Conference on the Human Environment, held in Stockholm, Sweden, in 1972, and the subsequent creation of the United Nations Environment Programme (UNEP) were two reflections[41] of that rising interest and growing concern.

During this decade, UNESCO's long-tem concern with the natural environment and its resources led to the launching of several large-scale international initiatives of scientific collaboration, in fields such as oceanography and hydrology. One such initiative focused on the non-oceanic part of that thin layer of living matter that envelops the planet Earth. The idea that the Organization ought to promote an intensification of international research efforts on the rational use and conservation of the resources of the biosphere dates to the mid-1960s, with UNESCO's hosting (in 1964) of the first session of the Special Committee for the International Biological Programme (IBP) and, a year later (September 1965), the first session of UNESCO's Advisory Committee on Natural Resources Research.[42]

In November 1966, the UNESCO General Conference, at its fourteenth session, asked the Director-General to convene, jointly with other interested international organizations, an intergovernmental meeting of experts in the field of 'ecological studies and conservation of natural resources.' The idea was destined to gain rapid momentum, in part fuelled by the enthusiasm and activities generated by the IBP.

THE 1968 BIOSPHERE CONFERENCE

The Intergovernmental Conference of Experts on the Scientific Basis for Rational Use and Conservation of the Resources of the Biosphere was duly convened

---

41   Other reflections of this rising concern included the MIT-sponsored Study of Critical Environmental Problems (SCEP, 1970), the Club of Rome's *The Limits to Growth* (Meadows et al., 1972), and a 1974 Scientific Committee on Problems of the Environment (SCOPE) study on critical environmental problems in developing countries.
42   UNESCO Document NS/201, Paris, 3 December 1965.

PART III: ENVIRONMENTAL SCIENCES

> **BOX III.4.1: 1968 – THE GREAT DIVIDE?**
>
> 'The year 1968 was that of the Great Divide. It marked the end as well as the zenith of the long post-war period of rapid economic growth in the industrialized countries. But it was also a year of social unrest, with student uprisings and other manifestations of alienation and counter-cultural protest in many countries. In addition, it was then that the general and vocal public awareness of the problems of the environment began to emerge.'
>
> — Introductory paragraph to the Foreword of the Council of the Club of Rome's *The First Global Revolution*, by Alexander King and Bertrand Schneider (1991).

in Paris in September 1968. The political base of the 'Biosphere Conference', as it came to be known, was very broad.[43] The conference was organized by UNESCO with the active participation of the United Nations, the Food and Agriculture Organization (FAO) and the World Health Organization (WHO), and in cooperation with IBP and the World Conservation Union (IUCN). Francois Bourlière, former president of both IBP and IUCN, was elected as president of the conference. Its secretary-general was Michel Batisse, at the time the director of UNESCO's Division of Natural Resources Research.

The conference was attended by 236 delegates from over sixty countries, and eighty-eight representatives of fourteen intergovernmental organizations and thirteen non-governmental organizations, coming from a wide variety of scientific fields, management and diplomacy. It was here that the now-familiar word 'biosphere' was introduced to the international vocabulary, having been previously familiar only to those who had read Vernadsky and Teilhard de Chardin.[44] More important, four years before the UN Conference on the Human Environment was held in Stockholm in 1972, this was the first worldwide scientific meeting at the intergovernmental level to adopt a series of recommendations concerning environmental problems and to highlight their growing importance. Indeed, one observer of the global environmental movement, John McCormick, has commented that the 'significance of the Biosphere Conference is regularly overlooked', and that 'the initiatives credited to Stockholm were in some cases only expansions of ideas raised in Paris' (McCormick, 1995).

Science journalist Daniel Behrman wrote that the full title of the conference was 'a proclamation in itself, it helped launch the then-radical assertion that we

---

43 Nowadays, it is frequent to see a whole list of organizations listed as co-organizers or co-sponsors of a major international meeting. Often the participation is emblematic, not substantive. That was not the case with the Biosphere Conference. For example, FAO detached one of its senior professional officers – forester René Fontaine – as Special Assistant to the Secretary-General of the Conference and Liaison Officer with FAO. He spent many months working as part of the Secretariat on the preparations, servicing and follow-up of the conference.

44 For overviews of the 'biosphere' as the specific, life-saturated envelope of the Earth's crust, see: Vernadsky, 1998; Samson and Pitt, 1999.

could go on using our planet only as long as we did not misuse it ... As often happens, UNESCO's focus on the subject gave new heart to those in various places who had been crying out in a rapidly vanishing wilderness. This was immediately seen in reports of individual nations to the conference. Some of the documents sound like the present-day youth protesters (who would never admit that the Establishment got in ahead of them)' (Behrman, 1972).

It is striking today, when scanning the report of the conference, to see how comprehensive and far-reaching the range of issues addressed was.[45] Perhaps the single most original feature of the Biosphere Conference was to have firmly declared that the utilization and conservation of our land and water resources should go hand in hand, and that interdisciplinary approaches should be adopted to achieve this aim. Twenty-four years before the UN Conference on Environment and Development (UNCED, in Rio de Janeiro, Brazil) – where this concept was to be recognized and advocated at the highest political level – the Biosphere Conference was the first intergovernmental forum to discuss and promote what is now called 'sustainable development'.

The conference provided the first occasion to review, at the global and intergovernmental level, the nature of the environmental problems facing humankind and the things which science and scientists could do to help solve these problems. In addition, the Biosphere Conference made a number of recommendations to governments and international organizations (UNESCO, 1969). The first of these twenty recommendations called for an 'international programme of research on man and the biosphere', indicating that such a programme should serve as a follow-up and an extension to the International Biological Programme, that the programme should be international and interdisciplinary and that it should take into account the particular problems of developing countries.

Some two months later, at its fifteenth session in November 1968, the General Conference of UNESCO supported in general all the recommendations of the Biosphere Conference and in particular invited the Director-General to prepare a plan for the long-term intergovernmental and interdisciplinary programme. This preparation involved, in the first place, close consultation with the Member States,

---

[45] Ten assessments were made as part of the preparations for the Biosphere Conference, presented and discussed at the conference and (after revision) included in the conference proceedings: Contemporary scientific concepts related to the biosphere; Impacts of man on the biosphere; Soils and the maintenance of their fertility as factors affecting the choice of use of land; Water resources problems, and present and future requirements for life; Scientific basis for the conservation of non-oceanic living aquatic resources; Natural vegetation and its management for rational land use; Animal ecology, animal husbandry and effective wildlife management; Preservation of natural areas and ecosystems, and protection of rare and endangered species; Problems of deterioration of the environment; Man and his ecosystems, and the aim of achieving a dynamic balance with the environment, satisfying physical and economic, social and spiritual needs.

PART III: ENVIRONMENTAL SCIENCES

the other interested institutions in the UN system, and IBP[46] and IUCN, in order to obtain preliminary views on the content and organization of the programme.

Subsequently, in November 1969, some fifty-five scientists from thirty-one countries and thirty-three specialists from international organizations came

BOX III.4.2: ON ECOLOGY

When, in 1869, the German zoologist Ernst Haeckel coined the term 'ecology', he could scarcely have envisaged the many dimensions this new discipline would take on in the last few decades of the twentieth century – scientific discipline and research approach, but also rallying call for political action, catalyst for social cooperation and trigger for economic change.

Ecology is first and foremost a distinct scientific discipline, initially defined by Haeckel as the study of the interrelations between organisms and their environment. Ecology is also an approach. Its primary focus on interactions between living things and their non-living environments has provided a pivot for cooperation among a host of disciplines spanning the natural and social sciences – from Earth sciences, such as geology and hydrology, to life sciences, such as genetics and physiology, to social sciences, such as anthropology and economics. It has also provided a stimulus in the development of a wide range of individual disciplines.[1]

Given this catholic agenda, it is scarcely surprising that 'ecology' provides the inspiration and underpinning of many collaborative efforts that address issues and problems related to the environment and its resources. These efforts range in focus and scale from local community action to political activism at national and regional levels, to research undertakings which engage the international scientific community.

It is arguable whether this spectrum of meanings of ecology is altogether a good thing for the science and scientific approach that bears its name. Although the social, economic and political connotations of ecology have brought the science into close touch with the broader society of which it forms part and have provided a means for putting scientific knowledge into action, scientific practitioners of ecology sometimes feel uneasy about their science being confused in the minds of some with politics. They are also uneasy about being asked to tackle problems and issues not amenable to ecological analysis.

That said, the science of ecology is reflected in a host of social issues relating to the human environment and the management of its resources. As an international programme with its roots in ecology, it is perhaps not surprising therefore that MAB's social fabric has many tints and textures.

1   Witness the title of one 1989 IHP publication – *Comparative Hydrology: An Ecological Approach to Land and Water Resources*.

---

46   The follow-up and extension of IBP within MAB was not to everyone's liking. For example, conservationist Max Nicholson referred to IBP's 'takeover by UNESCO' (Nicholson, 1989). And Frank Blair, Chair of US-IBP, commented bitterly on the perceived opposition of UNESCO officials to the proposed Programme for Analysis of World Ecosystems (PAWE), which he had piloted as a possible SCOPE initiative to complement the ecosystem work within MAB (see Blair, 1977, Ch. 4).

together to formulate the research themes and activities that would be proposed for the programme. In two weeks of intensive discussions and drafting, five working groups[47] prepared some ninety project proposals covering a wide spectrum of activities. These drafts were harmonized by a smaller group and streamlined further by the Secretariat. Eventually, after further consultation with the other organizations concerned, this resulted in a lengthy document[48] constituting the proposals for the Man and the Biosphere Programme (MAB).

As set out in that document, the programme would be 'an interdisciplinary one that emphasizes an ecological approach to the study of the relationships between man and the environment' (see Boxes III.4.2 and III.4.3). The proposed research programme consisted 'of thirty-one research themes, all of which are considered to be of global or of major regional importance, and to be of such a nature that an intergovernmental approach to their pursuit will be either essential or highly beneficial'. For convenience, the research themes were presented under four main headings:

- themes related to the natural environment, meaning environments that are little modified by man;
- themes related to the rural environment, meaning environments used primarily for agriculture, forestry, or other uses that do not involve major technological transformations of the landscape;
- themes related to environments affected by urbanization or subject to major technological modification by urban-industrial society; and
- themes concerned with pollution or related phenomena, as they affect the biosphere.

LAUNCHING OF THE PROGRAMME

The plan for the MAB Programme was submitted to the sixteenth session of the General Conference of UNESCO in November 1970. The General Conference unanimously decided to launch a programme to be known as MAB (see Box III.4.4) and to set up the machinery for the guidance, supervision and coordination of the new international research effort. Each country was invited to establish a National Committee to ensure maximum national participation in the international

---

47 The five working groups covered the following fields: Inventory and assessment of the resources of the biosphere; Systematic observations and monitoring; Research into the structure and functioning of terrestrial and aquatic ecosystems; Research into changes in the biosphere brought about by man and the effects of these changes on man; Education and other supporting activities. An account of the November 1969 discussions is given in *Nature and Resources*, Vol. 6, No. 2, pp. 2–6 (1970).
48 UNESCO Document 16 C/78. Paris, 6 October 1970.

> **BOX III.4.3: MAN'S PERSONAL FULFILMENT IN NATURE ... AND HIS RESPONSIBILITY FOR NATURE**
>
> If the MAB Programme is embedded in the ecological sciences, it is also one with strong ethical and humanitarian dimensions.
>
> Among those who played a key role in the advocacy of these dimensions within MAB was plant physiologist Magda Staudinger (1902–1997). Hers were the concluding words at a colloquium on 'Problems of the Rational Use and Conservation of the Resources of the Biosphere' organized by the German National Commission for UNESCO in Berchtesgaden in April 1968. In her presentation, she spoke of man as being part of nature and therefore having a biological conscience, and of knowledge giving rise to the development of a new partnership between man and nature, a real partnership comprising mutual giving and taking, not only one-sided taking.
>
> As a member of the German National Commission for UNESCO, Magda Staudinger pressed home such ideas at a succession of international meetings that led to the setting up of MAB – the Biosphere Conference of September 1968, the working groups of September 1969, the UNESCO General Conference of November 1970,[1] and the first session of the International Coordinating Council for MAB in November 1971.
>
> ---
> 1   See Rössler, 1992.

programme, to consider national priorities for participating and to ensure liaison at the international level. The first session of the International Coordinating Council for MAB was held in Paris in November 1971. Among its principal tasks, the Council defined the objectives and scope of the programme, and its main principles of action.

The very nature of the land-use and resource-management problems found in the real world – at the interface between human beings and their environment – meant that research within MAB had to cut across the natural and social sciences, if those problems were to be addressed properly. Within such a framework, the general objective for MAB adopted by the Council was no less than 'to develop the basis within the natural and social sciences for the rational use and conservation of the resources of the biosphere and for the improvement of the global relationship between man and the environment; to predict the consequences of today's actions on tomorrow's world and thereby to increase man's ability to manage efficiently the natural resources of the biosphere'.

The optimism of the period was reflected in the broad scope envisaged for MAB, and in the topics proposed within the thirteen (eventually fourteen) major international themes or project areas (see Box III.4.5) retained from an original list of nearly one hundred. One major group of project areas concerned human interactions with different ecosystem types or physiographic units: tropical and subtropical forest ecosystems, arid and semi-arid zones, temperate forest landscapes, mountains and tundras, inland waters and coastal zones, islands, urban systems. Processes and effects which may occur in all parts of the world, such as environmental perception and the environmental effects of large engineering works, were the concern of other project areas. One project area was particularly concerned with the conservation of natural areas and of the genetic material they contain.

BOX III.4.4: THE NAME OF THE PROGRAMME – MAN AND THE BIOSPHERE

The name of the programme is attributed to Barton Worthington, then scientific director of the International Biological Programme (IBP). The feeling was that such a name would give similar weight to human beings and the biosphere in which they live. As such, the new programme would seek in its very title to redress one of the perceived shortcomings in IBP, when human dimensions of biological productivity were largely confined to one of the eight sections of the IBP, that on Human Adaptability (Worthington, 1975). Worthington further suggested that the resulting acronym (MAB) would be a symbolically apt one, in recalling the fairies' midwife in Shakespeare's *Romeo and Juliet*, who visits men when asleep at night and delivers them of their dreams (Worthington, 1983).

The name Man and the Biosphere (MAB) was referred to explicitly in draft resolutions to the sixteenth session of the UNESCO General Conference in October–November 1970, which formally approved the launching of the programme. 'MAB' was subsequently adopted as the single internationally recognized acronym for the programme, even though the title in other languages would have given a different sequence of letters. Relatively quickly the acronym became widely adopted as a handy and not unattractive way of referring to the programme.

With the benefit of hindsight, the choice of the programme's name may not have been an altogether happy one, for at least two reasons. First, 'Man and the Biosphere' conveys the idea of man in juxtaposition with all the other components that make up the thin layer of life that envelops the planet, rather than being an integral part of the biosphere. In retrospect, it might have been more apt to call it Man in the Biosphere.

Certainly the approach advocated from the outset within the programme has been that of considering man within the context of his overall environment, rather than as a being apart. Hence the title ('Man in ecosystems') of a special issue of the *International Social Science Journal* (Vol. 34, No. 3, 1982), prepared to mark ten years of the programme.

Second, the use of 'Man' to refer to both men and women has aroused strong negative reaction, at least from some quarters. The counter-argument, as Duncan Poore notes,[1] is that the terms 'man' and 'mankind' are proper English usage for the human species (corresponding to the Latin *homo*); they are not intended to have any connotations of gender (Poore, 2000).

---

1 Duncan Poore is a classicist turned scientist, who played an influential role in the development of MAB in the 1970s, particularly as one of the principal scientific advisers to the United Kingdom delegation at early sessions of the MAB Council.

1970s: INTENSIVE PLANNING AND BURGEONING FIELD ACTIVITIES

Under the leadership of the founding director of the Division of Ecological Sciences, Francesco di Castri, the MAB Programme took firmer shape in the early 1970s. This was a densely eclectic period of creative elaboration of the scientific content and approach of the various project areas of MAB, with the ensuing proposals diffused in a series of MAB 'green' reports.[49] In many problem areas, the proposals were innovative and generated considerable enthusiasm and commitment in the scientific community. But given that there was no central source of funding for contributing field projects, the proposals were mostly framed in generalized terms. For each project area, proposals were multiple

---

BOX III.4.5: LIST OF MAB PROJECTS APPROVED BY THE MAB COUNCIL AT ITS FIRST SESSION (NOVEMBER 1971)

1. Ecological effects of increasing human activities on tropical and subtropical forest ecosystems
2. Ecological effects of different land uses and management practices on temperate and mediterranean forest landscapes
3. Impact of human activities and land-use practices on grazing lands: savannah and grassland (from temperate to arid areas)
4. Impact of human activities on the dynamics of arid and semi-arid zones' ecosystems, with particular attention to the effects of irrigation
5. Ecological effects of human activities on the value and resources of lakes, marshes, rivers, deltas, estuaries and coastal zones
6. Impact of human activities on mountain and tundra ecosystems
7. Ecology and rational use of island ecosystems
8. Conservation of natural areas and of the genetic material they contain
9. Ecological assessment of pest management and fertilizer use on terrestrial and aquatic ecosystems
10. Effects on man and his environment of major engineering works
11. Ecological aspects of urban systems with particular emphasis on energy utilization
12. Interactions between environmental transformations and the adaptive, demographic and genetic structure of human populations
13. Perception of environment quality
14. Research on environmental pollution and its effect on the biosphere[1]

---

1 Approved by the MAB Council at its third session (Washington DC, September 1974).

---

49 During 1972–77, over forty reports were produced in the MAB Report Series, resulting from various types of research planning meetings – including expert panels, working groups, task forces and regional meetings – as well as biennial meetings of the International Coordinating Council.

and wide ranging, rather than few and focused. Many subjects were flagged as important. Research topics were seldom described in terms of hypotheses to be tested using agreed methodology. Provision was not made to set up groups of specialists who would be responsible for the organization and coordination of activities in a particularly subject area. No time limits were established.

Not surprisingly, this planning process resulted in a very loosely knit network of field projects undertaken by countries, each project tending to focus on an environmental or resource management problem of priority interest at the local or national level. The result was a patchwork of projects dealing with

BOX III.4.6: MAB AND THE ECOSYSTEM APPROACH

'The relevance of MAB for ecosystem studies was threefold. First, MAB provided a reason to extend ecosystem studies into areas that had been neglected or inadequately treated in the IBP. This was especially notable for tropical rainforests and arid regions ... Second, MAB undertook to extend rapidly and improve on the conservation activity of the IBP. In this effort, MAB established a Biosphere Reserve Program, which identified natural or semi-natural areas worldwide where research could be undertaken and the environment protected. The idea of biosphere reserves was to allow research and some economic activities in a portion of the area, while leaving other portions as undisturbed controls.... Third, MAB extended ecosystem studies from natural landscapes to the human-built environment, leading to a revitalization of the subject of human ecology on ecosystem principles ... Thus, MAB led to a deepening and extension of the ecosystem concept – although it moved away from further theoretical evolution of the concept. Through MAB and the other activities discussed here, ecosystem studies gradually changed focus.'

--Ecologist **Frank B. Golley**, in *A History of the Ecosystem Approach in Ecology* (Golley, 1992).

complementary issues. Synthesis tended to be at the level of a given field project, rather than across groups of field projects. The primary preoccupation was with specific land-use and resource-management problems, and their underlying conflicts, in particular geographic contexts. For example:

- how to resolve the conflicting interests of the tourism industry and traditional agriculture and forestry in a mountain valley in the European Alps;
- how to develop sustainable systems of agro-silviculture in poor soil areas of a large island in South-East Asia, and provide a long-term source of livelihood for local people and migrants from more densely populated areas in the region;
- how to combine improved systems of animal husbandry and crop production in a semi-arid area on the fringe of the Sahara

PART III: ENVIRONMENTAL SCIENCES

desert, while continuing to support the social and cultural values of the populations concerned (di Castri et al., 1981).

As the 1970s progressed, there was gradual concentration within MAB on four areas of priority, at least in terms of the servicing by UNESCO of the programme: humid and subhumid tropical zones, arid and semi-arid areas, development of the biosphere reserve network, and urban areas considered as ecological systems. Nonetheless, substantive groups of field projects were undertaken around particular physiographic units, especially in mountain and island regions. The concerns of the process or impact themes (e.g. environmental perception) came to be integrated at the field level within projects having an ecosystem (e.g. tropical forest) or physiographic (e.g. urban systems) basis.

In terms of approaches and methods, the so-called 'ecosystem approach' to ecological research was an integral part of MAB from the beginning (see Box III.4.6), and could be considered as one legacy of its links with the International Biological Programme. In the sense that 'ecosystem approach' is used as a synonym for 'holistic', that approach remains an enduring and important part of MAB.[50]

But experience has revealed the limitations of using the ecosystem as the unit of research. It is not suitable for integrating the natural and social sciences, whose time and space frameworks tend to be very different. Moreover, the use of natural resources is often organized around the exchange of materials, energy and even populations between different ecosystems, which may complement each other in economic terms. Hence, the spatial unit of integrated, problem-oriented research within MAB tends to be the 'human use system', and its explicit recognition of the dimensions of space, time and perception (di Castri et al., 1980, 1981).

The role of systems analysis and modelling in ecological research was another methodological feature that received a fair amount of attention within MAB, at least in its early years. After the first session of the MAB Council, a number of ad hoc expert panels were convened to elaborate the scientific content of projects proposed under the programme and consider the methodologies and plans of studies that might be recommended to National Committees for the implementation of these projects. The first panel to be convened in April 1972 had as its charge to consider the role of systems and modelling approaches in the MAB Programme, in providing a conceptual framework and as a means for integration of projects, forecasting of change and optimization of resource

---

[50] This is reflected, for example, in the close association of the biosphere reserve concept and the World Network of Biosphere Reserves with the ecosystem approach that has been adopted by the Conference of Parties of the Convention on Biological Diversity (CBD) as the primary framework for action under the Convention. Thus, a booklet prepared by UNESCO in 2000 illustrates the twelve principles of CBD's ecosystem approach with examples from the World Network of Biosphere Reserves.

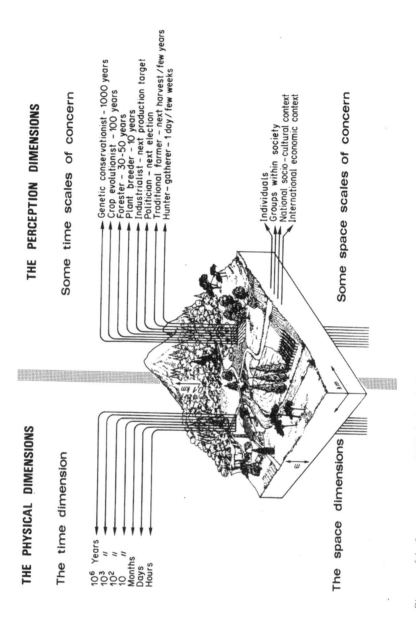

Diagram of the 'human use system', 1979. During the late 1970s and early 1980s in particular, UNESCO developed this concept to promote interdisciplinary work on resource management within MAB, as a way of linking biological and social systems. Image © UNESCO.

management. A preliminary stocktaking was undertaken by a reconvened second task force on systems analysis within MAB, which met nine years later in 1981.

One conclusion was that the use of systems analysis as an integrating tool had fallen short of expectations. Nevertheless, progress could be reported in the modelling of forest succession, human settlements, regional development and land-use schemes in marginal areas. Of particular significance had been attempts to use systems approaches in elaborating scenarios for the future development of particular areas. Two well-known examples were in central Europe, carried out by local scientific institutions and community associations and the International Institute of Applied Systems Analysis (IIASA), focused on land-use change in the Austrian alpine valley of Obergurgl and a simulation model of eutrophication in the shallow binational (Austria–Hungary) lake of Neusiedlersee/Lake Fertö.

In terms of the emergence of the biosphere reserve concept,[51] following an expert panel meeting for MAB Project 8 (held at IUCN headquarters) in 1973, a special task force was convened jointly by UNESCO and UNEP in 1974, which for the first time drew up a set of objectives and a set of characteristics for biosphere reserves (UNESCO, 1974). By and large, these are still valid today. They define and stress the multiple functions of biosphere reserves, covering to some extent the three basic needs of conservation, development and logistical support. The task force proposed a simple generalized zoning pattern for biosphere reserves, combining a core area, a delineated 'inner buffer zone' and an undelineated 'outer buffer zone', corresponding somewhat to what is now known as the 'transition area'. The task force also recommended that support be provided by UNESCO and UNEP to IUCN, to further develop a system for classifying natural regions and for facilitating the selection of representative sites for conservation (Udvardy, 1975).

On the basis of the task-force report, but mainly following their own interpretation of the new concept for their local needs, countries began to propose areas for designation as biosphere reserves. In some countries, this process attracted a high political profile and backing. One example concerned the mention of biosphere reserves in the final statement of the 1974 Brezhnev/Nixon summit meeting in Moscow (USSR), when the two countries declared that:

> Desiring to expand cooperation in the field of environmental protection ... and to contribute to the implementation of the Man and the Biosphere Programme of UNESCO, both sides agreed to designate, in the territories of their respective countries, certain natural areas as biosphere reserves for protecting valuable plant

---

51 For a more detailed account of the origins and development of the biosphere reserve concept, see the introductory chapter in UNESCO, 2002.

The Obergurgl Project. Preliminary dynamic model of the village of Obergurgl in the Tyrolean Alps, Austria, the location of intensive study by Austria-MAB and the International Institute for Applied Systems Analysis (IIASA) in the mid-1970s, as a microcosm of the problem of population and economic growth and diminishing natural resources. (Buba Himanowa, 1975). Four classes of components are (1) recreational demand, (2) population and economic development, (3) farming and ecological change, and (4) land use and development control.
Image © UNESCO.

At the international level, the Obergurgl project was first discussed at an expert panel meeting on MAB work in mountain areas held in Salzburg (Austria) in January 1973. Pictured in photo, three members of the UNESCO Secretariat: from left to right, Michel Batisse, Francesco di Castri and Gisbert Glaser.

and animal genetic strains and ecosystems, and for conducting the scientific research needed for more effective actions concerned with global environmental protection.

This declaration came as something of a surprise to the chancelleries of many countries, who had never heard of biosphere reserves.[52]

At the international level, the procedure for actual designation had not been defined. Overwhelmed with proposals, the MAB Coordinating Council felt that it was too delicate a matter to be handled in a large and open forum. At its fourth session in 1975, it delegated this task to its six-member Bureau, which designated a first batch of fifty-seven biosphere reserves in mid-1976. The main criterion used in the selection of approved biosphere reserves was their conservation role, together with the fact that they had some research facilities or history. The same thing happened in early 1977, when a second group of sixty-one biosphere reserves was designated. In fact, the Bureau adopted a very flexible approach, considering it sufficient when the areas proposed by the MAB

---

52 The inclusion of 'biosphere reserves' in the Brezhnev/Nixon summit declaration has its origins in the determination of Vernon C. (Tom) Gilbert, who was seconded to the MAB Secretariat by the US Parks Service in 1973–74. During his secondment, Tom Gilbert was critically involved in the development of the biosphere reserve concept (e.g. in the convening and follow-up of the 1974 task force on criteria and guidelines for biosphere reserves). It was an early example within MAB of the 'value added' role of specialists being detached by countries for relatively limited but not inconsequential periods (two to three years) in helping the UNESCO Secretariat in the development of work in strategically selected fields.

National Committees for designation appeared interesting for the conservation of ecosystems, had appropriate legal protection, and were the object of a reasonable amount of research work.

The process had been launched and was to continue over the following years, but at a slower pace. By 1981, 208 biosphere reserves had been designated in fifty-eight countries, with one publication in that year stating that 'Biosphere reserves form an international network of protected areas in which an integrated concept of conservation is being developed, combining the preservation of ecological and genetic diversity with research, environmental monitoring, education and training. Biosphere reserves are selected as representative examples of the world's ecosystems' (UNESCO, 1981).

In other words, in this first phase of implementation of the programme, the conservation role had been kept prominent, the logistical role minimal, and the development role largely forgotten. Almost all designated biosphere reserves were areas already protected, such as national parks or nature reserves, and in most cases the designation was not adding new land, new regulations or even new functions. Research work was conducted in these protected areas, but the research was in many cases of a rather academic character, not clearly related to ecosystem and resource management, and not addressing explicitly the relationship between environment and development. Moreover the linkages between biosphere reserves and the exchanges of information on this research remained very limited, and the international network was merely concentrated as a centralized UNESCO Secretariat function. Not only had a proper balance between the three central concerns of biosphere reserves not been reached, but they did not constitute a truly functioning network.

This situation did not prevail in all cases. Some biosphere reserves were created from scratch, adding not only new areas under protection but endeavouring to demonstrate the multiple functions of the new concept, including its development role (Box III.4.7). An example was Mapimi Biosphere Reserve in Mexico (designated in 1977), where scientists and managers began to experiment with the development function of biosphere reserves, under which the management of the reserve was increasingly expected to contribute towards meeting the economic needs of local people (Halffter, 1980, 1981). At other reserves, an existing national park was used as a starting point, but with new functions and areas added to it, which usually implied coordination between separate administrative units: an example here was Waterton Biosphere Reserve, centred on Waterton Lake National Park in Canada.

Another important step was the introduction of the interesting idea of 'clustering', developed at a joint United States–Union of Soviet Socialist Republics symposium on biosphere reserves in 1976 (Franklin and Krugman, 1979). This idea aims at accommodating the many situations where all the

> **BOX III.4.7: BORDERING ON THE LIMITS OF THE POSSIBLE**
>
> 'Science has the dreadful task of trying to unravel the paradoxical and incongruous processes that for thousands of years have placed man increasingly in opposition to and divided from nature; for man – as Ortega y Gasset tells us – "is made of such a strange substance that he is partly related to nature and partly not, at the same time natural and outside nature, in some way an ontological centaur of which one half is immersed in nature and the other half transcends it". This is why someone has said that such an immense task as that inaugurated by MAB can constitute a challenge that borders on the limits of the possible.'
>
> —**Valerio Giacomini** (1914–1981), Chair of the Italian MAB National Committee, in an article entitled 'Man and the Biosphere: an amplified ecological vision' (Giacomini, 1978).

functions of biosphere reserves cannot be performed in contiguous areas and where a regrouping and coordination of activities between several discrete areas is required. This is often the case when conservation in core areas has to be associated with integrated research and manipulative experimentation in other areas. An early example of this approach developed around Great Smoky Mountains National Park in the United States (part of what is now the Southern Appalachian Biosphere Reserve), which included Oak Ridge National Laboratory and Coweeta Hydological Station located outside the park. Another important development was the linking of a biosphere reserve with a MAB integrated pilot project for research, training and demonstration, as at Puerto Galera in the Philippines and Mount Kulal in Kenya. Efforts were also made to adapt the emerging biosphere reserve concept to particular regions (see Box III.4.8), with a regional meeting in Side (Turkey) for the Mediterranean, and a roving study tour of field sites in Australia and New Zealand, with special reference to techniques for selecting potential biosphere reserves (McAlpine and Molloy, 1977; Robertson et al., 1979; Davis and Drake, 1983).

Unfortunately, these early initiatives to 'regionalize' the emerging biosphere reserve concept did not lead to much effective action at the field level. In addition, by and large, the actual international list of designated biosphere reserves did not properly convey the innovative multifunctional approach embodied in the concept. Some thus held the view that, as a category of protected areas, biosphere reserves were not adding much to other existing categories and indeed were creating some confusion. Others considered that the initiative had to begin in some manner and stressed that the concept was a highly imaginative and valuable one, which had to be developed and applied in a more and more systematic manner, not only to supplement current efforts in nature conservation, but to build up the multiple functions of biosphere reserves.

BOX III.4.8: REGIONAL AND SUBREGIONAL COOPERATION AND NETWORKS

Planning at the regional and subregional level was one of the intrinsic features of MAB activities in the 1970s, with reports of over twenty regional and subregional meetings organized by UNESCO between 1973 and 1979 being published in the MAB Report series. In addition, groups of countries took the initiative in convening cooperation-meetings of various kinds (such as the MAB National Committees of the Socialist countries, with a first meeting in Moscow in March 1977). In terms of flexible arrangements for promoting regional cooperation over a stretch of time, the MAB Northern Sciences Network was set up in 1982, with the EuroMAB network following in 1987 and others during the 1990s in such regions and subregions as Africa, the Arab States, Latin America and the Caribbean, East Asia, South and Central Asia, South-East Asia, and the eastern Atlantic.[1]

These various networks are largely informal, in that each group of countries decides for itself its way of working, the nature of participation, the priorities to be addressed, and so on. The networks are not formal bodies set up under the aegis of the International Coordinating Council, and as such differ from arrangements within some of the other intergovernmental environmental undertakings of UNESCO. This said, UNESCO Field Offices in the different regions play a key role in animating and servicing the various MAB networks, including (in some cases) the hosting of network websites. Twinning arrangements[2] and transboundary cooperation are among the other forms of collaboration promoted within the World Network of Biosphere Reserves (UNESCO, 2003).

---

1 Further information on the various regional and subregional networks is accessible through the MAB website (http://www.unesco.org/mab/networks.shtml), which provides hyperlinks to the websites of individual networks.
2 For further information on exchanges between twinned biosphere reserves, see UNESCO, 2002, pp. 140–41.

1980s: STOCKTAKING AND REORIENTATION

Ten years after MAB was launched, the first overall assessment of the programme was carried out, by means of an international conference entitled 'Ecology in Practice' and an associated exhibit organized immediately prior to the sixth session of the MAB Council in September–October 1981. The overall conclusions were subsumed in a three-page general recommendation adopted by the MAB Council, which attempted to summarize lessons learned in the first ten years of MAB and to identify shortcomings and accomplishments. Guidelines for future development of MAB were put forward, including the need for continued concentration in activities and greater flexibility in working methods. More specific recommendations were directed to Member States, MAB National Committees, UNESCO, international intergovernmental organizations and the international scientific community, with a view to increasing the effectiveness of the contribution of MAB to measures for the improved management of the environment and its natural resources.

In its general recommendation concerning the status of MAB, on the tenth anniversary of its launching, among the points underlined by the Council was that MAB 'provides a highly significant and innovative international framework for scientific cooperation designed to improve the management of natural resources.' The Council also considered that after ten years the Programme had achieved a high degree of success, as indicated by:

- the number of countries (101), field projects (nearly 1,000) and individual scientists (more than 10,000) taking part in the Programme;
- more qualitative measures of accomplishment, such as the actual results of testing at the field level of new approaches to problem-oriented research on complex problems of land use and resource management;
- the important 'multiplying effect' – from tenfold to a hundredfold – of relatively modest amounts of funds invested in the MAB Programme, in attracting additional support for the Programme from national, bilateral and international sources;
- the extent to which individual field projects have succeeded in meeting one or more of the following criteria for evaluation of MAB activities: the production of new information of intrinsic value; the application of research results to resource management; the development of research capacities and national infrastructures; the building of bridges between disciplines and the breaking down of sectoral barriers; and the fostering of international cooperation.

Complementing the report and recommendations of the sixth session of the MAB Council (MAB Report Series 55 [UNESCO, 1983]), the two-volume book (di Castri et al., 1984) resulting from the Ecology in Practice conference provides a useful overview of the sort of activities that marked MAB's first decade, through its sixty-one chapters, grouped under six sectional themes:

- The search for sustained production systems in the humid and subhumid tropics;
- Scientific basis for the management of grazing and marginal lands;
- Providing a basis for ecosystem conservation;
- Ecological approaches for improving urban planning;

- Use of scientific information for environmental education purposes;
- Providing the types of information needed for decision-making on land management.

The conference included thematic syntheses, regional reviews and case studies, with an attempt made to flag some wider implications of field experience (see Box III.4.9).

Associated with the scientific conference was the protype of an 'Ecology in Action' poster exhibit, grouped around five themes. The exhibit was subsequently produced by UNESCO in three languages (English, French and Spanish), with further translations and adaptations in another twenty or so languages through arrangements with individual MAB National Committees and other collaborators. The exhibit had a high impact and is widely considered to be one of the most significant projects carried out within the MAB framework, in successfully stimulating efforts to communicate scientific information to lay audiences.

In confirming the value of biosphere reserves and other protected areas, the interlinked events in October 1981 of MAB Council, scientific conference and poster exhibit served to highlight the complexity of implementing the biosphere reserve concept in relation to the very diverse situations and concerns occurring in various parts of the world.

The need for a fresh look and a new impetus were to come in October 1983 from the first international Biosphere Reserve Congress held in Minsk, Belarus (USSR), jointly organized by UNESCO and UNEP in cooperation with FAO and IUCN. The congress took place at a time of major East–West political tension and could not achieve all expectations. However, on the basis of the very diverse experience gained in many countries, the congress was able to review the overall situation and to lay down general guidelines for the future (UNESCO-UNEP, 1984). It underlined the multiple functions that characterize biosphere reserves and explored how the network could function through complementarity and exchange of information. It developed proposals for research, monitoring, training, education and local participation.

On the basis of the work of the Minsk Congress, it became possible to draw up a world Action Plan for Biosphere Reserves (UNESCO, 1984), which, after considerable consultation, was adopted by the MAB Council in December 1984 and formally endorsed subsequently by UNEP, UNESCO and IUCN. The Action Plan marked an important stage in the evolution of the biosphere reserve concept, and provided a framework for developing the multiple functions of biosphere reserves and expanding the international network for the period 1985–89. The Action Plan comprised thirty-five actions grouped under the basic nine objectives drawn up at the Minsk Congress, including such challenges as improving and expanding the network, developing basic knowledge for conserving ecosystems and biological diversity, and making biosphere reserves more effective in linking conservation and development. Follow-up activities included the examination of the implementation of the Action Plan at the regional level, such as through a European Conference on biosphere reserves and ecological monitoring held in Ceske Budejvice (Czechoslovakia) in March 1986.

> ### BOX III.4.9: SCIENCE AND SOCIETY
>
> From its inception, MAB has sought to provide multiple links between science and society – between different types of ecological and human use systems, between scientists from a range of disciplinary backgrounds and (what are now called) different stakeholder groups.
>
> An early example of the sorts of issues that such linkages might entail was provided by a MAB field project in the late 1970s focused on pastoral management in the Waitiki Basin, New Zealand. The 1.3-million-ha Waitaki River Basin in New Zealand's South Island includes a range of ecosystem types, from mountain peaks to coastal plains. Grasslands in the basin were largely induced from forest by fire in Polynesian times and have been used for extensive pastoralism by Europeans for over a hundred years. In a case study of the Waitiki pastoral experience, three conclusions of Kevin O'Connor and co-workers bear repeating:
>
> - If a research programme into mankind's complex interactions with the biosphere looks at the past as well as the future, its progress is slower but its forecasting for action is likely to be more valid.
> - If such a research programme examines a wide range of interactions – as with water bodies, mountains, major engineering works, biological conservation, and grazing lands and irrigation – then a regional planning perspective becomes inevitable.
> - If such a research programme involves and interests local people, such people tend to take a new view of science and their environment, and also of themselves. It also induces changes in those doing the research (O'Connor et al., 1984, in: di Castri et al., 1984).

Among the specific proposals in the Action Plan was the creation of a Biosphere Reserve Scientific Advisory Panel, a group of independently nominated scientists charged with undertaking an intellectual reappraisal of the biosphere reserve concept in the light of the findings of the 1983 Congress and of future priorities. The panel – which met in Cancun (Mexico) in September 1985, and in La Paz (Bolivia) in August 1986 – reached agreement on a refinement of the biosphere reserve concept and on criteria for the selection of new biosphere reserves. It agreed at its last meeting that what distinguished the biosphere reserve network from other protected areas was the combination of three elements inherent in the biosphere reserve concept since its inception:

A poster of the 'Ecology in Action' exhibit, held in the early 1980s, presenting five major stages in the development of the science of ecology: from the study of single species and the conditions that control their lives (autecology, left), through studies of aggregates of mixed organisms called communities or biocoenoses (synecology) and research on the ecosystem and the biosphere, to recognition of man in the biosphere (far right). Image © UNESCO.

conservation of genetic resources and ecosystems; an international network of sites acting as a focus and base for research, monitoring, training and information exchange; and linking development to environmental research and education within the MAB Programme.

During the same period, a General Scientific Advisory Panel for the overall MAB Programme was organized jointly by UNESCO and ICSU, with two meetings in 1986 and 1988. Among the conclusions was that the MAB Programme continued to be dispersed over too large a number of subject areas. At the same time, there was a need to provide for innovations in the MAB research agenda, and four new research orientations were proposed, in such fields as economic investment and restoration of degraded land and water ecosystems. But the impact of the new research orientations was limited.

In effect, from the mid-1980s onwards, for more than a decade, the ability of the MAB Programme to take advantage of new and emerging opportunities was severely weakened by financial and technical consequences of the withdrawal of the United States and the United Kingdom from UNESCO, together with other factors. Symptomatic was UNESCO's substantive absence from discussions within the international scientific community on issues related to global climate change, at least in terms of terrestrial ecosystems (see Box III.4.10).

The impact of the withdrawal of the USA and the UK was described by Michel Batisse as follows:

> While those two countries nevertheless maintained their participation in the MAB Programme through active cooperation of their National Committees and continuing Biosphere Reserves and other projects, they were no longer contributing to the budget. More importantly, a number of people in the scientific establishment, and elsewhere, were no longer sure that cooperating with a UNESCO Programme was the right thing to do. At the same time, some scientists, being more interested in the cutting-edge of their disciplines than in the little-rewarding interdisciplinary efforts to solve land use problems, became more attracted by new, sophisticated research initiatives where striking results and clear synthesis could be expected (Batisse, 1993).

Hence, in the mid- to late 1980s, while the 'biosphere reserve' was gaining ground as a conceptual alternative to the 'national park' and other conventional protected areas, the solidarity that should have led the international community to consider biosphere reserves as priority sites – for testing and validating approaches to integrated conservation and development operations – was undermined.

BOX III.4.10: MAB AND GLOBAL CHANGE ISSUES

For MAB, the issue of 'global change' is a matter of 'what might have been': for some, it is a lost opportunity; for others, a question of the necessity of making a choice among conflicting priorities.

The issue goes back to the 1980s, when three interlinked sets of environmental issues – global change (and particularly climate change), sustainable development and biological diversity – entered into the scientific as well as political and public arena. Indeed, the scientific community played a leading role in drawing political and public attention to these issues, as part of their societal function of describing and understanding what is happening to the world in which we live.

UNESCO was involved in the early 1980s in discussions within the international scientific community on a fledgling international research programme on global change, for example through contributing to a seminal conference on Global Change organized as part of ICSU's twentieth General Assembly in Ottawa (Canada) in September 1984 (di Castri, 1985, in: Malone and Roederer, 1985). But for several reasons, UNESCO did not become a partner with ICSU in the launching in 1986 of the International Geosphere Biosphere Programme (IGBP). For MAB, some observers might argue that among the consequences was that a new generation of terrestrial ecologists 'was lost' to the programme, with the effects still being felt two decades later. Others might counter that not becoming involved as an active partner was an inevitable consequence of the oft-recognized and advocated need for concentration within MAB, rather than expansion into new areas.[1]

Later, several initiatives were undertaken to build bridges between MAB and global change issues and programmes. In February 1989, ICSU and UNESCO drew up a 'Memorandum of Understanding Concerning Cooperative Activities on Global Change'. Matching funds were provided by the US Government and UNESCO to ten cooperative activities between UNESCO and IGBP. They included work on soil fertility and global change, on savannah modelling for global change, and on geosphere-biosphere observatories, with the findings reported by ICSU's International Union of Biological Sciences (IUBS) in special issues of its bulletin *Biology International*.

In 1992, UNESCO-MAB and the Global Change and Terrestrial Ecosystems (GCTE) core project of IGBP, together with the Observatoire du Sahara et du Sahel (OSS), organized an international workshop at Ury, Fontainebleau (near Paris), on monitoring long-term changes in terrestrial ecosystems. The report of the workshop (published jointly as *MAB Digest 14* and *IGBP Global Change Report 26*) was entitled 'Towards a Global Terrestrial Observing System (GTOS)'. Subsequently, GTOS took shape under the aegis of five co-sponsors – WMO, UNESCO, FAO, ICSU and UNEP. UNESCO had the opportunity of taking a lead role in shaping and developing GTOS, but declined the opportunity. The resources for setting up a GTOS unit within UNESCO were not available.[2] Secretariat functions were

PART III: ENVIRONMENTAL SCIENCES

taken up by FAO, which set up a special GTOS unit within its newly created Sustainable Development Department.³

Since the formal launching of GTOS in 1996, UNESCO has continued to participate in its work, but in a secondary supporting role, through such channels as Biosphere Reserve Integrated Monitoring (BRIM), and cooperation with the International Long-Term Ecological Research (ILTER) initiative. Individual biosphere reserves are taking part in networked experiments and pilot monitoring schemes within GTOS, such as those on net primary productivity and terrestrial carbon. In the broader Sahara–Sahelian region, five biosphere reserves figured in the first phase of a regional monitoring scheme known as ROSELT (Reseau d'Observatoires de Surveillance Écologique à Long Terme). And more generally, GTOS itself forms part of the interlinked family of International Global Observing Systems (IGOS) and the even broader Global Earth Observation System of Systems (GEOSS).

Otherwise, global change has been addressed within MAB on a somewhat piecemeal basis. In mountain regions, plans took shape in 1989–90 for a collaborative study on 'The sustainable future of mountain communities: resources and tourism in the context of climate variability and change'. Those proposals were aborted, but more recently a series of assessments has been carried out on the implications of global change for the sustainable development of mountain biosphere reserves.

---

1   Other proximate factors contributing to UNESCO's non-participation as a partner in the IGBP included: (a) the constraints on UNESCO's ability to take on new cross-cutting initiatives by the existence of long-standing, open-ended intergovernmental initiatives on the environment (i.e. IHP, IGCP, IOC, MAB); (b) difficulties related to finance and credibility posed to UNESCO by the separation of the United Kingdom and the United States from the Organization in 1984–85; (c) the difficulties of a long-established open-ended multidisciplinary programme focused on resource use and environmental management at local and regional levels (i.e. MAB) in adapting to new sets of issues involving different space and time scales, paradigms, technologies and mind-sets; (d) the comparative lack in terrestrial ecology of large-scale networks of scientific cooperation characteristic for example of the meteorological and oceanographic research communities; (e) the resignation from UNESCO in 1983 of Francesco di Castri, who had been closely involved in the 1960s and 1970s as researcher and 'mover and shaker' in the major international programmes of scientific cooperation relating to the environment (IBP, SCOPE, MAB); (f) the opinion of some influential voices within the UNESCO Secretariat that 'global change' was a passing scientific whim, and was not of contemporary priority for an intergovernmental organization such as UNESCO, particularly given the difficulties that the Organization was facing at the time; (g) the disinclination of some influential members of the scientific community to become associated in new joint initiatives with UNESCO.
2   This is an example, one among many, of UNESCO being centrally involved in the early conceptual and launching phases of activities in a particular scientific domain, where the implementation is taken in hand by others with greater resources.
3   Within FAO, 'Department' is the equivalent of 'Sector' within UNESCO. The Sustainable Development Department of FAO was set up in January 1995, in response to the need to take a more holistic and strategic approach to development support and poverty alleviation. The Department serves as a global reference centre for knowledge and advice on biophysical, biological, socio-economic and social dimensions of sustainable development.

In spite of these difficulties, the biosphere reserve concept continued to gain attention and develop throughout the 1980s. Groups of scientists and managers gave attention to the application of the concept to particular types of physiographic units, such as coastal zones and islands. Links were explored with the emerging interests in 'integrated development and conservation projects'. The underlying philosophy and approaches of the biosphere reserve concept became much more widely appreciated within some parts of the conservation community, as a flexible and practicable approach to reconciling the needs of socio-economic development with those of conservation. Countries began to establish multiple-site biosphere reserves and to encourage voluntary linkages between large, ecologically delineated conservation units. Commentators underlined the ethical and spiritual dimensions of biosphere reserves, the need for efforts to promote public understanding and appreciation of the biosphere reserve concept, and biosphere reserves as 'innovations for cooperation in the search for sustainable development' (Engel, 1985; Francis, 1985; Kellett, 1986).

The mid- to late 1980s also saw some changes in the information materials produced within the MAB Programme. An occasional newsletter, *InfoMAB*, was introduced, designed primarily for those taking part in MAB, with twenty-four issues published between 1984 and 1996. A series of *MAB Digests* was started, with distillations of the substantive findings of MAB activities and overviews of recent, ongoing and planned activities within MAB in particular subject or problem areas. Nineteen booklets were produced from 1989 to 1998. The Man and the Biosphere book series was also launched, with the aim of communicating the results generated by MAB to a wider audience; twenty-eight volumes were published between 1989 and 2002. Included were reviews and monographs in particular technical fields (e.g. population and environment in arid regions, biohistory and the interplay between human society and the biosphere, agro-ecological farming systems in China) and syntheses of field projects (e.g. San Carlos in Venezuela, Gogol in Papua New Guinea, Třeboň in Czech Republic, collaborative studies on Nordic mountain birch ecosystems).

1990s: NEW LINKAGES, FURTHER CONCENTRATION ON BIOSPHERE RESERVES

The widening recognition of biosphere reserves during the 1980s was reflected in an observation by the MAB Council, at its eleventh session in November 1990, that the general interest in biosphere reserves had probably never been greater, even though the quality of the international biosphere reserve network (at the time, numbering 293 sites in 74 countries) was highly uneven and lacked credibility as an operational network.

It was within such a context that the MAB Council requested that an Advisory Committee on Biosphere Reserves be officially set up by UNESCO, in order to establish clear procedures for listing new sites and to consolidate the work of the international biosphere reserve network at the time when the overall MAB Programme itself was being reviewed in order to be adapted to the post-UNCED period.

The process associated with the Rio Conference on Environment and Development (1992) led directly or indirectly to a swathe of new frameworks for international cooperation on environmental issues and to a new landscape of organizations and programmes on environment and sustainable development issues. This called for a whole new constellation of responses from relatively long-established programmes such as MAB (see Box III.4.11). From the early 1990s to today, new poles of cooperation were developed with bodies such as the Convention on Biological Diversity (e.g. in respect to the Ecosystem Approach), the Convention to Combat Desertification (e.g. in respect to environmental education materials) and many others. There was also a renewal of cooperative arrangements with long-standing instruments and conventions such as those on the World Heritage and on Wetlands (see Box III.4.12).

Within MAB itself, there was a further concentration on the biosphere reserve concept and its implementation. Meeting for the first time in April 1992, the

BOX III.4.11: INTERNATIONAL CONNECTIONS AND COOPERATIVE PROJECTS

Given its catholic research agenda, it is scarcely surprising that many links have developed over the years between MAB and other international bodies. The need for such linkages and connections is formally recognized in MAB's statutes. It is reflected in the dozens of acronyms of organizations and programmes scattered throughout MAB reports and publications. It is also manifest in the range of topics that have formed points for concrete cooperation between UNESCO and other organizations working together with collaborating institutions at the national and regional levels.

Over the duration of the MAB Programme, many different types of collaborative projects have been carried out. In the 1970s and early 1980s in particular, MAB formed the focus of more than forty cooperative projects with UNEP, from project planning at regional and international levels to state of knowledge reviews, field projects geared to testing approaches to improved resource management, and training activities of various kinds. More generally, cooperative activities have ranged from relatively finely focused 'scientific partnership' projects over a three- to five-year period with a single or principal funding source, to broad frameworks for cooperation with multiple partners and sources of funding and complex networked activities spanning a decade and more. Among the issues addressed are South–South cooperation, biodiversity science, ethnobotany and the sustainable use of plant resources, tropical soil biology and fertility, survival of the Great Apes, and wetlands conservation.

> **BOX III.4.12: WORLD HERITAGE AND RAMSAR CONVENTIONS**
>
> Contemporary with MAB, the Convention on the Protection of the World's Natural and Cultural Heritage has turned out to be one of the principal international instruments for the conservation of nature. The Convention and the sites inscribed on the World Heritage List are serviced by the World Heritage Centre, set up by UNESCO in 1992 as an autonomous unit responsible for servicing the Convention. Prior to this, during two decades, the natural part of the Convention had been serviced by the Natural Sciences Sector.[1]
>
> The World Network of Biosphere Reserves and the list of World Heritage sites have each evolved to a point where they have strong complementarity with each other. (Though some sixty biosphere reserves are fully or partially World Heritage sites, biosphere reserves and natural heritage sites have fundamentally different purposes, objectives, legal status and management principles, and therefore should not be confused.)
>
> Another near contemporary to MAB is the Convention on Wetlands, popularly known as the Ramsar Convention (after the town in Iran where it was adopted in 1971). UNESCO is the depository organization for this convention, whose secretariat, known as the Ramsar Bureau, is located at IUCN Headquarters in Gland, Switzerland.[2] About sixty biosphere reserves include (or are part of) areas that are also inscribed on the Ramsar List of Wetland Sites of International Importance, with a Joint Programme of Work (signed in 2002).
>
> ---
>
> 1   Up to 1992, the cultural and natural parts of the Convention had been serviced by the Culture and Natural Sciences Sectors of the Organization, respectively, with each sector alternatively taking lead responsibility for the organization of successive sessions of the World Heritage Committee and its Bureau. In terms of the natural heritage, up till 1992 the staff servicing the natural part of the World Heritage Convention were the very programme specialists within the Division of Ecological Sciences responsible for the promotion of biosphere reserves.
> 2   In August 2003, Peter Bridgewater became secretary-general of the Ramsar Bureau, after previously serving as director of UNESCO's Division of Ecological Sciences and secretary of the MAB Council.

Advisory Committee for Biosphere Reserves noted the specificity of biosphere reserves and reviewed approaches for the future development of reserves, as well as the means for improving the quality of the network, addressing topics such as biogeographical coverage, zonation, management plans, the biosphere reserve nomination form, legal considerations, and the role of regional networks in promoting such perspectives as the societal dimensions of biosphere reserves.[53] The first meeting could not solve the many issues involved in a definitive way and came too late to have an impact on the wording of Agenda 21, adopted by the June 1992 Rio Conference. It was thus decided that some stronger specific action

---

53   An example was a regional EuroMAB working group on 'Societal Aspects of Biosphere Reserves: Biosphere Reserves for People', which first met at Kongswinter (Germany) in January 1995. See Kruse-Graumann et al., 1995.

was required at the international level, and the Advisory Committee devoted its efforts to this objective at its meeting in October 1993.

Crucial in this process was the International Conference on Biosphere Reserves, convened by UNESCO and hosted by the Spanish authorities in Seville (Spain) in March 1995. Among the outputs was the Seville Strategy for Biosphere Reserves, which reaffirms the nature and purpose of biosphere reserves and describes the criteria they must meet for formal designation. Thus, each biosphere reserve should fulfil the three complementary functions (conservation, development, logistical support). Depending on local conditions, a biosphere reserve will naturally fulfil these three functions to different degrees, but their combined presence is required in all cases. On the ground, each biosphere reserve should include three distinct territorial components (core area or areas, buffer zones, and a flexible transition area).

Subsequently, at its twenty-eighth session in November 1995, the UNESCO General Conference approved the Seville Strategy for Biosphere Reserves and also formally adopted the Statutory Framework of the World Network of Biosphere Reserves, which defines the principles, criteria and designation procedure for biosphere reserves, governs the general functioning of the World Network, and, critically, makes provision for a review of the state of biosphere reserves every ten years. Although not a formally binding text for states, the Statutory Framework applies to all biosphere reserves designated within the framework of the MAB Programme and, in effect, constitutes a set of 'rules-of-the-game' (Jardin, 1996).

The relevant resolution of the General Conference invited all Member States to take account of this text when establishing new biosphere reserves and requested the Director-General to ensure the full functioning and strengthening of the World Network in accordance with the Statutory Framework. In short, these two documents – the Seville Strategy and the Statutory Framework – provide the basic texts shaping and guiding the further development of the World Network of Biosphere Reserves and its component parts.

FROM SEVILLE (1995) TO PAMPLONA (2000) AND THEREAFTER

Since the Seville Conference in 1995, a major effort has been undertaken in many countries to review their biosphere reserves, in the light of the Seville Strategy and the Statutory Framework, at the individual site as well as national level. At the same time as this upgrading of long-established reserves, the immediate post-Seville period (1996–2000) has seen some 63 new additions to the World Network – an average of 12.6 new reserves per year, compared to the 186 reserves established in the first five years of designation, 1976–80 (for an average of 37.2 per year). These new reserves (designated during 1996–2000) adhere more closely to the desired multifunctional objectives. Such criteria figured among the considerations of an international expert meeting organized to review the first five years of implementation of the Seville Strategy (1995–2000). This 'Seville+5' meeting was held in Pamplona (Spain) in November 2000, and was based on the three levels of implementation of the Seville Strategy (international level, national level, site level).

The overall conclusion of the whole review process was that – with the Seville Strategy and the Statutory Framework – biosphere reserves have entered a new phase of development. The philosophy and concepts underpinning biosphere reserves have continued to spread into the broader international context, and protected areas are being considered as integral to socio-economic development.[54]

Most importantly, the Statutory Framework has provided a means to encourage those responsible for managing biosphere reserves to keep 'up to date' with the evolving concept, with the periodic review process acting as a mechanism for fostering sites of excellence for conservation and sustainable development (Price, 2002). The impact of the periodic review process is reflected in the expansion of such existing reserves as Omayed (Egypt), Babia Gora (Poland) and Braunton Burrows (United Kingdom), and the delisting of certain long-established biosphere reserves in such countries as Bulgaria and the United Kingdom. Another reflection is the further adaptation of the biosphere reserve concept to particular regions and countries and even sites, as in a review of lessons learned in putting the concept into practice in Latin America and the Caribbean (Jaeger, 2005).

In terms of networking between biosphere reserves, active areas of current development include encouragement to set up transboundary reserves and promote subregional and regional networks of various kinds. Networking in substantive technical areas includes initiatives on the development of quality economies and of integrated monitoring in biosphere reserves. Other topics of current concern include appropriation regimes and conflict prevention and resolution[55] and the role of sacred sites[56] in contributing to biological and cultural diversity.

---

54 This approach is reflected, for example, in the discussions on biosphere reserves during the World Conservation Congresses held in Montreal (October 1996), Amman (October 2000) and Bangkok (November 2004), as well as during the World Conference on Protected Areas held in Durban in 2003. It is reflected in various initiatives to link the biosphere reserve concept and the World Network of Biosphere Reserves with internationally agreed development goals, such as those encapsulated in the Millennium Development Goals (MDGs). It is also reflected in such recognitions as the award in 2000 of the UNEP Sasakawa Prize to Michel Batisse and in 2001 of the Prince of Asturias Award for Concord to the World Network of Biosphere Reserves.
55 'Appropriation regimes and management systems for biodiversity' was one of the MAB-organized workshops during the International Conference on Biodiversity, Science and Governance, organized in January 2005 at UNESCO House in Paris, under the aegis of the French Ministry of Research and the French Institute of Biodiversity (http://www.recherche.gouv.fr/biodiv2005paris) See also: Bouamrane and Ishwaran, 2005; Bouamrane, 2006.
56 An example is the International Symposium on 'Conserving Cultural and Biological Diversity:

STRUCTURE AND FUNCTIONING

As one of UNESCO's programmes of scientific cooperation at the intergovernmental level, MAB has its own structures at national and international levels, including national committees, an international policy-shaping body and a programme servicing secretariat.

MAB NATIONAL COMMITTEES

MAB is a nationally based programme of research, with MAB National Committees providing the mechanisms for animating and orchestrating national contributions to the programme. Optimally, MAB Committees bring together representatives and individuals from a range of disciplines and governmental and non-governmental institutions concerned with issues dealt with by MAB. They should be capable of mobilizing the human and financial resources needed for meaningful national participation in MAB, including participation in collaborative ventures at bilateral, subregional, regional and interregional levels.

In some countries, these conditions have been met more or less satisfactorily. In many more countries, however, there is a gap between theory and practice. MAB National Committees often lack resources, both human and financial. Some observers might argue that, in at least certain countries, MAB National Committees may serve as an obstacle rather than a facilitator to the participation in MAB of scientists in their country. Some Committees may have become ossified, beset by inertia and marginalized within the national scientific community and/or policy-making machinery.

INTERNATIONAL COORDINATING COUNCIL

Guiding and supervising MAB is the responsibility of the International Coordinating Council, composed of thirty-four elected Member States of UNESCO. As noted in the Council's own statutes, the individuals who represent the Member States on the Council are generally experts in the fields of ecology, natural sciences, agronomy, environmental and social sciences, and many are active in implementing MAB activities. They are designated by their countries. The Council normally meets once every two years.

At each of its sessions, the MAB Council elects its officers, comprising a Chairperson (see Box III.4.13), four Vice Chairs and a rapporteur. These persons comprise the MAB Bureau, which discharges such duties as the Council may lay upon it, and generally meets at least once between sessions of the Council.

The Role of Sacred Natural Sites and Cultural Landscapes', organized in May–June 2005 by UNESCO (MAB and World Heritage Centre) and the United Nations University (UNU) as part of the World EXPO 2005 in Aichi (Japan). See also Ramakrishnan et al., 1998.

BOX III.4.13: MAB INTERNATIONAL COORDINATING COUNCIL

Chairpersons: professional specialties and functions, and period of service
- Francois Bourlière, France (human physiology and vertebrate ecology, university professor), 1971–74.
- Arturo Gómez-Pompa, Mexico (tropical forest ecology, university professor), 1974–77.
- Ralph Slatyer, Australia (plant ecophysiology, Ambassador to UNESCO, senior scientific adviser to government), 1977–81.
- Balla Kéïta, Côte d'Ivoire (veterinary medicine, government minister), 1981–84.
- Gonzalo Halffter, Mexico (entomology, research institute director) 1984–86,
- Li Wenhua, China (forest ecology, research-institute director). 1986–90,
- Tânia Munhoz, Brazil (environmental science, national environment institute director), 1990–93.
- Tomas Azcarate, Spain (biology, environmental policy official), 1993–95.
- Peter Bridgewater, Australia (wetland ecology, ministerial-science adviser), 1995–98.
- Javier Castroviejo, Spain (conservation biology, government adviser), 1998–2000.
- Mohammed Ayyad, Egypt (arid zone botany, university professor), 2000–02.
- Driss Fassi, Morocco (geomorphology, university professor), 2002–04.
- Gonzalo Halffter, Mexico (entomology, university professor). 2004–present.

Secretaries: professional specialities and period of service[1]
- Francesco di Castri (terrestrial ecology, soil biology),1971–84.
- Bernd von Droste zu Hülshoff (forest science, conservation), 1984–92.
- Mohamed Skouri (agronomy, resource use in arid zones), 1992–93.*
- Pierre Lasserre (coastal-marine biology), 1993–99.
- Peter Bridgewater (wetland ecology, conservation), 1999–2003.
- Mireille Jardin (environmental law), 2003–04.*
- Natarajan Ishwaran (wildlife biology, conservation science and practice), 2004–present.

*Acting Secretary of MAB Council

---
1   Since the beginning of the Council, directors of the Division of Ecological Sciences have also served as secretaries of the MAB International Coordinating Council.

SECRETARIAT

The origins of MAB lie within the Natural Sciences Sector of UNESCO.[57] This institutional embedding has perhaps inevitably led to a bias in the disciplinary

---
57   From the outset, the programme and its progenitorial stages has been serviced by staff of what was successively known as the Natural Resources Research Division (1960–72), the Division of Ecology and Earth Sciences (1972–73) and then the Division of Ecological Sciences. In late 2004, the Division of Ecological Sciences became the Division of Ecological and Earth Sciences, thus reverting substantively to its earlier title.

background of the members of the MAB Secretariat, whose technical backgrounds have tended to be from various natural sciences disciplines, such as soil biology, forest science, agronomy, wildlife biology and conservation, entomology, genetics, mangrove biology, and so on.

That said, from the early years, the MAB Secretariat has included several geographers, who have played key roles in linking the natural and social sciences as well as in emphasizing crucial issues of scale in environmental research. During the first decade of the programme, staff members of the Social and Human Sciences Department contributed substantively to the planning and early implementation of MAB activities. In the early 1980s, two posts were transferred from the Social Sciences Sector to the Division of Ecological Sciences. And as of mid-2005, the Division includes two staff members whose disciplinary background is in ecological economics, with another in environmental law.

If the balance of disciplinary backgrounds of Secretariat members could be considered more satisfactory now than thirty years ago, the underlying challenge remains that of servicing a programme – which is intended to cut across the natural and social sciences and has a strong educational function – in an institution structured on programme sectors based on nineteenth-century disciplinary lines.

STOCKTAKING, QUALITY CONTROL, EVALUATION AND ASSESSMENT

The International Coordinating Council of MAB and its Bureau are responsible for overseeing the progress and development of MAB. International policy-guiding bodies such as the MAB Council tend – by their very make-up and nature – to be conservative in their judgements and advocacy. Some might consider that such bodies tend to be a brake on change rather than a motor for innovation, and that their assessments are necessarily biased and subjective, being more formalistic than real. For this reason, there may be a need for additional independent assessors in evaluation exercises, using the criteria that have been established for the planning, selection and evaluation of individual particular pilot projects and comparative studies.

Two programme-wide assessments were carried out in the 1980s as joint efforts of UNESCO and ICSU. The first assessment centred on the international conference and exhibit on 'Ecology in Action', in October 1981. The second assessment was carried out by the General Scientific Advisory Panel in 1986 and 1988. Major shortcomings identified by the Panel included: the dispersion of programme activities over too large a number of subject areas, problems related to the scientific quality of projects, a need for greater scientific coherence and for mechanisms for selection and effective evaluation of projects, continuing mismatch between MAB's goals and its resources, and lack of focus on regional and international problems with significant interregional impact. On the positive side, the panel considered that MAB has contributed to awareness and research in the interface between ecology and social sciences, leading to increased understanding and new solutions to land-use problems.

In addition to these two programme-wide assessments, an 'independent' evaluation was commissioned to be undertaken by the Scientific Committee on Problems of the Environment (SCOPE) in 1990–91.[58] Issues addressed included the phasing out of lower-priority projects, management strategies, publications, budget, intramural and extramural cooperation, education and communication. A postscriptum indicated that 'Much more effective evaluation of the Programme would be provided by a permanent review mechanism rather than by ad hoc evaluations as in the past, and with the present evaluation.'

IN GUISE OF A CONCLUSION

The MAB Programme was conceived in the late 1960s, at a time when there was an awakening interest and rising concern about what was happening to our living world and its resources. It was set up as interdisciplinary programme of scientific cooperation at the intergovernmental level, involving scientists from a wide range of natural and social science disciplines, with a focus on problem-oriented research on resource management and people–environment interactions. The programme has sought to involve the participation in field activities of the different groups of people concerned with land use and resource management – including scientists, educators, policy- and decision-makers, and local populations. It has strived to combine field research with training and education and with the communication and demonstration of results. In providing information and tools to convey understanding, a continuing concern has been that of reconciling the international character that a scientific programme should have with the specific national demands of development and the specificities of different ecological regions and cultural contexts. The programme was set up without a predetermined closing date. Some would argue that this very open-endedness was not helpful to MAB achieving the objectives set for the programme. Others might comment that the very nature of the issues addressed by MAB call for actions that are essentially open-ended, of the long term.

In the three and a half decades since MAB was launched, the world has changed apace. There have been many changes in the international firmament – in the instruments and frameworks that the international community has developed

---

58   UNESCO document SC.93/CONF.215/INF.2

for addressing environmental issues, in the regard in which the United Nations system is held, in the changing relations between the state and the international economic system and between government and community.

As an experimental programme of scientific cooperation, the MAB Programme has evolved over the years, with an increasing focus on the World Network of Biosphere Reserves. Over the last ten years or so, the process associated with the Seville Strategy for Biosphere Reserves and the Statutory Framework of the World Network has served as a means of improving the quality and credibility of biosphere reserves, and thereby of advancing the principles of ecosystem management and of sustainability. As one observer commented over two decades ago,

> A biosphere reserve is not just a pretty place, it's an idea and an approach to management. In an ideal world all protected areas would be managed in a 'biosphere reserve manner', with a zoning system which includes strictly protected core areas and buffer zones, institutionalized relationships with the surrounding land and people, management-related research and training programmes, and links with national and international monitoring programmes. In this sense, all of the world's protected areas may one day be 'biosphere reserves' as well, or at least managed in a 'biosphere reserve manner, (McNeely, 1982).[59]

EPILOGUE

This contribution is dedicated to Michel Batisse (1921–2004) and Francesco di Castri (1930–2005). With their creativity and talent, their rigour and charisma, they were decisive in the setting up and shaping of the MAB Programme and in mobilizing worldwide responses to its challenges. They both left an indelible mark, on MAB and also on the broader international communities that they enriched.

BIBLIOGRAPHY

Batisse, M. 1993. The silver jubilee of MAB and its revival. *Environmental Conservation*, Vol. 20, pp. 107–12.

Behrman, D. 1972. *In Partnership with Nature: UNESCO and the Environment*. Paris, UNESCO.

Blair, W. F. 1977. *Big Biology: The US/IBP*. US/IBP Synthesis Series 7. Stroudsburg, Penn. (USA), Dowden, Hutchinson & Ross.

---

[59] Jeffrey A. McNeely, senior scientist at the World Conservation Union, in a 1982 issue of the *IUCN Bulletin* (McNeely, 1982).

Bouamrane, M. (ed.). 2006. *Biodiversité et acteurs: des itinéraires de concertation* [Biodiversity and actors: itineraries of concertation]. Biosphere Reserves Technical Notes. Paris, UNESCO.

Bouamrane, M. and Ishwaran, N. 2005. Réconcilier les intérêts [Reconciling interests]. *Courrier de la Planète*, Vol. 75, pp. 55–57.

Buba Himanowa. 1975. The Obergurgl model: a microcosm of economic growth in relation to limited ecological resources. *Nature and Resources*, Vol. 7, No. 2, pp. 10–21. [Buba Himamowa is a pseudonym for the major authors of the article: F. Bunnell, S. Buckingham, R. Hilborn, G. Margreiter, W. Moser and C. Walters.]

Davis, B. W. and Drake, G. A. 1983. *Australia's Biosphere Reserves: Conserving Ecological Diversity.* Canberra (Australia), Australian National Commission for UNESCO.

di Castri, F. 1985. Twenty years of international programmes on ecosystems and the biosphere: an overview of achievements, shortcomings and possible new perspectives. In: T. F. Malone and J. G. Roederer (eds), *Global Change*, pp. 314–31. Cambridge (UK), Cambridge University Press.

di Castri, F., Baker, F. W. G. and Hadley, M. (eds). 1984. *Ecology in Practice.* Vol. 1: *Ecosystem Management,* and Vol. 2: *The Social Response.* Dublin, Tycooly International Publishing Company, and Paris, UNESCO.

di Castri, F., Hadley, M. and Damlamian, J. 1980. MAB: The ecology of an international scientific project: Insights from the Man and the Biosphere Programme. *Impact of Science on Society*, Vol. 30, No. 4, pp. 247–260.

——. 1981. MAB: The Man and the Biosphere Program as an evolving system. *Ambio*, Vol. 10, No. 2–3, pp. 52–57.

Engel, J. R. 1985. Renewing the bond of mankind and nature: biosphere reserves as sacred space. *Orion Nature Quarterly*, Vol. 4, No. 3, pp. 52–59.

Francis, G. 1985. Biosphere reserves: innovations for cooperation in the search for sustainable development. *Environments*, Vol. 17, No. 3, pp. 23–36.

Franklin, F. and Krugman, S. (eds.) 1979. *Selection, Management and Utilization of Biosphere Reserves.* Proceedings of USSR–USA Symposium, Moscow, 1976. Corvallis, Ore. US Department of Agriculture.

Giacomini, V. 1978. Man and the Biosphere: an amplified ecological vision. *Landscape Planning*, Vol. 5, pp. 193–211.

Golley, F. B. 1992. *A History of the Ecosystem Approach in Ecology.* New Haven, Conn. Yale University Press, pp. 162–63.

Halffter, G. 1980. Biosphere reserves and national parks: complementary systems of natural protection. *Impact of Science on Society*, Vol. 30, No. 4, pp. 269–77.

——. 1981. The Mapimi Biosphere Reserve: local participation in conservation and development. *Ambio*, Vol. 10, No. 2–3, pp. 93–96.

Jaeger, T. 2005. *New Prospects for the MAB Programme and Biosphere Reserves. Lessons Learned from Latin America and the Caribbean.* Montevideo, UNESCO. Spanish version published in 2005 by UNESCO-Montevideo as *South-South Working Paper 35.* http://www.unesco.org.uy/mab/documentospdf/wp35.pdf.

Jardin, M. 1996. Les réserves de biosphère se dotent d'un statut international: enjeux et perspectives. *Revue juridique de l'Environnement*, Vol. 4, pp. 375–85.

Kellett. S. R. 1986. Public understanding and appreciation of the biosphere reserve concept. *Environmental Conservation,* Vol. 13, No. 2, pp. 101–105.

King, A. and Schneider, B. 1991. *The First Global Revolution.* Council of the Club of Rome. New York, Simon & Schuster.

Kruse-Graumann, L., von Dewitz, F., Nauber, J. and Trimpin, A. (eds). 1995. *Societal Dimensions of Biosphere Reserves: Biosphere Reserves for People.* MAB Mitteilungen 41. Bonn, Germany, MAB National Committee.

McAlpine, J., Molloy and B. P. J. (eds). 1977. *Techniques for Selection of Biosphere Reserves.* Report of UNESCO Regional Workshop. Australia and New Zealand, 27 October to 7 November 1977. Canberra (Australia) and Wellington (New Zealand), Australian and New Zealand National Commissions for UNESCO.

McCormick, J. 1995. *The Global Environment Movement.* New York, John Wiley.

McNeely, J. A. 1982. *The IUCN Bulletin,* Vol. 13, Nos 7–9, p. 59.

Meadows, D. H., Meadows, D. L., Randers, J. and Behrens, W. W. 1972. *The Limits to Growth.* A Report for the Club of Rome's Project on the Predicament of Mankind. New York, Universe Books.

Nicholson, M. 1989. *The New Environmental Age.* Cambridge (UK), Cambridge University Press.

O'Connor, K. F., Costello, E. J. and Kerr, I. G. C. 1984. Pastoral production and grassland conservation in the Waitaki, New Zealand: an appraisal of their role in a river basin approach for land use planning. In: F. di Castri et al., *Ecology in Practice.* Vol. 1, pp. 344–67.

Poore, D. 2000. Setting the scene. In: D. Poore (ed.), *Where Next? Reflections on the Human Future,* pp. 3–11. Kew (UK), Royal Botanic Gardens.

Price, M. F. 2002. The periodic review of biosphere reserves: a mechanism to foster sites of excellence for conservation and sustainable development. *Environmental Science and Policy,* Vol. 5, No. 1, pp. 13–18.

Ramakrishnan, P. S., Saxena, K. G. and Chandrashekara, U. (eds). 1998. *Conserving the Sacred for Biodiversity Management.* New Delhi, Science Publishers.

Robertson, B. T., O'Connor, K. F. and Molloy, B. P. J. (eds). 1979. *Prospects for New Zealand Biosphere Reserves.* New Zealand Man and the Biosphere Report No. 2. Canterbury, Tussock Grasslands and Mountain Lands Institute, for New Zealand National Commission for UNESCO and Department of Lands and Survey.

Rössler, M. 1992. On 'Man and the Biosphere' and its history: 1971–1991. Unpublished manuscript. 6 February 1992. Paris, Division of Ecological Sciences, UNESCO.

Samson, P. R. and Pitt, D. (eds). 1999. *The Biosphere and Noonsphere Reader: Global Environment, Society and Change.* London and New York, Routledge.

SCEP (Study of Critical Environmental Problems). 1970. *Man's Impact on the Global Environment: assessment and recommendations for action.* Cambridge, Mass., MIT Press.

Udvardy, M. D. F. 1975. *A Classification of the Biogeographical Provinces of the World.* Prepared as a contribution to UNESCO's Man and the Biosphere Programme Project No. 8. IUCN Occasional Paper No. 18. Morges (Switzerland), IUCN.

UNESCO. 1969. *Use and Conservation of the Resources of the Biosphere.* Proceedings of the intergovernmental conference of experts on the scientific basis for rational use and conservation of the resources of the biosphere (Paris, 4–13 September 1968). Natural Resources Research Series 10. Paris, UNESCO.

——. 1974. *Task Force on Criteria and Guidelines for the Choice and Establishment of Biosphere Reserves.* Paris, 20–24 May 1974. MAB Report Series, No. 22. Paris, UNESCO.

——. 1981. *MAB Information System: Biosphere Reserves.* Compilation No. 2. Paris, UNESCO.

——. 1983. *Task Force on Methods and Concepts for Studying Man-Environment Interactions, Final Report.* MAB Report Series No. 55. Paris, UNESCO.

——. 1984. The Action Plan for Biosphere Reserves. *Nature and Resources,* Vol. 20, No. 4, pp. 11–22.

——. 2002. *Biosphere Reserves: Special Places for People and Nature.* Paris, UNESCO.

——. 2003. *Five Transboundary Biosphere Reserves in Europe.* Biosphere Reserves Technical Notes. Paris, UNESCO.

UNESCO-UNEP. 1984. *Conservation, Science and Society.* Contributions to the First International Biosphere Reserve Congress, Minsk, Belarus (USSR), 26 September–2 October 1983. Organized by UNESCO and UNEP in cooperation with FAO and IUCN at the invitation of the USSR. Two volumes. Natural Resources Research Series, No. 21. Paris, UNESCO.

Vernadsky, V. I. 1998. *The Biosphere.* New York, Copernicus, Springer Verlag.

Worthington, E. B. 1975. *The Evolution of IBP.* International Biological Programme Synthesis Series. Cambridge (UK), Cambridge University Press.

——. 1983. *The Ecological Century: A Personal Appraisal.* Oxford (UK), Clarendon Press.

PART III: ENVIRONMENTAL SCIENCES

ROCKY ROAD TO SUCCESS
A new history of the International Geoscience Programme (IGCP)
Susan Turner[60]

INTRODUCTION

'The original proposal was made in a letter to fifty leading international Earth scientists in 1964, but I had been thinking about it for two or three years before that, or even longer, following considerable involvement in the International Geophysical Year [1957–1958]'.

—H. J. Harrington to Ed Derbyshire (in correspondence, 17 July 2000)

AT a time when there was both the Cold War between the East and West and raging 'hot' wars in Asia and the Middle East, thirty-two men and one woman met in Paris in May 1973 to oversee the first official steps of a significant effort to solve world problems and create social cohesion. The International Geological Correlation Programme (IGCP) was on its way, after almost a decade of hard work, disappointments and setbacks, including avalanche and invasion. Mr John Fobes, then Assistant Director-General of UNESCO, proclaimed that the main tenet of IGCP was not only to promulgate both basic and applied geological research but also to achieve practical results. IGCP was to be a scientific research programme aimed not only at understanding the workings and history of the planet but also at improving man's environment and the search for natural resources. Another fundamental aspect of the programme: any one of the thousands of geologists around the world would have the opportunity to propose projects. This was the built-in 'grass roots' factor that would prove to be the lynchpin of IGCP's success.

As we enter the fourth decade of formal IGCP work, UNESCO's goal of placing geoscience in the service of solving problems for humanity is thriving, with projects on water and other resources, medical geology, and the understanding and reduction of natural hazards. The aim to be of practical use, which was a top priority from the very beginning, has come to fruition. In 2002, Ed de Mulder, president of the International Union of Geological Sciences (IUGS), provided statistics to illustrate the scope of IGCP by sampling a typical year in

---

60 Sue Turner was co-leader of UNESCO-IUGS IGCP 328; member of the IGCP Scientific Board (2000–2004); and currently (since 2001) part of the UNESCO Advisory Group of Experts for Geoparks.

the programme's thirty-year history: in 1996, some forty projects were running in 146 countries. On average, scientists from some thirty countries were involved in each project.

IGCP (renamed the International Geoscience Programme in 2003) was accepted as a joint initiative by the IUGS Council at the twenty-fourth International Geological Congress (IGC) at Montreal, Canada, in 1972, and subsequently at the UNESCO General Conference later that year (see von Braun, 1984). As implied by its name, the study of the Earth – or geoscience – requires global scope for investigations. As Sir Kingsley Dunham, a former IUGS president, noted in an address to the Geological Society, London (Dunham, 1985), the history of IGCP has 'probably involved more converging lines and deeper birth pangs than most'. In fact, each published version of the history of IGCP (see International Union of Geological Sciences, 1969; Skinner, 1992) has given a different interpretation of the programme's beginnings.

Thinking on a worldwide scale began with the foundations of the modern science of geology in the early nineteenth century and was catapulted by Wegener's Continental Drift hypothesis 100 years later (Oldroyd, 1996, 2002). Yet many geoscientists had to wait for the very successful International Geophysical Year (IGY) in 1957 (Crary, 1982; Wilson, 1961) to provide the organizational foundation for international cooperation to work on a grand scale. This research programme, however, mostly attracted physicists and atmospheric scientists; and many geologists felt there was much more to do in solid Earth science. Thus began attempts to create a global geological research effort. Launching such a programme within UNESCO was first discussed at the 1964 Paris General Conference.

On the other side of the world, the geologist H. J. 'Larry' Harrington, having been involved in Antarctic exploration during the IGY, was dreaming up a bold vision for a longer, more comprehensive global geological research programme (Harrington, 1991). He saw a particular need to investigate the relatively unknown southern hemisphere by an exchange of geologists from the northern hemisphere. Most resources and manpower were concentrated in the richer 'West', while the few scientists in developing and southern countries lacked the resources to study the classic geology of Europe. Because of the work he had done across the former 'Gondwanaland', Harrington – like many southern-hemisphere-trained geologists (see Le Grand, 1988) – recognized that much could be understood by testing the continental drift hypothesis in those places. He began to create a network by writing to many of the top geoscientists of the day, including the president of the International Council for Science(ICSU) and finally the IUGS secretary-general, with a proposal for a global effort in research. These overtures led, in 1965, to the formation – at the request of newly elected IUGS secretary-general, William Peter van Leckwijck – of an Australian ad hoc committee to develop the concept of an international geological research programme (Harrington, 1994*a*; Derbyshire, 2001; Turner and Harrington, 2002; Turner, 2003).

The young geological team in Australia then did several years of hard work to lay out the basic structure and principle aims of what has become the most successful of UNESCO-supported research programmes, producing a proposal that was forwarded to the IUGS in 1966. Recent historical research has brought to light the origins of the founding IGCP document (which first used the official title in March 1967), much of which had been forgotten (Turner, 2003). The IUGS took most of the Australian plan word-for-word for their submission to the twenty-third IGC. Unfortunately, that Congress, scheduled for Prague (Czechoslovakia) in 1968, was cancelled due to the Soviet invasion (Schneer, 1995; Winder and Schneer, 1971). A second testing phase by an interim planning committee followed, with a finalized IGCP document appearing in 1972.

The final design of IGCP came about in 1972, when collaboration between UNESCO and IUGS began in earnest, allowing the programme to run much as it does today – with UNESCO handling the operational and administrative details and the IUGS serving as the scientific guide, seeking the best geoscientists worldwide to serve as the arbiters. The first fifteen-member IGCP Scientific Board, under the chairmanship of Sir Kingsley Dunham (UK) – with E. M. Fournier d'Albe representing UNESCO, and Simon van der Heide, the IUGS – initiated some twenty-eight projects distilled from ninety-two proposals. For the intergovernmental UNESCO, linking up with a non-governmental scientific union for joint action on such a major research programme was a new departure.

By 2004, over 300 projects had been completed and 500 series-projects were being accepted. Much important work in the geosciences, especially at an international level and with the aim of solving problems of importance to humanity, has been carried out by IGCP participants. Numerous publications have been produced – with an annual budget that would not suffice for a typical Western household.

THE FIRST ATTEMPT

> The Australian planning committee was eight to ten people aged about 40 – 'a remarkable, young and productive group. They knew what needed doing and how it could be done effectively and cheaply, and time has proved that they were right'.
>
> —H. J. Harrington to Ed Derbyshire (in correspondence, 17 July 2000)

An agreement in principle to start a geological programme was formulated at the 1964 and 1965 UNESCO General Conferences, at the latter in conjunction with the IUGS. Advancement of geological knowledge was the aim, along with rational use of natural resources. These bodies had already taken some steps to promote standardization of terminology, nomenclature and cartographic methods, as well as encouraging research and exchange of scientific information on key geological problems, especially in developing countries. There was, for instance, at least one planned symposium and study expedition to correlate the granites of West Africa (UNESCO, 1965). These activities were directed towards the establishment of intercontinental and international correlations and comparisons. UNESCO and the IUGS had been looking for such a new all-encompassing programme to replace the Upper Mantle Project and bring geoscience to the fore again. After all, nearly a decade had passed since the IGY. Harrington's proposals to the IUGS – along with subsequent ones from the Australians – provided the basis. The IUGS took its time digesting the Harrington proposal sent in mid-1966. By the General Conference of November 1966, after two years of work by the Australian committee, there was a recommendation for an international research programme based on correlation. Van Leckwijck was there as an IUGS observer, promoting this as a joint initiative. Detailed consultations for the preparation of IGCP began in 1967.

The UNESCO-IUGS Coordinating Panel for the International Geological Correlation Programme (in 1967) comprised top geoscientists from around the world, heads of surveys, universities and academies. An official IUGS ad hoc Committee for IGCP was formed and met in Prague, 26–28 October 1967. This group worked on a draft IGCP proposal to be submitted to the next International Geological Congress (IGC).

Vladimir Zoubek, of the Czechoslovakian Survey, hosted the twenty-third IGC in Prague in August 1968. Around forty pre-conference fieldtrips began on 8 August; but just after the main conference began on 20 August, Soviet tanks rolled into the city (see Schneer, 1995). This led to a cancellation of much of the work planned, including the discussion of and voting on the 1968 IGCP proposal at the Council meeting. Van Leckwijck convened a rapid informal executive meeting at his home in Antwerp, Belgium, before the end of August, to decide what to do next.

POST-PRAGUE

'Of course, there was much in between Prague and Budapest, not official'.

—W. B. Harland to Ed Derbyshire (in correspondence, 2000)

The next planning meeting was held in Budapest, Hungary, on 11–16 September 1969, where the first *official* step to create IGCP was taken (Speden, 1970). This preparatory meeting of experts – convened by UNESCO in collaboration with IUGS – brought together nearly 100 geologists from thirty-five countries. Budapest was chosen because the Hungarian Geological Survey was celebrating its 100th anniversary. There, the IGCP Working Document SC/IGCP/3, with limited distribution, was produced (ibid., p. 25). The 'emphasis [was] on time correlation since relationships to Earth history are important in all branches of geology' (ibid., p. 5). The experts settled on six divisions – introduced with examples:

1. clarification of stratigraphic principles, terminology and procedures;
2. the application and evaluation of methods of time correlation;
3. standard definitions for the principal units of the worldwide chronostratigraphic scale;
4. the promotion of quantitative methods and data processing with respect to geological correlation;
5. the study of patterns in time and space of geological events;
6. the study of the genesis of economic deposits in relation to other events in Earth history.

New techniques were cited – radiometric dating, new concepts in biostratigraphy, 'automatic processing', and quantitative methods for greater precision and accuracy. The author of the UNESCO document predicted that new types of important correlation projects, especially those of an interdisciplinary nature or originating from circles outside the present organization, would be given proper attention. But did that happen? There was still no mention of the plate tectonics revolution underway. There was virtually no mention of the raging contemporary debate on continental drift; the 'old guard', who were mostly against any part of the drift hypothesis, dominated discussions. Speden, who gives virtually the only young 'outsider's' account, was rather horrified with their complacency. Clearly, so were others, because, by the final document, some redress was made, and the structure again encompassed much of the original document.

1972 – IGCP GETS UNDERWAY

The Intergovernmental Conference of Experts for Preparing an International Geological Correlation Programme was held at UNESCO House in Paris in October 1971, in pursuance of Resolution 2.321, adopted by the sixteenth UNESCO General Conference. Delegates from fifty-one Member States, UN organizations, and non-governmental organizations (NGOs) were welcomed by the Director-General René Maheu. This meeting had to consider the aims and objectives of IGCP, specifying procedures and a draft programme. Four scientific working groups each made reports, with a fifth dealing with the structure, coordination and execution of the planned IGCP. Initially a project was expected to run for eight years, with a review after the first five (UNESCO, 1971).

The IUGS president, addressing the Intergovernmental Conference of Experts for the preparation of IGCP in 1971, said that it was 'evident to all who are familiar with the scene that there is a great need to bring the data of the Earth sciences into a proper state of correlation on a fully international scale'. The delegates discussed the 'population explosion' and the increasing demand for consumer goods and energy in both the developed and developing worlds. Geologists were to answer this need by improving prospection methods, based on advancements in geological knowledge, with geological correlation taken in its broadest sense. Such geological knowledge was also supposed to aid rural and urban planning. Universally accepted standards and definitions, and, especially, international correlation were needed. Also needed was a long-term approach (as Harrington had first suggested) and flexibility, so that new directions could be introduced as required.

The dearth of geologists in some parts of the world or the scarcity of some specialists prompted the recognition that although basic education was not part of IGCP, more specialist workshops and field meetings would be important, especially in developing countries. A Board and Scientific Committees were recommended, along with national committees, a secretariat and projects with working groups; guidelines and procedures for each were set. Links were forged with other UNESCO programmes and NGOs, but none gave a financial commitment. The resulting 1971 IGCP draft programme document was, in the end, 'a judicious compromise', linking the basic science with the applied results needed by politicians and communities.

1972-2005

> 'Now we have to eat together the poor IGCP porridge cooked by others ... I hope we shall be able to do it'.
>
> —Vladimir Zoubek to Brian Harland
> (in correspondence, 23 June 1972)

Philip Abelson (1913–2004), IUGS president, and Simon van der Heide, secretary-general, brought the IGCP document to the Council meeting at the twenty-sixth IGC at Montreal. Glaessner, in his report to the Australian Academy of Science, happily announced the success and his own future involvement. In

PART III: ENVIRONMENTAL SCIENCES

late 1972, the General Conference of UNESCO finally signed off on the joint venture, considering rightly that the Earth sciences

> must develop general principles for analysis and understanding of regional phenomena which may have their clearest expression in widely distant parts of the globe and that geological correlation may lead to an evaluation of study methods and principles and, taken in a broad sense, may provide an important means for locating new resources and expanding those which are already known (UNESCO, 1972).

The revolution in the Earth sciences then underway (Hallam 1973; Le Grande, 1988) had made it possible to formulate realistic research programmes. The two sponsoring organizations approved statutes for the IGCP Board. The director-general and the president of the IUGS then appointed fifteen eminent geoscientists to undertake the task of formulating scientific committees and both fostering and assessing the first official projects.

From its inception, the founders set a framework of selected fields of prime importance (e.g. time, terminology, the Precambrian) (Box III.5.1). The Board imposed guidelines as necessary. A major outcome of IGCP in conjunction with the sub-commissions of IUGS has been the increased coverage and accuracy of the geological timescale (e.g. with the magazine *Episodes*). As younger geologists proposed projects and became members themselves of the Board and Scientific Committees, consideration of the correctness of the ideas of continental drift and plate movements took hold. National Committees were forged (some 102 in the first year or so), and the IGCP Board met for its first meeting in May 1973. Themes were outlined: Ordering the Past: refining the geological calendar; The Beginning: evolution of the ancient crust; Man's Home: his geological environment; Man's Needs: energy and minerals.

The main problem in the early years was lack of money. There were other minor difficulties with personal jealousies and conflicts. Some wanted to see the effort going into other research programmes, such as the successor to the Upper Mantle Project. But for the first time in history, thousands of geologists were cooperating all over the globe. More importantly, this included people from developing and southern hemisphere countries, who had mostly been excluded from nineteenth- and twentieth-century science.

In the first year, ninety-two project proposals were received, nearly thirty of which were approved in the first three divisions (Division IV missed out) (Petranek, 1974; *Geological Correlation*, 1973, Nos 1, 2). Geologists from Africa, as from many developing and smaller countries, were slow to contribute, and so the need for further publicity was realized. The mechanism of regional meetings (including those of the Board and Scientific Committees) was important, but financial constraints have always limited these avenues. 'Super' projects[61] exploited the increasing need for fertilizers and brought in many developing countries. A lot of hard work by thousands of geologists, in over eighty countries, ensued.

BOX III.5.1: IGCP's ORIGINAL PRIORITY AREAS OF RESEARCH

Following are the original four divisions of research defined by the planning committees – the first IGCP Board and Scientific Committees – in the early 1970s, which provided the forerunners for the working groups of the modern Board. Interestingly, Division IV has ceased to be a major theme, whereas the Earth processes deemed important in the original concept have moved to the fore in the Tectonics Working Group:

I.   Time and stratigraphy: principles, methods, definitions.
II. Major events in time and space and their environmental implications: patterns of geological processes.
III. Distribution of mineral deposits in space and time: relation of their formation to other geological processes.
IV. Quantitative methods and data processing: standardization, geomathematics.

Within these four Priority Areas for projects – on which the efforts and limited resources of IGCP were to be concentrated – in order to promptly achieve valuable results in fulfilment of its original aims, the following additional focus was set:

A.   Methods of time determination and correlation:
1.   evolution of the Earth's crust, with special reference to the Precambrian crust;
2.   the Quaternary Period, its significance for shaping man's environment.

B.   Sources of energy and minerals.

IGCP Guidelines now read as follows:

> IGCP pursues four broad objectives:
> - improving our understanding of the geoscientific factors controlling the global environment in order that human living conditions may be improved;
> - developing more effective ways to find and assess natural resources of minerals, energy and groundwater;
> - increasing understanding of geological processes and geological concepts through correlative studies of many locations around the globe;
> - improving research standards, methods and techniques of carrying out research.

---

61   For example, Project 32: Stratigraphic correlations between sedimentary basins in the ESCAP (Economic and Social Commission for Asia and the Pacific) region, and Project 156: Phosphorites of the Proterozoic-Cambrian.

PART III: ENVIRONMENTAL SCIENCES

The International Geological Correlation Programme's first international field workshop and seminar on 'Phosphorites of the Proterozoic-Cambrian', held near Mount Isa, Australia, 1978. Co-leaders: John Shergold (centre) and Peter Cook (far right).

MONEY

Financial stringency has been a hallmark of IGCP. It is said that UNESCO and IUGS dollars multiply fifty- or two-hundred-fold through the IGCP programme, if leaders are skilful (Seibold, 1988). Sources include UNESCO national commission and IGCP committee grants, either from government directly, or via a national geological survey or academy. Participants of those countries that have a grant scheme clearly benefit. Still, in 2005, the main problem was finding airfare to get to meetings on the far side of the world. As in 1972, research grants, both institutional and individual, bolster the actual efforts. However, if politicians want to see value for money, if not sheer magic, then they should examine most IGCP projects. Compared with the IGY (which had 50,000 scientists over three years, with thousands of dollars available to them), IGCP has survived – to put it colloquially – on 'peanuts'.

Anders Martinsson had witnessed the first birth pangs of IGCP and created *Geological Correlation* magazine, but by 1976 he was not happy with the financial arrangements. The same topic had dominated the backroom talk at the twenty-fifth IGC in Sydney (Australia) in 1976 (Turner, 2003). This led to considerable changes – with the Board and Scientific Committee travel reduced and, eventually (after the 1994 review), the abandonment of the Scientific Committees, in favour of a combined role for the Board. In general, the amounts given have not kept pace with inflation and in recent years have been severely cut.

THEMES

The men who finally forged the UNESCO-IUGS IGCP programme first identified certain priorities and themes that were uppermost in their minds in the late 1960s and 1970s (see Box III.5.1). These were: the importance of accuracy and definition of nomenclature to correlation, whether stratigraphic or otherwise; the identification of geological processes; the assessment and discovery of economic minerals; and the Precambrian. These themes are still reflected today, although the recent removal of the word 'correlation' perhaps shows a shift in emphasis away from traditional basic geological projects. Interestingly, an early document put forward to IUGS by the Australian ad hoc Committee in 1966 had as a key aim to test the continental drift hypothesis; and one of the first projects, IGCP 3, followed through with this aim; Earth processes in general are now a major underpinning for many projects. In 1987, a new sub-theme was added: Quaternary Geosciences and Human Survival. In addition, that year the Board was asked, by the assembly of General Conference, to consider 'new ways and future trends in mineral exploration'.

GEOSCIENCE SERVING SOCIETY

An official directive for a new priority came from the Board and UNESCO, after the first large evaluation in 1993, and project proposals that served society were to be received favourably. This emphasis reiterated the societal obligation for project aims, instituting the subtitle 'Geoscience in the service of mankind' (Brown, 1994).

As noted above, the projects had always had a practical and applied raison d'etre. Division III of the Scientific Committee had always been directed to immediate practical results on minerals and energy evaluation, including water resources. Over seventy out of some 300 projects (as of 2004) have had applied features. One of the most important early projects was the Economic and Social Commission for Asia and the Pacific (ESCAP), which covered Iran (Islamic Republic of) through Pakistan and India to the countries of the Western Pacific Ocean. Study of the sedimentary basins in this vast area was aimed at oil resources. ESCAP contributed to a rational exploration and exploitation through a uniform approach to understanding of the geological history and its representation in comparable maps and graphic sections, on the basis of a uniform standard of correlation. Remote sensing – now the subject of its own joint programme, as well as of current IGCP Project 474 (Depth Images of the Earth's Crust) – began in 1976, with Project 143 (Remote Sensing and Mineral Exploration), which kept pace with developing satellite technology.

PUBLICATION RATE

Scientific output rather than citation indices has been the baseline for an IGCP project, the production linked to actual IGCP meetings being most favoured: a symposium volume, set of abstracts and/or field guides and maps are the most common result. Many of these bear the hallmark pick hammer logo and can now be found on the internet. Compilation catalogues were produced by UNESCO (1980, 1984), and lists are provided in *Geological Correlation*. In 1988, E. Seibold, president of IUGS, boasted of 40,000 publications; good projects have on average around 1,000 publications over their five-year stint. Such productivity enabled the United Kingdom and the United States to continue their financial (and other) support of IGCP, even when their governments withdrew from UNESCO. The first IGCP websites began appearing in the early 1990s. A recent search brought up over 5,500 websites dedicated to IGCP projects.

REVIEWS AND EVALUATIONS

There have been regular summary reports on the scientific achievements of IGCP in *Geological Correlation* magazine (Box III.5.2) (see von Braun, 1984; Kaddoura, 1980) but no overall analysis on the way the knowledge has been disseminated to people, politicians or industries. The General Meeting 'Eight Years of the IGCP' (held at the twenty-sixth IGC in Paris) was meant to serve as a platform for comment, criticism and summaries of the experience of IGCP, according to UNESCO's Assistant Director-General for Natural Sciences, A. Kaddoura (lessons had been learned from the poor showing at the twenty-fifth IGC). Mr Kaddoura congratulated the perseverance and even stubbornness of geologists collaborating at the international level. 'Despite all the physical, political and psychological obstacles', they achieved 'what many only talk about: partnership, dialogue and consensus on the one hand and scrutiny, criticism and trial on the other'. The outcome was headway into new intellectual and physical terrains of the world. His advice:

> Continue! Be satisfied with what has been achieved, but do not allow room for complacency, neither with us (UNESCO/IUGS) nor yourselves. International appreciation is not measured by the modest sums appearing in our various contracts for geological research training, reviews and the like; it is rather the response elicited by the IGCP that we should like it to be measured against (Kaddoura, 1980).

He also emphasized the growing rift between those who have the knowledge and those without.

Skinner (1992) – after two decades calling for the first *independent* peer review – distilled the essence of IGCP, which had 'far exceeded the hopes of its founders'. He saw results falling into three major categories: development of better methods of finding and assessing resources; a better understanding of the geological events and Earth processes that affect human activities; and standardization of terminology and procedures in research, and development of new methods of correlation. We might add Harrington's first aim: to find ways of testing significant new hypotheses.

> **BOX III.5.2:** *GEOLOGICAL CORRELATION* MAGAZINE
>
> This A4 publication with a distinctive yellow cover has served as the 'mouthpiece' of IGCP. The first issue of the annual journal *Geological Correlation* was produced in 1973 as a joint enterprise with the International Union of Geological Sciences. With the formation of the joint cooperative research programme IGCP, it was decided at the first Board meeting in May 1973 that a publication of the annual meeting reports of the Board and the Scientific Committees, as well as all the annual reports of the projects, was needed. Dr Anders Martinsson (then in charge of the IUGS publications committee) took on the task of editing the first issues. The new periodical was designed to reach all the burgeoning National IGCP Committees as well as IGCP project participants. The well-known IGCP pick hammer logo graced the cover from the second issue onwards. In the first years, the issues included publication of the results of the projects, but later these were also issued as Special Publications.
>
> At high points, notably in 1992, the issue carried several feature articles by members of the Board, the majority reporting on the substantive findings and approaches of projects and activities contributing to UNESCO's IGCP programme. The magazine is published by UNESCO in English and French and has now completed thirty-two issues.

RESULTS

'Overall ... it has been a good thing'.

—Brian Harland to Ed Derbyshire
(in correspondence, 22 March 2000)

IGCP has become increasingly important in forging ideas, tests and concepts of global tectonics, and at solving problems that directly affect humanity, without obstacles based on political or national boundaries. The scientific work and publications (many thousands, from over 250 projects) that result from IGCP cooperative projects worldwide more than justify the continuance of any governmental support for the UNESCO-IUGS programme. Free-range, as opposed to mission-guided, research projects provide inspiration and motivate international cooperation, providing the greatest opportunity for scientists to take the lead in collaboration with colleagues in other lands and to support governments in furthering this collaboration. Both government and non-government support is still needed at all levels to promote geosciences; the recent major earthquake/tsunami events have underscored this. Basic geology underpins all activities related to primary industry. These themes of importance to the survival of humankind are uppermost in the current IGCP programme. Education in Earth Sciences is another new directive of the Earth Sciences and Water Divisions of UNESCO, which is being promoted through IGCP and UNESCO Assistance to Geoparks. These are still the only means by which many geoscientists get to travel to, and undertake research in, overseas countries. Despite satisfying all the original objectives and more, external publicity from UNESCO and IUGS has been almost nil. Martinsson early on (in 1974) complained of the lack of profile given to the programme. UNESCO preferred internal, controlled articles emphasizing key projects.

The less-scientific papers and articles regarding the work and progress of IGCP projects are useful and sometimes invaluable for the media, the politicians, and the historians of science, to see how the different project leaders brought their proposals to fruition. Many of the original proposals and publications are housed in UNESCO archives; others are scattered in archives around the world. Special thirtieth-anniversary compilations have been published by China, the Czech Republic and Hungary and national listings made (see Turner, 2003*b*). Reports by leaders on all projects are also available in *Geological Correlation*.

Since 2002, numerous documents concerning the origins of IGCP (Period 1, 1964–72) have been unearthed from libraries, archives and personal holdings. Many of the people who worked to initiate IGCP or were active in the early years have been contacted, with face-to-face interviews providing a first oral history. This early phase (Turner, 2003) offers an illustration of what Johnson and Buckley (1988) have called the 'closed politics' of the time – when the advice and influence of a small coterie (in this case, a few members of the National Committee of Geological Sciences of the Australian Academy of Sciences, and the IUGS Executive) 'massaged' the original proposal to suit their needs. These few saw the merits in the ideas and promoted the 'Southern Continents Project' proposal into the fully fledged IGCP draft document.

The planning phase 1969–74 saw attempts by the Board and Scientific Committees to control the conducted research, which almost strangled the progress. But, fortunately, after 1976, the grass-roots philosophy prevailed, and the tensions created by almost half the allotted money being spent on the assessment process dissipated. This continued in 1995, with the reduction from forty-three in the original Board plus Scientific Committees to sixteen (and then back to twenty in the current Board). A normal four-year time limit was set for members, instead of the previous six (e.g. in 1994). However, until Member States and governments invested direct grants, it was still difficult for many geologists to gain access to enough funds to tackle IGCP research, especially in the southern continents; the 'tyranny of distance' still prevails, as it did when Harrington first had his idea in the 1960s.

The years 1972–2004 saw a lot of hard work by thousands of mostly uncredited geologists. *Geological Correlation* provides the only unified history of project aims and cooperative outcomes and scientific achievements. The files at UNESCO are not so informative, and one would have to go back to individual leaders to see if archives were kept. Given the parlous state now of university archives, it is likely that in a few years nothing will remain. And, since from around 1993, so much has been done by email, such ephemera will not remain to record the day-to-day running of projects.

IGCP has been an extremely important endeavour – even an experiment – in cooperative science. There is still no other programme like it. It has shown that cooperation rather than competition builds good and ethical science, often with immediate results (the symposia, publications, maps and field guides – and even popular articles and websites – flow rapidly once a project has begun). More popular reporting of the work is still needed; but, with the advent of the World Wide Web, this is already happening.

In the thirty years since its inception, IGCP has made substantial contributions to our knowledge of the solid Earth – contributions that range from a detailed understanding of how and why deserts expand and retreat, to databases for the whole globe documenting the type and concentration of chemical elements of the Earth's surface that are essential for agriculture, industry and human health. One of the latest activities focuses on the impact of geology and the natural environment on human and animal health. At present, several IGCP endeavours cluster around 'climate change and desertification', and 'structure, tectonics and drifting continents', reflecting the fact that the Scientific Board does not commission projects, but holds fast to its tradition of operating in response mode.

THE FUTURE

The objectives originally laid down have not changed: IGCP projects must be relevant to major scientific and practical objectives of the programme; they must meet a worldwide, continental or regional demand; preferably be multi- or interdisciplinary; they require international action and facilitate common understanding between specialists from different countries and cultures; they result not only in long-term benefits to science and humanity, but also, whenever possible, yield short-term practical results that will increase capabilities within (especially developing) countries. IGCP has always been cost-effective and flexible and can continue to be so.

In general, for individual scientists the benefits of taking part in IGCP far outweigh the difficulties of long-distance and/or cyclical lack of funding and employment. Discontinuance of geology and (especially) palaeontology courses in universities, along with economic pressure to work only for large amounts of money, however, is leading to a generational gap, with fewer young workers coming up to participate. This problem was addressed in 2003, by offering a Younger Scientist's Programme – three-year projects of more limited scope to encourage younger people to try out their skills. The shortfalls also create difficulties for established scientists in gaining the necessary backing of their institutions to undertake such pro bono international work. Of three projects accepted in 2003, co-leaders of two are unemployed or self-employed scientists. In recent highly successful projects (as defined by the IGCP Scientific Board), all Australian leaders were retired or semi-retired. Therefore the financial backing provided by the national IGCP Committees, which enables people to participate in IGCP meetings, is essential. The annual funds at international or national level, however, have not increased in more than a decade and thus now have less 'buying power'. Interestingly, as many research programmes follow the principle of including developing nations – and conferences are often held in those countries – there is now increasing financial pressure, as those countries have reacted to the market economy by significantly increasing prices (for accommodation, field trips, etc.), compared to a decade ago.

In the long run, as pointed out by Seibold (1988), the most important correlation IGCP has fostered has been that of people. Literally thousands of human connections have been made that could never have been otherwise achieved. This is a social or cultural aspect of scientific endeavour that is sometimes forgotten. As project leaders well know, when you gather together a few or more scientists to 'brainstorm' an idea or a problem, so much more can be achieved than by letter or email. The key to the success of IGCP, and of its future role, is that it was built around and for people.

BIBLIOGRAPHY

Abelson, P. 1973. A new international programme. *Science*, Vol. 182, No. 4108, p. 119. [Editorial on International Geodynamics Programme]

Babuska, V. 1997. Report of the IGCP Secretary for 1996. *Geological Correlation*, Vol. 25, pp. 7–11.

Bally, A. W. and Bharadwaj, D. C. 1983. General stratigraphic projects: Summary. In: Bally and Hoover (eds), pp. 15–16.

Bally, A. W. and Hoover, L. (eds). 1983. IGCP – Science resources and developing nations: a review and a look into the future. In: *Geological Correlation*, special issue, September.

Bally, A. W., Bharadwaj, D. C., Cooray, P. G. and Cordani, U. 1983. Science, resources, and the developing nations: a review and a look into the future. In: Bally and Hoover (eds), 1983, pp. 11–15.

Bassett, M. G. (ed.). 1978. IGCP scientific achievements: 1973–1978. *Geological Correlation*, special issue, September.

Behrman, D. 1979. IGCP – A trip to planet Earth. *The Australian Geologist*, Vol. 24, p. 11.
Bouckaert, J. 1990. William van Leckwijck. In: C.C. Gillespie and F. L. Holmes (eds), *Dictionary of Scientific Biography*, Vol. 18, Suppl. 2. New York, Scribner's, pp. 948–49.
Brett, R. and Schmidt-Thomé, M. 1994. International geological science cooperation: the role of IUGS. *Episodes*, Vol. 17, No. 4, pp. 118–120.
Brown, M. 1994. IUGS and UNESCO review the International Geological Correlation Programme (IGCP). *Episodes*, Vol. 17, Nos 1 and 2, pp. 24–25.
Cook, P. J. and Shergold, J. H. 1982. Simply super. *UNESCO Review*, No. 6, March, pp. 15–18.
Crary, A. P. 1982. International Geophysical Year: its evolution and US participation. *Antarctic Journal of the United States*, Vol. 17, No. 4, pp. 1–6.
De Mulder, E. 2002 Report of the President of the IUGS. *Geological Correlation*, Vol. 30.
Derbyshire, E. 2000. The cheapest geoscientific show on Earth. *The Geoscientist*, Vol. 10, No. 2, February, p. 14.
———. 2001. The changing face of IGCP: a note from the Chair. *Geological Correlation*, No. 29, pp. 12–13.
Dunham, K. C. 1985. Presidential address. Proceedings of the London Geological Society for 1985. London, London Geological Society, pp. 1–5.
F. H. 1975. Obituary for William Peter van Leckwijck (1902–1975). Geological Society, London, Annual Report for 1975. London, London Geological Society. [Author probably Frank Hodgson]
Frick, C. 1994. North–South cooperation in geological science: what to expect in the twenty-first century. *Episodes*, Vol. 17, No. 4, pp. 123–25.
Geological Correlation. 1973. Nos 1–32. Paris, UNESCO.
Hallam, A. 1973. *A Revolution in the Earth Sciences: from continental drift to plate tectonics*. Oxford, UK, Clarendon Press, 127 pp.
Harrington, H. J. 1991. The scientific objectives of the IGY years. In: Branagan, Gibbons and Williams (eds), *The Geology of Antarctica, First and Second Edgeworth David Day Symposia, Abstracts*. Sydney, Australia, the Edgeworth David Society, University of Sydney, pp. 113–15.
———. 1994a. The Australian origins of the IGCP (International Geological Correlation Programme). Twelfth Australian Geological Convention, Perth, September 1994. *Geological Society of Australia, Abstracts*, No 37, pp. 166.
———. 1994b. The International (Geological) Correlations Project: how it started, what happened to it and why. In: D. F Branagan and G. H. McNally, *Useful and Curious Geological Enquiries beyond the World: Pacific-Asia historical themes*. Nineteenth International INHIGEO Symposium, Sydney, Australia, 4–8 July 1994. Sydney, INHIGEO, pp. 59–62.
Harrington, H. J., Dickins, J. M. and Campbell, K. S. W. (eds). 1966. International Union of Geological Sciences Proposed Southern Continents Project. Explanatory Note, prepared and sent 31 May 1966 (dated June). Canberra, Basser Library.
Harrison, J. M. 1978. The roots of IUGS. *Episodes*, Vol. 1, pp. 20–23. www.iugs.org.
Hedberg, H. D. 1969. Circular 23, International Subcommission of Stratigraphic Classification, p. 11–16.
Hollingworth, S. E. 1962. Our society and the geological sciences. *Quarterly Journal of the Geological Society*, London CXVIII, No. 472, December 1962, pp. 455–72.

PART III: ENVIRONMENTAL SCIENCES

International Geological Correlation Programme (IGCP). 1967. Report of the Ad Hoc Committee for the International Geological Correlation Programme. Prague, 26–28 October, 1967, 5 pp + 2 pp Appendix. Paris, UNESCO.

——. 1980. Agenda General Meeting: Eight years of the IGCP. 15 July 1980, Paris, Twenty-Sixth IGC. Document SC/GEO/541.13.1. Paris, UNESCO.

International Union of Geological Sciences (IUGS). 1969. History of the International Geological Correlation Programme. Document SC/IGCP/3, prepared for Budapest meeting in 1969. *IUGS Newsletter*, Vols 18–19.

Jackson, A. and Barraclough, D. 1998. Contemporary use of historical data. In: Good and Gregory (ed.), *Sciences of the Earth: An Encyclopedia of Events, People and Phenomena*. 2 vols. New York and London, Garland Press, Taylor & Francis, pp. 115–17.

Johnson, R. and Buckley, J. 1988. The shaping of contemporary scientific institutions. In: R. W. Home (ed.), *Australian Science in the Making*. Sydney, Australia, Cambridge University Press, pp. 374–98.

Kaddoura, A. 1980. Eight years of the IGCP. Address of Assistant Director-General for Science. *Geological Correlation*, Vol. 16, p. 7.

Le Grand, H. 1988. *Drifting Continents and Shifting Theories: the modern revolution in geology and scientific change*. Cambridge, UK, Cambridge University Press, 313 pp.

Martinsson, A. 1973. First session of IGCP Board, Paris, 22–25 May 1973. *Geological Correlation*, No. 1. SC/MD/37 IUGS-UNESCO-IGCP (September). Paris, UNESCO, 25 pp.

——. 1974. Editor's column: International Geological Correlation Programme. *Lethaia*, Vol. 7, pp. 171–72.

——. 1976. Editor's column: Stratification in international geology. *Lethaia*, Vol. 9, pp. 459–62.

Mason, A. 2002. International Geological Congress: Prague 1968. *Geological Society of New Zealand Historical Studies Group Newsletter*, No. 25, September, pp. 27–30.

Millbrooke, A. 1998. International Geophysical Year. In: G. Good (ed.), *Sciences of the Earth: An Encyclopedia of Events, People and Phenomena*, Vol. 2. New York and London, Garland Press, Taylor & Francis.

Oldroyd, D. R. 1996. *Thinking about the Earth: a history of ideas in geology*. Cambridge, Mass., Harvard University Press, 410 pp.

——. 2002. *The Earth Inside and Out: some major contributions to geology in the twentieth century*. London, Geological Society of London, 369 pp.

Petranek, J. 1974. The First Year of the International Geological Correlation Programme. Address to Board meeting (April in Vienna), dated 11 July 1974. *Geological Correlation*, Vol. 2, No. 4.

Reinemund, J. A. and Watson, J. V. 1983. Achievements of the international geological correlation programme as related to human needs. In: Bally and Hoover (eds), 1983, pp. 9–11.

Ronner, F. 1974. The meaning and purpose of IGCP. *Nature and Resources*, Vol. 10, No. 2, pp. 27–28. Paris, UNESCO.

Seibold, E. 1988. Address by E. Seibold, Acting President of IUGS. *Geological Correlation*, Vol. 16, p. 7.

Schneer, C. J. 1995. The geologists at Prague: August 1968. History of the International Union of Geological Sciences. *Earth Sciences History*, Vol. 14, No. 2, pp. 172–201.

Skinner, B. J. 1992. Scientific highlights of two decades of international co-operation at the grass roots level. *Episodes*, Vol. 15, No. 3, pp. 200–203.

Skinner, B. J. and Drake, C. L. 1987. An unacclaimed success story. *Geotimes*, No. X, November, pp. 11–13.

Speden, I. G. 1970. International Geological Correlation Programme. *Geological Society of New Zealand Newsletter*, No 28 (March), pp. 25–31. [Report of Budapest meeting for IGCP in April 1969]

Truswell, E. 1997. Stratigraphy in the service of society. Address to twenty-fifth anniversary IGCP Scientific Board. *Geological Correlation*, Vol. 25, p. 7.

Turner, S. 2003. History of Australian Involvement in UNESCO-IUGS IGCP: the early years. Report to the Australian UNESCO Commission, in fulfilment of a grant from Department of Foreign Affairs and Trade, Brisbane, Australia, 64 pp.

Turner, S. and Harrington, L. 2002. Thirtieth anniversary of IGCP. *The Australian Geologist*, Vol. 123, June, pp. 31–32.

UNESCO. 1964. Records of the General Conference, thirteenth session, Paris. 306 p. Paris, UNESCO, 13 C/Resolutions.

——. 1965. Geological sciences. *Nature and Resources*, Vol. 1, Nos 1/2, June.

——. 1967. Advisory Committee on Natural Resources Research, Second Session, Paris, 20–27 June 1967, International Programme for Geological Correlation. Document UNESCO/AVS/NR/235, Paris, 31 March 1967. 7 pp + Annex. Paris, UNESCO.

——. 1969. An international geological correlation programme. *Nature and Resources*, Vol. V, No. 4, December, pp. 2–6.

——. 1971. The International Geological Correlation Programme. *Nature and Resources*, Vol. VII, No. 4, December, pp. 4–9.

——. 1972. Final Report. Intergovernmental Conference of Experts for Preparing an International Geological Correlation Programme (IGCP). Paris 19–28 October 1971 UNESCO SC/MD/28, May, 35pp.

——. 1973. International Geological Correlation Programme: First Session of the Board. *Nature and resources*, Vol. IX, No. 3, July–October 2–6.

——. 1975. *Guidelines for Project Leaders in Geological Correlation*. No. 12, 5 pp. Paris, UNESCO.

——. 1992. *Focus*, No. 38 (June), special issue on IGCP. Paris, UNESCO, pp. 1–9.

——. 1994. *Nature and Resources*, Vol. 30, Nos 3 and 4, special issue on IGCP. Les geosciences et le partenariat: Le programme international de correlation geologique (PICG) au cours de la periode 1988–1993. Paris, UNESCO, 48 pp.

UNESCO AGI. 1980. IGCP Catalogue of publications 1973–1979. Foreword by IGCP Secretariat. Paris, UNESCO. 184 pp. [Covers Projects 1–164]

——.1984. IGCP Catalogue II. 1978–1982. Foreword by IGCP Secretariat. Paris, UNESCO, pp. i–xix + 790pp. [Covers up to Project 197]

Von Braun, E. 1984. Twelve years of IGCP. *Pangaea Geonews*, No. 2, June, 3 pp.

Wilson, T. J. 1961. *IGY: The Year of the New Moons*. Foreword by Lloyd V. Berkner. London, Michael Joseph Ltd, 352 pp.

Winder, C. G. and Schneer, C. J. 1971. Premature termination of the Twenty-third International Geological Congress, Prague, August, 1968. XII Congres International d'histoire des Sciences, Paris, 1968, Actes 1971, VII, pp. 56–58.

THE FINAL FRONTIER
UNESCO in outer space
Robert Missotten[62]

UNESCO's involvement in space activities dates back to the early 1980s. Initial activities were an outgrowth of a longstanding concern for the use of modern technologies as part of integrated approaches to natural resource management and environmental planning.

GEOLOGICAL APPLICATIONS OF REMOTE SENSING (GARS) PROGRAMME

THE Geological Applications of Remote Sensing (GARS) programme was launched by the Division of Earth Sciences and the International Union of Geological Sciences (IUGS) in 1984, after the successful implementation of the International Geological Correlation Programme (IGCP) project 'Development of a Methodology on the Use of Remote Sensing Data from Landsat Satellite for the Exploration of Mineral Resources' (1976–82). GARS aims at assessing the value and utility of remotely sensed data for geological research, and at assisting institutes in developing countries to acquire and apply modern technology in their research work.

The first phase of GARS was undertaken in selected countries in Africa (Botswana, Burundi, Ethiopia, Swaziland, Uganda, the United Republic of Tanzania, and Zambia) from 1985 to 1999. Research work focused on the improvement of lithologic mapping in heavily vegetated terrains by using multi-sensor data and Geographic Information Systems (GIS). The work resulted in the development of a methodology for baseline geological mapping for mineral resource assessment and exploration in inaccessible terrain.

The second phase of the GARS programme was carried out in Colombia from 1988 to 1994. It focused on landslide hazard mapping by using optical and radar data and GIS. As radar sensors see through clouds and rainfall, the fusion of satellite data allows the monitoring of landslides during rainy season. This methodology aims to greatly improve measures to mitigate the impact of landslides.

The third phase of GARS focused on natural hazards of geological origin such as earthquakes, volcanic eruptions, landslides and soil erosion. The Bulusan,

---

62   Robert Missotten has been a UNESCO staff member since the early 1980s and is currently Secretary of the UNESCO-IUGS International Geoscience Programme (IGCP) and Head of the Earth Observation Section of the Division of Ecological and Earth Sciences.

Mayon, Pinatubo and Taal volcanoes in the Philippines were used as test sites from 1995 to 2001 to develop a new methodology in volcanic hazard mapping, using optical, radar and thermal data and GIS. At the end of each phase of the programme, workshops were organized for local geoscientists with a view to transferring knowledge and technology, and training packages were produced in cooperation with its partners – BGR (Germany), CNES (France), CSA (Canada), ESA (France), ITC (Netherlands), JAXA (Japan), RMCA (Belgium) and NASA (USA).[63] The GARS programme is being continued through the IGOS Geohazards Theme programme.

IGOS GEOHAZARDS PROGRAMME

The Integrated Global Observing Strategy (IGOS) is a partnership of international organizations that are concerned with global environmental change issues. Its partners are: the Global Observing Systems (GCOS, GOOS, GTOS[64]), and the international organizations that sponsor them; the Committee on Earth Observation Satellites (CEOS); and the International Global Change Science and Research programmes. The current IGOS Themes include: Oceans, Carbon Cycle, Geohazards, Water Cycle, and Atmospheric Chemistry, with a sub-Theme of Coral Reefs.

The World Summit for Sustainable Development (South Africa, 2002) recognized that systematic, joint international satellite, airborne and *in situ* observation systems help nations to improve the preparation for and mitigation against phenomena of geological and geophysical nature. In 2003, a study on the Geohazards Theme was carried out by the British Geological Survey, the European Space Agency and UNESCO, resulting in the publication of a report that recommends actions based on the four strategic objectives: building global capacity to mitigate geohazards ; improving mapping, monitoring and forecasting based on satellite and ground-based observations; increasing preparedness; using integrated geohazards information products and improved geohazard models; and promoting global take-up of best practice in geohazards management. This report was adopted by the third Global Earth Observation Summit (Brussels, Belgium, February 2005) and included as an integral part of disaster management in the ten-year Implementation Plan of the Global Earth Observing System of Systems (GEOSS).

---

63 The Federal Institute for Geosciences and Natural Resources (BGR), Centre National d'Etudes Spatiales (CNES), Canadian Space Agency (CSA), European Space Agency (ESA), International Institute for Geo-Information Science and Earth Observation (ITC), Japan Aerospace Exploration Agency (JAXA), the Royal Museum for Central Africa (RMCA), and National Aeronautics and Space Administration (NASA).
64 The Global Climate Observing System (GCOS), the Global Ocean Observing System (GOOS) and the Global Terrestrial Observing System (GTOS).

GEOSS aims to achieve comprehensive, coordinated and sustained observations of the Earth system for improved monitoring of the state of the Earth, increased understanding of Earth processes, and enhanced prediction of the behaviour of the Earth system through *in situ*, airborne and space-based observations. GEOSS is expected to meet the need for timely, quality long-term global information as a basis for sound decision-making and to enhance delivery of benefits to society in the following initial areas: natural and human-induced disasters; environmental factors affecting human health and well-being; energy resources; climate variability and change; water resources; terrestrial, coastal and marine ecosystems; agriculture and desertification; and biodiversity.

> **BOX III.6.1: WATER RESOURCE MANAGEMENT**
>
> Space technology is being used to map water distribution and availability, measure the impact of droughts and floods, and collect information on how water is used in areas such as agriculture. In the framework of the International Hydrological Programme (IHP), a partnership has been developed with the European Space Agency's TIGER programme, which aims at supporting African development efforts with pertinent space-based information for water resources development and use. This cooperation – known as TIGER-SHIP (Space–Hydrology International Partnership) – focuses on integration into national institutional development and capacity-building programmes of space technology and space-based information, for better water resource management in Africa.

INTEGRATED MANAGEMENT OF ECOSYSTEMS AND WATER RESOURCES IN AFRICA

The application of remote sensing for integrated management of ecosystems and water resources in Africa has been the theme of the Science and Education Sectors' cross-cutting project for two biennia (2002–2003; 2004–2005). Intersectoral institutions are involved in this project – including ministries of water, land, forestry and coastal ecosystems; and universities, research centres, and NGOs in twelve African countries (Benin, Botswana, Côte d'Ivoire, the Democratic Republic of the Congo, Equatorial Guinea, Guinea, Mozambique, Niger, Nigeria, Senegal, South Africa and Zimbabwe).

MONITORING OF WORLD HERITAGE SITES AND BIOSPHERE RESERVES

An Open Initiative was launched by UNESCO in 2002 with several space agencies ESA, CSA, CONAE [Argentinean Space Agency], NASA, JAXA, etc.) to assist developing countries in monitoring and preserving World Heritage sites. Recently, activities are being extended to biosphere reserves. This has been extremely useful for UNESCO in demonstrating the capabilities that space technologies offer for conservation, as well as to create a basis of local capacity at site level to understand and make use of these technologies. The main goal is to provide local capacity for site managers to become operational in the use of space technologies, in order for them to use these in their daily work, and mainly to be able to report with data and facts in their reporting commitments, with respect to the World Heritage Convention and the Man and the Biosphere Programme (MAB).

SPACE EDUCATION PROGRAMME (SEP)

Participants of the World Conference on Science (Budapest, Hungary, 1999) called for the improvement of science education at all levels, through the development of new curricula and teaching methodologies, in response to changing educational needs and emergence of new technologies. Similarly, the third United Nations Conference on the Exploration and Peaceful Uses of Outer Space (UNISPACE-III, Vienna, Austria, 1999) underlined the need to enhance education and training opportunities in space science and technology and to raise public awareness of the importance of space science and technology to human security and development. Taking into consideration these recommendations, the Space Education Programme (SEP) was launched in 2002, putting space in the forefront and using it to make science subjects more exciting and to equip the younger generation with state-of-the-art technology brought about by the challenges and discoveries of the new millennium. SEP aims at enhancing space-related subjects and disciplines in schools and universities (particularly in developing countries) and their integration in the curricula, and at raising awareness of both decision-makers and the general public of the benefits of space activities for the well-being of their societies and the sustainable development of their countries. In focusing on the three space-related areas – namely, space science, space/aeronautic engineering, and space technology applications – SEP hopes to contribute to the preparation of the next generation of the space workforce, such as astronauts, astronomers, robotic engineers, structural engineers, computational scientists, physicists, and so on.

To achieve the above objectives, SEP develops, coordinates and implements outreach and capacity-building activities such as organization/sponsorship of teacher training courses, development of pedagogical materials, organization of space events, space technology application workshops, etc., in cooperation with space agencies and space-related institutions such as the European Space Agency, Eurisy, International Space University, NASA, National Space Society, Norwegian Space Centre (and NAROM), UN Office for Outer Space Affairs, etc. UNESCO is a member of the working groups on education and capacity-building of the Committee of Earth Observation Satellites (CEOS), Committee on Space Research (COSPAR), International Astronautical Federation (IAF), and Sociedad Especialistas Latinoamericana en Percepción Remota (SELPER).

An important feature of the Space Education Programme is the BRiSSU (Bringing Space to Schools and Universities) initiative, which reaches out to a maximum number of students and teachers/educators at different educational levels (primary, secondary, tertiary) in a certain country at a given time. This consists of organizing space education workshops in several areas in a single country, in which space experts give lectures and demonstration sessions on hands-on projects and best teaching practices, and share information on educational resources and existing networks. At the end of the workshops, a pilot national space programme is defined with the assistance and advice of the experts and submitted to UNESCO for assistance in building partnerships and identifying donor agencies. The national space programme becomes the blueprint for implementation of space activities in the country and eventual integration of space-related subjects in the curricula.

SAFETY FIRST
UNESCO's initiatives in natural disaster reduction
Badaoui Rouhban[65]

'Hazards have always been the lot of humanity', UNESCO Director-General Koïchiro Matsuura stated, speaking of the upcoming (18–22 January 2005) World Conference on Disaster Reduction in Kobe, Japan:

> Throughout history, societies have coexisted with the violent forces and assaults of nature and have had to adapt themselves to their environment. However, we should realize that, as a result of population increase and concentration, our societies are in some ways becoming more vulnerable and that our protective systems are not necessarily adapted to cope (UNESCO, 2003).

Natural disasters are increasing in terms of frequency, complexity, scope and destructive capacity. Indeed, urbanization, alteration of the natural environment, climate change, substandard dwellings and public buildings, inadequate infrastructure maintenance and grinding poverty in numerous communities have all exacerbated the risks of natural disasters. The early years of the twenty-first century alone have witnessed catastrophic earthquakes, tsunamis, heat waves, storms and floods that have ravaged lives and landscapes in the world's poorest nations as well as in the mightiest superpower.

But natural phenomena do not automatically have to cause disasters. Disaster reduction is both possible and feasible if the sciences and technologies related to natural hazards are properly applied. Today, there is more scientific understanding and technological know-how than ever before to anticipate the potential effects of a disaster before it strikes. Of all the global environmental issues, natural hazards present the most manageable of situations: the risks are the most readily identified, effective mitigation measures are available, and the benefits of vulnerability reduction greatly outweigh the costs.

Yet, while disaster relief captures the imagination of the public, disaster prevention often ranks relatively low on public agendas. Relief continues to be the primary form of disaster risk management. Resources spent on relief and recovery continue to account for 96 per cent of all resources spent on disaster-related activities annually, thus leaving a mere 4 per cent for prevention efforts.

---

65 Badaoui Rouhban serves as the focal point for UNESCO's programme on natural disaster reduction. He joined UNESCO's Natural Sciences Sector in 1983.

Cost–benefit analyses support the rationale of disaster-prevention-oriented actions. Some current studies estimate the cost-to-benefit ratio of investing in disaster warning systems, compared with the reduction of disaster-related economic damages, as between 1:7 and 1:10.

Clearly, there needs to be a general shift in emphasis from post-disaster reaction to pre-disaster action. The new emerging approach, to be spearheaded by UNESCO, should stress the merit of preventive measures through the design and dissemination of mitigation techniques, proper information, education and public awareness. As Kofi Annan, Secretary-General of the United Nations, has said: 'Building a culture of prevention is not easy. While the costs of prevention have to be paid in the present, its benefits lie in the distant future. Moreover, the benefits are not tangible; they are the disasters that did *not* happen' (UNESCO, 1999).

There are three measures which must be taken to cope with an existing natural hazard before it becomes a natural disaster:

1. risk assessment – which comprises the assessment of the hazard and of the vulnerability represented by the exposure of the natural as well as the man-made environment to it;
2. prevention – which corresponds to long-term reduction of vulnerability;
3. preparedness – which is achieved through temporary or short-term reduction of vulnerability.

An in-depth scientific study of natural phenomena, and their characteristics and distribution in space and time, is an essential prerequisite for a logical approach to the problems of risk reduction and calls for a multidisciplinary approach, involving cooperation between specialists in several sectors of science and technology. This idea of a multidisciplinary scientific approach is in fact to be found throughout UNESCO's activities relating to the Natural Hazards Programme.

Science and technology and their applications provide powerful means for limiting the effects of natural hazards. Studies of the violent forces of nature are made up of an orderly system of facts that have been learned from study, experiments, and observations of floods, severe storms, earthquakes, landslides, volcanic eruptions and tsunamis, and their impacts on humankind and its works. The scientific and technological disciplines that are involved include basic and engineering sciences, natural, social and human sciences. They relate to the hazard environment (i.e. hydrology, geology, geophysics, seismology, volcanology, meteorology and biology), to the built environment (i.e. engineering, architecture and materials), and to the policy environment (i.e. sociology, humanities, political sciences and management science).

UNESCO's activities in the study of natural disasters and protection against them date from the early 1960s. Originally concerned with basic seismology, these activities were later extended to the reduction of earthquake hazards and, still later, to other categories of natural hazards and their socio-economic aspects. Since 1960, UNESCO has brought an interdisciplinary approach to the study of natural hazards and the mitigation of their effects. These natural hazards include both those of geological origin (earthquakes, tsunamis, volcanic eruptions and landslides), and of hydrometeorological origin (storms, cyclones, floods, prolonged droughts, desertification and avalanches). Being at the crossroads of several sectors, UNESCO provides a unique intellectual setting, linking – within a single organization – the natural sciences with education, culture, communication and the social sciences. With its unusual broad mandate and breadth of expertise, UNESCO is able to integrate many of the essential ingredients for disaster reduction. The purposes of UNESCO in this field can be described as follows:

- to promote a better understanding of the distribution in time and space of natural hazards and of their intensity;
- to help set up reliable early warning systems;
- to devise rational land-use plans;
- to secure the adoption of suitable building design;
- to protect educational buildings and cultural monuments;
- to strengthen environmental protection for the prevention of natural disasters; and
- to enhance preparedness and public awareness through education and training.

And, when catastrophes do strike, to foster post-disaster investigation, recovery and rehabilitation.

UNESCO action is carried out through networking and the strengthening of regional and international coordination systems, direct partnership with Member States, field implementation of operational projects, reconnaissance and advisory missions, and preservation and dissemination of data, seminars and training courses. Constant cooperation is maintained with competent international and non-governmental bodies. Over the years, the Organization has made a significant contribution through a variety of scientific programmes.

SEISMOLOGY AND EARTHQUAKES

The United Nations Economic and Social Council (ECOSOC), at its thirtieth session in July 1960, invited the Secretary-General 'to seek the cooperation of the United Nations Educational, Scientific and Cultural Organization (UNESCO), the World Meteorological Organization (WMO) and other Specialized Agencies concerned in undertaking' studies of the ways of reducing the damage resulting from earthquakes. UNESCO's first activities, in collaboration with the International Union of Geodesy and Geophysics (IUGG), concerned the organization of 'seismological survey missions', which visited South-East Asia, South America, the Mediterranean and the Middle East, and Africa, between May 1961 and April 1963. The findings of these four missions formed the basis of the report submitted by UNESCO to ECOSOC at its thirty-fourth session.

An Intergovernmental Meeting on Seismology and Earthquake Engineering was held at UNESCO Headquarters from 21 to 30 April 1964. The report of this conference, the first large-scale international meeting on the subject, contained the main elements that would constitute UNESCO's seismology programme for years to come: the establishment of international and regional centres for the exchange and analysis of seismological data; the identification of seismically active zones and the quantitative assessment of earthquake hazard in those zones; the establishment of an international tsunami warning system in the Pacific Ocean area; field studies of the effects of large earthquakes, through the dispatch of expert missions to the affected regions as soon as possible after the event; the training of specialists in seismology and earthquake engineering; and studies of seismic activity induced by certain human activities, such as the impounding of large artificial reservoirs.

From 1963 onwards, UNESCO undertook responsibility, as Executing Agency of the United Nations Development Programme (UNDP), for several large-scale projects of training and research in seismology and earthquake engineering, notably in the Balkan region of Europe, in South-East Asia, and in Japan, Mexico, Romania and Yugoslavia. From 10 to 19 February 1976, an Intergovernmental Conference on the Assessment and Mitigation of Earthquake Risk was convened at UNESCO Headquarters. While the previous intergovernmental meeting was concerned mainly with the advancement of scientific and technical knowledge, this conference dealt with interdisciplinary aspects ranging from seismology to the application of scientific knowledge to risk reduction. The resolutions adopted by this conference, in so far as they called for action by UNESCO, have been reflected in the successive biennial programmes of the Organization up to the present time. They led to UNESCO's implementation of major regional initiatives aimed at earthquake risk reduction. These include: the regional seismological networks in South-East Asia and the Balkan region; the Programme for Assessment and Mitigation of Earthquake Risk in the Arab Region (PAMERAR); the UNESCO–US Geological Survey (USGS) programme on Reducing Earthquake Losses in the Eastern Mediterranean Region (RELEMR) and Reducing Earthquake Losses in the South Asia Region (RELSAR); and the Regional Seismological Centre for South America (CERESIS) in Lima, Peru. These endeavours have yielded accomplishments in the fields of earthquake risk assessment, the training of technical personnel, and hazard zonation, and in the creation of earthquake-resistant building codes.

UNESCO has also been involved in the creation and development of a number of important earthquake-related facilities and training centres: the Institute of Earthquake Engineering and Engineering Seismology (Skopje, Macedonia), the International Seismological Centre (Newbury, United Kingdom), the International Institute of Seismology and Earthquake Engineering (Tsukuba, Japan), the Centre of Earthquake Engineering in Algeria, and the International Institute of Earthquake Engineering and Seismology (Tehran, Islamic Republic of Iran).

At the request of the Algerian Government, and in the aftermath of the El Asnam earthquake on 10 October 1980, UNESCO has overseen the elaboration of the seismic microzonation of the region. This project drew conclusions with a view to the reconstruction of the disaster area, to reducing earthquake risk both in this region and in the rest of the country, and the updating of building codes and regulations in accordance with the regional earthquake risk.

While the achievements outlined above on earthquake risk have been led essentially by the Natural Sciences Sector, it should be pointed out that since 1975 other sectors of UNESCO have undertaken activities related to natural hazards. In the Culture Sector, the Division of Cultural Heritage initiated a programme aimed at the protection of cultural monuments against the effects of earthquakes and other natural hazards, and at the restoration of monuments damaged by such phenomena. In the Education Sector, particular attention has been paid to the problem of designing school buildings to withstand earthquakes, or strong winds in areas subject to tropical cyclones.

VOLCANIC ERUPTION HAZARDS

At the present time, about 800 volcanoes, most of them in developing countries, are known to be active throughout the world; another 500 are potentially active. Eruptions occur at a rate of about sixty per year. In UNESCO's developing Member States, there are not enough trained volcanologists to study and continually monitor all active or potentially-active volcanoes. Fortunately, most volcanoes, particularly the most violent ones, have long repose periods between shorter intervals of activity. Surveillance of these volcanoes may be entrusted to local volcano watchers, who need only be intelligent and observant individuals, with a few weeks' basic training in identifying phenomena that indicate approaching activity. With volcanoes showing signs of approaching activity, this basic surveillance needs to be supplemented by equipments and instruments, operated by volcanological technicians. A greater number of such technicians need to be trained, preferably at well-equipped observatories.

PART III: ENVIRONMENTAL SCIENCES

At the International Institute of Seismology and Earthquake Engineering, a strong-motion recorder furnished by UNESCO, within a UNESCO/UNDP Project aiming at training experts and developing advisory services for developing countries in seismic zones, Japan, 1969.

The Organization's volcanological and volcanic hazard activities fall into three categories: (1) missions to specific areas, usually at the request of local authorities, to advise on volcanic hazards and their mitigation; (2) meetings to review past volcanic activity in specific regions and the status of volcanological surveillance; and (3) projects attempting to identify the actions needed to improve volcanic surveillance and in the mitigation of volcanic hazards.

One important field mission involved the geochemical study of gases associated with the 1971 eruption of Tenequia volcano on La Palma in the Canary Islands (Spain). Another involved the evaluation of volcanic risk and monitoring on the islands of Saba and St Eustatius in the Netherlands Antilles, in 1977. Review meetings organized by UNESCO on volcanic issues took place between 1974 and 1981. Three regional seminars were held between 1974 and 1978 to examine work on active volcanoes. These seminars reviewed and evaluated the various monitoring and data interpretation methods. They made, *inter alia*, recommendations for reconnaissance groups to be sent to places where volcanic activity seemed imminent.

By far, the most ambitious of all UNESCO volcano-related projects was the UNESCO study of an International Mobile Early Warning System(s) for Volcanic Eruptions and Related Seismic Activities (IMEWS). The basic premise of the project was identified as 'the need to improve international systems of rapid response and mutual assistance to cope more efficiently with volcanic crises'. This study aimed at laying the principles for tasks to be performed in order to:

- create and operate an international system to dispatch mobile scientific teams to crisis areas;
- improve knowledge of early-warning precursors of impending volcanic eruptions;
- promote on-the-job training and experience of local scientists in volcano monitoring techniques; and
- encourage and assist local scientific institutions in important pre-crisis studies and preparations.

A future challenge for UNESCO is to convert an excellent feasibility study such as this into a reality.

TSUNAMIS

Tsunamis are sea waves generated by impulsive disturbances of the seabed. Earthquakes – such as the one that occurred beneath the Indian Ocean on 26 December 2004 – are responsible for most tsunamis. But other tsunamis causing great destruction have resulted from volcanoes (e.g. Krakatoa in 1883) or from landslides into or within the sea. Some tsunamis are destructive only on shores near their source; others also wreak damage many thousands of kilometres from their origin. UNESCO became, in 1965, the organization through which international cooperation was formally initiated. Through its Intergovernmental Oceanographic Commission (IOC), UNESCO convened a Working Group on the International Aspects of the Tsunami Warning System in the Pacific in Honolulu, Hawaii (USA), in April 1965. Eighteen Member States and agencies participated, examining such issues as the requirements for standardized tide and seismic data within the System, exchange of information, and the research needed for a greater understanding of tsunamis and for the improvement of detection systems.

Subsequently, the IOC established the International Coordination Group for the Tsunami Warning System in the Pacific (ICG/ITSU), in 1968, and started the development of the Tsunami Warning System in the Pacific (TWSP). The Group has a membership of twenty-six countries. Its main purpose is to assure that tsunami watches, warning and advisory bulletins are disseminated throughout the Pacific to Member States, in accordance with procedures outlined in the Communication Plan for the Tsunami Warning System. The TWSP is a network including national tsunami warning centres, regional tsunami warning centres, and the International Tsunami Information Centre (ITIC) in Honolulu. The ITIC supports ICG/ITSU by monitoring the activities of the TWSP, coordinating tsunami technology transfer among Member States, and acting as a clearinghouse for tsunami preparedness and mitigation activities.

The ITSU system makes use of the hundreds of seismic stations throughout the world to disseminate tsunami information and warning messages to well over 100 points (governmental contacts) scattered across the Pacific. The IOC is currently leading an effort to expand the existing system in the Pacific to the World Ocean, to ensure that appropriate warning systems are available in all regions of the world that are prone to tsunamis, including the Indian Ocean, the Caribbean and the Mediterranean.

LANDSLIDE HAZARDS

Various studies have been conducted by UNESCO on the cause and prevention of landslides, in particular the publication of guidelines on landslides hazards zonation. The UNESCO-UNEP project on the Protection of the Lithosphere as a Component of the Environment (1981–84) resulted in significant research being carried out on landslide mitigation. UNESCO is behind the recent establishment of the International Consortium on Landslides (ICL), a non-governmental and non-profit scientific organization promoting landslide research. It was created in 2002 with the support and encouragement of UNESCO.

The objectives of the Consortium are to:

- promote landslide research for the benefit of society and the environment, and capacity-building, including education, notably in developing countries;
- integrate geosciences and technology within the appropriate cultural and social contexts in order to evaluate landslide risk in urban, rural and developing areas, including cultural and natural heritage sites, as well as contribute to the protection of the natural environment and sites of high societal value;
- combine and coordinate international expertise in landslide risk assessment and mitigation studies, thereby resulting in an effective international organization which will act as a partner in various international and national projects; and
- promote a global, multidisciplinary programme on landslides.

The International Programme on Landslides (IPL) is an initiative of ICL that conducts international cooperative research and capacity-building on landslide risk mitigation, notably in developing countries. The protection of the Machu Picchu (Peru) from landslide hazards has been underway since the year 1999, under the aegis of UNESCO.

HYDROMETEOROLOGICAL AND FLOOD HAZARDS

Since 2002, UNESCO has taken a leading role in promoting the International Flood Programme, which aims at promoting an integrated approach to flood management in order to minimize the loss of life and the hardships such as loss of possessions that result from floods. This new initiative is set to integrate the scientific, operational, formal and public educational aspects of flood management, including the social response and communication dimensions of flooding (and related disaster) preparedness.

DROUGHT AND DESERTIFICATION

Studies on droughts and desertification are undertaken in the framework of the International Hydrological Programme (IHP) and the Man and the Biosphere (MAB) Programme. The arid land studies, including drought and desertification problems, are given a leading priority in MAB activities, with significant input to the Convention to Combat Desertification. Integrated pilot projects aimed at better understanding the functioning of arid and semi-arid ecosystems and at devising optimal land-use systems in drought-prone areas contribute to the combat against desertification, particularly in Africa (Sahelian countries, Kenya, Lesotho and Tunisia). A large number of scientific publications and guidelines tailored to decision-makers and the general public have emanated from these projects, including poster and audio-visual series.

AVALANCHES

The increasing use of mountain areas for leisure activities, particularly skiing, has led to a notable increase in the number of victims of avalanches. In an attempt to make the nature and causes of these complex phenomena better known, four regional training seminars have been organized. In cooperation with the International Commission on Snow and Ice, an *Atlas of Avalanches* – with photographs accompanied by an explanatory text (in five languages) clearly illustrating the many different types of avalanche – has been prepared and will shortly be published by UNESCO.

PROTECTION OF EDUCATIONAL BUILDINGS AND CULTURAL MONUMENTS

UNESCO helps Member States deal with the problems that sudden natural hazards present for school buildings. Emphasis is placed on practical advice on how to build schools that will be relatively safe if a natural disaster does occur.

In addition, all school construction projects benefiting from UNESCO technical collaboration are examined from the point of view of their vulnerability to natural hazards. The development of schools that can be used as a place of community refuge during – and as a relief centre after – a disaster is among UNESCO's preoccupations as well. In addition, prototype schools with emphasis on disaster resistance have been developed in certain countries to serve as models for large-scale projects.

Historic sites, monuments and other cultural works of art are liable to be affected by natural disasters. Important examples of this include: the 1950 earthquake that seriously damaged ancient buildings in Cuzco, Peru; the 1971 earthquake in Trujillo, Peru; the 1975 earthquake in Burma, in which temples in Pagan were severely damaged; the 1976 earthquake in Guatemala, in which Antigua was particularly affected; the April 1979 earthquake in Friuli, Italy, and the Republic of Montenegro; and the 2003 earthquake in Iran (Islamic Republic of), which levelled the citadel of Bam. In all these cases, UNESCO was able to send teams of experts to advise local authorities on restoration measures. UNESCO has also been called upon to assist in the safeguarding of works of art damaged by other natural disasters, such as the floods in the cities of Florence and Venice (Italy) in November 1966, and the violent cyclone which seriously damaged the Royal Palace of Abomey in Benin in April 1977.

COMMUNICATION AND INFORMATION

Activities in education and information in the field of disaster preparedness and prevention are carried out in the Communication and Information Sector, including studies conducted on the role of mass communication media in disaster situations. The *Disaster Information Kit for the Caribbean Media (6th Edition)*, produced by UNESCO in 1999, provides background information on natural and man-made hazards to help reporters when covering disasters and hazards in the region.[66]

POST-DISASTER INVESTIGATION FIELD MISSIONS

Various scientific, technical or socio-economic aspects of damaging or disastrous earthquakes, floods, volcanic eruptions, cyclones, etc., cannot be understood solely on the basis of documents and scientific papers and records. On-site observations are necessary, and field work has always been organized at national and international levels from the very beginning of seismological investigations.

---

66   This resource is available at: http://www.cdera.org/media/.

The complexity of the earthquake origin and of earthquake effects is such that every new piece of information resulting from a new event is extremely valuable for reducing future risks. Soon after a disaster, governments and their agencies must obtain an objective assessment of the impacts in the affected area, enabling them to organize immediate relief operations and to introduce the first rehabilitation measures. Complex, long-term scientific, technical and socio-economic studies must provide or complete basic data for the implementation of preventive measures of a legislative, economic and engineering nature.

It was at the initiative of UNESCO that first reconnaissance missions started to operate from 1962 onwards in earthquake- and other hazard-stricken areas. In the aftermath of natural disasters, and at the request of Member States affected, UNESCO's intervention is aimed at investigating and introducing transitional actions to draw lessons from the event, to propose and (sometimes) execute measures for reducing the impacts of the disaster as well as losses from any future event, and for developing human resources as a catalyst for recovery and national self-reliance. The purpose is also to fill the gap between emergency relief operations and long-term recovery and rehabilitation efforts.

Close to fifty missions have been dispatched since 1962. The number of UNESCO experts participating in individual missions varied between one and nine. Such missions have resulted in definite improvements in early warning systems and in disaster mitigation; useful information was transferred to governments. Furthermore, the scientific and technical findings of most missions improved substantially our knowledge of seismicity, and this knowledge was used on other occasions or in other regions. Consequently, UNESCO was better able to help national authorities in the repair, planning and new construction of school buildings, and also in avoiding mistakes already made elsewhere.

UNESCO'S PARTICIPATION IN INTERNATIONAL INITIATIVES

UNESCO played an active part in the International Decade for Natural Disaster Reduction (IDNDR, 1990–2000). It is currently an important partner in the International Strategy for Disaster Reduction, the successor of IDNDR. The Organization has also actively participated in both the United Nations World Conference on Natural Disaster Reduction held in Yokohama, Japan, in May 1994, and the World Conference on Disaster Reduction held in Kobe, Hyogo, Japan, in January 2005.

OVERALL IMPACT

UNESCO can be proud of its record in natural disaster reduction. Over the decades, there has been increasing cooperation in the field between international governmental and non-governmental organizations. Observation or warning networks have been completed, developed or improved; above all, they have been set up in places where they did not previously exist. Training has been made possible and effective by the allocation of fellowships and the organization of special courses. The establishment of international institutes has made a major contribution to the development of research and the dissemination of knowledge.

UNESCO possesses the necessary competence to refine its potential contribution to natural disaster reduction and deliver further achievable concrete results. A major effort will be needed on the part of the Organization to ensure an effective contribution to the implementation of the Hyogo Framework for Action 2005–2015 (HFA), which was adopted at the Kobe World Conference on Disaster Reduction. By tackling the issues of natural disaster reduction in the scientific and technological, and environmental and educational realms, UNESCO will continue to try to protect communities from the ravages of nature and the recklessness of many human offences against nature.

BIBLIOGRAPHY

UNESCO. 2003. *Address by Mr Koïchiro Matsuura, Director-General of UNESCO, on the occasion of the meeting on UNESCO CCT project on reduction of natural disasters in Asia, Latin America and the Caribbean.* UNESCO, 25 September 2003. DG/2003/126. Paris, UNESCO.

United Nations. 1999. *Report of the Secretary-General on the Work of the Organization.* General Assembly Official Records, fifty-fourth session, 31 August 1999, Supplement No. 1 (A/54/1). New York, United Nations.

OBSERVING AND UNDERSTANDING PLANET OCEAN
A history of the Intergovernmental Oceanographic Commission (IOC)
Geoffrey Holland[67]

IN writing this brief history of the Intergovernmental Oceanographic Commission (IOC), the author is haunted by images of the personalities and events that are missing from the text. In such a short précis of nearly fifty years of broad-ranging and often intense debates of important national, regional and global issues concerning the ocean, it is inevitable that most personal contributions, many important events, and even complete programmes have been omitted. For this he apologizes. It must also be recognized that all the IOC programmes are interrelated, and that the separation that has been chosen to clarify the historical development of many of the individual elements does not do justice to the integral nature of the total ocean programme.

THE FOUNDING

The eighth session of UNESCO in 1950 authorized the Director-General to promote the coordination of research on scientific problems relating to the oceans and marine biology, eventually leading to the formation of an international advisory committee in 1955. Over the same time period, the International Council for Science (ICSU) established a Special (later Scientific) Committee on Oceanic Research (SCOR). At its first meeting in 1957, SCOR initiated planning for an International Indian Ocean Expedition that UNESCO agreed to co-sponsor. It was clear that such cooperative efforts were needed to tackle ocean science projects, especially in areas where little regional capacity existed. In November 1958, UNESCO made the decision to convene an intergovernmental conference on oceanographic research.

The Intergovernmental Conference on Oceanic Research was held in Copenhagen, Denmark, on 11–16 July 1960. The principal recommendation was that an Intergovernmental Oceanographic Commission (IOC) be established within the framework of UNESCO. This recommendation was adopted by

---

[67] Geoffrey Holland: Canadian delegate to the Intergovernmental Oceanographic Commission (IOC) of UNESCO and many of its programmes since 1970; IOC Vice Chairman and Chairman (1993–99).

Resolution 2.31 at the UNESCO General Conference later the same year, together with approval of the initial Statutes and an Office of Oceanography to act as the IOC Secretariat. The justification for the birth of this new and valuable United Nations (UN) organization was based on the need for international cooperation in ocean research:

> The oceans, covering some 70 per cent of the Earth's surface, exert a profound influence on mankind and even on all forms of life on Earth ... In order to properly interpret the full value of the oceans to mankind, they must be studied from many points of view. While pioneering research and new ideas usually come from individuals and small groups, many aspects of oceanic investigations present far too formidable a task to be undertaken by any one nation or even a few nations (UNESCO, 1960).

By this act, ocean science took a major step forward in terms of intergovernmental political visibility. It is very likely that this growing awareness of the strategic importance of ocean science and information – and its usefulness in the resolution of a broad range of national, regional and global issues – had its origins in the application of ocean research to military operations during the war years. From its first session, it became obvious that Member States were looking to the new organization to be more than a meeting place to discuss ocean research and to plan cooperative oceanographic experiments. Throughout its lifetime, the IOC has continued to move towards the exploitation of ocean knowledge and information for the use and benefit of national governments and for collectively addressing regional and global problems. The thrust of discussions and programmes at governing body meetings has evolved over the years beyond ocean research per se, to the extension of the knowledge gained in tackling problems in areas such as coastal management, ocean health, climate change, ocean services and capacity-building.

THE FIRST SESSION

The IOC met for its first intergovernmental session in Paris at UNESCO Headquarters from 19 to 27 October 1961. By the end of the session, a total of forty states had become members of the Commission; several other countries sent observers to the meeting. The new organization had been established with the participation of existing UN organizations and other international and intergovernmental bodies with interests in ocean science. Many of these organizations would become important partners in the future programmes of the IOC. Among those present were representatives from the International

Atomic Energy Agency (IAEA), Food and Agriculture Organization (FAO), World Meteorological Organization (WMO), World Health Organization (WHO), Intergovernmental Maritime Consultative Organization (IMCO), International Civil Aviation Organization (ICAO), International Council for Science (ICSU), International Union of Geodesy and Geophysics (IUGG), International Association of Physical Oceanography (IAPO), Special Committee on Oceanic Research (SCOR), International Hydrographic Bureau (IHB) and the International Council for the Exploration of the Sea (ICES).

The chairman, A. Bruun (Denmark), and the two vice chairmen, W. M. Cameron (Canada) and Vice-Admiral V. A. Tchekourov (USSR), for the initial meeting were re-elected for a full term of office expiring at the end of the second session. Tragically, Bruun died in Copenhagen only a few months later (on 13 December 1961); and, according to the Rules of Procedure, Cameron assumed the Chair.

The initial session of the IOC has a distinct place in its history. The discussions highlighted many issues that would recur throughout the succeeding decades. Many of the future strengths and weaknesses of the organization became manifest and many of the personalities who would influence later years were present. The first secretary was Warren Wooster (USA), who also had played a large role in the Copenhagen Conference; also in attendance were K. Federov (USSR), D. Scott (UK) and M. Ruivo (Portugal), three scientists destined to become future secretaries of the Commission. In fact, Ruivo has been a constant force in the work of the IOC and, at the time of writing (forty-five years later), is one of its Vice Chairs. Two future IOC Chairs, George Humphrey (Australia), and N. Pannikar (India), were also participants, along with many of the leading ocean researchers of the time, including R. Revelle (USA), H. Lacombe and J. Cousteau (France), and G. Deacon (UK).

The major powers recognized the proven and potential strategic military implications of ocean research as well as the societal benefits. Many delegations had naval representatives, and the United States sent a delegation of twenty-eight, including two rear admirals and two senators. Even at this early stage, non-scientific politics were evident. The Soviet delegation expressed regret at the absence from the Commission of the lawful delegates of China, and protested the right of the representative of Taiwan to represent China at the meeting. Such political issues inevitably arose from time to time within the UN, and the meetings of the IOC were no exception. The question of Chinese representation, South African apartheid and many other instances of political differences, would disrupt future sessions of this scientific organization.

Nor were the politics reserved for governments alone. There were jurisdictional concerns among the interested UN agencies. In his welcoming speech at the inaugural session, the Acting Director-General of UNESCO, Mr

TABLE III.8.1: INTERGOVERNMENTAL OCEANOGRAPHIC COMMISSION (IOC) CHAIRPERSONS AND SECRETARIES, 1961–2005

Session of Assembly	Time of election	Chairpersons	Secretary
I	Oct. 1961– Sep.1962	Dr A. Bruun (Denmark)	Dr W. S. Wooster (USA)
II	Sep. 1962 – June 1964	Dr W. M. Cameron (Canada)	
III	June 1964 – Sep. 1965	Dr N. K. Panikkar (India)	Dr K. N. Fedorov (USSR) (from Sep. 1963)
IV	Sep. 1965 – Oct. 1967	Prof. H. Lacombe (France)	
V – VI – I Extr.	Oct. 1967 – Sep. 1969 Sep. 1969 – Nov. 1971 Nov. 1971 – Nov. 1973	Rear Adm. W. Langeraar (Netherlands)	Dr S. Holt (UK) (from Jan. 1970) Mr D. P. D. Scott (UK) (from Oct. 1972)
VIII IX	Nov. 1973 – Nov. 1975 Nov. 1975 – Nov. 1977	Dr G. F. Humphrey (Australia)	
X XI	Nov. 1977. – Nov. 1979 Nov. 1979 – Nov. 1982	Dr A. Ayala-Castañares (Mexico)	Dr M. Ruivo (Portugal) (from Jan. 1980)
XII XIII	Nov. 1982 – Mar. 1985 Mar. 1985 – Mar. 1987	Prof. I. A. Ronquillo (Philippines)	
XIV XV	Mar. 1987 – Sep. 1989 Sep. 1989 – Mar. 1991	Prof. U. Lie (Norway)	Dr G. Kullenberg (Denmark) (from Jan. 1989)
XVI XVII	Mar. 1991 – Mar. 1993 Mar. 1993 – June 1995	Prof. M. M. Murillo (Costa-Rica)	
XVIII XIX	June 1995 – July 1997 July 1997 – July 1999	Mr G. L. Holland (Canada)	(Executive Secretary) Dr P. Bernal (Chile) (from April 1998)
XX XXI	July 1999 – July 2001 July 2001 – June 2002	Prof. Su Jilan (China)	
XXII	June 2002 – Present	Dr D. Pugh (UK)	

René Maheu, stressed that it was not the responsibility of the IOC to examine problems in meteorology, fisheries and other areas that came under the purview of existing UN agencies, although he did instruct the Commission 'to cooperate closely with other institutions of the United Nations family, and all other competent intergovernmental and non-governmental organizations, respecting their various fields of competence, but working together with them to arrange meetings and other forms of useful collaboration'.

UNESCO gave the IOC an office attached to the Natural Sciences Department, an arrangement which allowed it to take advantage of the existing meeting facilities at UNESCO Headquarters and to use the administration and framework of an established UN organization. UNESCO also endorsed functional autonomy for the Commission, which was seen to be essential to the advancement of its work and programme.

Many of the topics and concerns that had the attention of the delegates at the initial meeting were to remain on the agenda throughout succeeding years. Marine pollution, capacity-building and the availability and exchange of data were high on the list of initial priorities and have remained so. Another recurring theme has been the almost constant plea for more resources to tackle the growing list of ocean responsibilities and challenges. The future usefulness of the Commission to organize regional and global ocean research programmes was demonstrated, as was its ability to facilitate cooperation in common concerns such as: organizing international research programmes, standards and formats for ocean observations, and data and the deployment of (and communication from) fixed and floating ocean platforms.

THE GROWTH AND DEVELOPMENT OF THE COMMISSION

POLICY DIRECTIONS AND STATURE

The IOC enjoyed a relatively high level of recognition within the United Nations system during the first decade of its existence. In December 1966, the UN General Assembly passed a resolution[68] requesting the Secretary-General to make proposals to ensure the most effective arrangements for an expanded programme of international cooperation, in terms of understanding the oceans and developing its resources. These proposals were to be made in cooperation with FAO and UNESCO, in particular with UNESCO/IOC. When considering this UN directive at its fifth session (in 1967), the IOC Assembly also examined a report, prepared by its advisory bodies, entitled 'International Ocean Affairs'. This report recommended that the Member States of the UN and its relevant

---

68   UN Resolution 2172 (XXI).

agencies give consideration to the establishment of a central intergovernmental oceanic organization to deal with all aspects of ocean investigation and the uses of the sea.

This session of the IOC recognized the need for additional financial support, but concluded that a major change to the existing organizational arrangements was premature. At the UN, a second resolution was passed in December 1968[69] endorsing the concept of a long-term and expanded programme of oceanographic research, which had been recommended by a Working Group set up by the Secretary-General, and which included the Chair of the IOC. The recommendation also urged Member States at the UN and relevant UN agencies to agree, as a matter of urgency, to broaden the base of the IOC so as to enable it to formulate and coordinate such an expanded programme. At the same session, the General Assembly adopted Resolution 2467 (XXIII), which welcomed the concept of an International Decade of Ocean Exploration (IDOE) and requested the IOC to coordinate this activity in cooperation with other organizations.

A special IOC Working Group, again acting with the input of advisory bodies, prepared a comprehensive outline of the Long-term and Expanded Programme of Oceanic Exploration and Research (known later under its acronym LEPOR). This plan was approved by the IOC in September 1969 and acknowledged by the General Assembly resolution[70] later that year. The UN instructed the IOC to keep LEPOR up to date, and the IOC established a group of experts to do this. For many years, LEPOR formed the basis for the scientific activities within the IOC.

In 1969, the IOC Bureau requested the Director-General of UNESCO to negotiate a formal basis of cooperation with other UN specialized agencies with interest in matters related to ocean science. The result of the negotiations was the establishment of a unique committee, called the Inter-Secretariat Committee on Scientific Programmes Related to Oceanography (ICSPRO). It consisted of the Executive Heads of FAO, WMO and the International Maritime Consultative Organization (now the International Maritime Organization). Membership was open to other UN agencies, and the UN Environment Programme (UNEP) joined in 1972. ICSPRO was chaired by the Director-General of UNESCO. The members of this high-level committee agreed to support the IOC activities through cooperation, provision of staff and assistance with publications and meeting facilities. A staff member was seconded from FAO, WMO and IMCO to facilitate the cooperation. In 1974, approximately one-quarter of the IOC staff salaries and operational funds were provided by the ICSPRO agencies. Unfortunately, by the mid-1970s, the financial constraints throughout the UN system became more apparent, and the ICSPRO arrangement faltered. Staff members were recalled

---

69 UN Resolution 2414 (XXIII).
70 UN Resolution 2560 (XXIV).

to their parent organizations, with the one exception of the staff officer position from WMO. ICSPRO continued to meet into the 1990s, but apart from a few special sessions, the level of representation was not kept at the executive head level. New arrangements at the UN and between agencies took up the burden of inter-secretariat cooperation. ICSPRO has never been formally disbanded, and therefore remains a potential vehicle for bringing ocean elements of the UN together when the need arises.

Many external forces would prove to have an influence upon the ocean community in general and the IOC in particular, one of the first being the political recognition of the importance of the environment and its place alongside the economy and health in dealing with human development. In 1972, the UN Conference on the Human Environment was held in Stockholm, Sweden, to draw attention to the planetary environment and the global issues that needed to be addressed by society. The oceans were not a large part of the agenda, but several recommendations focused on ocean pollution. The conference requested the IOC to create a programme for the investigation of pollution in the marine environment. This request was a reinforcement of an activity already commenced within the Commission as one of the major projects of LEPOR. In 1992, a second global conference on the environment was held in Rio de Janeiro, Brazil – the UN Conference on the Environment and Development (UNCED). This was a historic meeting, which would influence the evolution of environmental programmes over the succeeding years. In Rio, the oceans were to have more prominence. The conference produced an environmental agenda (Agenda 21), which included a chapter (Chapter 17) specifically dealing with oceans. It proposed that an integrated and comprehensive global ocean observing and information system be created to provide the information needed for oceanic and atmospheric forecasting, for ocean and coastal zone management by coastal nations, and for global environmental change research. Many recommendations were related to capacity-building, coastal protection and management, while other actions (prompted by global concerns) called for an examination of the effects of UV radiation and atmospheric carbon dioxide on the oceans. Agenda 21 provided a standard against which national and international environmental goals could be set and actions evaluated. The influence of Agenda 21 on the IOC and on other UN organizations concerned with the environment was substantial. Once again, the IOC had already taken steps to deal with many of these issues. As early as 1979, the Commission had established a Committee on Climate Change and the Ocean (CCCO), led by one of the pioneers of global warming studies, Roger Revelle, and it had been working on climate research for several years in close collaboration with WMO.

In 1995, governments agreed on a global plan of action, backed up by national plans, to address the protection of the marine environment from land-based sources of pollution. The Washington Agreement recognizes the need to address the pollution problems caused by the run-off of agricultural, industrial and human wastes into the vulnerable coastal waters. The 2002 World Summit on Sustainable Development (WSSD) took place to review the progress in the implementation of Agenda 21. The IOC undertook a proactive approach to this latest world conference on the environment. The Commission approved a message from its Member States to the Summit and submitted a document entitled 'One Planet, One Ocean' describing the Commission, together with other documentation explaining how its programmes related to the goals of sustainable development. The IOC was also a sponsor of a Global Conference on Oceans and Coasts at Rio+10, held at UNESCO Headquarters in December 2001. This conference had the aim of assessing the present status of oceans and coastal programmes and reviewing the progress made since UNCED, thus laying the groundwork for the upcoming Summit.

Another huge change in ocean affairs that has had significant impact on the programmes and policies of the Commission in the final decades of the twentieth century was the emergence of the 'new ocean regime'. This change began in earnest with the third UN Conference on the Law of the Sea, held in Caracas (Venezuela) in 1974, and it culminated with the Convention coming into force in 1994. Despite the enormity of the task, governments reached agreement on the greater part of the text in Caracas, although the final Articles on seabed resources were to take several more years of negotiations. Many of the Articles that concerned the IOC, however, were already in their final form, for example those on scientific research, marine pollution and technology transfer. The IOC was recognized within the Law of the Sea Articles as a 'competent international body'. This recognition gave the Commission a presence and prestige in ocean law and would be cited time and again as a proof of its stature and responsibilities. Unfortunately, the interpretation of what could or should be expected of the IOC in the light of these explicit and implicit responsibilities has proven to be a difficult task and one that has already been addressed by many studies, committees and expert meetings.

Over the years, the Commission has not neglected the need to maintain an up-to-date scientific basis for its programmes. In 1975, the Assembly agreed to establish a Scientific Review Board (SAB), originally chaired by H. Lacombe (France), to advise it on its research direction. The Board met in 1976 and 1977; an expanded version met in 1978, with C. Mann (Canada) as Chair. The Board recommended that a study be undertaken by the IOC, in collaboration with its advisory bodies, on ocean science up to the year 2000. This was to be the second assessment of ocean science to follow that which produced LEPOR twenty years earlier. An expert consultation was held in Villefranche-sur-Mer, France, 13-17 April 1982, with leading oceanographers of the time, and a report prepared and submitted to the twelfth Assembly the same year.

A similar exercise was organized twenty years later with a third assessment, again in collaboration with SCOR, and on this occasion the two organizations were joined by the Scientific Committee on Problems of the Environment (SCOPE). The latest assessment was carried out by scientists nominated by their peers and selected by the three sponsoring organizations. Participants met in Potsdam, Germany, in October 1999, and the resulting publication, *Oceans 2020*, was released in 2002.

A final word in this section must be devoted to the role of the Commission in promoting and coordinating activities in 1998. A resolution to designate that year as the International Year of the Oceans (IYO) was adopted in 1993 and endorsed by the twenty-seventh General Conference of UNESCO. The proposal was adopted by the UN General Assembly in 1994, and the following year the Assembly agreed[71] to undertake preparations for this event. The Commission itself had little additional funds to support the cooperative activities necessary, and initially some Member States were advocating that the IYO should not have any financial implications on the budget. Despite these difficulties, the executive secretary, Mr Kullenberg, and the deputy secretary, Mr Iliounine, managed to orchestrate a series of significant international and national events, and, by 1997, the Assembly was able to endorse a substantial effort that included: an IYO logo and flag; a theme ('Ocean, a Common Heritage'); a UNESCO pavilion at the Ocean Expo 1998 in Lisbon (Portugal); the creation of national committees; commemorative stamp issues; and numerous events organized by IOC bodies and programmes, national entities and other international organizations.

At the same Assembly, Member States approved the text and programme for the 'Ocean Charter', which, although not a legally binding document, represented a declaration of intent, a commitment to the future of oceans, marine areas and coastal environments. During 1998, IYO representatives from over eighty governments – including heads of state, ministers and ambassadors – were to sign the Ocean Charter.

MANAGEMENT

Any organization needs to have vigorous and expert managers at the helm. The IOC has been fortunate to have had a succession of active and dedicated secretaries. The increased responsibility and autonomy of the IOC within UNESCO has been recognized, and the position now holds the joint rank of IOC executive secretary and Assistant Director-General of UNESCO. The executive secretary has been supported by a permanent staff of professionals and

---

71   UN Resolution XVIII-3.

administrative personnel, many of whom have devoted a large portion of their careers to the Commission.

A valuable external source of support has been through the direct contribution of personnel and resources by Member States. In-kind support from Member States in hosting meetings or providing technical help has also been welcomed. From 1977, the IOC has operated a Trust Fund that can be used to accept financial support from Member States, either directed towards designated activities or programmes, or to be used at the discretion of the executive secretary for contracts or operational expenditures. The Trust Fund became of particular significance during the period when both the United States and the United Kingdom withdrew from UNESCO. Because of the autonomy of the IOC, both countries remained Member States of the Commission and made arrangements for funds to be transferred to the IOC Trust Fund in proportion to those that would have been forthcoming through the UNESCO budget. The arrangement benefited the IOC, as it received these substantial contributions without the usual administrative loss to UNESCO. When both countries returned to the UNESCO General Conference in 2000, the result was a net reduction in the IOC budget.

The elected officers of the IOC are not formally part of the management structure but are clearly an important part of the policy direction. When the Commission had a relatively small number of Member States, the Chair and two Vice Chairs constituted the IOC Bureau. At the third session of the Commission in 1964, a Consultative Council was formally established to assist the Bureau and the Statutes amended accordingly. The number of Member States continued to grow and, in 1969, the Statutes were again passed to the General Conference for amendment. By 1970, the Consultative Council had been reconstituted as an elected Executive Council, which included the Chair and four Vice Chairs.

The number of seats on the Executive Council was set at no more than one-quarter of the number of Member States, elected with due regard for geographical distribution. Originally this rule was interpreted as including the IOC Officers, but as the competition for seats on the Council became more intense, a more liberal interpretation allowed the restriction of one-quarter of the Member States to exclude these five positions and de facto adding an additional five seats. This structure continued, through a further amendment of the Statutes in 1987, until the latest revision in 2002, when the number of seats on the Council was set at an upper limit of forty, including the elected officers. Of course, the Statutes contained far more than the Articles referring to the conduct of the governing bodies, and the adopted changes over the years reflected the changes in the direction and policies of the Commission's work.

In 1960, the purpose of the IOC was stated simply as promoting the scientific investigation of the oceans, through the concerted action of its Member

States. In the latest Statutes, adopted in November 1999, the purpose has been expanded to include research, services and capacity-building, and the application of knowledge for the improvement of management, sustainable development, the protection of the marine environment and decision-making processes. The collaboration with related intergovernmental and international organizations is also specifically mentioned. A separate Article deals with the function of the Commission, which – while it still includes the coordination of ocean research programmes – now extends far beyond the responsibilities originally envisioned by IOC's founders. For example, Article 3.1c covers the response of the IOC, as a competent international organization, to the requirements deriving from the UN Convention on the Law of the Sea. The Rules of Procedure were also amended to be compatible with the latest changes to the Statutes; and the growing maturity of the organization and the broader representation of its Member States were reflected in the new voting procedures that specifically address the issue of geographical representation. The Commission now has five Vice Chairs, each selected from nationals of the respective five electoral groups, and the distribution of seats is mandated in an annex to the Rules of Procedure and kept under review. In a brief history of the IOC, it is not possible to list all the officers and personalities who influenced the development and impact of the Commission and its programmes.

In later years, the IOC was to experience difficulties that arose from its rather unique position within UNESCO. Unlike the UNESCO science programmes, the IOC possesses its own Statutes and Member States; it elects its own Chair, Officers and Executive Council. The IOC is responsible for recommending candidates for secretary to the Director-General for his consideration and has functional autonomy over its programmes. Nevertheless, although the number of Member States of the Commission was to approach 130 by the end of the century, the operating resources of the Commission take up a very small percentage of the UNESCO budget, and its programme has therefore remained of relatively minor consequence in the affairs of its parent body. Governmental representatives attending the respective governing body meetings of UN organizations are drawn from a variety of departments and ministries, with various responsibilities, and it is difficult to maintain coordination in the cases of overlapping programmes. For the Commission, this problem is compounded by having its programme and budget issues – as agreed by national representatives attending the IOC Assembly – passed for debate and approval at a subsequent meeting of the UNESCO General Conference, attended primarily by the same Member States, but with a very different representation of responsibilities and interests. It is no wonder that the general perception of the IOC is that of a science programme of UNESCO rather than being the UN entity representing the world's oceans.

THE IOC PROGRAMMES

CAPACITY-BUILDING

There have always been significant differences in the capacities and capabilities of IOC Member States in the understanding and use of the knowledge and information available from the oceans. As the number of Member States grew, the proportion of those needing scientific and technical assistance also grew, along with an increased focus on capacity-building within the programmes of the Commission. The ability of the IOC to respond to these needs has remained limited by the resources available.

The Commission quickly set up a Working Committee for Training, Education and Mutual Assistance (TEMA) and has undertaken many studies to address this problem. These actions have identified issues and requirements, but there have been few internal or external funds available to implement the many identified activities. The resources within the internal IOC budget have been sufficient to fund training courses and to assist with minor expenditures (such as travel expenses of scientists) but are insufficient to undertake larger initiatives. Occasionally, the internal funds have been supplemented by donations from Member States to the IOC Trust Fund, or by direct in-kind assistance with travel arrangements, hosting training courses or supplying associated personnel and materials. However, a few instances of more substantial support ably demonstrated what can be done with the help of funding organizations. For example, during the late 1990s, an African programme of regional cooperation in scientific information exchange was assisted by a generous contribution from the Flemish Government, supplemented by contributions from other Member States. This successful programme led to the establishment of national ocean data centres and a regional cooperative network in ocean knowledge and information. The programme has been endorsed by the countries in the region and is planned to expand further within Africa. The concept is being explored for use in other regions of the globe. Capacity-building is related to regional organizational strengths among Member States. Only a few of the regional bodies have succeeded in establishing permanent regional secretariat facilities. Two of these have reached sub-commission status (the Sub-Commissions for the Western Pacific and for the Caribbean, WESTPAC and IOCARIBE, respectively), while many others are at various stages of development and cover many ocean region areas of the world. It is hoped that these will eventually follow the development of WESTPAC, which became the IOC Sub-Commission for the Western Pacific in 1989. WESTPAC's roots can be traced back to a regional investigation of the Kuroshio Current in 1965, with twelve countries participating. A Regional Committee for the Western Pacific was established during the tenth Assembly in 1977. The initial emphasis on fundamental research subsequently evolved to include regional programmes that paralleled the IOC Programmes.

OCEAN DATA AND INFORMATION

The IOC, wishing to establish a continuing framework to build on the cooperation in ocean data exchange that had been generated in the International Geophysical Year and in 1960, set up a Working Committee to consider how this could be achieved. This Working Committee formed the basis for IODE, the International Oceanographic Data Exchange Working Committee which has continued to be one of the pillars of the IOC programme over the years. Its membership is based upon representatives from the National Oceanographic Data Centres of participating Member States and has therefore kept current with the methods and operational developments in ocean data. In recent years it has recognized and adapted to the innovations brought about by the electronic age and has expanded its role to include information exchange. It has been particularly successful in the establishment of data exchange and training programmes accessible to all, through the development of the IOC website and with the associated initiatives of Ocean Portal, Ocean Expert and Ocean Teacher. Another development over the years has been the recognition that the evolution of modern electronics is leading to the disappearance of the border between archived data and that available in 'real time'. Initially the IODE dealt only with the data archival; and data sets often took many years from the time of observation to their arrival at data centres, owing to the individual attention that had to be given to the quality control and verification of observations. Automated observations, quality control software, and electronic communications have revolutionized data management. Ocean scientists themselves have gradually moved away from the concept of individual data ownership to an awareness and acceptance of the benefits of rapid exchange and access to data sets of all sources and types.

The free availability and exchange of ocean data has long been a goal of the IOC. As knowledge and data are interpreted into information, the strategic and commercial value of the distributed products grows. The Commission began tackling this issue with the establishment, in 2001, of an intergovernmental Working Group on data policy (chaired by A. McEwen), the recommendations of which were approved by the Assembly in 2003.

The interest in the exchange of oceanographic data included a recognition of the need for improved bathymetric charts of the world ocean. This need had been identified as much as a century ago, when it led to the General Bathymetric Chart of the Oceans (GEBCO) project, established in 1903 by Prince Albert I of Monaco and accepted as a responsibility of the International Hydrographic Bureau (to become the International Hydrographic Organization, IHO) on its foundation in 1921. The GEBCO consisted of twenty-four sheets covering the whole globe, collected and assembled by various national hydrographic offices. Unfortunately, the programme ran into financial difficulties; the sheets could not be maintained and production decreased. In 1963, and again in 1964, the

IOC recognized the importance of bathymetric charts and urged its Member States to support the GEBCO efforts, emphasizing in particular the utility of scientific cruises in investigating doubtful soundings. The IOC programme in international ocean mapping began in 1969, with the compilation of the *Geological and Geophysical Atlas of the Indian Ocean*, taking advantage of the data collected through the International Indian Ocean Expedition. In 1971, the IOC included morphological charting of the sea floor as one of its eight major programmes of major importance and, in 1973 (in response to proposals from SCOR), approved the establishment of a joint IOC-IHO Guiding Committee for GEBCO. This committee set about its task, and by 1982 the fifth edition of the GEBCO was published. The role of ocean mapping at the IOC is a continuing challenge. The main goal is to provide decision-makers, scientists and students with information concerning the relief of the world ocean and its geological/geophysical parameters. The demands for new and more comprehensive maps grew rapidly over this period, especially from the Law of the Sea discussions on the extension of coastal state jurisdiction, advances in the development of coastal and offshore resources, and the need to improve the understanding and modelling of ocean processes.

At the same time, new technologies for obtaining data, digitization, and the development of electronic charts were revolutionizing the science. The IOC established the Consultative Group on Ocean Mapping (CGOM), as a primary subsidiary body of the IOC, to oversee its mapping programme. At its ninth meeting in Monaco in 2003, the Group recommended the creation of an IOC–IHO Ocean Mapping Board, which would bring the GEBCO and the regional International Bathymetric Charts (IBCs) under one management structure, thus continuing the close collaboration between the two bodies that has existed for forty years.

OCEAN SERVICES

Ocean data are by themselves of little use unless information is able to be extracted from them and useful products distributed in a timely fashion to decision-makers and managers of related activities. Although the truth of this precept was recognized at an early stage in the IOC's history, progress towards the goal has been slow and arduous. Among the reasons for this are the technical difficulties and time delays in obtaining the necessary ocean data, and the expense of setting up monitoring networks (especially as these would need to be in continuing operation). Moreover, there is a lack of recognition of the potential benefits of such a system, and a difficulty of demonstrating the benefits without the programmes being already in place. Yet, these difficulties are gradually being overcome; and the progress over the years towards viable ocean services can be mapped by the IOC's actions.

A tsunami warning system in the Pacific Ocean (ITSU) was established by the IOC in 1965, after two devastating tsunamis in that region, and the programme has been maintained since that time. The IOC took the lead in coordinating the efforts of the regional Member States and formed the International Coordination Group for the Tsunami Warning System in the Pacific (ICG/ITSU). The Pacific Tsunami Warning Centre, in Hawaii (USA), became the operational headquarters of the system in the Pacific. The service undertook to ensure cooperation among the Pacific Member States in terms of prediction, monitoring, civil defences and public awareness against these devastating ocean waves. Regional and local national systems are also part of the system. In recent years, other regions of the world have been petitioning the governing bodies of the IOC to expand the tsunami service into their regions. However, the relatively modest funds that would allow for this expansion have not yet been found. (On 26 December 2004, a devastating tsunami occurred in the Indian Ocean, causing tragic loss of life, with the result that the lack of warning networks in the region is now being addressed by governments.)

To a large degree, the ocean service activities originated with the formation of the Working Committee IGOSS (Integrated Global Ocean Stations System) in the first years of the Commission. IGOSS was a committee set up to address the availability and exchange of real-time data, it being accepted that IODE activities were addressing archived data only. The international exchange and use of atmospheric data were managed under the World Meteorological Organization and had been well established for many years. The production of weather maps and related information products was predicated on the constant and timely exchange of atmospheric observations using a Global Telecommunications System (GTS). A similar framework did not exist for ocean data. The shipping and fishing industries relied on atmospheric marine weather services, and the only large users of real-time subsurface data were the military, who were not anxious to share data. Many thought that the timescales in the ocean were so long as to render efforts to exchange data in real time unnecessary. Nevertheless, from the 1960s through the 1970s, the situation slowly changed. The more-sophisticated models for weather forecasting needed information on ocean–atmosphere exchanges, and longer-term forecasts needed data on the heat content of the ocean surface layers. WMO and the IOC began to cooperate with the interaction of the IOC-IGOSS with the WMO Panel on the Marine Aspects of Ocean Affairs. By the early 1970s, these two bodies were holding mutual planning meetings; and, in 1977, IGOSS became a joint IOC-WMO Working Committee. The union was not without its problems, however, as there were individuals on both sides who were resistant to change. For the IOC, the largest benefit was obtaining the use of the GTS to exchange ocean data, beginning with temperature versus depth (bathythermograph) data, and then expanding to temperature and salinity (TESAC) profiles. The collection of data was still slow, and often ocean data were not available for exchange until vessels reached port, making it necessary initially to define real-time ocean data as data transmitted within thirty days of observation. Advances in communication from sea – and developments in instrumentation to allow data to be collected automatically and from moving vessels – would eventually increase the quantity and timeliness of the ocean data available.

The successful involvement of IGOSS in the First GARP Global Experiment (FGGE) resulted in a large increase in the number of ocean messages exchanged, and it demonstrated the ability of the programme to observe and collect real-time ocean data. By the mid-1980s, the name of IGOSS was changed to the Integrated Global Ocean Services System, retaining its acronym but reflecting its growing maturity. The benefits of monitoring the ocean in real time and 'near real time' became more evident as advances in instrumentation, remote sensing, automation, online data management and electronic communications progressed, laying the groundwork for the establishment of an expanded role for the IOC in ocean services. IGOSS explored the exchange of many other parameters, one of these being sea level. Sea level monitoring can be used to plot tides, seasonal ocean changes such as the occurrence of El Niño, and rises due to climatic changes. In 1990, a Working Committee Global Sea Level Observing System (GLOSS) was formed to oversee ocean sea level measurements.

Despite the many activities within the IOC dealing with ocean services, there was a growing understanding that a more ambitious and comprehensive approach was needed. The Technical Committee on Ocean Processes and Climate (C/OPC), under the chairmanship of James Baker (USA), considered that understanding and forecasting climate change would require the existence of an ocean observing system similar to the World Weather Watch system underpinning weather forecasting, and it presented this vision to the twenty-first IOC Executive Council (1988). Support for this action was received from the Second World Climate Conference, which identified the need to establish a Global Ocean Observing System as the ocean component of the proposed Global Climate Observing System.

A strategy document was prepared by an expert group and, at the sixteenth IOC Assembly (1991), it was decided to undertake the development of a Global Ocean Observing System (GOOS), built initially on existing systems and operated by Member States for the needs and benefits of each. The Assembly noted that GOOS would be a highly complex and sophisticated undertaking and one of the most important programmes ever established by the Commission. The 1991 decision recognized that GLOSS and IGOSS were fundamental building blocks of GOOS, which was seen as including elements such as climate observations, MARPOLMON (marine pollution monitoring), coastal zone monitoring and regional programmes as 'modules' or 'subsystems' of the overall system, and would serve as the ocean component of the Global Climate Observing System. The free, open and timely exchange of data and information was seen as essential components, as were training, assistance and technology transfer.

The establishment of GOOS was considered to represent a 'new era in oceanography'. Following the instructions to build on existing systems, GOOS continued to promote the development of regional GOOS organizations. In Europe, the establishment of a EuroGOOS was spearheading the involvement of governments and industry in the provision of ocean services, while, on a smaller scale, the North-East Asia Regional GOOS (NEAR-GOOS) was also making progress. These successes spawned interest in other regions of the world. The first Regional GOOS forum in 2002 was followed by a second in 2004, and no less than fourteen existing or planned regional programmes were represented.

The obvious synergy between the IOC ocean services developments under GOOS and the climate and atmospheric services programmes under WMO led to a closer association of the two organizations. In 1999, the governing bodies of the two organizations, recognizing the increasing demand for integrated marine meteorological and oceanographic data and services, and the efficiencies achieved by combining the expertise and technological capabilities of the WMO and IOC systems, decided to establish the WMO-IOC Joint Technical Commission for Oceanography and Marine Meteorology (JCOMM). JCOMM is an intergovernmental body of experts providing the international, intergovernmental coordination, regulation and management mechanism for an operational oceanographic and marine meteorological observing, data management and services system.

As instruments improve, the possibility for data collection and the production of information products grows accordingly. One exciting programme that has been unfolding over the past few years is the Argo experiment. Argo is an international project to collect information on the temperature and salinity of the upper part of the world's oceans. It uses robotic floats that spend most of their life drifting below the ocean surface (some as deep as 2,000 m); every ten days, they rise to the surface, taking measurements on their ascent and communicating the data to a satellite. For 2006, the goal is to have 3,000 floats, producing 100,000 temperature/salinity profiles per year, and covering the world's oceans.

OCEAN SCIENCE

The IOC was founded as an intergovernmental science organization and continues to adhere to that role. An understanding of the oceans is fundamental to all ocean-related issues, and this understanding pervades all of the Commission's programmes. Major ocean research programmes coordinated by the IOC include involvement in the International Indian Ocean Expedition (1959–65), the

PART III: ENVIRONMENTAL SCIENCES

K. N. Fedorov (IOC Secretary, 1963–69) on board a research vessel in 1972, checking the CTD (Conductivity-Temperature-Depth) probe.

International Cooperative Investigations of the Tropical Atlantic (1963–64), the Cooperative Study of the Kuroshio and Adjacent Regions (1965–77), and the Cooperative Investigation of the Caribbean and Adjacent Regions (1967–76). Somewhat in contrast, the International Decade of Ocean Exploration (IDOE, 1971–1980) was adopted by the IOC to provide a general intensified effort on ocean research. Member States were invited to submit ocean research programmes to the IDOE as part of the decade. These needed to be multinational, serve exclusively peaceful purposes, actively involve scientists from other nations, and make the resulting data available. Even while these programmes were underway, planning was taking place on research activities driven by more specific objectives, such as weather, climate, ocean health, and fisheries.

OCEAN WEATHER AND CLIMATE

It is impossible to investigate the physical processes governing the atmosphere and the oceans as if they were two separate systems. Oceanographers and atmospheric scientists are well aware of the need to work together on weather and climate issues; and the IOC recognized the importance of this from its very first meeting. In 1965, this approach was formalized with the establishment of a Working Group on Ocean–Atmosphere Interaction. The need to involve WMO was recognized, and thus in 1967 the Commission dissolved the group in order to negotiate collaborative arrangements. WMO suggested that the Commission join with it and ICSU in a Panel on Ocean–Atmospheric Interaction, within the framework of the Global Atmospheric Research Programme (GARP), which the IOC accepted. The relationship between the IOC and WMO in terms of ocean observations under IGOSS (discussed above) was recognized as an integral part of this collaboration. The Commission was asked to arrange for oceanographic participation in the GARP Atlantic Tropical Experiment (GATE) and called on its scientific advisory body, SCOR, to identify the processes to be studied. GATE was successfully carried out in the summer and autumn of 1974, with the participation of about forty research vessels, and large numbers of buoys, moorings and aircraft. The results – which provided a valuable data set – were reported in numerous scientific papers, at several workshops, and led to a GATE Symposium, hosted in Kiel, Germany, in 1978.

This initial success was followed by an observational phase of the First GARP Global Experiment, to encompass an extended activity from December 1978 to November 1979. Once again, the Commission was tasked to support the associated oceanographic investigation, and again SCOR agreed to provide the scientific guidance. The IOC Assembly in 1975 established an IOC-SCOR task team to prepare a plan of IOC activities related to GARP that not only addressed the FGGE requirement but also looked at the future needs for understanding the physical basis of climate. In 1977, the report of the team was approved, and the Assembly assigned a high priority to FGGE for the next two years, recognizing that the study of ocean processes would need to be fully integrated into the understanding of climate variability.

In 1979, recognizing the importance of the ocean's role in global climate change, the IOC and SCOR formed the first Committee on Climate Change and the Ocean (CCCO), with Revelle as its chairman. The CCCO was to provide significant guidance to the Commission as its climate-related programmes evolved over the next few years. WMO organized an intergovernmental and interagency planning meeting on the World Climate Programme in 1980. One result was the establishment of a World Climate Research Programme (WCRP), under the direction of a Joint Steering Committee (JSC). In May 1982, the IOC was a co-sponsor of a conference in Tokyo (Japan) to study the need for large-scale ocean experiments under the WCRP. This conference recommended, *inter alia*, two major programmes: the Tropical Oceans and Global Atmosphere (TOGA), and the World Ocean Circulation Experiment (WOCE). Both came into being, with the IOC being a major sponsor. TOGA (1985–95) would be the forerunner to the development of the monitoring programme for the prediction of El Niño and its recognition as a driver of the seasonal global climate. WOCE (1990–97) would be the largest ocean experiment ever seen, involved the efforts of thirty countries, and yielded a data set that was essential for climate research and many other uses. In 1993, the IOC joined WMO and ICSU as a sponsor of the World

Climate Research Programme (WCRP). The CCCO was disbanded in favour of this more cooperative approach (with IOC representation on the JSC). The programme encompasses studies of the global atmosphere, oceans, sea and land ice, and the land surface – which together constitute the Earth's physical climate system. WCRP studies are specifically directed to provide scientifically founded quantitative answers to the questions being raised on climate and on the range of natural climate variability. Such research provides the basis for predictions of global and regional climatic variations, and of changes in the frequency and severity of extreme events.

OCEAN HEALTH AND BIOLOGICAL SCIENCES

The founding Conference of the IOC in 1960 recognized the importance of protecting the marine environment and requested governments to take, without delay, all steps in their power to prevent pollution of the oceans and seas by radioactive wastes and other harmful agents. In 1965, the Commission established an IOC Working Group on Marine Pollution to consider the IOC role in this area. In 1968, the Administrative Committee on Coordination of the United Nations (ACC) submitted a report on the state of marine science and its application for the Economic and Social Council of the United Nations (ECOSOC), outlining the responsibilities of the various agencies of the United Nations system in regard to the prevention and control of marine pollution. UNESCO and the IOC were given the responsibility for the coordination of scientific research and the evaluation of data on marine pollution. The IOC Working Group succeeded – in collaboration with SCOR and ACMRR – in preparing a clear definition of marine pollution and a classification of pollutants. They stressed the need for better coordination on these matters; and, in 1969, such coordination was exemplified by the agreement of IMO, FAO, UNESCO/IOC and WMO (with the approval of the ACC) to establish a joint group of experts on the scientific aspects of marine pollution (GESAMP), composed of distinguished scientists acting in their individual capacities.

During the early 1970s, WHO, the IAEA, the United Nations and finally (in 1977) UNEP, became co-sponsors. After adoption of UNCED Agenda 21, and in light of its direct bearing on the future work of all its sponsoring agencies, GESAMP changed its name to 'Group of Experts on the Scientific Aspects of Marine Environmental Protection'. GESAMP has produced about fifty technical documents, including three landmark assessments of the state of the marine environment in 1982 (published by the IOC), and in 1990 and 2001 (published by UNEP).

For the past twenty years, the IOC-IMO-UNEP sponsored GIPME (Global Investigation of Pollution of the Marine Environment) programme has been working on issues regarding contaminants in the marine environment. GIPME has tackled problems on the compatibility in methods, standards and intercalibration, designed the Marine Pollution Monitoring System (MARPOLMON) and undertaken the design of the Health of the Oceans (HOTO) module of GOOS.

A highly successful IOC programme on Harmful Algal Blooms (HAB) was established in 1992, owing to a growing concern with the increase in the global occurrences of these events. The programme office is located at the IOC, but two science and communication centres are hosted in Copenhagen (Denmark) and Vigo (Spain), together with coordination groups in several IOC regions. The HAB contributions to research, training and public awareness of causes and occurrences of these hazardous events have been significant.

The IOC recognized the need for scientific advice on fishery research and in 1962 designated the Advisory Committee on Marine Resources Research (ACMRR) of FAO as its advisory body on fisheries oceanography. Although the attention to this topic has been a constant part of the IOC agenda, especially under Ocean Sciences and Living Resources (OSLR), the programme has remained relatively small, the attention being directed towards biological oceanography programmes in such related programmes as Mussel Watch, Coral Reef Monitoring, Continuous Plankton Survey, and HAB, and international cooperation in programmes such as GLOBEC and JGOFS. In recent years, the scientific community has come to recognize that an examination of the relationship between biological and physical elements is crucial to the understanding and management of renewable marine resources. This combined approach to marine and environmental sciences is known as the 'ecosystem' approach, and it calls for greater attention to be paid to fisheries oceanography.

OCEAN MANAGEMENT

IOC Coastal Area Management activities date back to the 1980s, as many of the most pressing problems of Member States were related to the coastal waters. Initially these concerns were taken up by interdisciplinary scientific and social coastal activities within existing IOC and UNESCO programmes. However, in 1997, Integrated Coastal Area Management (ICAM) was adopted as an independent programme by the nineteenth session of the IOC Assembly, and the programme commenced the following year. The ICAM objective is to build marine scientific and technological capabilities in the field of Integrated Coastal Management, through the provision of reliable marine scientific data, development of methodologies, dissemination of information and capacity-building.

A great many other programmes in all aspects of ocean science have been initiated and completed over the history of the IOC. Some of the recent programmes include the IOC/World Bank Working Group on Coral Bleaching and Local Ecological Responses, initiated in September 2000; the International Ocean-Colour Coordinating Group (IOCCG), established in 1996; and the SCOR-IOC Ocean Carbon Dioxide Advisory Panel.

CONCLUSION

The IOC is now approaching its forty-fifth year. This chapter has charted the progress and growth of many of the most important elements of its programme. One could express disappointment that, although the oceans are such a vital part of the planetary existence, the Commission has not grown to occupy a larger and more significant role within the UN system. On the other hand, given its relatively small resources, the achievements of the IOC can be considered remarkable.

BIBLIOGRAPHY

UNESCO. 1960. *Declaration Adopted by the Intergovernmental Conference of Oceanic Research* (Copenhagen, 11–16 July 1960). UNESCO/NS/167, 7 October 1960. Paris, UNESCO, p. 6 (Appendices). http://unesdoc.unesco.org/images/0001/000177/017743EB.pdf

BUILDING BLOCKS FOR MARINE SCIENCE
A history of UNESCO's Marine Sciences Division (OCE)

Dale C. Krause, Selim Morcos, Marc Steyaert and Gary D. Wright, in consultation with Alexei Suzyumov and Dirk G. Troost[72]

THE BEGINNINGS

THE Division of Marine Sciences existed for only twenty years, from 1971 to 1991. But those two decades were marked by a strategic approach that yielded major achievements. UNESCO's Division of Marine Sciences was referred to internally as OCE, a name carried over from its predecessor, the Office of Oceanography. Its efforts were carried out jointly with the Organization's Regional Offices for Science and Technology (ROSTs). The crux of its mission was to help less-favoured countries to develop their marine science capabilities. The OCE also fostered cooperation with the international scientific community to develop research programmes, ultimately concentrating on coastal projects. Everything the OCE did employed an experimental approach, bringing together the natural and the social sciences. The result was a deeper understanding of how social development and scientific cooperation can nurture each other. The Division of Marine Sciences is no more. But its accomplishments continue to have an important impact.

Marine sciences had been included in UNESCO's science programme since the Organization's early days, and were marked by a steady evolution within the activities of the Secretariat and the marine scientific community. Though initially less recognized by UNESCO as compared to other branches of science – such as physics and biology – marine sciences received occasional

---

72 Dale C. Krause: UNESCO (1973-89), promoted to Director of the Division of Marine Sciences in 1980; Division's programme focused on building marine sciences in developing countries with help of scientific NGOs.
Selim Morcos: UNESCO Division of Marine Sciences (1972–89)
Marc Steyaert: Joined UNESCO in 1967 as Programme Specialist in Office of Oceanography. From 1975 to 1995, chiefly concerned with development and running of the Coastal Marine (COMAR) Programme. Head of Marine Science Related Issues, within UNESCO's Office of IOC and MRI (1990–94).
Gary D. Wright: Twenty years as UNESCO editor, eighteen of which in marine sciences.
Alexei Suzyumov joined UNESCO's Division of Marine Sciences in 1982.
Dirk G. Troost joined UNESCO's Natural Sciences Sector in 1980; Head of the Organization's Coastal Regions and Small Islands Platform since 1995.

support from the Organization in the early years in the form of training courses and fellowships to specialists (UNESCO, 1968). Such travel and exposure to the international community enabled a small number of outstanding scientists from the developing world and Eastern Europe to advance their careers and become leading oceanographers in their countries. (Konstantin N. Federov, second director of the UNESCO Office of Oceanography and a well-known physical oceanographer, was a UNESCO fellowship recipient.[73]) When the Intergovernmental Oceanographic Commission (IOC) was founded, several developing countries joined and participated, many of them because of their early awareness of the importance of the sea and the need to build their scientific capabilities, through their own resources or bilateral/multilateral support – including that of UNESCO.

OFFICE OF OCEANOGRAPHY

The IOC was established within UNESCO in 1960 by a decision of the General Conference, and became operational in 1961. This was accompanied by the creation of the Office of Oceanography in UNESCO to serve in part as the Secretariat for the Commission, as well as to consolidate and carry out the Organization's other marine science programme activities (see previous chapter).

As was stated by Rear Admiral W. Langeraar (of the Netherlands, chairman of IOC in 1971), 'the IOC Secretariat has to depend on the UNESCO Office of Oceanography for the majority of its staffing. Under the ... Statutes, the director of the UNESCO Office of Oceanography is the Secretary of the IOC and is working in both functions under the authority of the Director-General of UNESCO.' In fact, the director of the Office and some professional staff were working half-time for the IOC and half-time for the marine science programme of UNESCO (whereas the Food and Agriculture Organization [FAO] and the World Metrological Organization [WMO] seconded staff were working full-time for the IOC) (UNESCO-IOC, 1971).

By 1969, as noted at the sixth session of the IOC Assembly, sixty-seven Member States had joined the Commission, and its activities – such as cooperative investigations and IGOSS (Integrated Global Ocean Stations [later, 'Services'] System) – increased in number and sophistication. The Commission's Vice Chair, C. Frazer (Mexico) noted that 'most of the proposed studies could not be undertaken by about forty of the Member States: the developing countries.' Meanwhile, the production of IOC documents in a

---

73   This was a ground-laying function of UNESCO. Besides Federov (USSR), other examples of UNESCO fellowship holders who went on to become leading specialists are Marta Vannucci (Brazil), Selim Morcos (Egypt) and Aprilani Soegiarto (Indonesia).

timely manner in five languages placed a heavy load on the limited staff of the Office and on UNESCO's general services. René Maheu, Director-General of UNESCO at the time, was satisfied with the progress of the Commission, its recognition within the UN system and as the Secretariat for the Inter-Secretariat Committee on Scientific Programmes Relating to Oceanography (ICSPRO). However, he warned that the increased tasks required an increase in the resources available, and in 1971 asked for the addition of an item – on the relations between UNESCO and the IOC – to the agenda of the Bureau and Consultative Council meeting in Bordeaux (France) that year; on that occasion he participated in a lively discussion on how to cope with the needs of both UNESCO and the IOC.

In 1971, the IOC Assembly recommended that the Director-General 'reorganize the Secretariat of the Commission and UNESCO Office of Oceanography as separate entities, with the Commission's secretary reporting directly to him.' The recommendation was adopted, having been supported by a majority of Western states but opposed by the Union of Soviet Socialist Republics (USSR) and other Eastern European states (the developing countries were divided).

The rationale behind this recommendation was spelled out in a note by the chairman of the commission and by 'observations' of the Director-General. Both agreed on the separation and each one for reasons that supported their visions towards improved performance. This was not a simple administrative restructuring, but a management plan based on well-perceived targets and giving the IOC Secretariat and the UNESCO marine science programme distinct and separate functions. In his Long-Term Outline Plan for 1971–1976, Director-General Maheu wrote:

> UNESCO's specific programme for the promotion of the general advancement of marine science would continue ... All I wish to say is that I hope to be able to give more aid to Member States, particularly developing countries ... in strengthening their infrastructure and planning their oceanography activities in accordance with an integrated policy based on national development priorities (UNESCO-IOC, 1971, Add. 1).

For the chairman of the commission, on the other hand, the ICSPRO agreement among UN agencies gave the IOC a special place within the UN system. However, initially only two agencies (FAO and WMO) posted staff members to the IOC Secretariat. It was hoped that separating the IOC Secretariat from the rest of UNESCO would encourage other UN agencies to put more resources and staff members at the disposal of the IOC. These two compatible and complementary views, which prevailed in 1971, ensured the relative independence of the Division of Marine Sciences (as the Office of Oceanography was later renamed) for twenty years, until it was finally merged with the IOC Secretariat as the Office of the IOC and Marine Science Related Issues (IOC/MRI) in 1991.

MARINE SCIENCE FOR DEVELOPMENT

PROGRAMME MISSION AND APPROACH

Immediately after the separation of the two units (IOC and OCE), there were discussions with the Scientific Committee on Oceanic Research (SCOR) about what might be the approach of the Division. In 1974, SCOR produced a working paper on the basis of an extensive inquiry on the 'Promotion of Marine Sciences in Developing Countries' that formed one of the bases for the programme's strategy in the following years (ICSU, 1974). In 1980, as part of UNESCO's response to the 1979 UN Conference on Science and Technology for Development (Vienna, Austria), the marine science programme was reorganized with a visible concentration on coastal scientific problems under the coastal marine project COMAR.[74] At the same time, a major decentralization of activities occurred, giving rise to the strong cooperation of the Division with the UNESCO Regional Offices for Science and Technology (in the 1980s, after their having been strengthened by the addition of marine scientists) – namely in Cairo, Egypt (for the Arab States Region), Jakarta, Indonesia (for South-East Asia), Montevideo, Uruguay (for Latin America and the Caribbean Region), Nairobi, Kenya (for Africa) and Venice, Italy (for Europe).

INFRASTRUCTURE-BUILDING – GROWTH AND CHALLENGES

During 1974–82, the Law of the Sea Conference was taking place, which effectively put before governments the importance of marine science. In the conference there was much strenuous debate as well as tension among various groups. However, one positive aspect to emerge was that most of the developing countries decided to improve their marine science capabilities, both through their own resources and with the help of UNESCO, which resulted in a rapid growth of UNESCO's extrabudgetary marine science development programme, reaching a peak in 1981. One notable achievement (evident by the end of the 1980s) was that most developing countries now had established basic marine science capabilities, although they may have varied in size (from major laboratories to just a few scientists) and quality. UNESCO can take pride in having been associated with most of these developing countries in this achievement.

---

74 The Coastal Marine (COMAR) Programme, officially: the Major Interregional Project on Research and Training leading to the Integrated Management of Coastal Systems.

UNESCO developed a large and diverse publications programme in cooperation with the scientific community, which was strongly tied to the activities of the Regular Programme. The publications were designed to place scientific methodology in the hands of scientists: for advice, to plan research, to report on the results of working groups, workshops and conferences, and for activities in training and education. The books and documents grew out of the activities or were the purpose of the activity. These publications constituted major achievements of the marine science programme and were essential to its successful impact on scientists in both the developing and developed countries (see the summary on UNESCO publications in marine science at the end of this chapter.)

Over roughly three decades (from 1960 through the 1980s), the marine scientific community of the developing world and of the world as a whole grew by ten times (De Shazo and Krause, 1984).[75] Significantly, the relative growth in the developing countries began to exceed that in the industrialized countries. By 1983, the number of marine scientists in the developing world equalled the total number for the rest of the world in 1970. The potential and need for research in the developing world revealed itself to be much larger than it had been in 1972, as was confirmed by their actual accomplishments.

A primary strategy of the major regional initiatives was to use UNESCO's action as a catalyst to generate large extrabudgetary projects and national commitments. As shown in Figure III.9.1, the marine science extrabudgetary programme manifested high growth during the 1970s, initially doubling during each biennium – from US$480,000 (with only a few projects) in 1971–1972, to US$2 million (with thirteen projects) during 1977–1978. Then, suddenly, from 1979 to 1980, the funding jumped to US$21.6 million (with twenty-three projects).

A marked decrease was evident in the triennium 1981–1983, which was very different from the preceding one, with projects falling in value by two-thirds from 1981 to 1983. The funding for the triennium totalled US$22,269,000. For comparison to trends in other biennia, the triennium 1981–1983 can be divided into two artificial biennia, as follows: 1981–1982 (US$18,836,000) and 1982–1983 (US$10,294,000). Even this masks the abrupt change from US$11,975,000 in 1981 to US$3,433,000 in 1983.

---

75   This was reflected in the third edition of *International Directory of Marine Scientists,* published in 1983 by UNESCO in cooperation with the UN, IOC and FAO. After this edition (fairly exhaustive, and in much use throughout the world), no further global directories as such were published, at least not in printed form – although various directories by region or other groupings were compiled and made available. Of note, in this respect: in 1997, the IOC launched GLODIR, the *Global Directory of Marine (and Freshwater) Professionals,* a database (later renamed *OceanExperts*) containing information on individuals involved in all aspects of marine or freshwater research and management.

PART III: ENVIRONMENTAL SCIENCES

Figure III.9.1: Resources of the Marine Science Programme (in US$), 1961–1995.

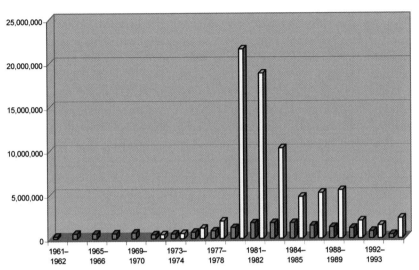

Dark bars = Regular Programme funds; light bars = extrabudgetary funds. 1961–1972: Office of Oceanography; 1973–1989: Division of Marine Sciences (OCE); 1990–1995: Marine Science Related Issues programmes. 1981–1983 was (exceptionally) a three-year budgetary period in UNESCO. For the sake of comparison, it has been transformed here into two artificial biennia: 1981–1982 and 1982–1983. Reminder: the above figures do not reflect IOC budgets.

Due to the global financial crisis, funding declined and hovered around US$5 million in the three biennia from 1984 to 1989 (US$4,768,000 in 1984–1985; US$5,211,000 in 1986–1987 with twenty-six projects; and US$5,506,000 with twenty projects in 1988–1989). A further decline to around US$2 million occurred in the three following biennia: 1990–1991, 1992–1993 and 1994–1995.

For comparison, the approved budget for the activities of the Office of Oceanography – covering both IOC and UNESCO (proper) activities – grew from US$269,200 in 1961 and 1962 to US$682,000 in 1969 and 1970. For the new marine science programme, the activities budget grew from US$491,000 for 1971 and 1972, up to US$2,426,000 for 1981–83. Afterwards there were numerous budget cuts throughout UNESCO as the consequence of the withdrawal of the United States, the United Kingdom, and Singapore, plus the global financial crisis.

Figure III.9.1 demonstrates that the extrabudgetary funds started at roughly the same level as the Regular Programme budget in 1971–1972, but soon and systematically exceeded it by several times over the following twenty years, reaching more than seventeen times the Regular Programme budget in the 1979–1980 biennium. This was a clear success in using Regular Programme funds in launching extrabudgetary projects.

In the years following this rapid growth (i.e. post-1989), the challenge to which the UNESCO marine science programme needed to respond was rather different in character from that of the previous two decades. On the one hand, there was the challenge to maintain and upgrade the marine science human resources and infrastructure that had been established, while ensuring their application to vigorous research programmes that were relevant to society's needs and/or to the scientific frontier. On the other hand, many less-developed countries and island states did not yet have a well-established marine science capability – and, in fact, most could not afford one. So the challenge became one of helping them develop their own expertise, while supporting their efforts to gain access to more extensive expertise elsewhere through networking approaches. Such a networking approach was then being followed, for instance, through the UNESCO–UNDP coastal marine projects in Asia and Africa – at the same time that the countries themselves were building up their capabilities.

OCE STRATEGY: GOOD SCIENCE IN A REGIONAL APPROACH

Supported by long experience, UNESCO's approach to marine science development was based on the building-up of three kinds of marine science capability: human resources, scientific infrastructure, and research programmes. If any one of these three components were missing, the whole development effort would fail. The regional projects were based on the participation of Member States, taking into consideration their common national priorities and their individual strengths, as well as what the scientific community of the specific countries and regions judged to be the relevant scientific frontiers. A further element of the UNESCO marine science strategy was that a project, programme or activity should be designed and developed from the beginning by the scientists and managers who would eventually implement it.

A good example of this approach was the investigation of the Physical Oceanography of the Eastern Mediterranean (POEM). POEM was developed in the region in 1983 through the collaboration of the very scientists who wanted to conduct the research, and who in due course were responsible for carrying out the activities. At that time, thanks to ongoing UNESCO-UNDP projects in several countries of the region, these countries' marine science capabilities had reached a satisfactory level of progress and hence the ground was well prepared for fruitful cooperation. Carried out over almost ten years, POEM included planning workshops, multiple-ship coordinated surveys, intercalibration exercises, data validation and research workshops. The results – mainly on the formation of intermediate and deep waters – were published in scholarly journals and led to a much-improved understanding of the Eastern Mediterranean, which up to then had been less known scientifically than the Western Basin.

As another example of its marine science strategy, UNESCO approached SCOR for advice on the mangrove ecosystem, and then took that advice to build up a large regional networking programme for research and training on such ecosystems in Asia and the Pacific, which stretched from Pakistan to Fiji. The United Nations Development Programme (UNDP) provided financial assistance (US$2.8 million) to the mangrove programme, and it was later shown that this was one of the best regional projects with which UNDP had ever been associated. Of course, that credit reflects back upon the people involved, who were mainly in the countries themselves – as well as to other associated scientists, among them Marta Vannucci – and who played a critical role, as well as the concerned UNESCO staff. In 1990, the mangrove project evolved into a non-governmental scientific society as a grass-roots initiative fostered by UNESCO and Japan. The non-governmental organization (NGO) eventually resulting from these activities was the International Society for Mangrove Ecosystems (ISME, based in Okinawa, Japan),[76] which carried on the types of work originated or previously carried out by UNESCO. This was a uniquely significant evolution that had the support of the scientists and countries, which allowed them to continue their cooperation after UNESCO was no longer able to help substantially.

In the Arab States, the regional approach of the marine science programme concentrated on the development of national capabilities, for example the establishment of several laboratories and university departments, as well as research vessels, involving ten countries and using (in the 1980s) US$13 million of extrabudgetary funding that had come from the Arab region. The flow of funds allowed UNESCO to recruit additional staff for a special operational unit created within the Natural Sciences Sector to manage the funds provided for the extrabudgetary projects. The future programme aim was to concentrate on developing subregional projects through which the national capabilities were built up. UNESCO can be credited with the fact that these national projects and the initial technical support of the Organization's marine science programme contributed to the creation of two active subregional organizations: the Regional Organization for the Conservation of the Environment of the Red Sea and Gulf of Aden (PERSGA, based in Jeddah, Saudi Arabia), and the Regional Organization for the Protection of the Marine Environment (ROPME, based in Kuwait).

As one example of the cooperative efforts between the Division, the IOC and relevant scientific NGOs, one can mention the support for the Joint Oceanographic Assembly (JOA) in Acapulco, Mexico, in 1988, organized by SCOR and its associated NGOs (as had been the case for previous JOAs). Given

---

76   ISME has a consultative status with the United Nations Economic and Social Council (ECOSOC).

that the Acapulco event was the first JOA to be held in a developing country, the number of participants from the developing world was high.

During the 1980s, there was much effort in UNESCO towards planning for the coming years, and in renewing and consolidating the programmes in response in part to global financial crises. With the departure of three Member States (USA, UK and Singapore) from UNESCO in 1984 and 1985, the amount of UNESCO funds diminished, so that each dollar of the limited funds available had to accomplish more than in the past. With the help of the UNDP-funded marine science regional projects, the impact was lessened. To further lessen the weakening effect, in the early 1990s (after the merger with the IOC Secretariat), the UNESCO staff responsible for the marine science programme developed a new financing mechanism in support of its various activities by inviting potential donors to directly co-fund the activities. These funds provided an important amplifying or booster effect to the programme elements.

THE MERGER

In 1990, the role and functions of the Division – its name being changed – were reorganized and placed under the administrative guidance of the IOC secretary. Important elements of the (former) OCE programme continued, though in a reduced and altered state. In the resulting merger – the Office of the IOC and Marine Science Related Issues (IOC/MRI) – these activities obviously constituted the MRI part of the programme. The MRI part was further restructured into three mutually supportive sub-programmes: the Coastal Marine (COMAR), Training and Education in Marine Sciences (TREDMAR), and Promotion of Marine Sciences (PROMAR) Programmes, with the main focus on the first two.

Under TREDMAR two major initiatives were launched: (1) the Floating University (Training-through-Research or TTR), and (2) a Global Faculty (which developed and exploited a computer-based set of training modules in remote sensing in the marine sciences and in coastal management). These efforts involved the complementary work of other UNESCO units. With the birth of newly independent states, UNESCO was conscious of their needs in marine sciences. To identify these needs, multidisciplinary advisory missions were sent in early 1990 (through the Division of Policy Analysis and Operations in cooperation with the IOC) to Namibia and Eritrea, the Palestinian Autonomous Territories (following the Oslo Accord) and war-torn Lebanon.

A CLOSER LOOK AT EXEMPLARY ACTIVITIES

COASTAL MARINE (COMAR) PROGRAMME

In the early 1970s, UNESCO realized that not enough attention was given to coastal areas and ecosystems – vital for the growing coastal population – such as mangroves, coral reefs, estuaries, marshes and coastal lagoons. By the mid-1970s, UNESCO joined with the Scientific Committee on Oceanic Research (SCOR) to initiate various working groups for the purpose of conducting thorough reviews of the status of knowledge and the need for further investigations on those coastal formations and ecosystems. Based on these results, at the end of the decade, UNESCO launched its 'Major Interregional Project on Research and Training leading to the Integrated Management of Coastal Systems' (COMAR). By the early 1980s and jointly with UNDP, UNESCO had developed a series of regional projects and networks in southern and South-East Asia, as well as in Africa and Latin America. In fact, COMAR was an expression of the profound evolution that took place, in the late 1970s, in the nature of the cooperation between the UN system (including UNESCO) and the Members States. Regional rather than bilateral cooperation had prevailed and the so-called 'country projects' were replaced by multinational efforts towards solving similar regional problems. This type of cooperative effort was particularly appropriate for the coastal environment, where similar characteristics prevail in many countries.

During 1983–87, through the UNESCO-UNDP 'Research and Training Project on Mangrove Ecosystems in Asia and the Pacific', considerable advancement of knowledge was made on the status of coastal ecosystems in the region. A special focus of the project was mangrove reforestation and management. The recommendations developed were addressed to several managerial levels, from local communities up to the level of governments.

For coastal Africa, as a part of UNESCO's cooperation with UNEP during 1982–85, COMAR addressed a specific regional problem – extensive coastal erosion in West and Central Africa. Shortly afterwards, the entire African coastal zone was the object of the UNESCO-UNDP Research and Training Project on Mangrove Ecosystems in Africa (COMARAF). This initiative united, for the first time, researchers from nearly all countries of the region in an attempt to understand the state of the coastal ecosystems and to provide for their better management. This successful project was headed by African scientist and former UNESCO (COMAR) grantee Salif Diop (Senegal).

In the Asian and African projects, training and South–South knowledge sharing were very important components. Regional groups of researchers, working together and sharing experiences, were established. The results were documented in a few dozen issues of a publication entitled *Research Bulletin*, as well as in country and regional reviews and reports. In Latin America, the COMAR project strategy was different. There the national capabilities were relatively strong. The Latin America and Caribbean COMAR projects were based mainly on the networking of national efforts. In general, the scientific community responded well. A good example is the ongoing project on coastal management in small islands of the Caribbean region, the Caribbean Coastal

Poster on the UNESCO/UNDP COMARAF project 'Research and Training on Coastal Systems in Africa.'

Marine Productivity (CARICOMP) programme, which was launched in the mid-1980s and has provided pertinent information for decision-makers.

INTERNATIONAL OCEANOGRAPHIC TABLES

Oceanographers measure salinity of seawater in different locations and depths and at different hours, seasons and years. They need a universal standard and a method to be able to compare their results. The Hydrographic Tables (1901) and Knudsen's method, based on the chemical measurement of halogen ions, served oceanographers from the beginning of the twentieth century until it was outdated by the more accurate measurement of electrical conductivity of seawater. In 1961, Roland A. Cox called on the newly established Office of Oceanography, asking for UNESCO's support to solve this universal problem. Beginning that year, the Organization co-sponsored the Joint Panel on Oceanographic Tables and Standards (JPOTS). The Division of Marine Sciences supported and helped coordinate the work of the Panel. As an OCE activity, UNESCO published the JPOTS findings, finally consolidated into a textbook manual, all of which were produced during the twenty-five years of the Panel's existence. The rich publishing programme of JPOTS benefited especially from the advice and council of Panel members Alain Poisson and Oleg Mamayev (Morcos, et al., 1990).

The results of the Panel marked the fruition of research work carried out by many distinguished institutions and scientists. The Panel adopted a new definition of salinity: the Practical Salinity Scale (1979), which also ensured the continued usefulness of salinity data collected since the beginning of the twentieth century. A new International Equation of State of Seawater (1980) replaced the traditional equations that describe the density of seawater as a function of temperature, salinity and pressure (depth). The culmination of this effort was the publication by UNESCO of two volumes of the *International Oceanographic Tables* in 1985 and 1987. The tables and/or equations are used nowadays by all oceanographers worldwide to ensure the universality and comparability of oceanographic data.

MARINE SCIENCE EDUCATION AND TRAINING

Curricula development

Throughout the years, the training and education programme supported national and regional capacity-building, especially in countries that were at the starting point of their development in the marine sciences. As a response to an interest expressed by the scientific community and IOC meetings in training and education, a workshop on teaching marine sciences at the university level was convened in UNESCO in 1973, at which university curricula in the main disciplines of oceanography were developed and recommended. This workshop was followed by a series of other workshops in which a number of curricula and syllabuses as well as sets of recommended guidelines were produced, published and distributed, such as for training marine technicians (to enable them to deal with imported oceanographic instruments), for secondary schools (early awareness-building), and for universities (courses in fishery sciences, ocean engineering, etc.). Regional recommendations (e.g. on the development of marine sciences in Arab universities), guidelines for research institutions (e.g. on the organization of marine biological reference collections) and inventories (such as one on innovative learning materials in marine sciences and technology) were also developed.

A major achievement of the programme was the preparation and distribution (in six languages) of the forward-looking study that resulted in the 1989 publication *Year 2000 Challenges for Marine Science Training and Education Worldwide* (UNESCO Report in Marine Science 52, Paris 1988). This was prepared on the basis of a world survey, an analysis of the survey, and a critical review and synthesis by a meeting of experts. It represented input from all over the developing and industrialized world. National and regional differences existed, but at the same time it revealed that there was – worldwide – a shared vision regarding certain kinds of problems and the directions that needed to be taken. The publication signalled those priority directions and the specific efforts that ought to be undertaken. UNESCO's marine science programme was adapted to reflect those priorities, as were programmes in Member States.

Floating university
In the mid-1980s, the Union of Soviet Socialist Republics (USSR) – in need of advanced student training in the geomarine field, due to preparations for deep-sea mining operations – proposed to UNESCO the joint launching of an international ship-based programme that would combine training and research. During 1988–90, the proposal was discussed and a plan was hammered out at various gatherings. 'Training-through-Research' (TTR) was finally launched in December 1990, at a meeting hosted by the Institute of Marine Geology (Bologna, Italy). The first international TTR cruise took place in the Mediterranean and Black Seas in the summer of 1991, on board the USSR's R/V Gelendzhik, headed by John Woodside and Michael Ivanov. The success (both scientific and social) of that cruise resulted in the co-sponsorship by the European Science Foundation and in further annual cruises as well as mid-cruise and post-cruise workshops and (later) international conferences. The participants represented a growing number of countries. By the end of 1995, some 300 students and young researchers from fifteen countries had been trained. TTR's study concentrated on a true research frontier: the poorly known interactive processes between the geosphere and biosphere in the deep ocean. In view of the programme's growing international recognition and contribution to the development of a universal

PART III: ENVIRONMENTAL SCIENCES

A selection of marine science publications issued by UNESCO from 1968 to 1997.

culture of peace and tolerance, that same year (1995) the project was designated as a UNESCO contribution to the celebration of the UN's fiftieth anniversary. Since 1996, the co-sponsorship of TTR has been carried on by the IOC.

TTR's achievements were based on two major elements: the availability, from the USSR (and later, the Russian Federation) of a large, well-equipped but relatively inexpensive research vessel; and the conduct of cutting-edge, pioneering research on the European and North African continental margins – thanks to which students were trained to the highest standards. However, what project can succeed without the dedicated efforts of a few individuals? Indeed, the co-founders of TTR were a tightly knit group of knowledgeable scientists from France, Italy, the Netherlands, Russia, Turkey and the United Kingdom. Their guidance provided the key to TTR's success, from data collection through analyses to the publication of the research results.

PUBLICATIONS AND PUBLIC INFORMATION

One of the major contributions of the Division was the production and broad distribution (mostly free of charge) of a wide variety of publications. These included documents, newsletters, monographs, bibliographies, and other reference works. These publications were in significant demand by scientists from around the world – but especially by those in developing countries, as it afforded them access to UNESCO's intellectual output, when most other sources were often prohibitively expensive.

A landmark reference series, UNESCO's Monographs on Oceanographic Methodology, was launched in 1966 with the publication of *Determination of Photosynthetic Pigments in Seawater* (69 pp., in English), the manuscript of which was produced by SCOR. In fact, the publication of the series followed a recommendation of SCOR to UNESCO in 1963. In this highly successful collection ('best-sellers' among UNESCO science titles), which was largely an OCE activity (from the OCE's creation, until the series was taken over in 1996 by the IOC), a total of eleven volumes were eventually published, providing guidelines and information for researchers – mainly in aspects of marine biology.

Well known for over three decades in the international marine science community were two document series: UNESCO Technical Papers in Marine Science and UNESCO Reports in Marine Science. The Technical Papers (sixty-seven volumes issued, from 1965 to 1994) informed the scientific community of recent advances in oceanographic research and on recommended research programmes and methods. They were mainly published as documents produced jointly with scientific NGOs, such as SCOR. The Reports series (sixty-nine volumes, 1977–96) served specific programme needs and reported on developments in projects conducted in the context of UNESCO's marine science related activities.

PART III: ENVIRONMENTAL SCIENCES

The quarterly *International Marine Science Newsletter* (IMS), launched in 1973, served as a valuable public information vehicle for UNESCO (both the Division and the IOC) and its partners in marine science. Its popularity in the international scientific community led to the extension of its production (initially in English only) to publication finally in all six UN languages (Arabic, Chinese, English, French, Russian and Spanish). In all, seventy-six issues were produced and distributed free of charge – with some 11,000 copies for each issue (all languages included) – informing practically all marine science and related institutions, governmental and non-governmental agencies (as well as other interested bodies and individuals) on a broad range of pertinent topics and events. Its production was stopped in 1996, with the discontinuation by the General Conference of the MRI part of the programme. The resulting loss of a valuable source of printed information and documentation was felt in the international community, especially in the developing world – as witnessed by the many letters of enquiry for several years thereafter. As well, a convenient medium for the visibility of UNESCO and its partners, with regard to their ocean-related activities, had disappeared. Although some of the IMS features were included in the IOC's website, many scientists in developing countries pointed out that they, at that time, were not well equipped to benefit from this electronically published information. Eventually, administrative and personnel changes effectively relegated to the IOC the responsibility for all publications of a purely oceanographic or marine scientific nature and content.

Throughout the OCE and MRI years, a number of other books were published (marine environmental bibliographies, publications on aspects of oceanographic history, ichthyological reference works, scientific diving codes and [in the earlier days, in cooperation with FAO] directories of marine scientists and marine science institutions etc.).[77] As well, the Division produced and mounted marine science exhibits and provided other types of public awareness materials for use, for example, at major relevant events. Among such events were JOA (Acapulco, 1988), the Earth Summit in Rio (UNCED, 1992), and other similar UN or other major conferences in Paris, Toronto, Yokohama, and Barbados.

In the autumn of 1995, the UNESCO General Conference decided to discontinue the Marine Science Related Issues (MRI) part of the programme,

---

77 Some highlight publications: (i) the UNESCO taxonomic work *Fishes of the North-Eastern Atlantic and the Mediterranean* (FNAM) was important to certain developing countries (UNESCO, 1984; 1986a; 1986b), and the *Checklist of Fishes of the Eastern and Tropical Atlantic* (UNESCO/JNICT, 1990). These works were produced jointly with the European Ichthyological Union; (ii) JPOTS references: the *Practical Salinity Scale 1978* and the *International Equation of State of Seawater 1980* (UNESCO, 1981a); *International Oceanographic Tables*, Vol. 3 (UNESCO, 1981b) and Vol. 4 (UNESCO, 1982); *Algorithms for Computing Fundamental Properties of Seawater* (UNESCO, 1983); and the JPOTS work on the thermodynamics of the carbon dioxide system in seawater (e.g. UNESCO, 1987), and *Processing of Oceanographic Station Data* (UNESCO 1991).

leaving the IOC Secretariat to carry on alone as the sole marine science arm of UNESCO. Still, thanks to the generations of scientists it inspired and the benefits it brought to so many developing nations, the Division of Marine Sciences, in ways both subtle and profound, lives on.

BIBLIOGRAPHY

De Shazo, Y. M. and Krause, D. C. 1984. *Marine scientists in the world.* Paris, UNESCO, 16 pp.
ICSU. 1974. Working Paper, SCOR Executive Meeting, Canberra, January–February 1974, 29 pp.
Morcos, S., Poisson, A. and Mamayev, O. 1990. Joint Panel on Oceanographic Tables and Standards: twenty-five years of achievements under the umbrella of international organizations. In: W. Lenz and M. Deacon (eds), *Ocean Sciences: their history and relation to man.* Proceedings of ICHO IV, Hamburg, 1987. Hamburg, Germany, *Deutsche Hydrographische Zeitschrift,* Ergänzungsheft, Reihe B, No. 22, pp. 344–56.
UNESCO. 1968. *Directory of UNESCO Fellows, 1948–1968.* Paris, UNESCO.
——. 1981a. *Practical Salinity Scale: 1978,* and *International Equation of State of Seawater: 1980.* UNESCO Technical Papers in Marine Science 36. Paris, UNESCO, 25 pp.
——. 1981b. *International Oceanographic Tables, Vol. 3.* UNESCO Technical Paper 39. Paris, UNESCO, 111 pp.
——. 1982. *International Oceanographic Tables, Vol. 4.* UNESCO Technical Paper 40, 195 pp.
——. 1983. *Algorithms for Computing Fundamental Properties of Seawater.* UNESCO Technical Paper 44. Paris, UNESCO, 53 pp.
——. 1984. *Fishes of the North-Eastern Atlantic and the Mediterranean (FNAM) Vol. I,* pp. 1–510. Paris, UNESCO.
——. 1986a. *FNAM. Vol. II.* Paris, UNESCO, pp. 511–1008.
——. 1986b. *FNAM. Vol. III.* Paris, UNESCO, pp. 1009–1474.
——. 1987. *UNESCO Technical Papers in Marine Science 51.* Paris, UNESCO, 55 pp.
——. 1991. *Processing of Oceanographic Station Data (JPOTS Manual).* Paris, UNESCO, 138 pp. (Co-published by UNESCO in Russian and Chinese.)
UNESCO-IOC. 1971. Suggestions on the Working Relations between the Commission and UNESCO, by the Chairman of the IOC. IOC (Intergovernmental Oceanographic Commission) seventh session, 26 October to 5 November 1971. UNESCO Document SC/IOC-VII/43 (30 September 1971). Add. 1: Extract from 'Long-Term Outline Plan for 1971–1976, presented by the Director-General' (UNESCO Document 16 C/4; IOC/B, 89 Add. 2).
UNESCO/JNICT. 1990. *Checklist of Fishes of the Eastern and Tropical Atlantic* (CLOFETA), 3 Vols. Paris, UNESCO, 1492 pp.

PART III: ENVIRONMENTAL SCIENCES

BREAKING DOWN BARRIERS, BUILDING BRIDGES
Coastal Regions and Small Islands (CSI) Platform
Dirk G. Troost and Malcolm Hadley

The world's small island developing states are front-line zones where, in concentrated form, many of the main problems of environment and development are unfolding. As such, they are the big tests for the commitments made at the 1992 World Summit.

— Kofi Annan, Secretary-General of the United Nations, at 'Barbados+5', September 1999.

THE Environment and Development in Coastal Regions and Small Islands (CSI) initiative, established in 1996, represents one of UNESCO's recent experiments in programme design and management. Conceived as a platform, CSI serves as a test bed to explore options, overcome barriers and demonstrate solutions.[78] From the outset, its goal has been to bring together people from different disciplines, backgrounds and sectoral affiliations to contribute to environmentally sustainable, socially equitable, culturally respectful and economically viable development in the world's highly vulnerable small islands and coastal regions. In addressing the challenge of transcending traditional obstacles to cross-sectoral, multidisciplinary cooperation, CSI has promoted three main areas of activity: field projects exploring complementary facets of a single shared problem; UNESCO Chairs and university twinning arrangements that pool cross-disciplinary expertise; and multilingual internet-based discussion forums. CSI has also served as the house-wide focal point for the Organization's inputs to the UN-wide process of contributing to the sustainable development of Small Island Developing States (SIDS), and lead unit for two cross-cutting projects.[79]

---

78 The CSI initiative partly builds on previous UNESCO work on coastal regions and small islands described elsewhere in this section (e.g. within COMAR, TREDMAR, PROMAR). In essence, this work lies outside and is additional to the work in these physiographic units (coastal regions, small islands) carried out within the special intergovernmental structures such as IOC, IHP and MAB. The fact that the successive administrative units responsible for COMAR and associated initiatives had no responsibilities for the servicing of special intergovernmental bodies has had certain implications: lower lobbying power and limited intergovernmental support in UNESCO's governing organs being offset by a greater flexibility and capability to innovate and a greater proportion of staff and financial resources being devoted to substantive activities (vs servicing of statutory organs).
79 Local and Indigenous Knowledge Systems (LINKS) http://www.unesco.org/links, and Small Islands Voice (SIV) http://www.smallislandsvoice.org.

COASTAL REGIONS AND SMALL ISLANDS IN FOLLOW-UP TO THE RIO CONFERENCE

In the immediate follow-up to United Nations Conference on Environment and Development, in Rio de Janeiro, Brazil, June 1992 (UNCED), there was considerable discussion within the UNESCO Secretariat on situations that might be particularly amenable to interdisciplinary, cross-sectoral treatment. Coastal regions and large-scale river basins were two such situations.

The idea of preparing a cross-cutting overview of UNESCO's work on small islands arose during informal discussions of an open ad hoc group within the UNESCO Secretariat which met on a dozen or so occasions between July and December 1992, in the immediate follow-up to the Rio Conference of June 1992. Among the topics considered by the informal group, chaired by Dirk G. Troost, was that of a certain discrepancy between the long-standing involvement of the Organization in certain fields and the lack of overviews of that work. Since its inception, UNESCO has had activities and programmes related to resource use and socio-economic development in particular types of biomes or physiographic units (coastal zones, forests, islands, urban systems, arid zones, mountains, etc.). In some cases, this involvement dates back to the early days of the Organization (e.g. the Arid Zone Research Programme of the 1950s). In several fields (e.g. coastal zones, islands), activities have been sponsored under the aegis of several sectors, programmes and administrative units. Generally, compilations and reviews of UNESCO's activities and their impact in these domains have been undertaken on a project-by-project or programme-by-programme basis, if at all. House-wide assessments are generally not available.

It was within such a perspective that suggestions took shape that an overview be prepared on UNESCO's work on small islands. Among other aims, this review would provide background on the Organization's past, ongoing and potential contribution to the issue of sustainable development in small island situations, raised during the UNCED process and subsequently in the lead-up to the Global Conference on Sustainable Development in Small Island Developing States (Barbados, April–May 1994). Such a review would also provide a better basis for others to take account of UNESCO's work in more substantive assessments of approaches to small island development.

In this light, an informal ad hoc task force met on a number of occasions during 1993, with a view to putting together a cross-sectoral overview of UNESCO's work on small islands. The idea of preparing such an overview was subsequently discussed and endorsed by the UNESCO Committee for the Follow-up to UNCED, at its meeting in Paris in April 1993. The review then became part of UNESCO's own internal preparations for the Barbados Conference, under the aegis of the Organization's Coordinator for Environment

PART III: ENVIRONMENTAL SCIENCES

Programmes. The responsibility for finalizing the 131-page report (UNESCO, 1994), and preparing it in camera-ready form using desktop publishing facilities, was entrusted to the Division of Ecological Sciences.

In a somewhat analogous but less detailed and intensive way, a sixteen-page brief for decision-makers was prepared on 'Coasts: Managing Complex Systems' (UNESCO, 1993), and a house-wide overview of UNESCO work on coastal areas presented to the World Coast Conference held in the Hague in November 1993 (Troost and Hadley, 1993). This overview provided a starting point for successive drafts of a compilation of UNESCO and UNESCO-related publications dealing with subjects pertinent to coastal regions and small islands, covering the period 1980–95.[80]

CSI PLATFORM: ESTABLISHMENT AND APPROACH

The intersectoral endeavour on Environment and Development in Coastal Regions and in Small Islands (CSI) was established by the General Conference at its twenty-eighth session in late 1995 in response to the recommendations of key United Nations meetings, such as those in Rio de Janeiro and Barbados. The establishment of CSI as an intersectoral 'platform' was based on the proposition that environment and development issues in coastal regions and small island settings 'require an integrated and interdisciplinary approach for their resolution, that UNESCO had the right mix of competencies to develop such approaches and that mutual benefits and efficiencies could be derived from bringing together UNESCO Sectors'.[81]

The CSI initiative was therefore an opportunity to develop an integrated coastal management approach, as well as an avenue for piloting intersectorality within the Organization. These two interlinked aspects of CSI each posed significant challenges in their own right. In addressing these challenges, activities have focused on three interactive modalities: university Chairs, field projects, and internet-based discussion forums.[82]

---

80   The compilation was published in 1997 as 'CSI info 2' (UNESCO, 1997). Information is grouped under various environmental sciences headings (Earth sciences, Ecological sciences, Marine sciences, Water sciences) as well as other sources (Basic sciences, Culture, Education, Social and Human Sciences, World Heritage Centre, World Solar Programme, UNESCO periodicals). As with the three publications mentioned just above, the scope of this compilation is house-wide and cross-sectoral.
81   From the report of an evaluation of CSI carried out in 2002 by a three-person evaluation team. Collectively, the evaluation team combined expertise in the social sciences, marine sciences and ecology, integrated coastal and natural resource management, and cultural and landscape planning. The evaluation consisted of a desk review of CSI documentation, semi-structured discussions with UNESCO staff at Headquarters and in selected field offices, visits to selected field projects and discussions with local stakeholders and a small pilot survey of recipients of the internet-based Wise Practices Forum. http://www.unesco.org/csi/intro/eval02.htm#eval
82   Note that additional scientific work on coastal regions and small islands has been undertaken within the frameworks of other UNESCO undertakings and programmes (e.g. IOC, IHP, MAB, IGCP) and is reported elsewhere in this book.

UNIVERSITY CHAIRS AND TWINNING ARRANGEMENTS

At the higher education level, the UNITWIN-UNESCO[83] Chairs programme serves as a prime means of capacity-building through the transfer of knowledge and sharing in a spirit of solidarity between countries. House-wide, as of late 2005, some 590 UNESCO Chairs and inter-university networks have been set up in over 120 countries. Several of these relate to coastal regions and small islands, with CSI directly involved in the establishment and subsequent liaison with three university Chairs relating to integrated management and sustainable development of coastal regions and small islands at: Cheikh Anta Diop University, Dakar, Senegal (Chair established in April 1997); the University of the Philippines-Diliman (July 2000); and the University of Latvia at Riga (late 2001).[84] Each Chair is closely linked with field projects, such as that in Latvia on municipal environmental management and public participation in the North Kurzeme coastal region. A number of university twinning networks have also been developed, with regional workshops on such topics as multi-stakeholder agreements as a tool for preventing and resolving conflicts in the use of coastal resources (Khuraburi, Thailand, November 2002) and ports and sustainable coastal management (St Petersburg, Russian Federation, May 2004).

FIELD PROJECTS

Technical and financial support has been provided to more than twenty field projects located in different regions of the world, which as an ensemble have been concerned with a wide range of issues[85] related to sustainable development in coastal regions and small island settings. Topics addressed include freshwater security, effects of flooding, impacts of a ship-breaking industry, underwater archaeology, stakeholder co-management in artisanal fisheries, municipal environmental management and public participation, and indigenous peoples and protected areas.

These field projects have provided the critical building blocks for the elaboration of Wise Coastal Practices, a programme designed for managing conflicts over precious resources. The various series of CSI publications – as well as website postings – have been used for the wide diffusion of the findings, conclusions and recommendations of the majority of the field projects, such as those on beach resources in the smaller Caribbean islands, management of the cultural heritage of Alexandria, Egypt, impacts and challenges of a large coastal industry (ship-breaking yard) in Gujarat, India, and the evolution of village-based marine resource management in Vanuatu.

---

83 UNITWIN is an abbreviation for the University Twinning and networking scheme.
84 UNESCO Chairs in Sustainable Coastal Development, http://www.unesco.org/csi/chairs_tw.htm.
85 CSI Field Projects, http://www.unesco.org/csi/pp.htm.

PART III: ENVIRONMENTAL SCIENCES

Several of the field projects represent long-term feedback-driven initiatives involving multiple partners and close links between research, education, public information and community action. An example is work on beach resources in the smaller Caribbean islands, whose roots date back to comparative studies (started in 1985) on managing beach resources and planning for coastline change in the Caribbean.[86] Methods were developed for the measurement of shoreline changes, training provided in environmental video production and broadcasting, and guidelines prepared and tested on what can be done in response to disappearing and degrading beaches.

Products of this work include a practical guide for beach users, builders, homeowners and other coastal stakeholders (Cambers, 1998), and a series of ten illustrated booklets on shoreline change in individual Caribbean islands.[87] The booklets represent the dedicated work of government agencies, non-governmental organizations, teachers, students and individuals. Together, they have carefully measured the changes in their beaches over a number of years, and have combined scientific research and monitoring with educational and environmental stewardship activities of various kinds. Each booklet combines generic and island-specific information, on such issues as natural and human forces that affect beach areas, national initiatives to monitor and manage changes, recommendations on wise practices for a healthy beach. Individual booklets have been prepared for Anguilla, Antigua and Barbuda, Dominica, Grenada, Montserrat, Nevis, St Kitts, St Lucia, St Vincent and the Grenadines, and Turks and Caicos Islands.

More recently, this work of monitoring and measuring beaches has been extended to small islands in other regions, including the Cook Islands, Palau and Seychelles. Within the Caribbean, links have been developed with the Caribbean Sea Project (CSP) – one of the flagship projects of UNESCO's Associated Schools Project Network, the aim of which is to heighten young people's effective response to the marine environment as a prerequisite for their positive action and to enable them to learn about the rich cultural diversity of the Caribbean region.

Among the components of CSP is 'Sandwatch', a joint initiative of two UNESCO sectors (Education and Natural Sciences), the UNESCO Office in Kingston, Jamaica, and the University of Puerto Rico Sea Grant Program. Objectives include: (1) reducing the level of pollution in the Caribbean Sea; (2) training schoolchildren in the scientific observation of beaches through field measurements and data analysis; and (3) assisting schoolchildren, with the help of local communities, to use the information collected to better manage the region's beaches. Participating countries and territories include Aruba, Bahamas,

---

86   Known locally by an old acronym (COSALC): http://www.unesco.org/csi/act/cosalc/summary_7.htm.
87   http://www.unesco.org/csi/act/cosalc/sandwatch1.htm.

Barbados, Cuba, Curaçao, Dominica, Grenada, Haiti, Jamaica, St Lucia, St Vincent and the Grenadines, Trinidad and Tobago.

ICTs AND MULTILINGUAL INTERNET DISCUSSION FORUMS

New information and communication technologies (ICTs) have been one of the driving forces in today's processes of globalization. Among other opportunities opened by ICTs are those allowing for the storing of large amounts of information in electronic format and making this information instantaneously accessible through the internet. CSI was set up at a time when such electronic services were becoming current. Not being weighed down by the baggage of the past, it was relatively easy for CSI to take advantage of the opportunities for information handling offered by ICTs. In this vein, the CSI website (and associated websites operated by CSI) provide access to considerable amounts of textual information.

Another aspect of ICTs providing new ways of sharing and exchanging information is internet discussion forums of various kinds. Within CSI, a first discussion forum was launched in May 1999 on Wise Coastal Practices for Sustainable Human Development (WiCoP).[88] A small team of moderators edit contributions before they are posted (in English, French and Spanish) on the forum site and in addition sent as email (thanks to the collaboration of Scotland On Line) to an ever-growing number of individuals connected with the forum.

Issues that have been addressed range from conflict prevention and resolution to approaches for coastal stewardship, from private sector investment in marine conservation to combining traditional and modern practices in coastal fisheries. Vulnerability and resilience in small islands was the focus of one 2004 discussion thread, which elicited substantial comment, reaction and controversy. Other lively debates have included those on 'Land purchase as an option for conservation' and 'Aid has failed the Pacific'. And more generally, the case studies and insights presented on the forum have proved valuable for learning, teaching and research purposes.

A second initiative – carried out as a cross-cutting project by the Natural Sciences and the Communication and Information Sectors – is Small Islands Voice (SIV), which seeks to provide the general public in islands with a 'space to speak and act'.[89] From early 2002, when the initiative was launched, considerable effort has been made to identify the key issues of concern to the general public in the Caribbean, Indian Ocean and Pacific regions, through opinion surveys, internet discussions, meetings and workshops – all facilitated by newspaper, radio and television coverage.

---

88   http://www.csiwisepractices.org.
89   Small Islands Voice: http://www.unesco.org/csi/smis/siv/sivindex.htm.

PART III: ENVIRONMENTAL SCIENCES

Three wide-ranging overviews of UNESCO's work on small islands, including activities in education, culture, social sciences, and communication and information, as well as the natural sciences.
Top: *Island Agenda* (1994, 131 pp.);
Middle: *Island Agenda* 2004+ (2004, 48 pp.);
Bottom: *Embarking on Mauritius Strategy Implementation* (2005, 6 pp).

BOX III.10.1: YOUTH VISIONING FOR ISLAND LIVING

Among ongoing activities, UNESCO is facilitating a means for young people to articulate how they want their islands to develop in the future, and how they plan to help make this happen. There are three main stages.

First, during the twelve-month period starting January 2004, preparatory activities among island youth included local meetings and discussions, fund raising activities, media promotion of the visioning activity, and web-based discussions through a special site operated by the international youth NGO TakingITGlobal.

Second, youth participants from island countries met in Mauritius, in January 2005, to discuss concerns, share information about activities, and shape their vision. This special event brought together ninety-six young people from thirty-one Small Island Developing States (SIDS) and six island territories with other affiliations. The conclusions were encapsulated in a four-page Declaration that includes commitments for follow-up action by the participants, and which was presented to the plenary of the main United Nations meeting. Proposed follow-up actions address three main themes:

1. Life and love in islands – island lifestyles and cultures (17 projects);
2. My island home – safeguarding island environments (15 projects);
3. Money in my pocket – economic and employment opportunities (11 projects).

Third, and most importantly – after the UN meeting, young delegates reported back to their local groups about the results of the Mauritius youth forum. Youth groups gave priorities to actions at a national and local level, and began implementation. Mini-grants are being made available to youth groups, based on a competitive selection process, in support of implementing their projects. A major challenge is that of involving poor, marginalized, disaffected youth in the overall process and in individual projects.

The whole Youth Visioning for Island Living event represents a partnership activity involving UNESCO (through the Coastal Regions and Small Islands Platform and the Bureau of Strategic Planning's Section for Youth) and the Mauritian authorities (particularly through the National Commission for UNESCO, the Ministry of Education and Scientific Research, the Ministry of Youth and Sports, and the Ministry of Social Security), as well as a range of other regional and international partners, including the Lighthouse Foundation, Indian Ocean Commission, Secretariat of the Pacific Community, Caribbean Community, UNICEF, and the international youth NGO TakingITGlobal.[1]

---

1 For further information on the Youth Visioning event – including the final Declaration, the list of participants, and country commitments for follow-up work – see the 'UNESCO at Mauritius' website: http://portal.unesco.org/islandsBplus10.

In a related fashion, the SIV global forum is serving as a 'small islands heartbeat' by promoting and profiling the opinions of ordinary people living in islands.[90] Every two weeks or so, over 40,000 islanders and people concerned with islands are exposed to a range of topical issues – spanning environment, development, society, economy and culture – via SIV global email postings.[91]

Topics profiled range from rethinking an archipelago's tourism strategy (initial posting from Seychelles) and exporting an island's spring water (St Vincent and the Grenadines) to road construction and its effects on people's lives (Palau), piracy of fishery resources in the South Atlantic (Ascension Island) and problems of solid-waste disposal (San Andrés archipelago). And in early 2005, contributors to discussions on 'Communities planning their futures in a post-tsunami world' provided many insights on the need and opportunities for communities to take the lead in planning their own destiny.

Further implementation of the Barbados Programme of Action (Barbados+10)
As mentioned above, in 1993–94 the Organization prepared a house-wide overview of its work on island environments, territories and societies, as one of its inputs into a meeting in May 1994 in Barbados on the Sustainable Development of Small Island Developing States (SIDS). In turn, preparation of this overview had been part of the immediate background to the setting up of the CSI Platform in 1995–96.

Nearly a decade later, in December 2002, the United Nations decided to undertake a thorough review of the Programme of Action adopted in Barbados. On its part, in October 2003, the General Conference adopted a resolution (32 C/Res. 48)[92] specifically addressed to the 'Sustainable Development of Small Island Developing States: further implementation and review of the Barbados Programme of Action (Barbados+10)'. The draft of this resolution was submitted by fifteen of UNESCO's Pacific Member States, supported by Member States in other regions, and was considered by each of the five substantive programme commissions of the UNESCO General Conference. As subsequently adopted by the Plenary of the General Conference, the resolution addressed the continued implementation of the Barbados Programme of Action and preparations for an international meeting with high-level participation (Mauritius, 10–14 January

---

90 Small Islands Voice Global Forum: http://www.sivglobal.org.
91 As for the WiCoP postings, the collaboration of Scotland On Line in sending out the moderated SIV discussions as email messages is gratefully acknowledged. Such dispatch to several tens of thousands of individual recipients (for WiCoP and SIV combined) depends on outside support. As of late 2005, UNESCO is not able to provide such a service.
92 Resolution 32 C/Res. 48: 'Sustainable Development of Small Island Developing States: further implementation and review of the Barbados Programme of Action (Barbados +10)'. http://www.unesco.org/csi/B10/32C-Res48.pdf.

2005), and reporting to UNESCO's governing bodies on the planning, outcomes and follow-up of the Mauritius meeting.

In responding to this resolution, the Director-General designated the CSI Platform as focal point for the Organization's participation in the Barbados+10 (B+10) review and Mauritius International Meeting (MIM) forward-planning process. In line with CSI's function to act as a testing ground for approaches to intersectoral cooperation, some of the features of the Organization's inputs to the B+10/MIM process might prove useful in planning future house-wide contributions to analogous large-scale UN activities. Thus, as part of the CSI piloted-contribution to the B+10/MIM process, UNESCO:

1. took active part in the dedicated Interagency Task Force set up by the United Nations, playing an instrumental role in the adoption of conference-call facilities as the modus operandi for the task force;
2. developed an interactive cross-sectoral website for the B+10/MIM process,[93] starting more than eighteen months before the main UN meeting took place and subsequently providing insights into the Organization's contribution to the follow-up of the Mauritius meeting;
3. participated in various UN and AOSIS (Alliance of Small Island States) preparatory activities for B+10/MIM;
4. contributed distinctively, in particular through the two above-mentioned internet-based discussion forums,[94] with over 40,000 (in late 2005) e-recipients, to the evolving SIDS agenda (e.g. in terms of incorporation in 2004 of dimensions such as cultural diversity and economic opportunities of island cultures, as well as the role of youth);
5. prepared overviews of recent and ongoing UNESCO activities relating to sustainable development in small islands, with special emphasis on Small Island Developing States (SIDS) (e.g. small-island dossiers in the *New Courier* and *A World of Science*, Decade of Education for Sustainable Development [DESD] Brief on SIDS);
6. undertook a substantive review of the Organization's recent, ongoing and planned activities relating to sustainable

---

93 UNESCO Implementing Mauritius Strategy: http://portal.unesco.org/islandsBplus10.
94 Developed within the Coastal Regions and Small Islands (CSI) Platform, the two discussion forums are Wise Coastal Practices for Sustainable Human Development (WiCoP, http://www.csiwisepractices.org) and Small Islands Voice (SIV, http://www.smallislandsvoice.org).

development in small islands, with special emphasis on SIDS, with a digest of that review prepared as a forty-eight-page illustrated booklet (UNESCO, 2004) for distribution in paper form at the Mauritius meeting in January 2005 and, later, in electronic format with many hotlinks;
7. held a dozen informal open planning meetings on a near-monthly basis, following a first such meeting in May 2003, alternating since May 2004 with meetings of the high-level intersectoral Working Group on SIDS,[95] with summary records of each meeting being sent out widely in electronic format and being made accessible through the dedicated website mentioned under (2) above;
8. used conference-call facilities as a means for regular, systematic discussions with colleagues in UNESCO field offices and other collaborators.

The Mauritius International Meeting took place in January 2005, with UNESCO's contributions focusing on a handful of activities consistent with the Organization's mandate and comparative advantage. Following proposals made by the Focal Point to United Nations Headquarters in New York, the Culture Sector took the lead in organizing a Plenary Panel on the 'Role of Culture in the Sustainable Development of SIDS'. As stated above, other UNESCO-led activities in Mauritius included 'Youth Visioning for Island Living', as well as a Small Islands Voice parallel event on Communities in Action, and another parallel event on Ocean and Coastal Management.

The principal negotiated outputs of the Mauritius International Meeting – the Strategy document and the political Declaration – call for action in many fields related to UNESCO's concerns, programmes and priorities. The underlying challenge is that of building bridges and networks to promote problem-solving actions that: (1) cut across societal sectors and institutional specialities, (2) mobilize key actors and constituencies (including youth), (3) generate effective momentum and impact, (4) are culturally sensitive and scientifically sound, (5) take advantage of the opportunities opened by modern information and communication technologies, and (6) promote the exchange of information and experience within and between regions and between islands of different affiliations.

Within such a context, SIDS programme planning and implementation increasingly calls for approaches that connect entities in mutually supportive ways – facilitating actions that are intersectoral, interregional and intergenerational

---

95 UNESCO Document DG/Note/04/07 (10 February 2004): http://www.unesco.org/csi/B10/DGnote04-07.pdf.

in nature, that will often and increasingly take advantage of the opportunities offered by the internet, with the whole process designed to contribute optimally to sustainable island living and development.

Underpinning all of CSI's activities is an emphasis on the building of capacities and various kinds of bridges and encouraging networks to promote effective collaboration between societal and organizational sectors, between regions and between generations. It is hoped that this ideal of cooperation and mutual understanding will create a new vision of, and commitment to, sustainable development in the world's small islands and coastal regions.

BIBLIOGRAPHY

Cambers, G. 1998. *Coping with Beach Erosion*. Coastal Management Sourcebooks 1. Paris, UNESCO. http://www.unesco.org/csi/pub/source/ero1.htm

Troost, D. and Hadley, M. 1993. UNESCO activities relevant to the management of the coastal areas and resources. In: *Proceedings of the World Coast Conference 1993*, Volume 1, pp. 351–56. CZM Centre Publication No. 4. The Hague, Coastal Zone Management (CZM) Centre.

UNESCO. 1993. *Coasts: Managing Complex Systems*. Environment and Development Brief 6. Paris, UNESCO Office for the Coordination of Environment Programmes. http://www.unesco.org/csi/intro/coastse.pdf

——. 1994. *Island Agenda: An Overview of UNESCO's Work on Island Environments, Territories and Societies*. Paris, UNESCO. http://www.unesco.org/csi/intro/islandagenda.htm

——. 1997. *UNESCO on Coastal Regions and Small Islands: titles for management, research and capacity-building (1980–1995)*. CSI info 2. Paris, UNESCO, 21 pp. http://www.unesco.org/csi/pub/info/pub1.htm

——. 2004. *Island Agenda 2004+: coping with change and sustaining diversities in small islands*. Paris, UNESCO. http://www.unesco.org/csi/B10/IslAgenda2.pdf

PART III: ENVIRONMENTAL SCIENCES

THREADS OF LEARNING AND EXPERIENCE
Recording and valorizing traditional ecological knowledge
Malcolm Hadley

IF recent years have seen a widening interest in the contribution of different knowledge systems to the understanding of environmental and natural resources management, UNESCO's interest dates back more than half a century. Assessments of traditional water use practices in arid and semi-arid zones featured in several symposia and reviews of research during the 1950s. In the humid tropics, an ethno-botanical survey in South-East Asia was carried out under the aegis of the UNESCO Office in Jakarta, Indonesia.

Somewhat later, from the early 1970s on, the Man and the Biosphere (MAB) Programme has contributed to work on traditional ecological knowledge, through a series of field studies on local and indigenous people and their decision-making and resource-management systems, carried out by national institutions participating in MAB in a variety of ecological and sociocultural settings. Projects encompassed such topics as: traditional medicinal plants in Caribbean and Mediterranean islands; the attempted re-creation of historical grassland management in a mid-latitude biosphere reserve (Vessertal-Thüringen Forest, Germany); the use of a traditional, community gathering – the Maori Hui – for discussing research and development plans in a coastal marine area in New Zealand; leadership, channels of communication and decision-making among pastoralists in northern Kenya, as ingredients for shaping extension programmes; understanding and adapting traditional energy-efficient agricultural systems (*chinampa*) in wetland areas in Mexico; people–forest interactions in East Kalimantan in Indonesia; and linking traditional production systems with the market economy in Peru.

Field studies such as these have been complemented by efforts to compile and diffuse technical information on traditional ecological knowledge. Examples include a preliminary survey of crop improvement techniques in tropical regions, the role of biosphere reserves in the conservation of traditional land-use systems of indigenous populations in Central America, the relations of African forest dwellers with their equatorial environment (with particular reference to food and nutrition), traditional reindeer and fisheries management in Fennoscandia, and the implications of traditional knowledge to the planning and management of biosphere reserves in South and Central Asia. Over its three and a half decades (1965–99), the quarterly magazine *Nature and Resources* addressed a range of issues related to traditional and local knowledge, with two whole issues (in 1994) devoted to 'Traditional knowledge in tropical environments' and 'Traditional knowledge into the twenty-first century', respectively.

Within the marine sciences, cooperation with the International Association for Biological Oceanography included a 1983 study on 'Traditional knowledge and management of marine coastal systems'. In the same year, a regional seminar on that subject in Asia and the Pacific was organized by the Regional Office in Jakarta, with follow-up materials including an anthology on traditional marine resource management in the Pacific.

In terms of collaborative work on ethnobotany, the 'People and Plants' initiative was launched in 1992 by the World Wide Fund for Nature (WWF), UNESCO-MAB and the Royal Botanic Gardens Kew (UK), with the aim of promoting ethnobotany and the equitable and sustainable use of plant resources. In cooperating with local, national and regional institutions of research, community development, higher education and park management, 'People and Plants' sought (over three four-year phases) to promote local community development programmes on the improved management of plant resources, to encourage greater involvement of local people in devising and implementing strategies for the conservation of biological diversity, and to increase local capacities and capabilities for research, training and management of natural resources. Written outputs included several series of methodological guidelines and information materials, including conservation manuals, handbooks and working papers. Generic issues addressed in the People and Plants conservation manuals include applied ethnobotany, invasive plants, plants and protected areas, valuation methods for woodland and forest resources, and certification and management of non-timber forest products. Complementing this work on ethnobotany, is work on the chemistry of natural products, with a series of Asian symposia on medicinal plants, spices and other natural products, and an *International Collation of Traditional and Folk Medicine: North-East Asia*.

LINKING BIOLOGICAL AND CULTURAL DIVERSITY
Local and Indigenous Knowledge Systems (LINKS) project
Douglas Nakashima and Annette Nilsson[96]

THE environmental knowledge of local and indigenous peoples is now widely recognized as an essential building block for sustainable development and the conservation of biological and cultural diversity. Emerging on the international scene at the 'Earth Summit' (Rio de Janeiro, Brazil, 1992), and through the Convention on Biological Diversity – whose Article 8(j) incites States Parties to 'respect, preserve and maintain knowledge, innovations and practices of indigenous and local communities' – the domain has rapidly gained prominence and momentum.

UNESCO launched the Local and Indigenous Knowledge Systems (LINKS) project in 2002 as one of a new generation of cross-cutting programmes to heighten interdisciplinary and intersectoral action.[97] Contributing to the Millennium Development Goals of poverty eradication and environmental sustainability, the project aims to empower local and indigenous peoples in biodiversity management by advocating full recognition of their unique knowledge, experience and practices. LINKS is led by the Natural Sciences Sector's Coastal Regions and Small Islands Platform and involves all five UNESCO programme sectors, as well as the Field Offices in Apia (Samoa), Bangkok (Thailand), Dhaka (Bangladesh), Hanoi (Viet Nam), Montevideo (Uruguay) and Moscow (Russia). From its inception, the project has hosted and benefited from the yearly support of indigenous youth interns sponsored by Canada.

The LINKS project was in part born out of the debate generated by UNESCO's inclusion of indigenous knowledge on the agenda of the World Conference on Science (Budapest, Hungary, 1999). Placing scientific and indigenous knowledge side-by-side triggered considerable discussion about the status and validity of knowledge. Some questioned whether it was appropriate for the world's scientists to give recognition to these 'other knowledge systems' (Nakashima, 2000). Responding to these concerns, the International Council for Science (ICSU), with the support of UNESCO-LINKS, produced a

---
96 Douglas Nakashima heads the UNESCO LINKS project.
Annette Nilsson worked as a consultant with the LINKS project at UNESCO in 2005.
97 See the LINKS website at: www.unesco.org/links.

report on 'Science, Traditional Knowledge and Sustainable Development', which was launched in Johannesburg (South Africa) during the World Summit on Sustainable Development, and which defines and differentiates science, traditional knowledge and pseudoscience (ICSU, 2002).

Since its inception, LINKS has combined field-based action with efforts to raise awareness and build dialogue among indigenous knowledge holders, scientists and the public at large. A number of field projects have been established. In the Bosawas Biosphere Reserve in Nicaragua, the Mayangna people have requested that their knowledge be recorded in the form of an 'encyclopaedia of nature' that would serve both to educate their children and to affirm their status as knowledgeable managers of their lands and resources. Work has begun in the community of Arandak, with an initial focus on elucidating the economically important category of 'things of the water', which encompasses numerous types of fish, but also extends to turtles. Another Latin American project is underway with the Mapuche-Pewenche people in Chile. Working with the Asociación de Comunidades Mapuche Pewenche Markan Kura, a local indigenous non-governmental organization (NGO), this project focuses on the many facets of Mapuche knowledge. It is centred on the monkey puzzle tree (*Araucaria imbricata*), a keystone species for the ecological system, and a core element in Mapuche society. Other LINKS field projects have been initiated with the fishers of Charan District, Bangladesh; the Cree First Nation hunter-trappers of subarctic Quebec, Canada; the Even and Koryak herders and hunters of Kamchatka, Russia; and the Pacific Island peoples of Vanuatu, Solomon Islands and Palau. In the framework of the latter field project, LINKS launched the first volume in its publication series Knowledges of Nature. Entitled *Reef and Rainforest: An environmental encyclopaedia of Marovo Lagoon, Solomon Islands*, the volume (by Edvard Hviding, 2005) provides a meticulous documentation of Solomon Islander knowledge of reef and land topography, and of marine and terrestrial animals and plants. Containing more than 1,200 Marovo terms, with definitions in both Marovo and English, it is destined for use in local schools – where there is a dearth of indigenous language materials – and provides a first basis for dialogue between scientists and the Marovo peoples.

In addition to empowering local and indigenous communities in biodiversity governance, through their recognition as knowledge holders, the LINKS project also seeks to maintain the vitality of local knowledge within communities. Indigenous and local rural peoples are often marginalized by mainstream society. In the formal school system, this results in the exclusion (and even denigration) of local knowledge, values and worldviews. The resulting alienation, loss of identity and self-esteem is devastating for indigenous youth and for the society as a whole. In several of its field projects, LINKS seeks to strengthen ties between elders and youth in order to reinforce the transmission of indigenous knowledge.

One approach targeting youth has been the use of new information and communication technologies – such as multimedia CD-ROMs – as a vehicle for conveying traditional knowledge. The LINKS CD-ROM series thus far includes two interactive CD-ROMs: the first on Aboriginal Australian life-worlds, entitled *Dream Trackers: Yapa art and knowledge of the Australian desert* (Glowzewski, 2000), and the second (launched at the end of 2005) on Pacific Islander knowledge of ocean navigation and the arts of canoe construction and sailing, entitled *The Canoe is the People: indigenous navigation in the Pacific* (Glowzewski, 2005).

With bow and arrow, a Mayangna of Arandak, Nicaragua, applies his ecological knowledge and technical know-how to catch fish, the community's main source of protein. The Mayangna project has been supported financially and scientifically through the Local and Indigenous Knowledge Systems (LINKS) initiative since October 2003, with Mayangna knowledge being recorded and compiled in the form of an 'encyclopaedia of nature'.

Despite broadened support, the issue of local and indigenous knowledge continues to be contentious and is plagued with stereotypes and misconceptions. To build awareness and promote dialogue and mutual understanding, the LINKS project has organized international seminars and workshops, and prepared publications on the theme. In the framework of the World Water Forum (WWF), LINKS organized workshops on water and indigenous peoples at the second WWF in the Hague (the Netherlands) in 2000, and at the third WWF in Kyoto (Japan) in 2003. At the World Summit on Sustainable Development (Johannesburg, South Africa, 2002), LINKS organized a session with ICSU on 'Linking Traditional and Scientific Knowledge for Sustainable Development'. A further international seminar, 'NGOs, Indigenous Peoples and Local Knowledge', was organized with the Centre National de la Recherche Scientifique (CNRS, France) at UNESCO Headquarters in 2003, leading to the publication of a thematic issue of the *International Social Science Journal* (ISSJ) on 'NGOs and the Governance of Biodiversity' (Roué, 2003). An earlier issue of ISSJ was produced on 'Indigenous Knowledge' (Agrawal, 2002). Most recently, LINKS organized, in cooperation with CNRS – and in the framework of the international conference on 'Biodiversity: Science and Governance' – a workshop on 'Sustaining Biological and Cultural Diversity: The challenge of local knowledge, practice and worldviews'.

Rural populations and indigenous peoples are often excluded from governance processes, such as decisions pertaining to access, use and management of land and resources. Yet such decisions are critical for their economic, social and cultural well-being. By promoting local and indigenous knowledge systems, the LINKS project argues for the rights of local and indigenous peoples not only to participate in these processes, but also to shape them to their own needs and aspirations.

BIBLIOGRAPHY

Agrawal, A. (ed.). 2002. Indigenous knowledge. *International Social Science Journal*, No. 173. (In six languages.)

Glowzewski, B. 2000. *Dream Trackers: Yapa art and knowledge of the Australian desert.* LINKS Interactive CD-ROM series, No. 1. Paris, UNESCO. (In Walpiri, English and French.)

———. 2005. *The Canoe is the People: indigenous navigation in the Pacific.* LINKS Interactive CD-ROM Series, No. 2. Paris, UNESCO.

Hviding, E. 2005. *Reef and Rainforest: an environmental encyclopedia of Marovo Lagoon, Solomon Islands.* LINKS Knowledges of Nature series, No. 1. Paris, UNESCO, 252 pp.

ICSU (International Council for Science). 2002. *Science, Traditional Knowledge and Sustainable Development.* ICSU series for Science and Sustainable Development, No. 4. Paris, ICSU, 24 pp.

Nakashima, D. 2000. What relationship between scientific and traditional systems of knowledge?: science and other systems of knowledge. In: A. M. Cetto, *Science for the 21st Century: a new commitment.* Paris, UNESCO, pp. 432–44.

Roué, M. (ed.) 2003. NGOs in the governance of biodiversity. *International Social Science Journal*, No. 178. (In six languages.)

A GIFT FROM THE PAST TO THE FUTURE
Natural and cultural world heritage
Bernd von Droste zu Hülshoff[98]

THE cultural and natural heritage of every nation is a priceless possession, a precious gift that has been inherited from the past and is to be kept in trust by the present generation for generations yet to come. Any loss or serious impairment of this heritage is a tragedy, because these gifts are irreplaceable. The greatest heritage comprises those cultural and natural properties generally regarded to be of exceptional worldwide significance and value. For a variety of reasons, this world heritage is being damaged or lost at an alarming rate. In recognition of this crisis, the Member States of UNESCO adopted the Convention concerning the Protection of the World Cultural and Natural Heritage on 16 November 1972, at the sixteenth General Conference, known as the World Heritage Convention.

Today the Convention is considered the most successful international legal instrument in the field of heritage conservation, whether natural or cultural. It has been ratified or accepted by 180 states. Under the Convention, properties deemed to have outstanding universal value are identified by being placed on the World Heritage List. Each State Party is invited to nominate properties located on its territory for the World Heritage List. There are very strict criteria, for both cultural and natural properties. Following the World Heritage Committee's twenty-ninth session in Durban, South Africa, in July 2005, there were 812 sites on the World Heritage List, located in 137 States Parties. Of these, 160 had been inscribed on the basis of the natural criteria as elaborated in the Committee's operational guidelines; 628 were inscribed under the cultural criteria; and another 24 were listed under both sets of criteria as so-called mixed sites.

The major objective of the Convention is, however, to ensure protection and conservation of properties on the World Heritage List and to assist States Parties in this task, if so required, by using resources of the World Heritage Fund derived mainly from the contributions by states adhering to the Convention.

---

[98] Bernd von Droste zu Hülshoff was Director of UNESCO's Division of Ecological Sciences and Secretary of the MAB Programme from 1984 to 1992; subsequently he built up UNESCO's World Heritage Centre, which he directed from 1992 to 1999 (to retire at level of Assistant Director-General). Currently, he is UNESCO's Special Advisor for World Heritage.

THE ORIGINS OF THE CONVENTION

According to Michel Batisse, former Assistant Director-General and director of environmental programmes of the Natural Sciences Sector at UNESCO, the World Heritage Convention was born from the convergence, at the international level, of two movements: the first, mainly fuelled by cultural advocacy, was led by the International Council of Monuments and Sites (ICOMOS) and the Cultural Sector of UNESCO; the second was directly associated with the International Union for the Conservation of Nature (IUCN) and UNESCO's Natural Sciences Sector.

CULTURAL HERITAGE

When the governments of Egypt and the Sudan appealed to UNESCO for help to rescue the Abu Simbel temples from the rising floods of the Nile at the Aswan dam in 1959, the Organization decided to launch an international safeguarding campaign, the first and most spectacular of its kind. The so-called Nubian campaign of UNESCO was an unquestioned success and was soon followed by an increasing number of other preservation campaigns starting with Venice after the great flood of 1966. There was a growing conviction that an international convention, accompanied by an assistance fund, should be established for the protection of monuments, buildings, and sites of global interest. The necessary process was initiated by UNESCO's Culture Sector with the help of the newly established ICOMOS. A draft of the convention was circulated in the middle of 1971.

NATURAL HERITAGE

On the environmental side, the idea of a World Heritage Trust was first proposed by a White House conference on international cooperation in Washington DC (USA), in 1965. The conference called for the establishment of 'a trust for the world heritage that would be responsible to the world community for the stimulation of international cooperation efforts to identify, establish, develop and manage the world's important natural and scenic areas and historic sites for the present and future benefit of the international citizenry'.

The world heritage trust idea was vigorously promoted by Russell E. Train, chairman of the US Council of Environment and Quality. The proposal for the World Heritage Trust was eventually submitted to IUCN, which took the necessary steps to prepare a draft convention along these lines. This draft, which was ready by the beginning of 1971, was immediately supported by the United States Government. US President Richard M. Nixon expressed the wish that the convention be adopted on the occasion of the centennial celebration of Yellowstone National Park (USA).

PART III: ENVIRONMENTAL SCIENCES

Galapagos Islands, Ecuador, one of the first 'natural sites' inscribed on the List of the World Natural and Cultural Heritage, 1978. The 'Convention concerning the Protection of the World Cultural and Natural Heritage' was adopted by UNESCO's General Conference in November 1972 and came into force in December 1975 after ratification by twenty states.

Thus two independent movements were leading to two parallel international instruments. Because the proposed IUCN convention covered both natural areas and major historic sites, it threatened to overlap with the proposed cultural UNESCO-ICOMOS convention. This danger was brought to light in the course of the preparatory process for the UN conference on the Human Environment, held in Stockholm, Sweden, in June 1972.

The movement towards a single convention was initiated at the third meeting of the conference's preparatory committee in September 1971. At its fourth meeting, in March 1972, a delicate negotiation was conducted to reconcile the wishes of all parties concerned and, in particular, those of environmentalists, who were urging that the conservation of nature at long last be given the same attention and weight as cultural preservation. Eventually, it was agreed by the preparatory committee that a single text should be worked out for adoption by the General Conference of UNESCO. This highly desirable compromise resulted in cultural and natural properties being given equal treatment under the new convention. Addressing conservation of the masterpieces of human endeavour with that of the wonders of nature in a single legal text was a profoundly innovative move.

IMBALANCES AND UNDER-REPRESENTATION – A GROWING CONCERN

The Convention came into force on 17 December 1975. A World Heritage Committee was appointed from among the States Parties, and in 1976 the advisory bodies IUCN and ICOMOS were asked to produce proposals for criteria for the inscription of sites on the World Heritage List. In 1977, the Committee adopted operational guidelines for the Convention's implementation based on the proposals of the advisory bodies. The outcome was that the Committee, at its second session in Washington DC, inscribed the first sites on the list in 1978. Among the natural sites were Yellowstone National Park, and the Galapagos Islands (Ecuador).

In retrospect, it is interesting to note that already at this second session in Washington DC (1978), Committee members got involved in a lengthy discussion as to whether the number of nominations per country and per year should be limited. At that time, some members were considering an overall ceiling for the World Heritage List of about 100 properties. The Committee, mindful of producing a balanced first list, decided that States Parties should be limited to nominating two properties per year. Unfortunately, this decision was not maintained in later sessions.

In terms of application of the World Heritage criteria adopted in 1977, the Washington session of the Committee in 1978 set high standards. The Committee was fully aware of its great responsibility for setting the tone for the future. It decided to enter only twelve properties on the World Heritage List. In doing so, it gave equal weight to cultural and natural properties, and – to the extent possible – took into account that sites were situated in different cultural and ecological regions of the world.

During its first years, the Committee's attempts to achieve a geographical balance in the composition of the World Heritage List were doomed to failure, since the coverage by States Parties of the different regions of the world was heavily skewed. Initially, countries from Western Europe and North America, the Arab region and Latin America had joined the Convention. Most Eastern European and Asian countries followed in the late 1980s and early 1990s, and only recently has the void in some parts of sub-Saharan Africa, and the Pacific and Caribbean regions been filled. Nevertheless, during its thirty years of existence, the World Heritage List – albeit unbalanced – has grown beyond the imaginations of those who prepared the ground for it, mainly at the price of a Eurocentric composition.

Already in 1978 in Washington DC, the Committee was preoccupied by an under-representation of the natural heritage domain. This did not only concern the number of nominations for the World Heritage List, but also the composition of the World Heritage Committee. In order to give more weight to natural heritage, it was decided that the Chair should alternate from year to year between experts coming from the cultural and natural heritage fields. Furthermore, it was recommended that States Parties should be represented not only by a cultural expert, but also by a natural heritage expert. Some of the decisions of the World Heritage Committee at its recent session in Suzhou, China (2004), still address the above concerns, however in the context of an even more unbalanced distribution of World Heritage properties.

Although the Convention states (in Article 14) that UNESCO provides the Secretariat for World Heritage, initially no special provision was made for this by the Organization. I was a Programme Officer in the Ecological Science Division at the time, and the part of the Convention dealing with natural heritage was simply added to my job description – 'to act as one of the Secretaries to the World Heritage Committee'. The main responsibility remained to assist in the implementation of the Man and the Biosphere Programme (MAB). The Director-General, Amadou-Mahtar M'Bow, decided that Anne Raidl (of the legal standard section/Culture Sector) and I (of Natural Sciences) should alternate yearly in our responsibilities for organizing the World Heritage statutory meetings, and take care of the preparation and implementation of their decisions.

Unfortunately, during the first years we both had to perform secretarial duties for World Heritage, without especially assigned staff by UNESCO. From 1984 onward, I was asked to direct the Division of Ecological Sciences – with the implementation of UNESCO's MAB Programme as my main duty. This gave me the opportunity to involve staff of this division in World Heritage, when appropriate. To my disappointment, the Natural Sciences Sector tended to ignore the World Heritage Convention – which was considered non-scientific, elitist and not development-oriented. For these reasons, priority was given to the development and strengthening of the international biosphere reserve network. Anne Raidl was, in 1986, promoted to the post of director of the Cultural Heritage Division. As such, she was able to implement the cultural part of the World Heritage Convention in close cooperation with the cultural campaigns of UNESCO.

WORLD HERITAGE SITES AND BIOSPHERE RESERVES

The Secretariat for the natural part of the World Heritage Convention was, until 1992, provided by the UNESCO Division of Ecological Sciences, which was also responsible for MAB and its international network of biosphere reserves. From the first day of implementation of the World Heritage Convention, many wondered what precisely the two international conservation concepts had in common and what their differences were – particularly since many biosphere reserves figured in the World Heritage List (e.g. Yellowstone Park, the Galapagos Islands). Indeed, as of 2005, more than 70 biosphere reserves are found among the 160 natural World Heritage sites. It would seem that the two have much in common; and there is, indeed, a case of overlapping concern. Their differences and points in common can best be highlighted by comparing the criteria established for selecting suitable sites for each programme.

One major difference between the biosphere reserves and World Heritage sites is that the former are established in order to preserve ecosystems representative of the world's terrestrial and aquatic biome and their subdivisions, the ecosystems, whereas the latter serve the protection of outstanding cultural and natural properties of universal importance. Therefore World Heritage sites must fulfil criteria of uniqueness and universal value, whereas biosphere reserves must meet criteria of representativeness and naturalness. However, both require conservation in their ecological integrity, adequate long-term legal protection, management planning and monitoring.

Biosphere reserves are distinct from any other category of protected areas. Other protected areas may have conservation, research, or education/training as their prime purpose; but biosphere reserves depend upon having a combination of all three. Biosphere reserves form an integral part of the MAB research programme by focusing on the relationship between humans and the biosphere, and on natural ecosystems conservation and research. They are linked to other MAB project areas and are internationally coordinated by the intergovernmental structure of MAB. Biosphere reserves provide a natural baseline area giving scientific information against which to measure changes in the environment.

Although the World Heritage programme is not a research programme, the World Heritage Committee has in many cases listed biosphere reserves (or parts of them) as World Heritage sites, as requested by the States Parties in whose territories they are located. The main selection criterion for natural World Heritage sites is Criterion No. X: 'The area should contain the most important and significant habitats for *in situ* conservation of biological diversity, including those containing threatened species of outstanding universal value from the point of view of science or conservation'. So far, about 100 natural World Heritage sites are listed based on this criterion. As stated above, nearly half of them are at the same time wholly, or in part, biosphere reserves. One reason for this is that both the international network of biosphere reserves and the natural World Heritage sites are prioritized by attempting to integrate biodiversity hotspot areas.

In what follows, we will focus on natural World Heritage sites in danger. Most of these are irreplaceable in their significance for biodiversity conservation. In addition, many play an important role within the international network of biosphere reserves.

NATURAL HERITAGE IN DANGER

Under the Convention, States Parties are obligated to protect their properties (and to refrain from actions damaging the sites of other countries). If a site from the World Heritage List nevertheless becomes endangered, it may be inscribed on the List of World Heritage in Danger, which calls attention to its plight and makes the country in which it is located eligible to receive support from the World Heritage Fund. The number of sites included on the Danger List has increased considerably over time; at present there are thirty-three, the majority in the category of Natural Heritage.

What are the means available to the Committee to ensure that a State Party does not endanger the values for which the property was placed on the World Heritage List? The Committee engages in a series of moves to encourage a State Party not to pursue actions which threaten a property inscribed on the List:

- The Committee can initiate the procedure to inscribe the property in question on the World Heritage in Danger List, in accordance with Article 11(4) of the World Heritage Convention, if the conditions for such an inscription are fulfilled.
- Based on Article 13 of the Convention, the Committee may not grant international assistance to the State for the property concerned and even suspend or withdraw international assistance already approved.
- The Committee may request actions on behalf of other States Parties.
- The Committee must report to the General Conference of UNESCO on the breach of the Convention.
- The Committee may request for an advisory opinion from the International Court of Justice in accordance with Article 96(2) of the Statute of the International Court of Justice.
- The Committee may also consider the removal of the property in question from the World Heritage List. This is an action not foreseen by the Convention, but which should be considered as inherent to the spirit of the convention.

The World Heritage in Danger listing has been handled by the World Heritage Committee in a rather inconsistent way during the last thirty years. Some seriously threatened natural World Heritage sites – most of significant value for biodiversity conservation – may serve as examples of this. Four different cases can be distinguished, as outlined below.

World Heritage in Danger listing requested by the State Party
There are numerous examples of Natural Heritage sites being placed on the World Heritage in Danger List at the request of the State Party concerned, such as: the Ngorongoro crater (the United Republic of Tanzania), Djoudj National Bird Sanctuary (Senegal), Manovo-Gounda St Floris National Park (Central African Republic), Ruwenzori Mountains National Park (Uganda), Garamba National Park (Democratic Republic of Congo). These sites (all located in Africa) were inscribed on the List not only because they were exposed to serious threats, but also in the expectation of increasing their chances of receiving international assistance from the World Heritage Fund and other sources (e.g. the Global Environment Facility or bilateral aid).

World Heritage in Danger listing by the Committee, with consent
Mount Nimba Nature Reserve (in Guinea/Cote d'Ivoire), which was inscribed on the Danger List by the Committee with the consent of Guinea in 1992, became the first example of mining interests clashing with World Heritage conservation concerns. The conflict was resolved by removing from the World Heritage site an enclave for mining.

In 1995, US environmental groups brought to the attention of the World Heritage Committee the planning of a mining project close to the World Heritage site of Yellowstone National Park that could have caused major damage to the park's outstanding natural resources. The Committee responded by fielding a fact-finding mission to Yellowstone, which confirmed that the mining project (although located a good distance from the site) could indeed cause major damage to the park. Yellowstone was therefore inscribed on the Danger List, with the consent of the US Government. (At a later stage, the mining company was compensated for abandoning its mining lease.)

World Heritage in Danger listing, without consent
Only since 1992 has the World Heritage Committee taken it upon itself to inscribe natural sites on the Danger List against the wishes of the country to which the property in question belongs. In 1992, India's Manas National Park was listed as in danger by unilateral decision of the World Heritage Committee. In 1993, the World Heritage Committee placed the Everglades National Park in the USA on the Danger List; this had not been requested by the USA, which consequently abstained from voting.

Other sites – such as Ichkeul in Tunisia, Sangay National Park in Ecuador, Simien National Park in Ethiopia, or natural World Heritage sites in the Democratic Republic of Congo – were included on the List of World Heritage in Danger by unilateral decision of the World Heritage Committee, causing considerable friction between the Committee and the States Parties concerned.

The Danger List – a last resort
Edith Brown Weiss, professor of international law at Georgetown University (Washington DC), groups international legal and institutional strategies to encourage compliance with international agreements into three categories: (1) 'sunshine methods', such as monitoring, reporting, on-site inspection, access to information and NGO participation; (2) positive incentives, such as special funds for financial and technical assistance, training programmes or access to technology; and (3) coercive measures, in the form of penalties, sanctions and withdrawing of membership.

The chief strategy adopted by the World Heritage Convention is clearly based on the 'sunshine method': no sanctions are foreseen. This strategy – which relies on the so-called 'reputation factor' in order to induce compliance – works most effectively when there is a culture of compliance with norms. A good example of the effectiveness of the sunshine method is the following. In 1995, IUCN reported serious threats to the Galapagos Islands. The Committee embarked on a procedure for a World Heritage in Danger listing, which Ecuador wished to avoid. Ecuador agreed to a fact-finding mission to the Islands, which confirmed that the situation was serious. Finally, the Committee was able to enter into lengthy negotiations with the government, which led to a drastic change in conservation policy and action. In light of these new commitments, the Committee decided not to inscribe the Galapagos on the World Heritage in Danger List.

PREPARING THE GROUND FOR MONITORING AND REPORTING

A debate on monitoring the state of conservation of World Heritage properties was triggered by the United States. In 1982, the US National Park Service submitted a report to the World Heritage Bureau on the state of parks in the USA, which, in the Service's opinion, could serve as a model for reporting on the state of conservation of World Heritage properties.

Despite the evident advantages of knowing more about the state of conservation of World Heritage properties, commitments to monitor have been slow to arrive in the World Heritage community. Consensus around the need for, and value of, monitoring has been difficult to achieve – primarily because of different perceptions of the purpose of monitoring. I remember for example that the debate in 1982 was profoundly influenced by a translation error in our documents: in the French-language version, 'monitoring' had been (mis)translated as 'controlling', which ruined the whole discussion. Although there was stiff opposition (in particular from some French-speaking developing countries, owing to a misunderstanding of the notion of monitoring), monitoring and reporting of some World Heritage properties – notably in the natural heritage realm – proceeded on a voluntary basis, mainly at the initiative of IUCN. ICOMOS joined in later, by presenting to the Committee a number of so-called 'reactive monitoring reports' on cultural properties.

In fact, no consensus was reached until 1997, when the eleventh General Assembly of States Parties to the World Heritage Convention suggested to the General Conference of UNESCO 'to activate the procedures in Article 29 of the Convention and to refer to the World Heritage Committee the responsibility to respond to the reports'.

HOW TO DEFINE AND WHERE TO PLACE CULTURAL LANDSCAPES?

The 1980s also laid the groundwork for 'cultural landscapes' to become eligible for the World Heritage List. The term 'cultural landscape' is absent in the Convention, and it took more than twenty years before the concept of cultural landscapes was finally defined and accepted by the World Heritage Committee, in 1993. Why did it take so long?

In 1987, the United Kingdom submitted the Lake District as a test case for the cultural landscape issue, which demonstrated the need to revise the operational guidelines of the Convention. These were clearly not in line with Articles 1 and 2 of the Convention. A breakthrough occurred at a meeting in La Petite Pierre, France, in 1992, at which 'cultural landscapes' were finally defined. Three categories for World Heritage purposes were established: a clearly defined landscape, designed and created intentionally; agricultural landscapes, of exceptional harmony, of the works of man with nature; and an associative cultural landscape.

In 1993, Tongariro National Park in New Zealand, entirely owned by the Maori, became the first cultural landscape included in the Word Heritage List. Today about fifty-one cultural landscapes have been included on the list, mostly agricultural landscapes used for wine- and tobacco-growing, rice cultivation, and pastureland. The protracted debate on the value of cultural landscapes in the Committee helped to reconnect culture and nature. These links were later further explored by the Global Strategy meeting 'Linking Nature and Culture', which was held in Amsterdam in 1998.

THE ESTABLISHMENT OF THE WORLD HERITAGE CENTRE IN 1992

The challenge of intertwining nature and culture brings me to what is close to my heart – namely the establishment of the World Heritage Centre in 1992. The unresolved problem of establishing a true World Heritage Secretariat on its own became increasingly critical with the rapid increase of States Parties, and the number of properties on the World Heritage List and the World Heritage in Danger List.

The so-far bicephalous World Heritage Secretariat in the distinct sectors of Science and Culture was not any longer able to cope with the tasks at hand. In December 1991, my colleague Anne Raidl did not see any solution and resigned. Left alone with an obsolete situation, I approached the Director-General, Federico Mayor, with the unresolved problem of establishing a true World Heritage Secretariat. In February 1992, after three confidential meetings, I convinced him to establish such a secretariat under the designation 'World Heritage Institute' (later the name was changed to World Heritage Centre), mainly by employing the following arguments:

- The World Heritage Secretariat has to give an example for the concept of unity of the Convention by linking nature and culture in one organizational set-up.
- States Parties need to have a clear orientation as to whom to address requests to and who is accountable as Secretariat to the Convention.
- Since World Heritage is an intersectoral issue, the Secretariat must be empowered to coordinate and harmonize approaches within the Organization under the direct supervision of the Director-General.
- World Heritage has grown to such a dimension, significance and visibility for the Organization and its Member States that a larger professional permanent Secretariat has to be put in place.
- Twenty years of work under the Convention have to be reviewed, and major challenges for future strategic planning have to be met. Opportunities for providing greater visibility for the Convention should not be missed.
- The forthcoming commemorative and forward-looking World Heritage session in Santa Fe, New Mexico (USA, 1993), needs to be carefully prepared, since this should transmit a positive image of UNESCO worldwide.
- The World Heritage is UNESCO's foremost symbol of its unique broad and long-term mission.

In a written note (February 1992), I stated: 'The first requirement is to bring together the so-far separated Secretariat and unite it in daily work at one physical location, close to both parent sectors, but functioning in an autonomous way'.

The Director-General, Mr Mayor, not only adopted my proposal without any substantial modification but also took rather quickly the decision to establish the

World Heritage Centre, which began its work in June 1992 under my direction, with staff from the Culture and Natural Sciences Sectors of UNESCO.

Today, the World Heritage Centre is indeed one of the most highly visible symbols of UNESCO's mission to promote peace and understanding through education, science and culture. It could be said that, through its identification, protection and preservation of cultural and natural heritage around the world, the Centre serves all of humanity.

BIBLIOGRAPHY

Batisse, M. 1992. The struggle to save our World Heritage. *Environment*, Vol. 34, No. 10, pp. 12–20.

Batisse, M. and Bolla, G. 2003. L'invention du 'patrimoine mondial'. *Les Cahiers d'Histoire 2*; Associations des anciens fonctionnaires de l'Unesco, 101 pp.

Brown/Weiss, E. 1998. The five international treaties: a living history. In: E. Brown Weiss and H. K. Jacobson (eds), *Engaging Countries, strengthening compliance with international environmental accords*. Cambridge, Mass., MIT Press, pp. 89–172.

Francioni, F. 2002. The destruction of the Buddhas of Bamiyan and international law. A study commissioned by UNESCO. Unpublished document.

Franck, E. 2001. Legal advice on questions concerning the World Heritage Convention. Unpublished document.

Slatyer, R. O. and von Droste, B. 1984. The origin and development of the World Heritage Convention. *MONUMENTUM*, Special Issue 1984, pp. 3–7.

Smith, G. and Jakubowska, J. 2000. A global overview of protected areas on the World Heritage List of particular importance for biodiversity. Cambridge, UK, UNEP World Conservation Monitoring Centre, 15 pp.

Train, R. E. 1973. An idea whose time has come: World Heritage Trust. *Nature and Resources*, pp 25.

US National Park Service. 1980. *State of the Parks 1980: a report to the US Congress*. US Department of the Interior, 57 pp.

Von Droste, B. 2004. Three decades of world heritage implementation work. Workshop in Siena (Italy): 'The legal tools for World Heritage Conservation' (in press).

Von Droste, B. and Ishwaran, N. 2004. World Heritage in Danger: how it can build on political support. Proceedings of the fifth World Parks Congress, 2003 (in press).

Von Droste, B., Plachter, H. and Rössler, M. (eds). 1995. *Cultural landscapes of universal value – components of a global strategy*. Jena, Germany, Gustav Fischer Verlag, in cooperation with UNESCO, 464 pp.

Von Droste, B. and Robertson, J. 1984. Biosphere reserves and World Heritage sites: relationships and perspectives. Contributions to the First International Biosphere Reserves Congress, Minsk. *Natural Resources Research*, XXI, Vol. I, pp. 242–45.

Von Droste, B., Rössler, M. and Titchen, S. (eds). 1998. Linking nature and culture. Report of the Global Strategy Natural and Cultural Heritage Expert Meeting, Amsterdam, the Netherlands, 25–29 March, 237 pp.

Van Hooff, H. 1995. The monitoring and reporting of the state of properties inscribed on the World Heritage List. *ICOMOS Canada Bulletin*, Vol. 4, No. 3.

ACTING TOGETHER
Promoting integrated approaches, and the paradigm shift from environmental policy to sustainable development from Stockholm 1972 to Johannesburg 2002

Gisbert Glaser

INTRODUCTION

With UNESCO playing an instrumental role, worldwide environmental concern emerged in the 1950s and 1960s. Over the following forty years, UNESCO made a major contribution to the paradigm shift from dealing with environmental protection as a sectoral issue to taking a holistic approach based on the principles of sustainable development. The combined efforts to this end of the Organization's major scientific programmes – and the challenges in getting these programmes to work jointly in influencing an international environmental agenda – have regularly helped redefine UNESCO's environmental mission. The paradigm shift was also evident in policy-oriented interagency cooperation throughout the United Nations system. In order to ensure a coherent UNESCO action, the Natural Sciences Sector, in particular, has devoted a significant amount of time and energy over the past three decades to coordinating environmental programmes. But for what purpose and by which means is, of course, the question.

Hence, this chapter will dwell on both the changing objectives and the changing mechanisms of coordination. Since the late 1960s, with the 'explosion' of UNESCO's environmental activities, there was a growing need for: (1) coordination among the environmental sciences programmes internally within the Natural Sciences Sector; (2) coordination of the environment and, later, sustainable development related activities of all programme sectors of the Organization; and (3) coordination with the environment and sustainable development-related activities of other organizations in the UN system, in particular with the United Nations Environment Programme (UNEP).

The bulk of UNESCO's activities on the environment and on sustainable development are in the Natural Sciences Sector. Yet, the comparative advantage of UNESCO in the fields of environment and sustainable development goes well beyond the scientific field. Among the Organization's comparative advantages are the following. First, its environmental sciences programmes address most major component parts (terrestrial and coastal ecosystems, freshwater, oceans, and the Earth crust) of the Earth system (except the atmosphere). UNESCO is thus the UN organization with the largest portfolio in environmental sciences, giving it a great institutional advantage in launching multidisciplinary research on the interactions among these components of the environment (and on the functioning of the Earth system as a whole). Second, UNESCO can bring to bear its other programme sectors – social science, education, culture and communication – to join the natural sciences programmes in developing 'intersectoral' approaches in support of environmental protection and sustainable development. Third, UNESCO is the lead organization in the UN system in environmental education and education for sustainable development. Thus, it is often said that it has a unique opportunity to ensure that education on the environment and sustainable development is based on the best available information provided by its own scientific programmes. Fourth, all UNESCO programmes are placed at the interface of intellectual work and government. The intergovernmental context gives UNESCO's programmes in the environmental sciences their specific niche – a legitimacy as active agents in solving society's problems.[99]

UNESCO'S KEY ROLE IN THE EMERGENCE OF AN INTERNATIONAL ENVIRONMENTAL AGENDA AND IN THE EARLY WORK OF UNEP

There is no doubt that UNESCO's early scientific programmes in the 1950s and 1960s – dealing with arid zones and humid tropics, oceans and freshwater – contributed significantly to the emergence of the modern era of environmental concern. The discovery by scientists of the growing scale of human impact on the natural environment – as well as the experience of people worldwide of local air and water pollution, soil degradation, clear-cutting of forests, improper use of land and water resources, etc. – translated in the 1960s into growing environmental concerns on the part of the public, governments, NGOs and multilateral agencies. The first phase in UNESCO's environmental sciences activities culminated in the 'Conference on rational use and conservation of

---

99 In this chapter – following a widely used terminology in the international science arena – *multidisciplinary research* refers to research bringing together several disciplines of the natural sciences, whereas *interdisciplinary research* refers to combining natural and social sciences disciplines. The latter is required to understand environment–development interactions (coupled biophysical and socio-economic systems). *Intersectoral activities* are activities combining contributions from different 'sectors' (in UNESCO terms, all its 'programme sectors': natural sciences, social sciences, education, culture and communication). However, very often in UNESCO documents, the term *interdisciplinary* has been used indiscriminately as a synonym for *intersectoral*. When referring in this chapter to the titles of UNESCO activities – as used in UNESCO documents when these were launched and undertaken – it should become clear in each case, from the context provided, what type of activity is meant.

PART III: ENVIRONMENTAL SCIENCES

the resources of the Biosphere' (the Biosphere Conference) held at UNESCO Headquarters in Paris in September 1968. This was the first-ever global intergovernmental meeting of scientific experts on the topic of the changes affecting the planet's terrestrial and coastal ecosystems worldwide.

Referring to the results of the International Biological Programme (IBP) of the International Council for Science (ICSU), the Biosphere Conference was the first to address the fundamentally different scales, rates, kinds and combinations of changes taking place in nature due to human activities. IBP had focused on a better understanding of the functioning of the world's major ecosystems but stopped short of research on the human drivers of ecosystem changes and the question of how to manage ecosystems more rationally. Through its insistence that utilization and conservation of land and water resources should be seen as mutually supportive perspectives, rather than as essentially opposed, the Biosphere Conference established a precursor concept of 'sustainable development'.

The conference called for the launching by UNESCO of an international, interdisciplinary research programme to enhance knowledge for a more rational use of natural resources and a better conservation of the biosphere. At a time when its budget was still growing from one biennium to the next, the Organization had no difficulty following up on this. The launching of the major international scientific cooperation programme 'Man and the Biosphere' (MAB), in late 1970, represented another very significant UNESCO contribution to the development of an international environmental agenda and the emergence of international environmental institutions.

The Biosphere Conference, like all preceding UNESCO projects related to environmental and natural resources research since 1950, was an initiative taken by Michel Batisse, who in 1968 was Director of the Division of Natural Resources Research in the Natural Sciences Sector. Shortly after the UNESCO Conference, the United Nations came under pressure to organize its own conference, focused on the need to promote environmental protection policies worldwide and international environmental cooperation. Philippe de Seynes, UN Deputy Secretary-General for Economic and Social Affairs, requested UNESCO Director-General Renée Maheu to 'rent' Batisse to the UN for preparing the Report of the UN Secretary-General to the General Assembly, with a proposal to organize the UN Conference on the Human Environment. The proposal was adopted and the invitation by Sweden to hold this conference in June 1972 in Stockholm was accepted. The UN was so impressed by the work of Batisse, that de Seynes included him on top of his list of the candidates for the position of secretary-general of the Stockholm Conference.

However, Michel Batisse's main interest had in the meantime shifted towards preparing the first session of the MAB International Coordinating

Council (Paris, November 1971). This session, held just ten months before the UN Conference on the Human Environment – and aimed at defining MAB's programmatic scope and cutting-edge interdisciplinary research agenda – was very timely in providing an important UNESCO input into the preparations of the Stockholm Conference.

All UNESCO programmes in environmental sciences made a contribution to the Stockholm Conference process. This was facilitated by a restructuring in 1970 of the Secretariat units in the sector responsible for environmental sciences. For the first time, UNESCO institutionalized coordination among its environmental sciences programmes through the creation of a Department of Environmental Sciences, headed by a principal director (Walther Manshard, 1970–74), comprising at that time the Division of Natural Resources Research, the Office of Hydrology, and the Office of Oceanography.

In Stockholm, in June 1972, some argued that the follow-up of the conference should be entrusted to UNESCO, given the Organization's record in environmental sciences and, in particular, its launching of the Man and the Biosphere Programme. However, the prevailing view was that UNESCO and the planned UN environment secretariat – which, in December of that year, would be dubbed the United Nations Environment Programme (UNEP) – should join forces and become the main drivers of international action in the field of environment. This view was strongly advocated by governments at the first session of UNEP's Governing Council in June of 1973. Many delegates referred to UNESCO's complementary role in the field of environment and specifically to the new MAB Programme. Maurice Strong, the first UNEP executive director, in his concluding remarks:

> acknowledged with approval the strong and unanimous desire of representatives that UNEP's programme should be carried out as far as possible by using existing capacities, and that UNEP should not create and conduct its own programmes in a way which might compete with the activities of specialized agencies. He added that the role of specialized agencies was a key element of the entire approach to the environment, and that the UNEP was a programme of the whole United Nations system, the Governing Council being responsible for giving it a common outlook and direction (UNEP, 1973).

The UNEP Governing Council, at its first session in 1973, defined three general policy objectives. The first was the need to enhance knowledge 'through interdisciplinary study of natural and man-made ecological systems'. This objective, even in its wording, corresponded exactly to those of the MAB Programme. The second objective focused on the need for an integrated approach, and the third dealt with the need for technical assistance to developing countries, and 'education, training and free flow of information and exchange of experience'. In conformity with the decisions of its Governing Council, UNEP, during its first decade, did not embark on programmes of its own for meeting its objectives on environmental research, training and education. Instead, it used existing capacities by supporting and joining (primarily) UNESCO's activities in these fields. It also supported relevant activities in a few other international organizations, such as those of the newly created Scientific Committee on the Problems of Environment (SCOPE) of ICSU.

In its decision to establish UNEP, the UN General Assembly also decided to establish a UN Environment Fund, as part of the new organization. During its first decade, the Environment Fund spent a major share of the money pledged by the UNEP Governing Council on joint UNESCO-UNEP activities, aimed at enhancing environmental knowledge, training and education (projects worth tens of millions of US dollars). In UNESCO, the UNEP funds for these joint projects in environmental sciences and for the UNESCO-UNEP International Environmental Education Programme (IEEP) represented, of course, very welcome extrabudgetary resources.

However, most of the joint activities receiving support from the Environment Fund during 1973–82 represented de facto UNEP cooperation in UNESCO programme activities in the Work Plans of the Science and Education Sectors, respectively. They could justifiably be called an extension of UNESCO's Regular Programme with UNEP money. The Natural Sciences Sector endeavour that benefited most from this cooperation was the MAB Programme. The significant additional financial resources from UNEP were a major reason for the rapid and successful development of MAB. As the joint UNESCO-UNEP project on Microbiological Resources Centres (MIRCENs) demonstrates, 'non-environmental sciences' Divisions in the Natural Sciences Sector also benefited from the Environment Fund.

1982–1986: PAINFUL CHANGES FOR UNESCO's ENVIRONMENTAL PROGRAMMES

At the time of the 1972 Stockholm Conference and the creation of UNEP, countries did not have a dedicated national institutional set-up for dealing with environmental problems. National delegations to the conference (and, later, to the UNEP Governing Council sessions) consisted mainly of representatives of ministries responsible for research, and of the national agencies responsible for freshwater, oceans and protected areas. The establishment of environment ministries called for by the Stockholm Conference did not materialize overnight. However, by the time of the UNEP Governing Council's tenth session in 1982, about two-thirds of UN Member States had a ministry of the environment. Consequently, the UNEP Governing Council had become primarily a global forum of environment ministries. There was often no longer any direct link to the ministries responsible for science and technology, which had been so beneficial to UNESCO for many years.

UNEP's new constituency – the environment ministries – seized the opportunity of the tenth session to substantially modify the organization's terms of reference. Its mandate to support environmental research – which figured so prominently in its 1972 constitution – disappeared in the 'Basic orientations of the United Nations Environment Programme for 1982–1992', adopted at the session. With regard to the scientific knowledge needed for environmental policy development, the original orientation of supporting environmental research was replaced by the new emphasis on scientific assessments. The decision referred to on 'basic orientations' states in this regard: '(a) Stimulate, coordinate and catalyse monitoring and assessment of environmental problems of worldwide concern' (UNEP, 1982). The two other major areas that were present in UNEP's 1972 mandate – i.e. development of policies and environmental law, and promotion of activities in the fields of information, education, training and institution-building especially for developing countries – were retained.

In addition, the original idea of UNEP's supporting the environmental programmes of other UN agencies had, to a large extent, disappeared. In its place, the notion of a UNEP programme, to be implemented by the organization itself, took centre stage. Consequently, the Environment Fund disbursements focused from that point onward on financing UNEP's own activities and projects submitted by environment ministries from developing countries.

As a consequence, UNEP's support for joint projects with UNESCO in the different fields of the environmental sciences dried up rapidly after 1982. As UNEP funds had sometimes represented an increase over the regular budget of 100 to 200 per cent, this sudden shortfall of extrabudgetary resources represented a major setback for further expansion and for ongoing activities of UNESCO's programmes in the environmental sciences. However, given that environmental education had been retained as a priority, UNEP's support for the joint UNESCO-UNEP activities in environmental education, through IIEP, continued unabated for a number of years.

Another major setback for UNESCO's programmes in the environmental sciences came in 1984, when two of its most important Member States – the United Kingdom and the United States – ended their membership in UNESCO. These programmes, which had just lost their major extrabudgetary source of support, now lost one-third of their core regular budget. Of course, this major reduction of the regular budget affected all UNESCO programmes. Those of us who had joined the Organization in the 1970s (or earlier) will always make the distinction between the golden years of steadily growing budgets and staff before 1984, and the sudden shortfall of 30 per cent of UNESCO's regular budget thereafter. The Organization has never really recovered from these major budgetary and, in particular, staff cuts. The US and UK, however, remained members of the Intergovernmental Oceanographic Commission (IOC); and their governments and UNESCO agreed that scientists from these countries would continue to cooperate with the other UNESCO undertakings in the environmental sciences.

A third development in 1982–86 led to ICSU launching the International Geosphere-Biosphere Programme (IGBP). UNESCO's environmental sciences programmes were concerned that the ICSU-driven IGBP preparatory process did not reach out to them. On the other hand, as soon as the IGBP research agenda took shape, most senior UNESCO programme managers acknowledged that it was complementary to the Organization's programmes. Moreover, they were realistic in judging that it would have been very difficult to sell to UNESCO's governing bodies the practical usefulness of research on the functioning of the Earth system and on global change. In particular – in the difficult budgetary context – it would have been impossible for the Natural Sciences Sector to obtain the necessary financial resources from the General Conference. Hence, there was no move made to propose to ICSU a co-sponsorship of the new IGBP programme.

ICSU was probably not unhappy about this. Scientists involved in the IGBP preparatory process let it be known that they wished to avoid IGBP getting linked to the complex intergovernmental structures of UNESCO's programmes (particularly now that, with the exception of the IOC, they functioned without US and UK membership). The launching of IGBP in 1986 as an ICSU multidisciplinary programme – with no UNESCO involvement – led to a situation in which many leading environmental scientists (particularly from industrialized countries) paid less attention to UNESCO's programmes.

PARTICIPATION IN UN SYSTEM-WIDE COORDINATION ON ENVIRONMENT, 1972–1992

By the time of the UN Conference on the Human Environment in Stockholm in 1972, several UN organizations other than UNESCO and The World Meteorological Organization(WMO) had begun to develop and implement environmental activities. Consequently, one important recommendation made by the Stockholm Conference concerned the need to ensure efficient coordination of environmental programmes throughout the UN system. In defining the functions of the UN Environment Programme to be established, governments at the conference charged it with a UN system-wide coordination function in the area of environment.

To this end, the UN General Assembly (in Resolution GA 2997, adopted in December 1972) at the same time established an Environment Coordination Board (ECB) – a Board of Heads of UN agencies having environment-related programmes. As UNESCO at that time was the agency with the largest environmental portfolio in the UN system, the UNESCO Director-General or his representative (Michel Batisse) were key figures in the work of the Board. UNESCO's governing bodies, the international coordinating bodies set up specifically by the Organization for guiding the implementation of MAB, the International Hydrological Programme (IHP) and the International Geological Correlation Programme (IGCP), and the IOC General Assembly, all called regularly for enhanced cooperation with UN system partners and for the active participation of the Natural Sciences Sector in environment-related UN system-wide coordination efforts.

Having ECB as a senior-level coordination body to address issues of policy and strategy related to coordination in the field of the environment proved to be useful. However, it quickly became clear that there was a need for more-focused coordination mechanisms at the level of specific topics, such as ecosystems conservation, desertification or marine pollution. Hence, UNESCO's environmental sciences programmes and UNEP often acted jointly in proposing such mechanisms to their partners in the UN system. For example, UNESCO and UNEP together called for the establishment of the Ecosystem Conservation Group (ECG), launched in the mid-1970s to harmonize approaches to ecosystem and biodiversity conservation, and to coordinate international networks of protected areas. UNESCO's active involvement in the ECG became particularly important in the light of the rapid development of both the biosphere reserves network and the natural World Heritage sites.

In 1978, the overall functions of the ECB at the executive heads' level were taken over by the newly established Administrative Committee on Coordination (ACC). For the purpose of the actual environmental programme coordination within the UN system, UNEP established a UN system-wide committee of the Designated Officials on Environmental Matters (DOEM), chaired by the UNEP executive director. The 'designated officials on environmental matters' from UNESCO were Michel Batisse (up to 1983), Sorin Dumitrescu (1983–89) and the author (Gisbert Glaser, 1990–92).

DOEM tasks included the preparation of the so-called System-Wide Medium-Term Environment Programme (SWMTEP), whose objective was to achieve a coordinated priority setting and forward-planning of all environmental activities and programmes across the UN system. Consequently, all Natural Sciences programmes related to environment had to be included in these six-year plans, and each draft SWMTEP was submitted to the UNEP Governing Council for approval. Although SWMTEP in the end had no real influence on UNESCO's programmes, it proved useful for programme managers to be able to report to UNESCO intergovernmental bodies that the Organization's environmental sciences programmes were part of SWMTEP. Governments felt they had proof on paper of UN system-wide coordination on the environment.

UNESCO's RESPONSE TO THE BRUNDTLAND REPORT: COORDINATION OF ENVIRONMENTAL PROGRAMMES

By the mid-1980s, concern about global environmental problems had reached a new dimension. It had now become such an important issue that for the first time elections could be won or lost on it. Consequently, the environment was now included on the political agenda of heads of state and government. In addition, mitigation of the effects of global environmental change and prevention of further degradation had been incorporated into national strategic planning in many industrialized countries. Governments had also come to recognize that international cooperation was needed if solutions were to be found for these problems. The environmental programmes of the UN organizations became, to a certain extent, caught up in this political 'elevation' of environmental concerns.

In 1987, the World Commission on Environment and Development (WCED, usually referred to as the Brundtland Commission[100]) presented its Report, 'Our Common Future', to the UNEP Governing Council and (later) to the UN General Assembly. The Brundtland Report established the concept of 'sustainable development'.[101] The Commission concluded that the present path of development worldwide was unsustainable, not only in environmental terms but also in terms of addressing poverty in developing countries. Moreover, current production and consumption patterns of both developed and developing countries were unsustainable. In response to the Brundtland Report, the UN General Assembly in 1988 decided on a process of preparing for a UN Conference on Environment and Development, to be held in 1992 at the level of heads of state, the first-ever Earth Summit.

In 1988, governments (under the aegis of UNEP) started to negotiate a UN Biodiversity Convention and a Framework Convention on Climate Change. In the same year, governments in the WMO Assembly decided to convene (in 1990) the Second World Climate Conference (SWCC), and to call on WMO's

---
100 After its chairperson, Ms. Gro Harlem Brundtland, Prime Minister of Norway.
101 Although both the International Union for the Conservation of Nature (IUCN) and UNESCO's MAB Programme had advocated sustainable development since the early 1970s, they usually used a different terminology.

partner organizations in the UN system to co-sponsor it. By now, the general public and politicians in many countries worldwide had started to become concerned about the possibility of global warming (another issue where trade-offs between environmental protection and development needed to be considered). Consequently, the SWCC was to have a ministerial-level segment, including a session with several invited heads of state and government – a novelty for a meeting of climate experts.

Federico Mayor, who became Director-General in November 1986, felt it was a moment of strategic importance for UNESCO to gain recognition as an international leader on environment-development matters. He saw the Organization as ideally placed to provide the science for environment-related policy-making, as well as to promote integrated approaches to environment and development. He also considered that such a positioning of UNESCO would help convince the United States and the United Kingdom to rejoin the Organization. In addition, there was always some concern that other UN organizations (such as UNEP and/or WMO) might be able to take advantage of this new situation to the detriment of UNESCO's programmes.

July 1988 saw an unusual series of high-level meetings in which UN organizations addressed the challenge of a coordinated response to the Brundtland Report. The most important was the Oslo Conference on Sustainable Development (9–10 July), organized by the Government of Norway, bringing together the UN Secretary-General and the heads of many of the organs and agencies of the UN system, as well as a few members of the WCED, under the Chair of Gro Harlem Brundtland. Previously, Federico Mayor had proposed to the UNEP executive director, Mustapha Tolba, to jointly convene at UNESCO a one-day preparatory consultation of heads of UN organizations planning to attend the Oslo Conference. This meeting took place at UNESCO on 8 July 1988, with six organizations in attendance. Moreover, in trying to define how UNESCO could best respond to the Brundtland Report, the Director-General – following a proposal by Sorin Dumitrescu, in his capacity as Designated Official on Environmental Matters – sought the advice of an 'Advisory Panel on Environmentally Sound and Sustainable Development'. The Panel, consisting of twenty-one experts, met at UNESCO Headquarters from 4 to 6 July 1988 under the chairmanship of Mansour Khalid, Vice Chair of the Brundtland Commission.

The Panel concluded that sustainable development should become a central goal permeating all UNESCO activities in education, science, social sciences, culture and communication. UNESCO's major scientific programmes most directly concerned (as well as the joint UNESCO-UNEP undertaking on environmental education) should be reoriented so as to allow a more explicit contribution to sustainable development. UNESCO's impact at the country level should be enhanced by conducting a series of Interdisciplinary Field Projects

PART III: ENVIRONMENTAL SCIENCES

on Sustainable Development, in which the Organization would bring to bear an integrated approach through a coordinated effort of the different programmes in environmental sciences, as well as of relevant programmes in the other sectors. The Panel identified UNESCO's competence in sustainable development in the following areas: oceans (IOC), freshwater (IHP), resources of the biosphere and land use planning (MAB), Earth sciences (IGCP), natural hazards, basic sciences, applied microbiology, engineering education and technician training, new and renewable energies, educational policies, environmental education, protection of cultural and natural heritage, cultural dimensions of development.

The Panel's report was the first submitted to UNESCO's governing bodies, addressing specifically the issue of UNESCO's comparative advantage in environment and sustainable development. Implementation by the Organization of most of the Panel's strategic and programmatic recommendations only began after the UN Conference on Environment and Development (UNCED) in 1992. Overall, it took the Organization eight years to get sustainable development recognized as a central goal for all programme sectors.

Among the difficulties to be expected in the Secretariat, the Panel explicitly pointed to the lack of inter-programme and intersectoral coordination. It felt that interdisciplinary and intersectoral activities had proven difficult to implement within the existing sectoral structure of the Organization's programmes and Secretariat. To overcome this constraint, the Panel recommended that the Director-General exercise personal supervision of an internal intersectoral approach, using whatever mechanism he considered most appropriate to coordinate and monitor UNESCO's activities for sustainable development.

Even the Designated Official on Environmental Matters, Sorin Dumitrescu, had no official mandate to coordinate UNESCO's environmental programmes house-wide. Within the Natural Sciences Sector it was, of course, his responsibility to ensure some coordination, in particular as the coordination function of the head of the Department of Environmental Sciences had disappeared in 1982 (when the Director-General Amadou-Mahtar M'Bow did away with the 'departmental' layer of authority in all programme sectors). Mr Dumitrescu, who became Assistant Director-General for Natural Sciences in late 1988, agreed with the Director-General that the time had come to create intersectoral coordination capacity for a fully coordinated cross-sectoral response to this new situation, and for promoting intersectoral activities for environment and sustainable development.

Mr Dumitrescu also wanted to ensure that any such structural response recognized the lead function of the Natural Sciences Sector in the Secretariat for environmental activities. In order to create capacity within the Sector for: (1) managing inter-programme environmental issues within the Sector (e.g. the Second World Climate Conference); (2) providing both leadership in and support for intersectoral coordination; and (3) strengthening UNESCO's role in interagency coordination within the UN system, Mr Dumitrescu obtained the agreement by the Director-General to appoint an intersectoral environmental focal point within the Sector, who would report to the Assistant Director-General of Natural Sciences.

The focal point function was officially established in January 1989. The author, Senior Programme Specialist in the Division of Ecological Sciences, who had already helped Mr Dumitrescu throughout most of 1988 in 'environmental affairs' (i.e. those affairs going beyond the scope of any particular environmental sciences programme), was appointed to this position. For the next eighteen months, the Environmental Focal Point unit, linked to the Assistant Director-General's office, assisted the Assistant Director-General, and through him the Director-General, in all 'environmental' matters concerning either the whole of the environmental sciences programmes or the whole of the Organization.

In March 1989, the Director-General announced the creation of an Intersectoral Committee for Coordinating Environmental Activities, under the chairmanship of the Director-General himself. The Environmental Focal Point was appointed secretary of the Committee. The Committee served as an instrument for the Director-General to ensure that all programme sectors and the general offices concerned gave due attention to his strategic vision of the role of UNESCO with regard to environment and sustainable development. It analysed strengths and weaknesses of the Organization's relevant programmes and debated how best to capitalize on its intersectoral comparative advantage. In terms of hands-on coordination, the Committee during 1989 addressed in particular UNESCO's intersectoral contribution to the preparations of the Second World Climate Conference and, more importantly, to the UN Conference on Environment and Development. Moreover, the Intersectoral and Interagency Cooperation Project on Environmental Education and Information became a standing item on the agenda of Committee meetings.

The Medium-Term Plan for 1990–1995 (adopted by the General Conference in November 1989) included a new set of priority objectives for the Organization. Among these was, for the first time, the 'environment'; sustainable development, however, was not on the list. Indeed, despite all the attention that the Brundtlandt Report had received, UNESCO's constituencies – as well its governing bodies, and those of other UN organizations – continued to put 'development' and the 'environment' in two very different 'boxes'. Outside of the community of advocates, the concept of sustainable development had not made much headway. There was intellectual inertia and a resistance to accepting the practical consequences of a new intersectoral paradigm. Yet, under the heading 'environmental programmes,' UNESCO's scientific programmes started to respond to the Brundtland Report, by putting greater emphasis on the links between environment and development.

In February 1990, the Director-General issued a forty-six-page executive note on major structural reforms of the Secretariat with a view to implementing the new Medium-Term Plan. One new element introduced in the Secretariat structure was the establishment of several intersectoral 'coordinators', at director level and above. Also included was a coordinator of environmental activities.

In fact, the Director-General decided to create a new Bureau for Coordination of Environmental Programmes (COR/ENV), as part of the Natural Sciences Sector, headed by a Coordinator of Environmental Programmes, at the rank of Assistant Director-General. The Bureau became operational on 1 July 1990 with the arrival of Francesco di Castri as Coordinator.[102] The author was appointed Deputy Coordinator. The main responsibilities of the Coordinator, and ipso facto of the Bureau, included ensuring enhanced interaction and coordination of UNESCO's environmental programmes, both within the Natural Sciences Sector and across all programme sectors of the Organization, thus improving programmatic coherence and cohesion. The immediate priority for the Bureau was to organize a coordinated input of UNESCO into the UNCED preparatory process (Intergovernmental Preparatory Committee and its three Working Groups, numerous Working Parties of the UNCED Secretariat) and into the conference itself.

The executive note outlining the responsibilities of the new Bureau COR/ENV was written in such a way as to avoid any 'conflict' with the intersectoral coordination responsibility regarding UNESCO's policy/activities related to socio-economic and human development, entrusted to the Social Sciences Sector. Indeed, a pragmatic division of labour was respected by both sides. One of the main objectives of the Bureau COR/ENV was moving UNESCO's environmental programmes to focus more and more on environment-development links. However, given obvious political and professional sensitivities, no efforts were made by the Bureau to lobby for all UNESCO's development–related studies and activities to become oriented towards sustainable development.

UNITED NATIONS CONFERENCE ON ENVIRONMENT AND DEVELOPMENT, 1992

Throughout 1991, numerous activities were launched within UNESCO in preparation for the UN Conference on Environment and Development (UNCED), with the Bureau COR/ENV taking the lead on several. The Director-General, in late 1991, agreed to set up an ad hoc Intersectoral Working Group

---

102 UNESCO was fortunate that Francesco di Castri, an ecologist of wide international reputation, who had been the first Director of the Division of Ecological Sciences from 1974 to 1984, accepted to rejoin the Organization at this critical moment for the future of its environmental sciences programmes.

on UNCED to finalize UNESCO's input into the conference. For many, it would become the finest example during their careers in the Organization of successful cooperation between a great number of programme specialists (across disciplinary and sectoral lines) in the Secretariat, with the highest degree of professionalism, team spirit and personal responsibility.

UNESCO was active and visible on several fronts at UNCED – held in Rio de Janeiro in June 1992 – putting it in a unique position compared with other UN organizations, which were present at only one or two venues. It was, of course, present at the meetings of the actual Intergovernmental Conference. The statement to the plenary was delivered by Mr E. Portella, Deputy Director-General, on behalf of the Director-General. The evening before, Mr Portella decided to deliver what he called 'his own speech'. It was a memorable night for a few colleagues, including the author, who worked in his hotel room up to the early morning hours, trying to strike a balance between what the Organization needed to say, in our view, and what Mr Portella wanted to say in his home country, Brazil.

The second front was a Joint Government of Brazil/UNESCO Scientific Meeting on Environment and Development. This unique event brought together Brazilian and international specialists (natural and social scientists) in ten round tables (several with more than 400 participants) devoted to key issues of UNCED. UNESCO also organized an international symposium on 'Man, the City and Nature' and presented a large, very professional exhibit on the themes of UNCED and how UNESCO was addressing them. While the environmental sciences programmes shouldered the largest share of this exhibition, the other programme sectors – and UNESCO's role as a champion of promoting interdisciplinary and intersectoral approaches – also featured prominently.

The Bureau COR/ENV was able to publish within ten months a comprehensive kit of documents on Environment and Development in UNESCO, including a number of Environment-Development Briefs, aimed at communicating scientifically sound information about selected environment and development issues of global relevance, as a basis for informed action by decision-makers. Distribution of these documents at UNCED contributed significantly to the great visibility of UNESCO at this summit meeting.

Agenda 21, adopted by the summit, represented significant progress in linking environment and development, in comparison to the Stockholm Conference in 1972. Yet, it fell short of applying consistently a broad and integrated (environmental, social and economic) sustainable development approach. Two important new conventions were endorsed by UNCED and signed by many countries on this occasion: the Biodiversity Convention, and the Framework Convention on Climate Change.

PART III: ENVIRONMENTAL SCIENCES

UNESCO poster on the environment – 'Pour un monde meilleur, développement de l'homme en harmonie avec la nature' – 1992, year of the United Nations Conference on Environment and Development.

UNESCO had been helpful 'behind the scenes' in preparing Agenda 21 and the two Conventions. Indeed, the 'handwriting' of the Organization's environmental sciences programmes – often in cooperation with ICSU – can be seen in a number of Agenda 21's chapters. UNESCO (COR/ENV) and ICSU prepared the first draft of Chapter 35 on Science for Sustainable Development; the Education and Natural Sciences Sectors were instrumental in preparing Chapter 36, on 'Promoting education, public awareness and training'.

INTERDISCIPLINARY AND INTERSECTORAL INITIATIVES AFTER UNCED

The reorientation of the activities of IHP, IGCP, IOC and MAB, in response to UNCED and in line with UNESCO's commitment to contribute to the implementation of Agenda 21 is presented in the chapters dealing with the history of these programmes. It should be noted, however, that UNCED influenced other parts of the Natural Sciences Sector outside of the environmental sciences, notably the Division of Engineering Sciences, which henceforth attached a high priority to the promotion of renewable energy technologies and launched the World Solar Programme.

A major thread running through all chapters of Agenda 21 was the recognition that effective implementation of Agenda 21 would depend not the least upon development and application of integrated approaches. Given UNESCO's unique ability to link disciplines and research focused on specific thematic areas (such as biodiversity or freshwater) in the natural and social sciences with education, culture and communication, it could be argued that the Organization had a special responsibility after UNCED to demonstrate leadership in enhancing interdisciplinary and intersectoral approaches. Needless to say, Francesco di Castri and the author felt strongly that the Organization's response to UNCED must include such a direction, alongside the necessary reorientation of its existing environmental programmes.

Adnan Badran, who had taken over as Assistant Director-General for Natural Sciences in January 1990, worried that any new activities outside the scope of the existing programme structure of the Sector might weaken these programmes. Such differing views left the final decision to the Director-General and, as a last resort, to the Executive Board. Federico Mayor's position was clear. He lent full support to most submissions made by di Castri (until December 1992) and the author (Glaser). He was keenly interested in launching post-UNCED intersectoral activities on environment and development, both for their own sake and to herald a shift towards more cross-sectoral activities in the Organization. There was also relatively strong support in the Executive Board for testing whether UNESCO was capable of changing the 'ivory tower' mentality of its programme sectors and of the major programmes. Of course, Badran and the Division directors had reason to be concerned about the budget for the existing programmes. Hence, there was general agreement in the Sector, and in the Secretariat in general, that new interdisciplinary and intersectoral activities should be promoted primarily through enhanced cooperation and coordination of existing programmes, and that specific 'structures' for such activities should be created only if they were considered more cost-effective and more promising in terms of successful implementation.

Immediate UNCED follow-up was, of course, an issue for UNESCO's main undertakings in the environmental sciences, IGCP, IHP, IOC, and MAB. Moreover, the Board at its first session following UNCED authorized the Director-General, 'within the existing budget and staff resources' of the ongoing biennium 1992–1993, to launch four innovative intersectoral activities in the following areas: (1) production of information materials with a view to improving knowledge on interrelated environment–development issues; (2) assistance in reorienting training programmes and academic institutional functioning; (3) studying and promoting diverse approaches to sustainable development; and (4) developing and testing integrated approaches to the conservation and sustainable use of biodiversity.

The promotion and coordination for the implementation of these activities was entrusted to the Bureau for Coordination of Environmental Programmes (COR/ENV), within the Natural Sciences Sector. Many in UNESCO thought that the Bureau COR/ENV would disappear after UNCED (and more particularly after the retirement of the Coordinator, Francesco di Castri, at the end of 1992). However, the Director-General considered the Bureau essential in its support of several important UNESCO objectives: to champion interdisciplinary and intersectoral approaches to sustainable development by improving cooperation and coordination between disciplines and sectors in-house, and to ensure a coherent link between UNESCO's relevant programmes and activities (in all programme sectors) and international policy-making on environment and sustainable development. Also considered important was providing leadership in key areas of the environmental sciences and in environmental and sustainable development education. Consequently, the Director-General updated the Bureau COR/ENV's mandate in early 1993, and the author was appointed Coordinator for Environmental Programmes/Director of the Bureau.

In addition, Federico Mayor established a small 'Committee for the Follow-up to UNCED' (reporting to the Director-General) to provide guidance for UNESCO's follow-up to UNCED, in general, and for the implementation of the four 'intersectoral activities', in particular. Francesco di Castri served as chairperson of the Committee, which had four additional members, with expertise and/or experience in social sciences, systems science, environmental education, and as 'permanent secretary' of a ministry of environment in a large developing country. The director of COR/ENV acted as secretary.

At the twenty-seventh session of the General Conference in 1993 (the first after UNCED), UNESCO Member States fully supported these new initiatives. It was a General Conference that clearly launched UNESCO on the path of greater 'interdisciplinarity' (meaning intersectorality), including enhanced cross-sectoral cooperation on environment and sustainable development. For the first time, UNESCO's Programme and Budget 1994–1995 (27 C/5) included the term 'sustainable development' in the title of a Major Programme. However, in 'UNESCO language', sustainable development continued to be regarded as a primarily environmental policy objective only. Hence, the wording normally used now was 'environment and sustainable development', replacing 'environment and development'.

In Major Programme II.3, 'Environmental sciences and sustainable development', for the first time a sub-programme was included on 'Coordination and promotion of interdisciplinary and interagency cooperation'. This represented the activities and the budget of the Bureau COR/ENV. Included among its activities was the responsibility (with some seed money) for organizing/coordinating the continued implementation of the four 'intersectoral activities' referred to above. This sub-programme continued until 2001.

In response to both UNCED and the UN Conference on Population and Development (in Cairo, Egypt, 1993), the biennium 1994 also saw the launching of the interagency and interdisciplinary project 'Environment and Population Education and Information for Human Development' (EPD). This project included the joint UNESCO-UNEP International Environmental Education Programme (IEEP). One idea behind establishing EPD was to ensure a better link between UNESCO's environmental sciences programmes and IEEP. The Bureau COR/ENV was tasked to facilitate or coach this intersectoral cooperation, which proved a difficult task. Very few of EPD's activities were in the end cooperative activities with any of the environmental sciences programmes. The same was true of the 'Education for a Sustainable Future' Project – which was the continuation of EPD, under a different name and without IEEP – starting from the biennium 1996–1997. IEEP ceased to exist after UNEP withdrew from the programme in 1995, due to a slashing of its budget after UNCED.

The Coordinator of Environmental Programmes proposed to the Secretaries of the four environmental sciences programmes – IGCP, IHP, IOC and MAB – to organize (on the occasion of the 1993 General Conference) a meeting, for the first time, of the chairpersons of their scientific coordination councils/governing bodies. The objective was to move beyond basic coordination among these programmes to greater programmatic interaction, and to develop an enhanced common identity. The chairpersons reacted with enthusiasm to this opportunity of interaction and tasked the four Secretaries (to be facilitated by COR/ENV) to develop an initial set of cooperative activities. The meeting resulted in a 'Joint Statement by the Chairpersons' to the General Conference, acknowledging that up to that time 'by and large the four environmental science initiatives have been carried out separately from each other, as well as from [environmental activities in] the other sectors of the Organization'. The General Conference, in turn – pleased that the chairpersons pledged the readiness of their programmes to enhance interdisciplinary cooperation among them and to participate actively in relevant intersectoral activities of the Organization – officially endorsed the Statement.

Since then, joint meetings of the chairpersons have been held at every General Conference. Since 1995, the chairperson of the UNESCO social sciences programme 'Management of Social Transformations' (MOST) has regularly joined his/her colleagues of the environmental sciences programmes. This joining of the social sciences has made the meetings even more valuable for proposing and monitoring interdisciplinary and intersectoral activities among the programmes involved. The meetings have continued to produce a Joint Statement addressed to the General Conference and implicitly to the Director-General and the senior staff of the Organization. Over the successive biennia, interdisciplinary cooperation among the five programmes has increased. Nevertheless, overall, such cooperation continues to remain quite limited.

During the 1994–1995 biennium, Federico Mayor pursued an ambitious project idea of his own: for UNESCO to prepare regular 'state of the planet' reports containing the best available interdisciplinary information on different aspects of the state of the planet – in a language that would be understood by high-level decision-makers (starting with heads of state and government). The Bureau COR/ENV was in charge of preparing this. Although the project was included in the following biennial Programme and Budget (28 C/5), it quickly became clear from the resources allocated to it that such a major project could not be made a priority in UNESCO, given the needs of the existing scientific programmes. Obviously, it would also have required considerable cooperation from partners both within and outside the UN system. Potential partners, however, basically withheld their support and made it clear that they were not in favour of UNESCO leadership in this case.

GLOBAL CONFERENCES: ENSURING AN INTERSECTORAL UNESCO INPUT

After UNCED in 1992, and the Cairo World Conference on Population and Development in 1993, several other important world conferences took place in the period 1994–99 that required coordinated input by UNESCO's environmental sciences programmes, as well as by the other programme sectors. UNESCO, first and foremost the Director-General himself, attached great importance to the Organization's involvement in these conference processes, given the important role of sciences – as well as of social sciences, education, culture and communication – in the issues concerned. In each case, inter-programme input was required from the Natural Sciences Sector, as well as intersectoral input from all programme sectors.

Particularly important contributions were expected from the environmental sciences programmes to the preparations and proceedings of the UN Conference on the Sustainable Development of Small Island Developing States (Barbados, September 1994). The conference, convened at the level of heads of state and government, resulted in the Barbados Plan of Action on the Sustainable Development of Small Island Developing States. Beyond the need to coordinate the contributions of the IOC, the Divisions of Water Sciences and of Ecological Sciences, and Marine Science Related Issues, all other programme sectors (notably Education, and Communications, Information and Informatics) were also concerned. COR/ENV was tasked to organize this cross-sectoral UNESCO input. The sizeable intersectoral UNESCO delegation, led by the Coordinator of Environmental Programmes, mounted a large exhibit in Bridgetown, giving examples of the Organization's activities in support of small island developing states undertaken by all programme sectors.

After the conference, the Director-General reacted favourably to the proposal by the UNESCO Committee on UNCED Follow-up to include in the next Medium-Term Plan, and the Programme and Budget for 1996–1997 (28 C/5), a new initiative to promote interdisciplinary scientific activities and intersectoral cooperation on coastal zones and small islands in the context of sustainable development. After approval by the General Conference in November 1995, this led to the establishment (in January 1996) of the Coastal Zones and Small Islands Unit (CSI) in the Natural Sciences Sector, as a platform for organizing intersectoral activities in this field. Governments of the small island developing countries enthusiastically welcomed this initiative, and they have ever since provided strong political support to this intersectoral 'project' in one of the most vulnerable types of ecosystems.

In 1996–97, COR/ENV once again had the responsibility of arranging for coordinated UNESCO contributions to the preparation and proceedings of three major global conferences. The first was the World Conference on Food and Sustainable Development (Rome, Italy, 1996), an initiative taken by the Food and Agriculture Organization of the United Nations (FAO). Because of their focus on science for improving natural resources and ecosystems management, UNESCO's environmental sciences programmes have always had a strong link to FAO's work. Another area of close UNESCO-FAO cooperation relevant to the conference was the promotion of rural education. The other two global conferences which received major UNESCO input were the Special Session of the UN General Assembly on the Review of the Implementation of Agenda 21 (Rio+ 5), held in New York (USA) in September 1997, and the Global Civil Society Forum on the same subject (Rio de Janeiro, June 1997). Federico Mayor led the UNESCO delegations to all three events.

In June–July 1999, the 'World Conference on Science for the Twenty-first Century: A New Commitment' (Budapest, Hungary) – convened jointly by UNESCO and ICSU – provided an excellent opportunity to highlight progress made by UNESCO's environmental sciences programmes in bridging the science–policy gap, and in their commitment towards science for sustainable development. A major part of its programme was focused on 'science, environment, and sustainable development' and drew extensively from input by UNESCO's environmental sciences programmes, the social sciences programme on Management of Social Transformations (MOST), as well as on ICSU's interdisciplinary global environmental change research programmes. Among the significant contributions made by COR/ENV towards the preparations of the Budapest Conference was an initiative to bring together for the first time in a special interactive dialogue session the chairpersons of the five UNESCO scientific programmes (including MOST) and of the four global change research programmes sponsored by ICSU. The 'Science Agenda – A Framework for Action' document adopted by the World Conference on Science also included an agenda on the way forward in the natural, social and engineering sciences in the service of sustainable development.

It would be too long a story to tell of the implementation of the 'interdisciplinary activities' launched in 1993 (referred to above), or of the impact of the UNESCO Committee on UNCED Follow-up, which continued until the end of 1998. In brief, the 'interdisciplinary activities' had a mixed outcome. While the Committee proved useful (particularly in its early years), the ambitions of its Chair – and shared by the Director-General and a few others in the Secretariat (such as the author) – did not come to full fruition. The reasons for this are manifold, but include: 'psychological' non-adaptation to interdisciplinary and intersectoral cooperation by some colleagues concerned; the fact that it implied additional work, while the existing workload was already heavy for many, if not most, colleagues; and insufficient financial incentives. The bureaucratic way in which UNESCO's programme sectors traditionally interact also played a major role.

UN SYSTEM-WIDE COORDINATION REGARDING SUSTAINABLE DEVELOPMENT AFTER 1992

With the shift from environmental protection to sustainable development emerging as a major outcome of UNCED in 1992, governments recommended that UN system-wide coordination of environmental activities be superseded by a coordination of all efforts in the UN system aimed at meeting sustainable development objectives and at making a joint contribution to the implementation of Agenda 21. The Administrative Committee on Coordination (ACC), comprising the heads of all UN organizations, followed this recommendation, establishing in 1993 the ACC Interagency Committee on Sustainable Development (IACSD), bringing together the senior officials responsible for sustainable development activities in the various UN organizations. The chairmanship was entrusted to the UN under-secretary-general for Economic and Social Affairs, Nitin Desai.

During its first five years, the IACSD was very active, but it became less efficient in its remaining four years. A core group of the committee (including the chairman and the senior representatives of FAO, ILO, UNESCO, WHO and UNEP) acted as a sort of de facto 'Bureau' of the IACSD, interacting intensively between meetings; and decisions by the committee were basically decisions agreed upon by this core group. The IACSD was in charge of preparing the background documentation for the first nine sessions of the intergovernmental UN Commission on Sustainable Development (CSD) and fulfilled this function to the full satisfaction of the governments participating in CSD. UNESCO's scientific and educational programmes on environment and sustainable development were able to make major contributions to the work of CSD, both via the IACSD and directly. An important side benefit for UNESCO was that its visibility, as a key international player in supporting sustainable development goals, was much enhanced during these years, in particular outside the traditional UNESCO networks.

The IACSD introduced the mechanism of 'task managers' within the UN system for the different topical areas (chapters) in Agenda 21. The UN organizations entrusted with 'task manager' functions were responsible for coordinating the activities in their area of responsibility across the UN system (under the supervision of the IACSD), as well as for preparing progress reports for submission to CSD. UNESCO was made 'task manager' for both science for sustainable development (Chapter 35 of Agenda 21) and education for sustainable development (Chapter 38). Obviously, the task manager function was normally entrusted to the 'lead organization' in the UN system in a particular field.

The IACSD ceased to exist in 2001, after a decision of the ACC in late 2000. The UN system-wide input to the preparations of the World Summit on Sustainable Development (WSSD) in 2002 was organized by the CSD Secretariat in the UN Department of Economic and Social Affairs, without the benefit of the support of a dedicated interagency coordinating body. UNESCO – and the other UN organizations most concerned – did not get a chance to inject their ideas into the preparations of WSSD in the same prominent manner as they had been able to do during the UNCED process in 1992.

PART III: ENVIRONMENTAL SCIENCES

WORLD SUMMIT ON SUSTAINABLE DEVELOPMENT (WSSD) IN 2002

The decision to convene, in 2002, a World Summit on Sustainable Development (WSSD), ten years after UNCED, was taken by the UN General Assembly in 2000. The Summit was held in August–September 2002 in Johannesburg, South Africa. As part of this ten-year review of the progress achieved in the implementation of Agenda 21, UN organizations, including UNESCO, were invited to review their own contributions to this end. However, as stated above, this (as well as other input of UN organizations) remained as 'background documents' exclusively for the Commission on Sustainable Development, which acted as the intergovernmental preparatory committee for WSSD.

In UNESCO, the beginning of the WSSD preparatory process in 2001 coincided with the arrival in March of Walter Erdelen, a bio-geographer, as the new Assistant Director-General for Natural Sciences. After the restructuring of the Secretariat by the Director-General in July 2000, Koïchiro Matsuura in early 2001 decided on further structural amendments. In all programme sectors, he introduced again the position of a deputy assistant director-general. In the Natural Sciences Sector, he promoted the director of the Division of Water Sciences, Andras Szöllösi-Nagy, to this rank, giving him the additional responsibility and title of Coordinator of Environmental Programmes.

It also coincided with the preparations by UNESCO of its Draft Programme and Budget for 2002–2003. This Draft (31 C/5) reflected a hesitation concerning the priority to be given to sustainable development: it was left unclear whether it was an organization-wide priority objective or not. Only the Natural Sciences Sector stressed its continued focus on sustainable development. It is in this context that the UNESCO General Conference, at its thirty-first session in late 2001, adopted for the first time a general resolution on sustainable development. The proponents of the draft resolution were not satisfied with the Draft and wished to ensure that sustainable development would remain an Organization-wide priority.

In this resolution, the General Conference highlighted that UNESCO's most important contributions to sustainable development had been made in the environmental sciences programmes and in education. It also 'noted a global shift in emphasis from a focus on environmental concerns to the more holistic approach of sustainable development which centres on environment, society and economy, and their interrelationships, as well as the eradication of poverty and changing wasteful consumption and production patterns' (31 C/Res.40). This reflected remarkable progress made by the Organization's national constituencies in their awareness of the full scope of the sustainable development goal.

In the resolution, the General Conference also invited the Director-General to examine whether to make sustainable development a new cross-cutting theme for the whole of UNESCO's programmes in the future. Koïchiro Matsuura, immediately after the Johannesburg Summit, announced that adding sustainable development to the two existing cross-cutting themes of poverty eradication and access to the knowledge society was not the best way forward within the Organization. He proposed to make sure that contributing to sustainable development would become part and parcel of the design of each and every results-based programme activity of the Organization, thus joining other important 'cross-sectoral issues' in UNESCO, such as human rights and intercultural dialogue.

After a slow start, by the end of 2001 the Director-General decided to establish an Intersectoral Task Force on Sustainable Development to promote and coordinate UNESCO-wide preparation for the Johannesburg Summit, composed of high-level UNESCO staff from all programme sectors and central administration. UNESCO's delegation in Johannesburg was led by the Director-General, who was invited to be part of the official programme of high-level round tables, giving him an opportunity to address the Summit on behalf of the Organization.

Moreover, the Director-General participated in several UNESCO-organized side events. One major event in which he took part was the Round Table on Cultural Diversity and Biodiversity for Sustainable Development, co-organized with UNEP, and chaired by President Jacques Chirac of France. Another was a seminar entitled 'Educating for a sustainable future: actions, commitments and partnerships', organized jointly with the Ministry of Education of South Africa, and in cooperation with the NGO-UNESCO Liaison Committee. The Natural Sciences Sector partnered with the Government of South Africa and ICSU in co-organizing several sessions of the WSSD Forum on Science, Technology and Innovation for Sustainable Development, which took place in parallel to the intergovernmental proceedings of WSSD.

The Johannesburg Plan of Implementation adopted by the WSSD included numerous recommendations aimed at a better harnessing of science and technology for sustainable development that were directly relevant to UNESCO's environmental sciences programmes. Moreover, the Summit recommended to launch a UN Decade of Education for Sustainable Development (2005–2014) and proposed that UNESCO become the international lead organization for the Decade. The Decade was officially launched by Director-General Koïchiro Matsuura and Mrs Nane Annan, representing the UN Secretary-General, in New York in March 2005. The Decade will be a challenge for the Natural Sciences Sector, as all UNESCO scientific programmes – and, in particular, the environmental sciences – are called to enhance significantly their activities in education and training in support of sustainable development.

PART III: ENVIRONMENTAL SCIENCES

Many inside and outside UNESCO were taken by surprise when the Director-General dissolved the Intersectoral Task Force on Sustainable Development only weeks after the Johannesburg Summit. What about coordination of UNESCO's contribution to the follow-up of WSSD, particularly the Johannesburg Plan of Implementation? The explanation given was that the College of the Assistant Directors-General would be ensuring this coordination. By the end of 2002, Andras Szöllösi-Nagy had stepped down as house-wide Coordinator of Environmental Programmes, explaining he was unable to continue assuming this responsibility while his main function as director of the water programmes required all his energy. He also cited a lack of any specific means for the purpose of coordination. Obviously, these decisions must be seen in the context of limited human and financial resources. However, the consequences are also clear. At present, coordination of UNESCO's environment and sustainable development-related activities receives little attention.

A MIXED HISTORY OF SUCCESS AND FAILURE

It has been stated in the Introduction to this section on the history of the environmental sciences in UNESCO that, overall, the story of the Organization's environmental sciences programmes is undoubtedly one of success, with impressive achievements to show for each individual programme. However, when trying to evaluate whether UNESCO succeeded over the years in capitalizing on its specific interdisciplinary and intersectoral comparative advantage, the author comes to a more mixed conclusion.

One particularly positive development during the last forty years of UNESCO's environmental sciences programmes has been their becoming more and more relevant to policy. Up to the 1960s, and to some extent during the 1970s, when most of the current programmes were first designed and launched, their objectives were almost exclusively science-driven, focused on enhancing knowledge of the natural environment and on better understanding newly occurring changes in the environment. Today, societal needs and policy priorities drive the development and priorities of these programmes as much as strictly scientific considerations. Moreover, these programmes are now recognized as providing scientific information to many intergovernmental forums, beyond UNESCO, mandated with implementation of international legal instruments and policy formulation related to environment and sustainable development.

MAB was the first programme to make policy relevance a stated objective, from its beginning in 1972. The other programmes followed this direction gradually. MAB had the advantage of being launched as an interdisciplinary programme – involving both the natural and social sciences – undertaking integrated research on a changing natural environment and the socio-economic drivers of this change. The other programmes, while remaining basically natural sciences programmes, have now come to interact with social sciences networks focused on their topical areas (such as freshwater). Next to MAB, IHP has moved furthest in this direction and now includes programme elements that integrate the environmental, social and economic pillars of water-based sustainable development.

The list of the successes achieved would be incomplete without mentioning the very impressive contribution that all of UNESCO's environmental sciences programmes have been making year by year to human and institutional scientific capacity-building, particularly in developing countries. It is true that, in the environment and sustainable development-related sciences, the North–South knowledge divide has been widening during the last decades. But this fact in no way diminishes the immense value of the training and retraining of tens of thousands of scientists and of establishing and strengthening scientific institutions. UNESCO has been instrumental in setting up interdisciplinary institutions that have by now become fully independent, such as the International Centre on Integrated Mountain Development (ICIMOD) in Kathmandu, Nepal. Others have become an integral part of the Organization, such as the UNESCO-IHE Institute for Water Education in Delft, the Netherlands, whose object is the training of professionals and the strengthening of relevant institutional capacities in developing countries.

When it comes to efforts at capitalizing on intersectoral cooperation between the natural and social sciences, education, culture and communication within UNESCO, the road has been full of obstacles, and concrete cooperation limited. In fact, such intersectoral cooperation remains one of the most difficult aspects of UNESCO's work almost fifty years after it was first called for in the area of environment and, later, sustainable development. The activities of the intersectoral project on Environment and Development in Coastal Regions and Small Islands (CSI) since 1996, and the small number of other ad hoc intersectoral activities that have been organized regularly during most biennia (including the current one) are no proof to the contrary, as successful as a few of these may have been. The overall experience during five decades – reflected in the current situation – has been one of a major missed opportunity for the Organization. Interdisciplinary research that integrates environmental, social, economic and health sciences is currently spearheaded by international scientific programmes outside UNESCO. The Organization was not able to retain the leadership in this domain that it established during the first ten years of MAB.

For an Organization acting as a leader at the global level in the sciences for sustainable development, and in education for sustainable development, it must be considered a serious shortcoming that even after forty years of its having scientific and educational programmes side by side, UNESCO has not succeeded in ensuring concrete cooperation between them. The science programmes have never been – and are not today – instrumental in feeding the best available science into the Organization's relevant educational programmes, as has been so often called for during these four decades.

There is, finally, another general 'failure', which should, however, not be blamed solely on UNESCO and its programmes related to environment and sustainable development. Why has there not been much more significant progress in applying the impressive body of knowledge accumulated over the last sixty years in the implementation of projects and policies in the countries which are in need of the best science available? Application at the country level remains low worldwide. The process leading from knowledge generation to knowledge application is often unacceptably slow. There is, most importantly, a woefully insufficient political will to correct the current unsustainable path of human development.

In conclusion, moving towards sustainable development will require a much better harnessing of the sciences and education. In fact, the educational and scientific tasks to be tackled remain huge. UNESCO is called to continue being an indispensable and leading international player, if these momentous efforts are to be undertaken. There remains the hope that UNESCO will do better in living up to its unique potential of combining natural and social sciences for sustainable development, forging close cooperation between dedicated scientific and educational programmes, and enhancing the roles of culture and communication in efforts aimed at a more sustainable future.

BIBLIOGRAPHY

UNEP. 1973. *Report of the Governing Council of UNEP on the Work of Its First Session*, as submitted to the UN General Assembly, p. 16.
——. 1982. *Report of the UNEP Governing Council on the Work of Its Tenth Session*, as submitted to the UN General Assembly, at its thirty-seventh session, p. 35.

PART IV
SCIENCE AND SOCIETY

PART IV: SCIENCE AND SOCIETY

INTRODUCTION
THE SHOCK OF THE NEW
The contemporary age begins

Jacques Richardson[1]

THE six decades of science between 1945 and 2005 were to prove even more dramatic than the two preceding sixty-year periods (i.e. from around 1825 until the end of the Second World War). This fertile period witnessed enormous leaps in our knowledge of nature; indeed, future historians of science may well judge it to be the most exciting era ever in the evolution of the natural and engineering sciences (including also the advent of the informational and environmental sciences).

The unifying ideas of electromagnetism advanced by Clerk Maxwell, the work of Poincaré and Einstein (among others) in relativity, and of countless others in atomic theory (physicists and mathematicians); the classification of the elements by Mendeleyev and creation of the managed laboratory by Liebig (both chemists); Darwin's discoveries in evolution, Mendel's in genetics and Haeckel's in the compatibility of living things (all fruits of biology) – this was progress in extravagance. Knowledge grew, expanding into fields until then unassociated with the natural sciences (e.g. philosophy and economics).

During the twentieth century, the proceedings and documents of the League of Nations' International Institute for Intellectual Cooperation occasionally expressed the need for concern regarding the relationship between evolving science and social transformation. One of the precepts underlying the 1945 San Francisco Conference (the UN Conference on International Organization, which established the United Nations) was the reconstruction of war-torn countries, so that they might make up for lags in research and higher education and in the preparation of a new generation of scientists, engineers and technicians. Much of the rest of the world had not yet been touched by science at all, fast-evolving technology was still on the horizon, and the education for both had yet to be expanded into what would become known as the 'developing world'.

The Second World War was to prove an astonishing agent of change in other respects, too, with the invention of radar (some claim this science-based

---

1   Jacques Richardson: From 1972 to 1985, Head of UNESCO Science and Society Section, and Editor of journal *Impact of Science on Society*. Consultant to UNESCO, UNDP and national governments on environmental management in West Africa and Ethiopia (1986–2003).

technology won the war) as part of the sweeping advent of 'big science'. The science and technology of the 1940s also brought the world the energy of the atom – for evil and for good – and a stunning realization of the many life-saving applications of biochemistry, as well as the power of information theory.

The Nuremberg Code, which came out of the war crimes trials, led to the adoption in many countries of 'informed consent', to prevent the potential outrages of biomedical experimentation. It was a ground-breaking (if not outright revolutionary) development in the evolution of professional ethics, affecting specialists whose work was based on scientific and technological change. Professional ethics would soon extend to welcoming into the world research community scientists from former colonies, as well as the (sometimes grudging) inclusion of women to professions long considered masculine domains.

> **BOX IV.1.1: HOW ONE HUMAN CHANGED OUR EVERYDAY WORLD**
>
> Norbert Wiener (1894–1964) gave the term *feedback* 'modern meaning ... He was the first to perceive the essence of ... information' (Conway and Siegelman, 2005). Wiener worked with biologists and neurophysiologists to crack the communication codes of the human nervous system, then with the engineers who incorporated these codes into the circuits of the first programmable 'electronic brains'. Later, Wiener led the medical team that created the first bionic arm controlled by the user's own thoughts.

'Ethics' in its scientific connotations came late to UNESCO, largely at the insistence of one of its directors-general and of a consultative committee. Yet, just as the 'science for peace' of the Organization's first decades ceded to 'science for development' from the 1960s onward, the conscience-driven ethical obligations finally took hold in the 1990s, as UNESCO's scientific and technical mission further adjusted itself to a new age and a broadening societal outlook.

In this section on Science and Society, we have opted for the following order of subjects: science and technology policies, scientific research and industrial applications, science statistics, science and society, women in science and engineering (as seen from UNESCO), and science and ethics.

BIBLIOGRAPHY

Conway, F. and Spiegelman, J. 2005. *Dark Hero of the Information Age: in search of Norbert Wiener.* New York, Basic Books.

PART IV: SCIENCE AND SOCIETY

The UNESCO travelling science exhibition 'New Materials', on plastics and alloys, 1952.

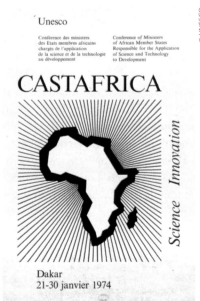

Poster UNESCO, Conference of Ministers Responsible for the Application of Science and Technology to Development in Africa' (CASTAFRICA), Dakar (Senegal), 1974.

HELPING HANDS, GUIDING PRINCIPLES
Science and technology policies
Jürgen Hillig[2]

NEEDS TO BE MET

THE term 'science policy' first appeared on an organization chart at UNESCO in 1963, when the Research Organization Unit became the Science Policy Unit. It was a time when the storms of the Cold War had quieted enough for the Organization to turn its attention to another epochal political transformation: the crumbling of colonialism. Many newly independent developing countries were emerging on the international scene. By 1965, when the Science Policy Unit was made into a Division, UNESCO was fully engaged with the fast-shifting international environment that recognized the increasingly important role of science and technology in national development.

UNESCO's first moves in addressing issues that would form the body of the Science and Technology Policy Programme actually occurred in the late 1950s. The Director-General, in his introduction to the Programme and Budget for 1959–1960, emphasized that 'in the natural sciences, the major problem with which UNESCO will cope is the general lack in many countries of a knowledge of the fundamentals of natural science and of the scientific method, and the shortage of trained scientists and technicians'.[3]

These countries were lacking not only an appropriate scientific and technological potential but also the capability to manage the development and application of science and technology. Initiatives to deal with these issues were already under way in member states of the Organisation for Economic Co-operation and Development (OECD) and, to a lesser extent, in members of the Council of Mutual Economic Assistance (CMEA).[4] UNESCO's relevant activities were grouped under the heading 'General Problems of Scientific Research', with National Research Councils as privileged partners. A first meeting of directors of National Research Councils was held in Milan, Italy, in 1955. No distinct secretariat unit dealing exclusively with science and technology policy existed at that moment.

---

2   Jürgen Hillig: UNESCO Division of Science and Technology Policies, then Coordinator for Natural Sciences Sector, 1967–88; Director, Regional Office for Science and Technology, Jakarta, 1989–94; Director, then Assistant Director-General, Division of Decentralization and Field Relations, 1994–97.
3   UNESCO Programme and Budget for 1959–1960, p. IX, para. 16.
4   In Eastern Europe, CMEA was better known by its initials in Russian, SEV.

The year 1959 marked a turning point when the United Nations (UN) and UNESCO jointly began preparations[5] for a 'Survey of the Main Trends of Inquiry in the Field of the Natural Sciences', with UNESCO coordinating a study directed by physicist Pierre Auger of France. Its results were published in 1961, when a United Nations Economic and Social Council (ECOSOC) resolution called for a UN Conference on Science and Technology to be held, with the organizational support of UNESCO, in 1963.

Activities labelled 'contribution to scientific research' were grouped, in 1962, within a separate Research Organization Unit, the precursor of the Science Policy Division. The Programme and Budget for 1961–1962 took into account the need for UNESCO to extend its previous activities (information regarding national research organizations, surveys of national research councils, case studies on research careers) so that data on national research policies and organization could be regularly reviewed, analysed and disseminated. For the first time, UNESCO proposed analysis on:

- main objectives of science policies and principal research and development programmes supported nationally;
- legal and administrative structures of national scientific research bodies;
- financial support given to scientific research with special emphasis on governmental expenditures; and
- human resources engaged in research, and methods to encourage scientific careers.

Building on these activities by recognizing the growing interest of governments in science as a vital part of their economic, social and political life, the programme for 1963–1964 (now called 'Aid to National Scientific and Technological Development', with sections concerning 'Information on Science Policy of Member States' and 'Aid to Technological Research') outlined a broader and more ambitious range of activity:

- information on the organization and financing of research;
- surveys and studies on Member States' national science policies;
- meetings of experts to recommend objectives and methods in national science policy; and
- help to Member States to improve and develop their institutions of national science policy.

---

5   UN General Assembly Resolution 1260 (XIII).

The conception and general thrust of these activities were modified and strengthened, as were other parts of the natural sciences programme,

> so as to bring them into line with the broad principles embodied in the ten-year plan which the Programme of the General Conference approved [based] on the report of the working party appointed to examine the 'Survey of the Main Trends of Inquiry in the Field of the Natural Sciences', prepared at the request of the United Nations General Assembly. An attempt was made to reflect ... the priorities which UNESCO was invited to accord to its activities by the Economic and Social Council.[6]

DEFINING ROLES

The year 1963, marked by the United Nations Conference on the Application of Science and Technology for the Benefit of the Less-Developed Areas (Geneva, Switzerland) was to be crucial in determining responsibility for science and technology within the framework of the UN and its system of specialized agencies. As a result, UNESCO strengthened its activities in science and technology policy among Member States. Since then, however, there has remained ambiguity concerning the extent of UNESCO's mandate, as opposed to that of the UN and of its Office for Science and Technology (OST, created in 1963) and the establishment (also in 1963) of the United Nations Advisory Committee on the Applications of Science and Technology (UNACAST). The latter would become a major player in this field in the years to come.

During 1960–65, other actors besides the UN and UNESCO became increasingly active in science policy activities. Most prominent was OECD's Directorate of Scientific Affairs; it initiated a series of monographs on the science policies of its member states, together with methodological work on the definition and measurement of 'research and development'. In 1964, the OECD countries accepted statistical definitions, laid down in the so-called *Frascati Manual* (OECD, 1964), which allowed member states to standardize their statistics in this sphere.

This evolution paralleled the reinforcement of UNESCO's efforts, through its Office of Statistics, to develop methodological tools to collect, analyse and disseminate data on the growing scientific and technological work by its Member States. Benefiting in part from the work done by OECD, UNESCO's own contribution was twofold; the Organization:

---

6   UNESCO Draft Programme and Budget for 1963–1964 (12 C/5).

PART IV: SCIENCE AND SOCIETY

- reconciled concepts and methodologies used by OECD countries and those used by the centrally planned economies of CMEA countries; and
- promoted the collection of science and technology statistics in developing countries.

One result was to upgrade, by 1969, the small unit in the Office of Statistics responsible for scientific statistics: it became the Division of Science Statistics. To this unit were assigned comprehensive studies for the development of an international standard for science statistics, in an effort to reconcile (as far as possible) differences in concepts, definitions and classifications as a common basis for presenting science statistics at the world level. In the following years, this Division (then situated within the Communication Sector) became closely associated with the broader activities of the Science Policy Division (within the Natural Sciences Sector). Much of the data required for the preparation of the Regional Conferences of Ministers Responsible for Science and Technology – organized in subsequent years by UNESCO – was provided by the Division of Science Statistics.

Two more biennia, from 1963 through 1966, were required to develop and refine the methodological instruments and the collection and analysis of relevant information in order to embark on large-scale initiatives. There was, for instance, the series of regional Conferences of Ministers Responsible for the Application of Science and Technology to Development (CAST), which extended over a little more than twenty years, beginning with CASTALA in Santiago de Chile (Chile) in 1965 and ending with CASTAFRICA II in Arusha (the United Republic of Tanzania) in 1987.[7]

With these regional conferences, UNESCO underscored its claim to overall responsibility within the UN system for science and technology policy, echoing specifically the United Nations initiative in 1963 to convene the first UN meeting on the application of science and technology to the 'least developed areas'. Coordinating all these conferences proved to be critical in the evolution of UNESCO's science and technology policies, both methodologically and practically, and benefited Member States over a period of two decades. Functionally, the Science Policy Division had about two years to organize a major conference, while simultaneously effecting the follow-up of previous encounters, with limited resources in both staff and funding.

---

7 Analogous conferences included: CASTASIA (Delhi, India, 1968), MINESPOL (Paris, 1970), CASTAFRICA (Dakar, Senegal, 1974), CASTARAB (Rabat, Morocco, 1976), MINESPOL II (Belgrade, Yugoslavia, 1978), CASTASIA II (Manila, the Philippines, 1982), CASTALAC II (Brasilia, Brazil, 1985).

The same years saw the advent of an impressive series of 'Science Policy Studies and Documents', beginning in 1965 with a report (in French only) on *La politique scientifique et l'organisation de la recherche scientifique en Belgique*. These initiatives underscored UNESCO's efforts to assume responsibility within the UN system for matters of science and technology policy. A *Directory of the Main National Scientific Research Organizations* (1963–64) provided insight into their operational and administrative structures, while specific studies on national science policy emphasized expenditures for and the main aims of national

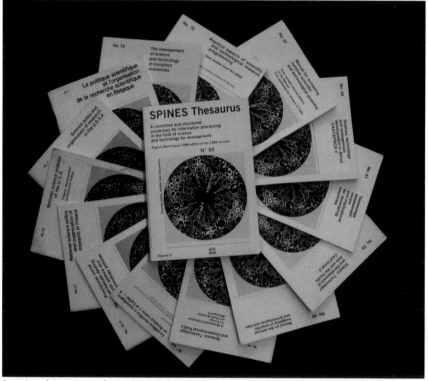

Samples of the seventy-four volumes of the Science Policy Studies and Documents series, published by UNESCO, 1965–94

research. The Work Plan for these years included the preparation of a handbook of research organizations and their statistics, and a multilingual glossary of terms related to science policy problems.

The period 1965–1966 must be singled out for the unprecedented rise of more than 50 per cent in budgetary allotment for the Natural Sciences. And again, in 1967–1968, the Natural Sciences received the highest percentage of increase in funds that benefited all programmes – including activities in science policy and the planning of technological progress. The Director-General, René Maheu, wished with this dramatic shift in emphasis to underline the need to give to science the same impetus as had been given to education in 1960.[8] This recognized that science policy and the planning of technological progress play the same fundamental roles as educational planning.

GROWING IMPORTANCE

The quickly developing and expanding science policy programme also commanded increasing resources. The small unit of two professionals (in 1961) grew to a full-fledged Division, with six professionals by 1967–1968, commanding a regular budget of US$300,000, besides US$400,000 in extrabudgetary funds for technical assistance. At the peak of its development (around 1977–78), the Division had a staff of ten professionals and managed a regular budget of US$1,751,400. Regional Field Offices were highly active in the expanded programme. National Commissions were urged to associate themselves closely with the new programme by providing data on governmental machinery for science policy and on national networks in this field. They also presented an inventory of potential and priorities in scientific and technological research.

At the time of the Regular Programme and Budget for 1969–1970, work on science policy was no longer exploratory – it was, by then, playing a basic role in the Natural Sciences Sector's activities. Emphasis was placed on initiatives to promote internally generated technological progress as opposed to intensive technological imports. It was also a time of particularly close cooperation with UNACAST, as this unit prepared a world plan of action for development. This plan was intended to provide, for the first time, a rational framework for long-term collective action – based on a common order of priorities – throughout the entire UN system.

UNESCO itself 'did not fail to take this opportunity of setting [its work] … in the twofold context of inter-agency coordination and long-term planning'.[9] The Organization contributed to the preparatory phases:

- Stage I, current and future programme of the UN system;
- Stage II, needs of the developing countries in nine sectors as determined by UNACAST; and
- Stage III, overall approach, aggregation and problems across the globe.

---

8 Introduction to the Draft Programme and Budget for 1965–1966, para. 89.
9 Introduction to Draft Programme and Budget for 1969–1970, para. 102.

Moreover, UNESCO was assigned an important system-wide responsibility for science policy and a spectrum of scientific and technological institutions constituting a country's infrastructure in this area. The World Plan of Action (1971) remained valid until 1979, when it was supplanted by a Programme of Action adopted by the UN Conference on Science and Technology for Development (held in Vienna, Austria, 1979).

While interaction with ECOSOC and its commissions and committees helped UNESCO give its science and technology programmes broad exposure and recognition, the interaction also posed problems of coordination and attribution of responsibility for certain activities. This was not a problem limited to science and technology; in his introduction to the Draft Programme and Budget, the Director-General stated that while 'the diversification of [ECOSOC] organs is inevitable if there is to be functional development', raising the number of organs charged with 'the same task leads to confusion and thereby reduces efficiency'.

During the 1970s and in later years such problems were frequently addressed by UN agency heads before the Administrative Committee on Coordination. The problems became more acute as the Vienna Conference of 1979 approached, a meeting in which UNESCO believed it was not given a role corresponding to its unique mandate within the UN system for science and technology. While UNESCO contributed to the conference's preparations as required, the Director-General, Amadou-Mahtar M'Bow, specified for the delegates (on 21 August 1979) those mechanisms likeliest to permit efficient implementation of whatever the conference should decide. He stressed the polycentric character of the UN, by virtue of its Charter, and declared that no specialized agency should be subordinate to another or to the UN itself. Cooperation is usually based on agreements, the one between UNESCO and the UN dating from 14 December 1946.

The Director-General recalled ECOSOC's mandate to study economic and social questions and to make corresponding recommendations to the General Assembly and to the specialized agencies. He underlined the general and sectoral actions of these agencies essential to the overall effort to mobilize the international community so that science and technology might serve all nations – especially the most underprivileged. The speech was unusual for its candour, reminding delegates of the problems inherent in the effective distribution of responsibilities among the different agencies.

TROUBLES ARISE

The Director-General reported critically on the Vienna Conference to the Executive Board during its session in the fall of 1979. He regretted the shortcomings of its preparatory process, the difficulties in informing delegates of the comprehensive information prepared by UNESCO, and the absence

PART IV: SCIENCE AND SOCIETY

of clear recognition of specialized agencies' roles in conference follow-up. He welcomed the establishment of the Intergovernmental Committee on Science and Technology for Development as a plenary unit of the General Assembly (reporting through ECOSOC), a committee that would invite specialized agencies to participate in its work. He regretted, however, that the committee would not be assisted by an inter-agency secretariat. Instead, a secretariat would be set up using the resources of the UN's Office for Science and Technology. These arrangements were intended to facilitate the ambitious Programme of Action voted by the conference, concentrating on three objectives:

- Target Area A, strengthening the science and technology capacities of developing countries;
- Target Area B, restructuring the existing pattern of international scientific and technological relations;
- Target Area C, strengthening the role of the UN system in the field of science and technology and providing increased financial resources.

The Programme of Action created a paradoxical situation since UNESCO's activities in this field were already heavily engaged in the three Target Areas. The Vienna Conference had come at a time when international concern for science and technology, in general, may have reached its peak. Another paradox was demonstrated by the subsequent failure to implement the global financial arrangements recommended by the conference. The proposed long-term arrangements to fund a scheme for Science and Technology for Development were, in fact, never implemented; even the Interim Fund, to be sustained by voluntary contributions, fell far short of the US$250 million targeted for 1980–1981. And, despite impressive new machinery established by the conference, the international community was quickly diverted by issues it considered more pressing.

As for UNESCO, the results of the conference required adjustments. Yet a new organizational, administrative and financial set-up did not hinder the intensity and scope of its Science and Technology Policy Programme. The process of growth and reinforcement came to an end only with the decline in financial resources, in part because of the withdrawal of the United States from UNESCO (1985), and in part because of an increasingly critical attitude of Member States towards these activities. This culminated in a decision not to include policy conferences in the Third Medium-Term Plan (1990–1995). A gradual decline of the remaining science and technology policy activities followed, including a halt in 1990–1991 to Regional Conferences of Ministers. Activity was discontinued altogether in 1992, for the sake of programme concentration called for by the General Conference.

A selection of books and reviews on science and society, produced or supported by UNESCO between the 1960s and 1980s.

AMBITIONS AND ACCOMPLISHMENTS

Major conferences and cooperation at the international level, although important, had not been the only concern of the programme in science and technology policies. Myriad other actions fashioned UNESCO's programme between the late 1960s and the late 1990s: development of normative instruments, methodological approaches, technical assistance projects at country and regional levels; symposia and training workshops; fellowships; clearing-house activities and publishing. The Regular Programme and Budget for 1969–1970 proposed a gamut of services, ranging from training in research planning, administration and statistics to budgeting techniques; clearing-house services embraced the collection, analysis and dissemination (through UNESCO's Field Offices) of data on national scientific and technical human resources as well as current infrastructure and research programmes of existing institutes. These activities were the basis, too, of UNESCO-provided consultative services to the UN's regional Economic Commissions.

The programme also developed methodological studies, with particular regard for the scientific and technological potential needed for technical assistance. Help to Member States in the planning of science policy through short-term staff and consultant missions covered:

- assessment of national scientific and technological potential;
- surveys of current trends in scientific research resources and programmes;
- establishing or improving governmental structures for science policy, preparation of relevant legislation;
- studies on efficiency of research, productivity of research workers;
- formulating national science plans: infrastructure, priorities;
- establishment of budgets for science policy and its integration within the state budget; and
- creating bilateral links between institutions of scientifically advanced states and similar units in developing countries.

While initially financed under Regular Programme resources, these activities benefited increasingly from funding through the UN Development Programme (UNDP). One example is the range of activities carried out in 1969–1970 in Cameroon, the Democratic Republic of the Congo, Indonesia, Senegal and Tunisia.

While similar in structure, the programme for 1971–1972 highlighted consultations and contacts with international non-governmental organizations, for instance: the ICSU Committee on Science and Technology Policy in Developing Countries, the Pugwash Continuing Committee on Science and World Affairs, and the World Federation of Scientific Workers. Anticipating the value of computerized management and information processing, a thesaurus of key words (the *SPINES Thesaurus*, used for the storage and retrieval of documents and their publication) was launched, together with a comparative glossary of terms for use by science policy specialists.[10] During the same period, work continued on the *World Directory of National Science Policy-Making Bodies* and on the *International Recommendation on the Status of Scientific Workers and Technicians*. By then, too, the volume of planned technical-assistance projects financed from extrabudgetary sources (e.g. UNDP) had grown to a respectable US$470,000.

When he introduced the Regular Programme and Budget for 1973–1974, the Director-General referred to a dilemma confronting UNESCO's science programme since its inception. On the one hand, he said, there existed the relatively imprecise terms by which UNESCO's competence in scientific matters was defined in its Constitution; on the other, there was a problem of meeting two conflicting requirements. These were (1) the need to cover a large number of scientific fields and subjects tending towards dispersion, and (2) the inevitability of

---

10 See Nos 33, 39 and 50 of the Science Policy Studies and Documents series (UNESCO, 1974, 1976, 1988).

selection and limitation – given scarce resources – that leads to concentration. The science and technology programme for this (and future) periods not only escaped unscathed from the veiled threat of restricted resources, it continued to expand – largely because of the budgetary effort required to organize four (of nine) Regional Conferences of Ministers Responsible for Science and Technology in 1970–78.

The planned financing of UNDP-supported projects, for example, rose to US$856,000 in 1973–1974. During the following biennium, the Natural Sciences Sector's budget once again grew faster than that of the Education Sector – with the additional resources added over-proportionately to the environmental sciences. Science and technology policies benefited, nevertheless, from overall growth in budget. During the same biennium, Field Science Offices were assigned responsibility for periodic assessments of developing countries' institutional needs in science and technology, this by taking into account the timing of UNDP country programming. UNESCO's science policy unit (by then called the Division of Science Policy Programming and Financing, DSPPF) served as the focal point of this exercise.

By 1975–1976 the Natural Sciences Sector enjoyed above-average budgetary growth. The DSPPF carried a particularly heavy charge, with preparations for two ministerial conferences (CASTARAB and MINESPOL II), follow-up for CASTASIA and CASTAFRICA (including the establishment of a Special Fund for African R&D), and contributing to global science and technology policy intended for the UN system through ECOSOC (and its Committee on Science and Technology) and UNACAST.

During the same period, clearing-house activities related to the dissemination of information on national science and technology policies, and the organization of research in Member States, was reinforced. Preparatory work was initiated to establish the Science and Technology Policy Information Exchange System (SPINES). One notable event was the adoption by the General Conference (during its eighteenth session) of the 'International Recommendation on the Status of Scientific Workers', a standard-setting instrument of special importance. With this recommendation, UNESCO aimed to define, consolidate, project and promote a category of personnel which, while essential to the advancement of science and its application to development, did not enjoy a status commensurate with its importance in many countries.

A GOOD SIX YEARS

Beginning in 1977, UNESCO aligned itself with UN practice by introducing six-year Medium-Term Plans as a framework within which the traditional biennial programmes/budgets would function. The first of these plans, covering the years 1977–1982, included (as its Objective 4.2) the 'Promotion of the formulation and application of policies and improvement of planning and financing in the fields of science policies'. The plan lists three principles of action:

- the desirability of countries being equipped with decision- and policy-making bodies;
- the priority of three factors in scientific and technological development: financial resources, conditions in which scientific work is done, international cooperation; and
- the use of modern scientific-management methods (especially data processing).

Based on these considerations, targets for 1982 were set, with three objectives:

- Promotion of policy formulation, planning and financing in science and technology at national level. Desired impacts: improvement of policy-making machinery; strengthening of staff, in particular specialists responsible for policy analysis; using modern analytical tools for planning and evaluation.
- Promotion of international cooperation with a view to policy formulation, planning and financing. Desired impact: all Member States should have gone through at least one round of review of their policies and the evaluation of regional programmes of cooperation.
- Participation in the formulation of overall science and technology policy for the UN system. Desired impact: contribution to the convergence and harmonization of programmes conducted by the various UN bodies in order to optimize its own contribution to the development process in Member States.

These aims became the conceptual basis and framework for the formulation of programmes and budgets for the years 1977–1982.

During that sexennium, the follow-up of CASTARAB (Rabat, Morocco, 1976) included exploration of the setting up of an Arab Fund for Scientific and Technological Research. Major efforts were also made to cooperate closely with the UN Secretariat in preparation for the Vienna Conference on Science and Technology for Development (1979), in spite of the previously mentioned difficulties. Equally important were preparation and follow-up in regard to the second MINESPOL meeting (Belgrade, Yugoslavia, 1978) and preparation of CASTASIA II (Manila, the Philippines, 1982). As of 1977, the strengthening of aid to Member States called for a restructuring of action in science and technology, drawing as much as possible on science policy studies and the experience acquired through international cooperation. UNESCO's Regional Offices for Science and Technology were assigned greater responsibility for gearing this effort much more to the specific needs of countries or regions.

The importance attributed to science planning during those years was highlighted by the decision (1979–1980) to undertake a feasibility study on establishing an International Institute for the Planning of Scientific and Technological Development. While the plan did not materialize, progress emerged with the implementation of a major 'International Comparative Study on the Organization and Effectiveness of Research Units' (ICSOPRU), designed to stretch over several biennia. Meriting mention in this respect was the publication *Method for Priority Determination in Science and Technology* (No. 40 in the Science Policy Studies and Documents series).

With UNESCO's change in alignment of its programme cycle, there occurred an exceptional, three-year programme-and-budget exercise for 1981–1983. The General Conference commissioned a report on the impact of regional intergovernmental Conferences of Ministers Responsible for Science and Technology Policies (CAST and MINESPOL), a study that would become the rationale a few years later to discontinue these meetings. Set up as a follow-up to CASTASIA II (1982), the Science and Technology Policy Asian Network (STEPAN) proved quite useful to Member States and participating institutes. They even continued to use STEPAN as a forum after the discontinuance of the programme of science and technology policy. And – as if the looming budget squeeze of the coming years was anticipated – this triennium was particularly rich in (1) analyses of national experience and information exchange relating to science and technology policies;[11] (2) contributions facilitating formulation of such policies at national and regional levels; and (3) initiatives for the training of personnel needed in the planning and management of national development in science and technology. Specific programme actions included:

- methods to identify priority programmes in science and technology;
- evaluation and choice of national or foreign technologies;
- organization, management and efficiency of research units;
- planning of resources allocated to research;
- stimulating demand ('pull') for technological progress; and
- training of staff at the level of national science and technology policy bodies.

---

11   See especially the *Manual on the National Budgeting of Scientific and Technological Activities* (UNESCO, 1984).

PART IV: SCIENCE AND SOCIETY

THE LEAN YEARS

The six years covered by the Second Medium-Term Plan (1984–1989) witnessed the withdrawal of the United Kingdom, Singapore and the United States from UNESCO. This was a time of transition, when the science and technology programme lost considerable momentum, not only because of reduced financial and personnel resources, but also because of Member States' changing priorities. The decision to discontinue regional ministerial conferences signalled both doubts about their cost effectiveness and a loss of political support. Indeed, organizing one of these conferences extended over two to three years and entailed significant cost, not including staff and overhead expenses and the considerable cost incurred by the host nation together with those of other countries involved. These same years saw an increase in bilateral cooperation between Member States or specialized institutions in different countries. Soon the new possibilities of electronic communication would further democratize and facilitate exchanges and access to information.

Thus, still looking back to major science policy events, such as the regional ministerial meetings and the UN Conference on Science and Technology for Development and its follow-up Programme of Action, the Medium-Term Plan of the time fell short of being explicit about science and technology policy. We have been able, however, to cite key issues at the time (e.g. strengthening national research potential and improving its infrastructure, broadening international cooperation). New priorities – such as the dissemination of novel technologies in informatics, applied microbiology and other biotechnologies, and renewable energy – reflected shifts in emphasis. The Plan recommended, moreover, that activities pertaining to the application of science and technology to development should be implemented through reinforced Major Regional Projects – especially those projects adopted by the General Conference of 1981.

During the years of the same Medium-Term Plan, the science and technology policy budget fell by about 3 per cent; its major, new activities were the holding of CASTALAC II (Brasilia, 1985) and CASTAFRICA II (Arusha, 1987). Plans to organize MINESPOL III and CASTALAC III were abandoned for the reasons previously stated. There was strengthening, on the other hand, of activities relating to the *SPINES Thesaurus*, intended to facilitate development of national databases on science and technology.

As for 1988–1989, the only major proposal was that of creating an International Scientific Council for Science and Technology Policy Development. The Council's purpose was to advise UNESCO on the orientation of its work in training and research related to science and technology policy, and its first meeting took place in Paris in 1988. In the same period, the Science Policy Studies and Documents series expanded to include a *World Directory of National*

Science and Technology Policy-Making Bodies (No. 59, 1984), a *Manual for Surveying National Scientific and Technological Potential* (No. 67, 1986), and a *World Directory of Research Projects, Studies and Courses in Science and Technology Policy* (No. 70, 1990).

CHANGING WITH THE TIMES

Aware of the widening gap between industrialized and developing countries, the Third Medium-Term Plan (1990–1995) emphasized the need for UNESCO to

- respond to the basic priority of transferring science and technology fully and rapidly to developing countries;
- promote the emergence of a scientific and technological culture in developing countries through international cooperation;
- work out strategies for the development of science and technology; and
- address ethical issues arising in various fields of science and technology.

Science and technology policy, as such, was no longer singled out as a specific objective. The stress was, rather, on 'assisting Member States in working out strategies for science and technology development' and on 'providing advisory services and facilitating policy-discussion forums at the regional and subregional levels'. Activities of this nature were thus to be found in the Draft Programme and Budget for 1990–1991. However, in the following biennium (concentrating on programme and resources), only a task entitled 'Management of science and technology development' remained. This offered advisory aid to 'some ten' Member States, support for two existing regional networks in Asia and Latin America, assistance to Eastern European nations in restructuring their management systems for science and technology, and improvement of regional database and other information dissemination.

During 1994–1995 and, in fact, for the six-year period of the Medium-Term Plan for 1996–2001 science and technology policy activities came to a halt. Yet the promotion of the advancement of science and technological knowledge – together with the acceleration of the sharing and transfer of such knowledge – remained an objective of the Plan. Specific science and technology policy activities were no longer, however, a means to achieve this objective.

It is not surprising, therefore, that the World Conference on Science held in Budapest (Hungary) in 1999 did not have 'science and technology policy' as its main focus. Convened by UNESCO in cooperation with the International Council for Science (ICSU) and other partners, this conference adopted

A selection of books on science and society produced or supported by UNESCO during the 1980s and 1990s.

unanimously two statements: a 'Declaration on Science and the Use of Scientific Knowledge', and the 'Science Agenda – Framework for Action'. Among the results expected through follow-up by the various parties were

- development of guidelines for decision-makers and legislators in conducting policy reviews, and
- formulation of national science and technology policy and management.

As a result, the Regular Programme and Budget for 2000–2001 bore a major budgetary provision of US$900,000 for conference follow-up and, to do this, maintained a Policy Analysis and Operations Division within the Natural Sciences Sector.

The Medium-Term Strategy 2002–2007 singles out, as a major objective of UNESCO, 'Promoting principles and ethical norms to guide scientific and technological development and social transformation'. The Strategy recognizes that science's contribution cannot be based only on research and accumulated knowledge. Its ethics must be justified by relevance and effectiveness in addressing the needs and aspirations of societies. Ethics also calls for participation by all societal groups in decision-making and strategy-definition: making effective use of research and innovation. All this highlights the relationship between research, education, technological innovation and their practical benefits; briefly, it means adapting science policy itself to the needs of society.

As a consequence, UNESCO today fosters international cooperation through the adoption of science and technology policies that put science squarely in the service of social requirements, with the following expected outcomes:

- improved science policy analysis and formulation of national policies, budgets and legislation, and harmonization of science and technology within regional and subregional integration;
- strengthened endogenous research and operational capacities for national science and technology systems, including the creation of regional centres of excellence;
- enhanced public participation in policy formulation; and
- improved understanding by parliamentary science and technology committees and senior policy-makers.

CONCLUSION

'Challenges emanating from globalization and from the trends in many areas are becoming ever more complex [and are] often driven by scientific and technological insights and breakthroughs' (Medium-Term Strategy, 2002–2007). These challenges carry manifold institutional implications within UNESCO, one of which has been realized through the reorganizing (since 2002) of the Policy and Operations Division, known until 2005 as the Division of Science Analysis and Policies, and now known as the Division of Science Policy and Sustainable Development. UNESCO has thus re-established, in the early twenty-first century, science and technology policy among its programmes, underlining its long-standing claim to competence in this crucial field at world, regional and national levels.

BIBLIOGRAPHY

OECD. 1964. *The Measurement of Scientific and Technical Activities: Proposed Standard Practice for Surveys of Research and Experimental Development* ('Frascati Manual'). Paris, OECD.

UNESCO. 1974. *Science and Technology Policies Information Exchange System (SPINES)*. Vol. 1: *Feasibility Study*. Vol. 2: *Provisional World List of Periodicals Dealing with Science and Technology Policies*. No. 33 (Vols 1 and 2) in the Science Policy Studies and Documents series. Paris, UNESCO.
——. 1976. *SPINES Thesaurus: a controlled and structured vocabulary of science and technology for policy-making, management and development* (Vol. I: Rules, conventions and directions for use; Vol. II, Parts 1 and 2: Alphabetical structured list; Vol. III: 34 Terminological graphic displays). No. 39 in the Science Policy Studies and Documents. Paris, UNESCO.
——. 1984. *Manual on the National Budgeting of Scientific and Technological Activities: integration of the 'science and technology' function in the general state budget*. No. 48 in the Science Policy Studies and Documents series. Paris, UNESCO.
——. 1988. *SPINES Thesaurus: a controlled and structured vocabulary for information processing in the field of science and technology for development* (rev. edn). No. 50 in the Science Policy Studies and Documents series. Paris, UNESCO.

PUSHING AND PULLING
Scientific research and industrial applications
Jacques Richardson

IN the late 1980s, UNESCO's Director-General initiated investigation into the potential of the Organization to serve as interlocutor and catalyst between the research potential of universities and the applied-research needs of various industrial sectors, a push–pull relationship[12] intended to serve both the North–South and South–South systems of technology transfer. There was heavy pressure from the outset to use as a model for research–industry integration the experience of the Massachusetts Institute of Technology (MIT, in the United States), which, decades before (under wartime conditions), had established MITRE (MIT Research), an initiative in private enterprise.

UNESCO engaged a full-time consultant to: (1) catalogue what was being done in this sphere, and what still needed to be done; and (2) estimate how effectively UNESCO might offer its offices to both academia and manufacturing and industrial services – while taking care not to tread on the work of the UN Industrial Development Organization (UNIDO) and the World Health Organization (WHO, with special reference to the biotechnological applications).

The efforts of UNESCO's consultant led to the establishment of the University–Industry Science Partnership (UNISPAR), approved by the General Conference at its twenty-seventh session in 1993. UNISPAR stimulates exchanges in information, research, training, and certain services and exchanges: creating technological infrastructure, assisting academics in obtaining industrial experience, and advising industrial managers how to engage university research facilities for business purposes. The project has enjoyed notable success in Africa, especially Nigeria and the United Republic of Tanzania. Propelled by a dynamic of its own – largely through workshops organized according to objectives – and functioning (to a limited degree) in tandem with UNIDO, this initiative remains one of the key projects of UNESCO's recently revived unit concerned with science and technology policy.

The concept behind this effort is of particular interest because of the involvement of the industrial elements of society with a UN organization. For in its preoccupation with the academic and cultural worlds, UNESCO has (historically) overlooked, or even disdained, the collaborative possibilities and potential financing in the broad field of the applications of science and technology.

---

12   Push–pull: 'Push' here refers to the influence of new basic scientific knowledge on possible applications of this knowledge, and 'pull', to the expectations of technologists concerning what new knowledge the research laboratory might be plumbed for, for improvements of an incipient or emerging technology.

CRUNCHING NUMBERS
Science and technology statistics at UNESCO [13]
Ernesto Fernández Polcuch [14]

THE EARLY YEARS

STATISTICS at UNESCO were initially collected and published by the UNESCO Statistical Service. In 1952, a Statistics Division was established in the framework of the Department for Social Science. Before producing actual science and technology (S&T) statistics, in the 1950s, UNESCO had set up multiple directories on national science councils, scientific and scientific-cooperation institutions – both national and international – adopting a model usually followed by countries to keep track of their scientific and technical resources (Godin, 2001*b*).

The first attempt to systematically measure resources was carried out in 1960, with the collection of existing data from various countries. Building on this experience, in 1964–65 a formal questionnaire was designed and sent to Latin American and, later, Asian countries, requesting data on the total number of scientists, engineers and technicians by field of specialization and sector of employment, and information on current expenditures for research and experimental development by sector. A new questionnaire incorporating counts for 'manpower' engaged in research and development (R&D) was later sent to the remaining countries of the world. The survey was carried out by the new Division of Science Statistics of the UNESCO Office of Statistics, created in 1965, and its results were published in the 1968 UNESCO *Statistical Yearbook*, heralding the beginning of regular statistical data collection on science and technology at UNESCO.

In parallel to data collection activities, UNESCO devoted important efforts to developing international standards. After first establishing international standards in statistics on education (1958) and culture (1964), it was the turn for science and technology as well as development activities in 1966. At the same time, several other organizations were getting involved in standardization, particularly the Organization for Economic Co-operation and Developmen (OECD), which had published its *Frascati Manual* in 1964 (OECD, 1964). After conducting several meetings

---

[13] This chapter was prepared by the UNESCO Institute for Statistics. The authors wish to acknowledge the contribution of two fundamental documents: Barré, 1996; and Godin, 2001*c*.
[14] Ernesto Fernández Polcuch: Programme Specialist since 2002 at UNESCO Institute for Statistics, after serving as consultant to UNESCO in the same field.

and bringing together numerous experts, the 'Recommendation Concerning the International Standardisation of Statistics on Science and Technology' was adopted by the UNESCO General Conference at its 1978 session.

INTERNATIONAL STANDARDIZATION

The Recommendation and the subsequent 'Manual for Statistics on Scientific and Technological Activities'[15] and 'Guide to the Collection of Statistics on Science and Technology'[16] addressed the need for standards that could be applied in all Member States, both those with advanced S&T statistical systems and those whose systems were still under development. Standards were also to be applied in the 'various types of socio-economic systems' in the world at the time, particularly 'market' and 'centralized' economies.

The Recommendation provides a conceptual framework for 'scientific and technological activities' (STA), comprising 'research and experimental development (R&D)' activities, but also 'scientific and technological education and training at broadly the third level (STET)', and 'scientific and technical services (STS)'. Furthermore, the concept of 'national S&T human resources' was introduced, covering the complete stock of scientists, engineers, and technicians in a country, active or 'potentially' active. This concept was later abandoned by UNESCO, but was reintroduced in 1995 by OECD in its *Canberra Manual* (OECD, 1995). The concept of scientific and technological activities would become the basis of UNESCO's philosophy of science and technology measurement, because it was understood to be more relevant to the conditions of developing countries, where 'proportionally more resources are devoted to scientific activities related to the transfer of technology and the utilization of known techniques than to R&D per se' (UNESCO, 1972, in: Godin, 2000).

UNESCO took on an enormous task (Godin, 2001c). For the Organization, standardizing science and technology statistics meant two things: first, extending OECD standards to all countries; second, extending standards to activities beyond R&D – that is, to all scientific and technological activities (STA). The two problems were intimately related, since the particular characteristics of 'centralized economy' countries at the time included the fact that R&D was not considered separately from 'science' – a term that also included training, design, and museums (Freeman and Young, 1965). Therefore, UNESCO chose to extend its measurement proposals to the broader concept of STA.

Following the 1978 Recommendation, measurement of STA was to be carried out in two stages. The first was limited to measuring R&D activities and

---

15   UNESCO Document ST-84/WS/12.
16   UNESCO Document ST-84/WS/19.

human resources in S&T (HRST), in a fairly aggregated manner. The second stage would include experimental approaches to measuring STS and STET, as well as a deeper level of detail in the measurement of R&D and HRST. The first stage was swiftly launched through a worldwide survey based on the 'Statistical Questionnaire on Scientific Research and Experimental Development' in 1981, focusing on HRST and R&D.

In order to prepare the second stage, a 'Guide to Statistics on Scientific and Technological Information and Documentation (STID)'[17] was published, after being tested in seven countries. The guide states that 'STID statistics aim at measuring all possible aspects of information activities: the level of generation, how information is collected, processed and disseminated, and how it is used'. The principal items to be measured were the institutions and individuals performing these activities, the amount of financial resources and physical facilities available, and the quantity of users. Surveys were directed initially to 'collectors, processors and disseminators' of STID, which usually meant libraries, as well as editing, publishing, printing, consulting and advisory services and enterprises. 'Producers' and 'users' of STID were expected to be included in a later stage of the project.

The measurement of STID was seen then as a first step in the extension towards covering the full range of STA. Eventually, however, UNESCO concentrated the measurement activities solely on R&D, which was easier to locate and to measure, and had the virtue of being 'an exceptional contribution to science and technology'. Hence, while UNESCO pushed for broadening the measurement, it simultaneously argued for the centrality of R&D.

At the same time, only a few countries were actually collecting data on STID. Since these activities were not always deemed important, the purpose of measuring them was not obvious, and there were difficulties in interpreting the definitions (UNESCO, 1985). One could argue that, while the policy questions regarding the extension of the measurement beyond R&D were significant, it was very difficult to find a path to convert them into practical measurement priorities and relevant instruments. STID measurement, in particular, did not manage to provide significant answers to science policy questions.

THE DECLINE AFTER THE MID-1980s

It is widely agreed that, after the mid-1980s, S&T statistics at UNESCO entered a period of constant decline, suffering greatly from the significant budget reduction that affected the Organization as a whole. The first cause of the decline was the marked decrease in human and financial resources available for these

---

17   UNESCO Document ST-84/WS/18.

activities from 1980 to 1993. However, as pointed out in an evaluation of the area conducted in 1996, the causes of this decline were manifold (Barré, 1996).

Another cause may be traced to the difficulties facing Member States in completing the UNESCO S&T statistics questionnaires, especially since many countries stopped conducting S&T surveys during the 1980s and the early 1990s. This decline in national S&T statistics was partly due to a loss of relative weight of S&T policies in the framework of national policies, now devoted principally to 'structural adjustment' rather than to promoting (endogenous) development (Albornoz and Fernández Polcuch, 1996). A third reason identified for the decline of S&T statistics was related to shifts in the priorities of UNESCO's Natural Sciences Sector. By the end of the 1980s, most S&T policy programmes were discontinued, and international S&T statistics lost one of its most significant users.

The heterogeneity of UNESCO Member States also conspired against its success, resulting in an 'absence of a community of views' (Godin, 2001c). By contrast, the OECD, which was very active and successful in this field, had a consistent membership of industrialized countries with similar levels of development and economic structures (Godin, 2001a). This common understanding – together with strong and committed leaderships from participating experts and the OECD Secretariat – helps to explain the success of NESTI (the OECD Working Group of National Experts in S&T Statistics).

All these factors led to a slowdown in S&T statistics activities at UNESCO. At the same time, definitions and classifications used by OECD and European Union countries became the internationally accepted norm – especially after the fall of the centralized economies made many of the UNESCO S&T statistics classifications (such as sectors of performance) obsolete – and many developing countries adopted OECD methodologies in order to obtain valid benchmarks. The reduction of the UNESCO S&T statistics programme to minimal levels of activity only widened the gap between the more developed countries with a well-oiled S&T statistics system and those countries that did not attract the interest of other international organizations and therefore received no attention in this area.

1992–2003: IDENTIFYING PROBLEMS, RETHINKING STRATEGY

From 1992 on, UNESCO's Division of Statistics S&T programme entered into 'consultation mode'. Several case studies were conducted from 1992 to 1994, in order to survey country needs and current practices in this field, followed by a series of meetings of experts, and various documents prepared by consultants. S&T data collection did not take place in 1994 and 1995. In 1996, an external evaluation was conducted by Rémi Barré – a renowned expert and head of the French Observatoire des Sciences et de la Technologie – who proposed strategies for improving the scope, quality, reliability and relevance of S&T statistics (Barré, 1996). The conclusions of this evaluation pointed out the need for the following activities:

- Establish renewed relations with countries, frequently on a regional basis.
- Share the work with other international organizations.
- Subcontract the construction of indicators involving external databases.
- Raise funds and external competencies to carry out the S&T statistics programme.
- Urgently harmonize classifications and definitions.

The evaluation document proposed three scenarios for the future: Option 1: Start 'at minimum', to prepare future activities; Option 2: Produce standard statistics and indicators; and Option 3: Establish UNESCO as a major player in the field. Mostly compelled by circumstances, the Division of Statistics chose Option 1 and advanced only in the cooperation with other international organizations, and (most importantly) in the harmonization of classifications and definitions. From 1998 on, UNESCO S&T questionnaires adopted the OECD classification of sectors of performance, which had been the major source of disagreement between the two organizations. At the same time, the UNESCO questionnaire borrowed other features from OECD, becoming significantly more complicated to complete. It also shifted focus towards more detailed data on R&D expenditure, rather than on human resources, which used to be the traditional UNESCO strength.

With the formal establishment, in 1999, of the UNESCO Institute for Statistics (UIS), and its subsequent move to Montreal (Canada) in 2001, a new turning point for S&T statistics at UNESCO arrived. Data collection was once again interrupted after the 2001 survey, and an intense process of consultation was launched, reaching experts and users from around the world. The intense collaboration with UNESCO's Natural Sciences Sector – and in particular the Division for Science Analysis and Policy, and the Regional Office for Science and Technology in Latin American and the Caribbean (ROSTLAC) – played a decisive role in the success of this consultation.

REPOSITIONING UNESCO AS A MAJOR PLAYER

The strategy prepared from the results of the consultation (UNESCO Institute for Statistics, 2003) sought to reposition UNESCO as a principal player in the field, very much in line with 'Option 3' of the 1996 evaluation. As short-term priorities, the improvement of data collection on 'input indicators', as well as further work on human resources in S&T (with particular attention to such issues as 'science education', 'gender', and 'brain drain') were set. Medium-term priorities included work in the field of innovation indicators. Longer-term priorities concerned output indicators and methodological development of issues such as measuring the social impact of science. One of the main characteristics of the new UIS strategy in S&T statistics is a clear commitment to capacity-building, as well as the recognition of the need to foster demand and use of S&T statistics and indicators in the framework of establishing active partnerships with other international organizations.

Based on this strategy, the 2004 S&T statistics survey was launched. A questionnaire was sent to countries not covered by partner organizations (such as OECD, Eurostat, and the Ibero American Network on Science and Technology Indicators), designed with a bias towards information on R&D personnel rather than expenditure, using classifications harmonized with OECD. It included a large section on metadata (in order to assess the quality of the data), as well as on the identification of a contact in each country that would permit direct exchange of information and participation in networks. This was also the first UNESCO statistical questionnaire to be available for completion online via the internet. Capacity-building activities were initiated through the planning of workshops to discuss methodologies and 'good practices' of data collection in sub-Saharan Africa, and southern Asia. These activities will no doubt help to increase response rates and quality of data.

The inquiries into methodological development saw the drafting of an Annex to the OECD/Eurostat *Oslo Manual* (OECD/Eurostat, 1997) on measuring innovation in developing countries (in cooperation with various countries), as well as the start of a project to develop an internationally harmonized survey on the Careers of Doctoral Holders, in cooperation with the OECD and Eurostat, and funded by the US National Science Foundation (NSF). Some preliminary work on measurement of the social impact of science has also been initiated, in cooperation with academic partners from the Institut national de la recherche scientifique (INRS, Quebec, Canada) and Georgia Tech (United States).

A new opportunity for UNESCO's S&T statistics programme has appeared on the horizon. Its constant interaction with UNESCO's Natural Sciences Sector signals the return of a strong internal demand for S&T data. International partners are strengthening the links with the UNESCO Institute for Statistics. Renewed interest and high participation of developing countries in recent activities might also signal that perhaps this time UNESCO will manage to recover its historical leading role in S&T statistics.

BIBLIOGRAPHY

Albornoz, M. and Fernández Polcuch, E. 1996. Latin American Science and Technology Indicators. *Research Evaluation*, Vol. 6 (December), pp. 209–13.

Barré, R. 1996. *UNESCO's Activities in the Field of Scientific and Technological Statistics*. Document BPE-97/WS/1. Paris, UNESCO.

Freeman, C. and Young, A. 1965. *The R&D Effort in Western Europe, North America and the Soviet Union*. Paris, OECD, pp. 27–30, 99–152.

Godin, B. 2000. *Neglected Scientific Activities: the (non) measurement of related scientific activities*. Montreal (Quebec), Observatoire des sciences et des technologies (OST).

——. 2001a. *Taking Demand Seriously: OECD and the role of users in S&T statistics*. Montreal (Quebec), OST.

——. 2001b. *The Number Makers: a short history of official science and technology statistics*. Montreal (Quebec), OST.

——. 2001c. *What's So Difficult About International Statistics?: UNESCO and the measurement of scientific and technological activities*. Project on the History and Sociology of S&T Statistics, Paper No. 13. Montreal (Quebec), OST, 21 pp.

OECD. 1964. *The Measurement of Scientific and Technical Activities: Proposed Standard Practice for Surveys of Research and Experimental Development* ('Frascati Manual'). Paris, OECD.

——. 1995. *The Measurement of Scientific and Technological Activities: Manual on the Measurement of Human Resources Devoted to S&T* ('Canberra Manual'). Paris, OECD.

OECD/Eurostat. 1997. *The Measurement of Scientific and Technological Activities: Proposed Guidelines for Collecting and Interpreting Innovation Data* ('Oslo Manual'). Paris, OECD

UNESCO. 1972. *Considerations on the International Standardization of Science Statistics*. Document COM-72/CONF.15/4. Paris, UNESCO, p. 14 (cited in Godin, 2000).

——. 1985. Meeting of Experts on the Methodology of Data Collection on STID Activities. 1–3 October 1985. Background Paper; Document ST-85/CONF.603/COL.1. Paris, UNESCO.

UNESCO Institute for Statistics (UIS). 2003. *Immediate-, Medium- and Longer-Term Strategy in Science and Technology Statistics*. Montreal (Quebec), UIS.

CLOSING THE CULTURAL GAP
Science and society
Jacques Richardson

AS UNESCO's earliest programmes were taking shape, its monthly magazine the *Courier* commented (in July 1948) that scientists joining the Organization should bring 'with them ... a familiarity with some of the implications of science in the modern world'. The General Conference of 1949 authorized, furthermore, the launching of a journal to be called *Impact of Science on Society*, a quarterly periodical that took form in French and English in 1950. The editorship of the new journal was haphazard, however: no more than a task added to the principal duties of a succession of civil servants, usually those concerned with science policy. The result was a product of uneven quality, often coming out late. In 1967 came the appointment of a full-time editor. This post lasted until 1986, when the assignment was relegated once again to part-time responsibility.

The combined circulation of the journal's editions never exceeded 30,000 per issue. A strenuous effort to find major co-publishers (more than thirty were contacted, during the years 1976–84) failed because of the prevailing attitude among the most successful of specialized publishers that the dissemination of scientific information should be made to precisely identified audiences and not be left in intergovernmental hands.

After the Second World War, C. P. Snow, a molecular physicist and novelist, identified the existence of a culture gap between those trained in the natural and engineering sciences, on the one hand, and, on the other, those trained in the humanities. This gap between what Snow named the 'two cultures' – where practitioners of each discipline were ignorant about the other, making communication between them difficult – was far from disregarded by UNESCO. For both groups and for others sensitive to a world beginning to undergo morphological change in the politico-economic sense, the beginnings of decolonization were creating yet another category: the scientific 'have-not' status of the slowly nascent new nations.

Over the years, the journal *Impact* increased its circulation by publishing in several languages. By the 1980s, it was published in the six official languages, as well as Korean. The edition in Arabic, however, posed endless problems in distribution in the Arabic-speaking countries outside of Egypt, where it was co-published with the UNESCO centre there. The Russian edition, never more than a selection of articles, managed to avoid topics controversial in the Soviet context of the time. Efforts to publish in German and Japanese met with rebuffs from official quarters. Furthermore, a Portuguese-language edition, arranged by the

Office of Publications to appear in Brazil, came to nought when the would-be publisher absconded with funds collected for subscriptions. By the time of the Regular Programme and Budget for 1971–1972, publication of *Impact* in four languages was costing a modest US$52,500 per biennium.

The publication of *Impact* continued until the early 1990s, when it was converted into the biennial *World Science Report*. After a refreshingly universal coverage of new developments in the scientific life in its first volume, the *World Science Report* became a statistical compilation of 'scientometry' (distribution of the world's scientists, technologists and science teachers; levels of funding) in its succeeding volumes. Publication was halted in 2000 'by Executive-Board decision', then resumed with the publication of the *UNESCO Science Report 2005* in December 2005. Meanwhile, the Organization launched in October 2002 the quarterly magazine *A World of Science*. This is a first-class magazine of twenty-four pages in colour, which reports broadly on what UNESCO is doing in the fields of science and engineering round the world, carries editorials and is published on time in French and English. The journal is distributed free of charge and has an online edition.

Parallel efforts to popularize science through the mass media began through what is now the Communication component of UNESCO, with the creation in 1957 of a network of International Centres for Advanced Training in Journalism. The first of these schools was established at the University of Strasbourg (France), directed by jurist Jacques Léauté and journalist Francis Cook, both of French nationality. A workshop on the problems and methods of science journalism was not held until 1966, an indirect result of which was the creation in Paris, later that year, of the International Science Writers Association (ISWA), a non-governmental organization (NGO). The Strasbourg Institute, which in the end was not duplicated elsewhere, ceased to exist around 1970.

ISWA, with a worldwide membership of several hundreds, is still affiliated with UNESCO and proved its dynamism at a UNESCO-ICSU (International Council for Science) World Conference on Science held in Budapest (Hungary) in 1999, during the combined European Scientific Open Forum and European Science Foundation meeting of August 2004 in Sweden, and at a world science writers' congress held in Montreal (Canada) in October 2004. The initiative now seems to be in the hands of NGOs. Project 2.111.1 of the Regular Programme and Budget for 1973–1974 authorized (rather verbosely) study of the Human Implications of Scientific Advance and Promotion of the Public Understanding of Science and Its Relationship to Society. This activity consolidated the journal *Impact* and management of both the Kalinga Prize for the Popularization of Science and the UNESCO Science Prize honouring excellence in research in developing countries. The awarding of these and other international science prizes continues today: there are now a half-dozen in all (see Annex 5 for full list). The

1. UNESCO Kalinga Prize for the Popularization of Science, 1953. Fourth from left: Luther Evans, Director-General of UNESCO. Fifth from left: Julian Huxley, laureate, former Director-General of UNESCO.

Three images 'illustrating' UNESCO's implication in the popularization of science

2. Popularization poster 'Handle with Care/Fragile', UNESCO's Man and the Biosphere Programme (MAB), 1986.

PART IV: SCIENCE AND SOCIETY

3. Children's art exhibition, part of the campaign 'Children's Views on Science in the Twenty-First Century', 2001.

same activity authorized the preparation of a book provisionally entitled 'Science in the '70s'. An additional objective, called 'Misuse of Science by Society', was left unstaffed and unfunded and never came to fruition.

Subsequent editions of the Regular Programme and Budget authorized expansion of the scope and languages of *Impact*. Considerable experience was offered by UNESCO via consultants to science museums, especially in the Latin America and Caribbean region. The book originally to be entitled 'Science in the '70s' was finally published in 1977, signed by the former Assistant Director-General of Natural Sciences, who had directed its compilation. Portions of the book proved to be controversial, and the work appeared finally only in English (published by UNESCO),[18] Italian and Spanish (co-published with private firms).

The Regular Programme and Budget of 1976–1977 made the first reference to 'ethics' (para. 4124), stating that the systematic dissemination of 'objective information on the long-term significance of ... discoveries would be progressively introduced'. This was accomplished, in effect, through *Impact* and several book titles, as well as through the multiplier effect of professional workshops in science popularization. These problem-solving seminars were organized in several countries (Bulgaria, China, Egypt, Tunisia and Yugoslavia, among them) during the late 1970s and until 1984, under the aegis of the programme then called, more simply, 'Science and Society'.

---

18   Published (in English) as *The Scientific Enterprise, Today and Tomorrow* (Buzzati-Traverso, 1977).

In 1976, the programme co-published *Solar Energy: The Awakening Science* (Behrman, 1976), based on the personalities and proceedings of a world alternative-energy conference held at UNESCO. At the same time, UNESCO sought to create a semi-popular journal chronicling scientific discovery throughout the world; however, a major effort to enlist a co-publisher failed, again because specialized houses did not endorse this kind of dissemination by an intergovernmental entity. Preparatory to the UN Conference on Science and Technology for Development (Vienna, 1979), the Organization commissioned a 273-page analytical essay, *Science and the Factors of Inequality: lessons of the past and hopes for the future*, by Charles Morazé of the Parisian Institut des Sciences Humaines (Morazé, 1979). The volume, a moralistic tribute to the sociology of natural science, was published in French, Spanish and English.

The annual meetings of the Director-General's Advisory Group on Science and Technology for Society (1981–83) resulted in – among other things – the publication (in English only) of the well-received *Science and Technology for World Development* (Clarke, 1985). Other didactic volumes produced during the same period included a manual on the use of cinematography in the world of science; a collection of essays on the societal consequences of scientific and technological creativity; and *UNESCO:Why the S?*, an accessible survey of science at UNESCO, by David Spurgeon (UNESCO, 1985). By arrangement with a US publisher, UNESCO also produced a series of volumes based on themes appearing in *Impact*: technology transfer, models and modelling, law and resources of the sea, natural hazards, and creativity and invention. Published under license from UNESCO, these books generated revenue for the Organization.

Today, there is no longer any Science and Society programme. Over the decades, UNESCO's scientific and technological management showed little enthusiasm for popularization of the societal implications of science and technology, perhaps because the intrinsic value of science as a cultural pursuit was insufficiently pushed. One result of the joint UNESCO-ICSU conference held in Budapest in 1999, however, was the formation among a number of NGOs of a working group (complete with webpage links) concerned with science and ethics.

BIBLIOGRAPHY

Behrman, D. 1976. *Solar Energy: The Awakening Science*. Boston, Mass., Little, Brown.
Buzzati-Traverso, A. 1977. *The Scientific Enterprise, Today and Tomorrow*. Paris, UNESCO.
Clarke, R. 1985. *Science and Technology in World Development*. Oxford, UK, Oxford University Press.
Morazé, C. 1979. *Science and the Factors of Inequality: lessons of the past and hopes for the future*. Paris, UNESCO.
UNESCO. 1985. *UNESCO: Why the S?* Paris, UNESCO, 63 pp.

PART IV: SCIENCE AND SOCIETY

TAPPING AT THE GLASS CEILING
Women, natural sciences and UNESCO
Renée Clair[19]

LET me begin with a telling anecdote. In 2002, the Women in Physics Conference in Paris was attended by more than 300 scientists from sixty-five countries. When asked about the situation of women in the field, a round-table participant assured the audience that all was well, and that in his firm women had the same career opportunities as their male colleagues. Less than three years later, on 13 January 2005, UNESCO launched the International Year of Physics with an international conference on 'Physics of the Future'. How many women were present at the inaugural session? None. How many at the closing session? None. And how many women were there among the eleven lecturers? Only one: Mirian Sarachik, a North American laureate of the L'Oréal-UNESCO Prize for Women in Science.

I can already hear the cries of reproach in response to this anecdote. Why, certain readers will ask, are women scientists so impatient? Look at the undeniable progress that has been made in only the past two generations. Women scientists, like women in other professions, should wait for the unrelenting advance of sexual equality to establish a balance between men and women in science. They just need to be more patient.

Although the involvement of women in science varies according to discipline, country, and level of industrial development, the fact remains that it is far from generally prominent. And while, at UNESCO, the concern for the low-level of female participation emerged relatively late, one may conclude that the overall situation has improved with time and effort. Indeed, women are growing more and more numerous among science students and research personnel. Increasing numbers of women are choosing this career path, especially in the life sciences. What is troubling, however, is the decreasing number of women higher up the scientific professional ladder. Some argue that this is because family responsibilities take priority over the women's careers, or that women do not find the more responsible and powerful jobs attractive. In short, their situation has nothing to do with any sort of dissuasion in the scientific realm. Certainly, it is not due to any possible discrimination. There is a strong denial of the fact that, when it comes to scientific professions, the 'glass ceiling' is still one of the most impenetrable.

---

19   Renée Clair joined UNESCO in 1997, charged with the special project on Women, Science and Technology. She helped launch the L'Oréal-UNESCO Prize for Women in Science, and since 2002 has been Executive Secretary for this prize. From 1993 to 1997, she was Counsellor of the French National Commission for UNESCO, coordinating a quadrilingual book on the scientific education of girls.

SOME STATISTICS AS TESTIMONIAL TO INEQUALITY

In 1997, when Marianne Grunberg-Manago, the first woman to be elected president of the French Academy of Sciences, received me in her laboratory at the Institute of Physical-Chemical Biology, in Paris, she suddenly asked her secretary, 'How many men and women have we in the laboratory?' She added that she had never asked herself the question.

Without going into the history of the role of women in science, let me cite a few revealing cases. Marie Curie was awarded two Nobel Prizes: the first in physics in 1903 for her work on radioactivity, the second in chemistry in 1911 for her discovery of polonium. Yet she was refused membership in the French Academy of Sciences. Her daughter, Irène Joliot-Curie, who won the Nobel Prize in Chemistry in 1935 for her work in artificial radioactivity, also failed to make her mark in the august halls of French science. It was not until 1988 that the first woman, Marie-Anne Bouchiat, joined the Academy, and 1995 when a woman, Ms Grunberg-Manago, was named president.

Of the 190 members of the French Academy in 2004, only fourteen were women, although five of the twenty-four members elected that year were women – a strong proportion and an encouraging sign. In the United States, in 1999, there numbered 118 women among 1,904 members of the National Academy of Science; in the Netherlands, one woman among 237 members of the Royal Netherlands Academy of Arts and Sciences; and, in the United Kingdom, forty-three women among 1,185 members of the Royal Society of London. Of the 503 Nobel Prizes given in science between 1901 and 2004, two in physics, three in chemistry and seven in physiology or medicine were awarded to women (including the two to Ms Curie). The most recent went to Linda Buck (shared with Richard Axel) in 2004 for research on the functioning of the olfactory system. Not a single woman has been awarded the Fields Medal, the highest distinction in mathematics and one comparable to the Nobel Prizes.

AND WHAT HAPPENS AT LESS PRESTIGIOUS LEVELS?

According to a study done in 2004 at the world level, women represent more than 55 per cent of undergraduate students in more than half the countries, whereas in 60 per cent of the countries women represent less than 45 per cent of graduate students in the scientific disciplines. In more than half the countries, women number less than 35 per cent among research personnel. In the private research sector of thirty-five countries that are either industrialized or in the process of industrialization, the spread is considerable: from only 10 per cent of women researchers in Japan to more than 50 per cent in Argentina, with an overall average of 30 per cent. In the United States, the salaries of women with a first degree in science or engineering are 35 per cent lower than those of equivalent males; with a doctoral degree, they are 26 per cent lower.

So, are women scientists impatient? No, but they are firm about having their rights and talents recognized, determined to change mentalities – not only to protect their interests but because they believe in the essential role played by science in culture and development. All the major international conferences of the past thirty years have declared that there can be no durable development without the participation of everyone, men and women, in support of science as a force of progress. What about UNESCO in this respect, a body of the United Nations created in 1945 with very specific missions concerning the promotion of science and equality between men and women? Let us take a look at the record.

PRIOR TO THE FIRST WORLD CONFERENCE ON WOMEN (MEXICO, 1975)

UNESCO's Charter indicated clearly that the new organization proposed to contribute to the maintenance of peace by strengthening collaboration among nations through education, science and culture. A few lines further on, it is specified that this will be done 'without distinction of race, sex, language or religion'. Thus, from the moment of its creation, UNESCO has had a mission to reinforce scientific cooperation at the international level, monitoring such cooperation in order to reject all discrimination, especially between men and women. Putting this into practice, UNESCO would be guided (as an entity subject to the vagaries of history) by the evolution of thought as it encountered the spectacular acceleration of the 1970s.

UNESCO's initial activities meant to ensure the equality of opportunity for women began, basically in terms of education, during the 1950s. This hardly involved the science programme. Then, impelled by several non-governmental organizations, UNESCO participated in 1967 in a first effort; this was in Chile, and concerned the access of girls to scientific and technical careers. The project was supported financially by UNESCO until 1974. During the first thirty years of its existence, however, the results of such efforts were inconclusive. Should one deduce that UNESCO, much in the image of scientific societies of the time, paid little attention to 'women's' issues?

Mentalities change. During the early 1970s, a number of countries experienced social turmoil that was centred on the question of the equality between men and women. The International Women's Year World Conference held in Mexico in 1975 – focusing on the condition of women in the judicial, economic and political spheres – adopted a World Plan of Action. The participation of women in scientific and technological development was raised in only one of its paragraphs, which concerned the means used to urge women to take up scientific and technical

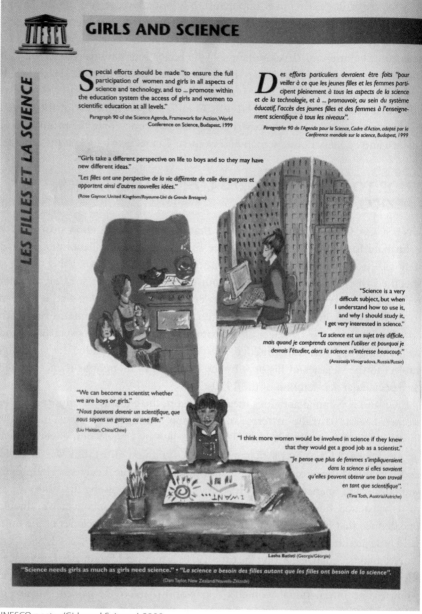

UNESCO poster 'Girls and Science', 2000.

professions. This recommendation may strike us today as a rather timid one, yet it must be emphasized that this was the first time that such a proposal was presented at a global level – incontrovertible progress, given the preoccupations of the time.

We note that, when UNESCO's journal *Impact of Science on Society* devoted an entire issue to the role of women in science (also in 1975), its editor was upbraided by a number of women scientists and engineers both outside and within UNESCO, who asked: 'Why set women further apart? They have, already, sufficient problems surviving in the world of science and technology.' The conference in Mexico led the General Assembly of the United Nations to adopt the UN Women's Decade: Equality, Development and Peace (1975–1985).

Although, as early as 1973, the role of women in science had become a formal responsibility of UNESCO's unit charged with human rights, the Natural Sciences Sector itself contributed nothing that the Director-General could call attention to in his biennial report for the period. One result of this was, for the first time, specific mention in UNESCO's programming of the issue of women. The first Medium-Term Plan for 1977–1982 contained a reference to an augmented role for women in the development process, an objective shared among the education, social-science and communication-information programmes – but still without the involvement of the natural sciences programmes. Nonetheless, the Plan's aims included analysis of the discrimination inherent in scientific education and professions, together with projected statistical studies on the subject. There was little substantive follow-up to the scheme, however, an inertia that persisted until the first half of the 1980s. Then a series of three studies on women engineers became the basis for further elaboration of a programme on women in science in general.

At the World Conference of the International Women's Year (Nairobi, Kenya, 1985), on the results of the UN's decennial effort to promote the role of women in development overall, emphasis was laid on the need for access by girls and women to education and training in science, mathematics and engineering, especially in all of their peaceful uses. A new aspect of the issue emerged at that time: the need to evaluate the effects, both real and potential, on science and technology of all factors influencing the integration of women into different economic sectors – e.g. women's health, income and status – and to ensure that women benefit fully from modern technologies. The Nairobi meeting sought, in effect, to accord women both a new status and a greater importance in UNESCO's programme.

The second Medium-Term Plan (1984–1989) featured the continued participation of UNESCO in the UN's Women's Decade, reaffirming the priority given to women. The Plan foresaw (within a horizontal Major Programme called the Condition of Women) two programmes intended to reduce certain observed disparities: one to eliminate discrimination based on sex, and the other on the equality of girls and women in the educational process. The latter had a sub-programme on the promotion of equality of access by girls and women to technical and professional education. Furthermore, the Major Programme on the natural sciences had a sub-programme on the training of engineers and technicians that sought to increase participation by women. October 1984 saw, for instance, the convening of a seminar in India on the role of women in the assimilation and diffusion of technological innovation in developing countries.

This Medium-Term Plan sought increased integration of the 'feminine dimension' in all programmes and activities. What did such a recommendation signify? It meant that – while awaiting perfect equality between men and women – all of the Organization's programmes and intended activities were to be reviewed in the light of particular difficulties encountered by women in respective areas, and that a strategy of differentiation be adopted. An effort was made to enumerate the women participating in seminars, conferences, committees and other assemblies of the Organization. A woman coordinator was named to guide this policy on the feminine condition. The third Medium-Term Plan (1990–1995) reaffirmed the priority accorded by UNESCO to women, adding youth to priority and interdisciplinary programmes, together with their integration throughout the sectors.

An evaluation of programme execution for 1994–95 showed that the few activities conceived specifically for women had often suffered from budget cuts, that follow-up on particularly well-received actions had not been made, while integration of women seemed to progress within several projects. As for the participation of women in the scientific programmes, the Man and the Biosphere Programme (MAB) was mentioned as one in which training included 13 per cent women, whereas women had participated in only 5 per cent of the research scholarships awarded. In sum, one observes that despite a better recognition of the situation of women in science, the Organization was undertaking little noticeable action in the field. Things were still at the talking stage.

AFTER RIO DE JANEIRO (1992) AND BEIJING (1995)

Two world conferences had a considerable impact at the international level on the subject of women in science: the UN Conference on the Environment and Development (Rio de Janeiro, Brazil, 1992) and the Fourth World Conference on Women (Beijing, China, 1995). In its Agenda 21, the Rio Summit emphasized for the first time the role of women in sustainable development, more precisely in the management of ecosystems and the preservation of the environment. Of the forty chapters of Agenda 21, thirty-four refer to women, in matters such as deforestation, the management of mountainous regions, agriculture, biodiversity, biotechnology, mountains and coastal zones, freshwater, natural catastrophes, and the like. The Summit's main recommendations concerned gender data, the education of women and their integration within all decision-making mechanisms.

The Fourth World Conference on Women was an important landmark, demonstrating a new awareness, on the global scale, of the inequalities persisting between men and women twenty years after the Mexico conference. In its Action Programme, the conference ranked among twelve critical areas for the future the unequal access to education (science and technical teaching were judged 'discriminatory') as well as the disparities between men and women in the fields of natural resources management and environmental conservation. Did these world events have repercussions on policy at UNESCO? They were, in my estimation, decisive.

In its Mid-Term Strategy (1996–2001), UNESCO retained women among the Organizations' priorities, while listing alongside them three other 'priority groups': youth, the least-developed countries, and Africa. An essential change was the replacement of the intersectoral programmes by the 'special projects' organic to each of the Organization's Major Programmes. Among these, two special projects amplifying themes featured at Rio and Beijing within the Major Programme on natural sciences were 'Women, Science and Technology', and 'Women, Water Supply and the Use of Water Resources'. Contrary to previous activities, subject to the fortunes of the biennial budget, the special projects had clear objectives, a definite budget (one protected from unexpected cuts), a specific duration (allowing a strategy to take form) and a responsible staff member named by the Organization (leading to better accountability).

The special project 'Women, Science and Technology' – launched by UNESCO at the NGO Forum at Beijing in 1995 – featured a workshop on the issue. The same year, UNESCO's (previously published) *Second World Science Report* helped considerably to emphasize the need for attention to the role of women in the world of science. What were the results of this special six-year project (ended in 2002), of which certain activities are still under way? Among the most important were: the organization in 1998–99 of six regional forums on Women, Science and Technology; UNESCO's participation in the World Conference on Science (Budapest, Hungary, 1999); the creation of five UNESCO Chairs on the theme of Women, Science and Technology; support for numerous women scientists and engineers; and one of UNESCO's main partnerships with the private sector, the L'Oréal 'Women in Science' project.

Since 2000, the UNESCO-L'Oréal partnership has awarded prizes worth US$100,000 each to five outstanding women from five different regions – Africa, Asia, Latin America, Europe and North America – working in the life sciences, or alternatively in materials science. Furthermore, fifteen fellowships (worth US$20,000 each), are attributed annually to young women aspiring to conduct laboratory research of their own choice. And, since 2003, a number of national projects have been launched under the supervision of National Commissions for UNESCO in tandem with L'Oréal's national divisions.

L'ORÉAL-UNESCO Awards for Women in Science, created in 1998; first awards ceremony at UNESCO Headquarters. Left to right: Gloria Montenegro, laureate; Myeong-Hee Yu, laureate; Béatrice Dautresme, Director-General of Helena Rubinstein (representing L'Oréal); Federico Mayor (biologist), Director-General of UNESCO; Grace Oladunni L. Taylor, laureate; Pascale Cossart, laureate.

L'ORÉAL-UNESCO Awards for Women in Science, 2006. Left to right: Koïchiro Matsuura, Director-General of UNESCO; the five laureates, Esther Orozco, Pamela Bjorkman, Habiba Bouhamed Chaabouni, Christine van Broeckhoven, and Jennifer Graves; and Sir Lindsay Owen-Jones, Chairman and CEO of L'Oréal.

UNESCO's special project on water has also met with much success. It made possible the training, in a Mauritanian village, of some forty young women in the popularization of techniques related to research on, and the proper use of, water, as well as on sanitation and finding and exploiting water sources. This project unfortunately was not extended beyond its first four-year term, which ended in 2000.

And what about the aftermath to the World Summit on Sustainable Development (Johannesburg, South Africa, 2002)? In its Medium-Term Strategy (2002–2007), UNESCO emphasizes priority programmes rather than priority groups: specific activities have progressively disappeared from the scientific programmes. The end of special projects and the financial penury still affecting UNESCO's programmes has reduced budgets drastically. The reputed 'feminine dimension', also known as the 'gender perspective' – which ought to have been integrated within the Organization's programmes and activities – has never worked, although it is always amply used textually as the ultimate (and improbable) lifesaver. And this remains the case even after the World Summit on Sustainable Development at Johannesburg, where everyone expected a more solid stance concerning the Rio propositions.

And yet, the impact of numerous world conferences since the late 1980s has had its effect at UNESCO, with increasing attention paid to the need to enhance the position of women in the scientific and engineering professions. The results obtained, while far from the best possible, augur well for foreseeable progress in the decades to come.

BIBLIOGRAPHY

European Commission. 2000. *Science Policies in the European Union: promoting excellence through mainstreaming gender equality.* http://www.europa.eu.int/comm/research

———. 2002. *National Policies on Women and Science in Europe.* Helsinki Group on Women in Science. http://www.europa.eu.int/comm/research

———. 2003. *Women in Industrial Research: a wake-up call for European industry.* http://www.europa.eu.int/comm/research

———. 2004. *Increasing Human Resources for Science and Technology in Europe.* Report presented at the EC Conference, 2 April 2004, by José Maria Gago. 192 pp. http://www.europa.eu.int/comm/research/

Hartline, B. K. and Michelman-Ribeiro, A. (eds). 2005. *Women in Physics: Second IUPAP International Conference on Women in Physics* (Rio de Janeiro, Brazil, 23–25 May 2005). AIP Conference Proceedings, Vol. 795. Melville, NY, American Institute of Physics, 260 pp.

UNESCO. Reports of the six UNESCO Regional Forums: Women, Science and Technology. http://www.unesco.org/science/wcs/meetings/list.htm

UNESCO Institute for Statistics. http://www.uis.unesco.org

QUESTIONING AUTHORITY
Science and ethics
Jacques Richardson

E**THICS,** one now perceives, came into its own within UNESCO's endeavour only slowly; the process required nearly fifty years to take form. Thanks in part to a Director-General (Federico Mayor) who was scrupulous about the moral implications of the fast-developing biotechnologies, UNESCO created in 1993 the International Bioethics Commission (IBC, with thirty-six experts) under the aegis of the Social and Human Sciences Sector and in close collaboration with the Natural Sciences Sector. The IBC's main aims were to emphasize scientific rigour, and the ethical and juridical interpretations of this rigour throughout general biology, genetics and medicine. This led to the founding, in 1998, of the Intergovernmental Bioethics Committee (also composed of thirty-six experts) as a specialized, ethical forum representing governments. This committee's work, translated through diplomatic negotiations leading to international agreements, was solidified publicly in 1999 by a resolution of approval adopted by the United Nations General Assembly.

In 1997, largely through the efforts of the IBC, the International Declaration on the Human Genome and Human Rights (the human genome and its implications for both good health and disease) was proclaimed; it was also formally adopted by the UN General Assembly. This was followed, in 1999, by the adoption of the International Declaration on Human Genetic Data. Both instruments were intended to protect, to the maximum extent possible, the rights of the individual, as knowledge in the life sciences broadens and deepens.

In 1998, the World Commission on the Ethics of Scientific Knowledge and Technology (COMEST) was created. Headed by Pilar Armanet Armanet (Chile), COMEST is responsible for

- facilitating technical dialogue between scientists and decision-makers,
- providing specialized counsel to decision-makers, and
- giving early notification to governments of risks inherent in evolving science and technology.

Among its first areas of involvement have been: freshwater supply and distribution, energy, new information and communication technologies, and extra-atmospheric research and the manned/unmanned exploration of space. Of particular note is the work of the physician Alain Pompidou on ethical issues

PART IV: SCIENCE AND SOCIETY

related to space exploration, an analysis of which was published in 2000 under the auspices of the European Space Agency and UNESCO.[20]

It is still too early to know what UNESCO's new activities in scientific ethics might achieve worldwide. But it is notable that the 'science and society' initiatives entrusted to UNESCO scientists and engineers for a good half-century met with less success than did the analogous activities of the humanists of the Organization, in the ten years between 1995 and 2005.[21] The initiatives in bioethics – and in the overall ethics of scientific investigations and their technological applications – originated in the UNESCO sectors concerned with social sciences and culture, rather than in that concerned with research and engineering. There may, indeed, be 'two cultures' still at work within UNESCO itself.

BIBLIOGRAPHY

Pompidou, A. 2000. *The Ethics of Space Policy*. Paris, The European Space Agency and UNESCO.

---

20 A. Pompidou's *The Ethics of Space Policy* offers an unusual perspective on professional behaviour relating to astrophysical research and technologies (Pompidou, 2000).
21 It will be very interesting to see how a new, parallel activity at UNESCO, 'Economics and Ethics', fares over the coming years.

HARD TALK
The controversy surrounding UNESCO's contribution to the management of the scientific enterprise, 1946–2005
Bruno de Padirac[22]

UNESCO'S programme of science and technology policy emerged during the 1960s, grew during the 1970s and 1980s, then disappeared during the 1990s, only to reappear in part early in the twenty-first century. What are the reasons for the eclipse of this programme?

In this brief chapter, we shall take a broad view of the concept of 'science and technology policy', where it shall be understood to include:

- research in epistemology, in order to have a better understanding of the scientific enterprise (the 'science of science');
- methodological studies to define the tools that aid the decision-making process in science policy;
- support for the elaboration of governmental policy intended to utilize national scientific and technological potential or resources (managing the scientific 'supply' – the 'science policy' programme in its strictest sense);
- the democratic expression of society's hopes and needs regarding scientific explanations and technological solutions (arousing the societal 'demand', otherwise called 'science and society');
- the dissemination of research results by facilitating scientific and technological information exchange (the UNISIST and SPINES[23] programmes); and
- the building of consensus regarding the legal principles and ethical values that should govern the scientific enterprise and the use of its results (the 'ethics of science' programme).

---

22   Bruno de Padirac: UNESCO, Programme Specialist in Science and Technology Policy Division (1976–90); Director of Cyberspace Law and Ethics Task Force (1999–2002); Director, Science Policy Studies and Information (since 2002).
23   The United Nations Information System for Science and Technology (UNISIST), and the Science and Technology Policy Information Exchange System (SPINES).

PART IV: SCIENCE AND SOCIETY

UNESCO's activities concerning science and technology policy – in this broad sense – have been assigned to different components of the Secretariat over the last forty years, regardless of whether they should be part of the mandate of the Natural Sciences Sector from a conceptual viewpoint.

The main objective of the science and technology policy programme was to allow developing countries to acquire the ability to (independently) participate in the expansion of scientific knowledge (especially that relating to their specific social, economic and environmental situations), and to select and adapt those technologies which would contribute to their endogenous development. This programme thus ran up against the views of governments and the transnational firms of certain developed countries, which saw within the same countries resources and markets to be exploited. It is not insignificant that the term *endogenous* – referring to a nation's capacity to decide freely its own future – was replaced during the 1980s by *sustainable* (in French, *durable*). 'Sustainable' is meant to signify 'being for the common good of humanity'; yet this 'common good' is too often determined only by those market laws which are controlled by the richer countries. It is noteworthy that the unit currently responsible for a large portion of this programme is called the Division for Science Policy and Sustainable Development.

To an outside observer, this programme might have seemed (during the 1970s and 1980s) very much under the influence of concepts applied by socialist countries in the central planning of science policy. UNESCO encouraged the adoption of a cybernetic model of national science system comprising only public institutions and state control – whereby programmed research was linked to a national development plan (see illustration). It should be said in UNESCO's favour that, as an intergovernmental organization, its preferred channels of action (especially during the period in question) were in the public sector, and that the scientific potential of most developing countries (except in Latin America) was concentrated in that sector. In fact, underlying this programme was the mixed-economy model as found in France; for instance, the relationship between government, universities (enjoying a broad functional autonomy), and both private and public enterprises played a prominent role.

The fall of Communism, nonetheless, deprived this programme of firm support by the socialist countries and by developing nations operating through the Group of 77. The programme was also viewed with much misgiving by scientists themselves, who invoked both the independence of the scientific enterprise and the neutrality of science. They accepted being subsidized by public funds, but without yielding to the control of the use of such funding and thereby surrendering to accountability. Some Member States, in the name of laissez-faire – and endowed with strong scientific potential, mainly in the private sector – systematically attacked UNESCO's science and technology policy programme during each General Conference of the 1970s and 1980s. It is perhaps only a

SIXTY YEARS OF SCIENCE AT UNESCO 1945–2005

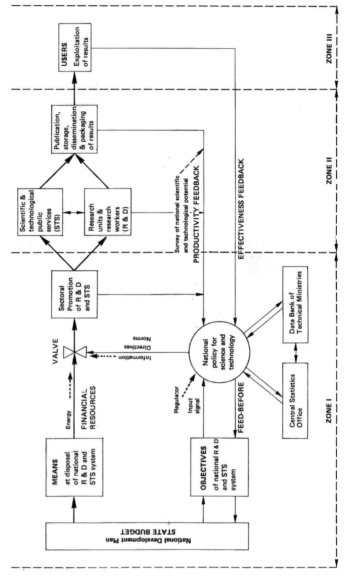

Cybernetic Diagram of the National Science System, illustrating the machinery and structures for science policy formulation, and their links to institutions implementing that policy. This diagram, promoted by UNESCO, was very widely used for information, advisory and teaching purposes in science and technology policy during the 1970s and 1980s.

coincidence that this programme disappeared when UNESCO began to give priority to the return of the United Kingdom and the United States (after their departure in the mid-1980s). Yet it cannot be denied that the programme had, and continues to have, considerable conceptual and operational impact on the policies of developing countries.

The fate reserved for one of these activities – the system of information exchange regarding science policy known as SPINES (Science and Technology Policy Information Exchange System) – is symptomatic. Beginning in 1972, UNESCO conducted an activity to facilitate the management and international exchange of documents and data having a direct impact on scientific and technological policy – at the level of both government and institutions involved in scientific research or technology transfer. A feasibility study published in 1974 dealt with setting up an international SPINES scheme, comprising several national or regional units responsible for information input/output, and a central group responsible for computerizing these data. This was to be in accordance with the INIS (International Nuclear Information System), as developed by the International Atomic Energy Agency (IAEA). The central processing entity was to be financed by Member States, on a pro rata basis of their contributions to UNESCO's regular budget.

After about ten years of deferred action, notably innumerable consultations with experts (albeit those in favour of the system) and a pilot programme wilfully underfunded, the SPINES project was abandoned in 1983 by the General Conference. What was especially revealing here were the comments of delegates heard in the conference corridors, to wit, 'Why would we finance the system and furnish our documents and data gratuitously to Third World countries at no benefit to ourselves? Let's trade the information (scientific data) for petroleum!' Despite this, SPINES influenced many developing countries, which used some of its features to create or reinforce their national systems of scientific and technical information.

After the abandonment of the SPINES scheme, one of its by-products, the *SPINES Thesaurus* (a controlled and structured vocabulary of 11,000 terms and 78,000 semantic relations for information processing in all fields of science and technology) continued to be disseminated; it was adapted into several languages (English, French, Portuguese, Spanish) and made up-to-date until the end of the 1980s (UNESCO, 1988). Thereafter, its many users fended for themselves. Certain countries and national or international institutions dropped the *SPINES Thesaurus* once UNESCO relinquished its programme in science and technology policies. Others still make use of it today in different forms, asking UNESCO to bring it up-to-date and adapt it to the internet.

Another controversial aspect of the overall science and technology policy programme is the development of legal principles and ethical values governing the scientific undertaking. While declaring its own particular ethical mandate within the UN system, UNESCO has produced few international legal instruments concerning the ethics of science. Exceptions were, in 1974, the 'Recommendation on the Status of Scientific Researchers' and, recently, two instruments concerning bioethics. These are little known, however, and rarely invoked by the actors and beneficiaries of science. And yet the Organization has known the agitation of numberless debates leading only to vague conclusions (or none at all) in regard to ethics. Does there exist, for example, a universal ethics acceptable to all cultures and applicable to all disciplines? Again, what is the difference between scientific ethics and technological ethics? And so forth.

Ethics touches upon delicate issues, both complex and controversial, such as respect for life or freedom of information. In no UNESCO document will one find the term 'ethics' defined. UNESCO, burdened by ideological debates and intimidated by a lack of consensus on regulating the new information technologies and the internet, left it to the Council of Europe to elaborate the Convention on Cybercrime, which, after four years of expert work, was open for signature on 23 November 2001.

The World Commission on the Ethics of Scientific Knowledge and Technology (COMEST), created by UNESCO in 1998, never dealt with bioethics (a concern of specific intergovernmental bodies) and seems to have strayed into issues that are not, strictly speaking, ethics of science issues (e.g. the precautionary principle, water ethics, energy ethics). COMEST itself, after seven years of work, has turned out no international legal instrument. One wonders if an intergovernmental organization whose decision process is founded on consensus among more than 190 Member States – representing many cultures – can define and adopt clear, normative texts that are demanding and constraining in fields whose economic stakes are considerable. The International Declaration on Human Genetic Data, adopted by UNESCO in 2003, distinguished itself by not uttering a word on the commercialization of such data.

And yet the role of the Organization is to bring together the points of view of different actors and beneficiaries of the scientific enterprise on complex – and thus often controversial – questions that require transborder, transcultural and universal responses, so as to arrive, if possible, at a world consensus that would be built into international legal instruments. Instruments of this kind (declarations of principle, international recommendations, conventions, and the like) provide a framework for national legislation that states can develop, and for codes of conduct that can be adopted by researchers, industrialists and service-providers as self-regulation. Questions of the moment that are treated by numerous other international organizations, or those already reaching some sort of common accord, do not need UNESCO's intervention.

One can hope that not too much water will have flowed under the bridge before the Organization recovers one of its constitutional priorities: a solid programme in science and technology policy – one contributing to progress in the understanding of scientific phenomena, as well as to better governance over the scientific response to societal demand. In the best-case scenario, such an effort should help to ensure individual fulfilment and common prosperity in an interdependent and peaceful world.

BIBLIOGRAPHY

UNESCO. 1988. *SPINES Thesaurus: a controlled and structured vocabulary for information processing in the field of science and technology for development* (rev. edn). No. 50 in the Science Policy Studies and Documents series. Paris, UNESCO.

WHAT IS TO BE DONE?
A few conclusions
Jacques Richardson

IN the area of 'science and society' – with the exception of the eminently successful undertaking in science and technology policy (itself interrupted in the 1990s, to be resumed a decade later) – the investment in resources, the target definition, and the management of objectives have not been at the level of the other science programmes of the United Nation's main scientific agency. Consequently, both the thrust and the intensity of UNESCO's activities in 'science and society' have been limited in scope, inconsistent, and not always successful.

The Natural Sciences Sector has been supervised since its inception by natural scientists. These managers were often specialists so preoccupied by the need to succeed in the areas of advanced education and research – or in those of UNESCO's priorities of application (largely of an environmental nature) – that

Captain Jacques-Yves Cousteau at UNESCO Headquarters, Paris, 1993. When, more than forty-five years ago, Cousteau and his teams embarked aboard *Calypso* to explore the world, the general public did not yet know about the effects of pollution, overexploitation of resources, and coastal development. The films of *Calypso's* adventures drew the public's attention to the potentially disastrous environmental consequences of human negligence.

they gave little recognition to the great variety of societal implications of scientific progress. The statistics of science, for example, often proved deceptive as to the true strength and merit of the efforts of industrializing countries, and thus failed to seize an intensity of interest on the part of senior managers at Headquarters or in the field.

The Natural Sciences Sector's managers also considered as nuisances and distractions the efforts to publicize the danger points (especially the menaces or threats) represented by scientific and technological advance (nuclear energy and biochemical warfare being vivid examples). Another obstacle concerned the means of getting the word out: an audiovisual production or a print publication requires an editorial and production staff, a means of distribution, and a capacity to evaluate results leading to self-correction. With the exception of one or two training manuals, however, UNESCO has never excelled at the distribution of its publications, and its competence in evaluation has often proved limited or annoyingly self-serving.

As a consequence of all these factors, the area of ethics came late to UNESCO's science programmes, and, likewise, concern for the participation of women in science and engineering is still struggling to attain its deserved legitimacy. Scientific research as a 'feeder' of the processes of industry – which should have long ago been one of the Organization's *raisons d'être* in the world of science – remains a goal undefined and undirected on the far confines of its undertaking.

All in all, more could have been attempted during the past sixty years in terms of both quantity and quality. It is therefore in this area of 'science and society' that the UN's sole 'cultural organization' might, and should, try much harder in the future.

PART V
OVERVIEWS AND ANALYSES

PART V: OVERVIEWS AND ANALYSES

IMPRESSIONS: FORMER ASSISTANT DIRECTORS-GENERAL FOR NATURAL SCIENCES

A tremendous privilege
Abdul-Razzak Kaddoura[1]

IT was a great honour and privilege for me to serve the science programmes of UNESCO. I feel a deep emotion as I look back to the years 1976–88, when I was Assistant Director-General for Natural Sciences.

Although it may sound presumptuous for someone associated with the Organization to say so, it is nonetheless true, and right, to state that UNESCO is a great idea and a good embodiment of it. It is a noble endeavour to root its existence in the sacred cause of peace, and to proclaim that knowledge's acquisition and transmission are the best defence of this cause.

The finalization and publication of this volume in 2005 and 2006 is well timed. This period is not only the sixtieth anniversary of the foundation of UNESCO; it has other claims to fame as well. For one thing, 2005 is the World Year of Physics, which was officially launched at a conference held at UNESCO Headquarters from 13 to 15 January 2005. About 1,000 participants, including 500 students from over 70 countries attended the conference. Eight Nobel laureates addressed it. It was a good illustration of the symbiosis between the world of science and UNESCO. The World Year of Physics is a celebration of the 100th anniversary of Einstein's *annus mirabilis* of 1905, when (at the age of 26) he published, in *Annalen der Physik*, his seminal papers on relativity, radiation, molecular dimensions, and Brownian motion. These papers ushered in the new era of physics that has profoundly affected science, technology, and human life throughout the twentieth century.

It is gratifying that this celebration of Einstein's achievements should start at UNESCO. Let us also recall, with pride, that UNESCO commemorated, in 1979, the centenary of Einstein's birth with a series of events at the Organization's Headquarters in Paris, at Einstein's birthplace (Ulm, Germany) and in Switzerland, where he was residing in 1905. The official meeting at UNESCO House was attended by many scientists, including several Nobel Prize winners,

---

1 Abdul-Razzak Kaddoura served as Assistant Director-General for Natural Sciences at UNESCO from April 1976 to November 1988. See Annex 3: 'Heads of Natural Sciences at UNESCO'.

among them the legendary physicists Paul Dirac and Abdus Salam, who gave keynote speeches. Professor Dirac, who was then in his late seventies, spoke beautifully for about thirty minutes without the help of a single note. A recording of his talk became a much sought-after item. These events are an instance of the determination of UNESCO to exist in symbiosis with the international scientific community, and to seek the advice, help and criticism of active scientists in the elaboration, implementation, evaluation and termination of its programmes. UNESCO's association with the international scientific bodies, the national academies of science, and the International Council for Science (ICSU) was only the most conspicuous aspect of a multifaceted and constant twinning. I pay a heartfelt tribute to the thousands of scientists, engineers and technicians who have selflessly contributed to UNESCO's scientific programmes. Their incentive was their attachment to scientific cooperation and their belief in the ideals of UNESCO.

The year 2005 was also exceptional, tragically and gloriously, as the year following the Asian tsunami and the wonderful human solidarity it aroused. It illustrated man's vulnerability to the forces of nature, but also his nobility to his fellow man. UNESCO is, of course, at the heart of this predicament and opportunity. The risk presented by earthquakes and their aftermath calls for international scientific and technical cooperation to develop a better understanding of the phenomena involved, and to seek better methods of protection and mitigation. Earthquakes are, of course, but one of several natural disasters that can befall man with various magnitudes and probabilities. Many, if not most, of these threats are global in character, and can be addressed only by a collective human endeavour. Environmental degradation, water scarcity or floods, and climate change, are the most conspicuous and likely in the near future. But there are also other, less likely, but more potentially devastating dangers lurking in the future. To name many of these would be tedious and dispiriting, but two stand out because of what would be their horrible consequences: the use of nuclear weapons, and the impact of a very large meteorite. The dangers that humanity faces at present led the (UK) Astronomer Royal, Sir Martin Rees, to state, in a recent book, that the odds of mankind surviving the present century are about 50/50. This bleak picture can be improved only through concerted action of the international community, based on better knowledge of the risks and the ways of mitigating these.

It is in this context that I recall one of my regrets about the scientific programmes of UNESCO. This concerns earthquakes. UNESCO has had a good programme in this field. Within its framework, an international conference had been organized, in the late 1970s, after the terrible earthquake in China, which caused hundreds of thousands of deaths and untold damage. So the Organization was well equipped, had additional support become available, to respond favourably to the proposal of Frank Press, who had served as the president of the US National Academy of Sciences, as well as scientific advisor to the US president. As an eminent geophysicist, Press had recognized the importance of establishing a worldwide network of seismometers to obtain a continuous tomography of the Earth. Many countries, particularly developed ones, have of course installed such instruments, but the coverage is far from being uniform throughout the different regions. An international programme of this kind would have had great value, not only (from a scientific point of view) in the study of the Earth's interior, but also in the mitigation of the effects of earthquakes. Unfortunately, it was not created, mainly because of concerns that it could also be used for matters affecting national security.

UNESCO can be proud of the fact that its marine science programmes – implemented both by UNESCO directly, and through its Intergovernmental Oceanographic Commission (IOC) – have been instrumental in launching studies and cooperation on the oceans, whose influence on the climate is paramount. The IOC was so highly appreciated by Member States that when the United States withdrew from UNESCO, in the 1980s, they chose to remain a member of the IOC. I was always opposed to the attempt by some to sever the links of the IOC with UNESCO, believing their association to be beneficial to both.

The year 2005 was also one in which Bill Gates, Microsoft's chairman, announced new initiatives in his support for science and technology (but, of course, he has now been doing this for a number of years). I recall this because one of my deep regrets is that the informatics programmes at UNESCO were not more important. UNESCO had recognized, very early on, the impact of computers, which is why it established the Intergovernmental Bureau for Informatics (IBI), as early as the period that saw the Organization found CERN (the European Organization for Nuclear Research). Our great misfortune then was the absence of a Bill Gates, who could have given a great impetus if we had known how to approach him. Finally, it was in 2005 that the Kyoto agreement came into effect. This major international response to a great risk – inherently international and potentially ominous – was shepherded by the United Nations, and rightly so, because of its high political content.

I wish to end these reminiscences by paying tribute to the three directors-general whom I got to know personally. Mr René Maheu, a great intellectual and an admirable international civil servant, established the Science Sector in the form I found when I joined the Organization. Although I did not serve under him, I was strongly attached to him, and he chose me, in various capacities, for important international assignments. Mr A. M. M'Bow appointed me as Assistant Director-General for Natural Sciences. I constantly treasure his leadership and friendship. A man of great integrity, tireless activity and profound wisdom, he led the Organization during turbulent times and always highly valued and strongly supported its scientific programmes. I served twice under Federico Mayor, the first time when he was Deputy Director-General, and the second when he headed the Organization. He was the first scientist to become Director-General since the founding father, Sir Julian Huxley. His outstanding qualities, as a specialist and as a man, made him a worthy successor of the great leaders who formed the galaxy of directors-general of UNESCO.

I should also like to pay a warm tribute to the scientists who preceded me, or succeeded me, as heads of the science programmes of UNESCO. I have been privileged and honoured to know all of them personally, except the current Assistant Director-General. I think all would agree that we owe a special debt of gratitude to the first among us, the man who started the science programmes. Professor Joseph Needham, whom we invited to a meeting of the Natural Sciences Sector of UNESCO, has become a part of the history of science by his monumental treatise on the history of Chinese science.

Last, but not least, I express my deep gratitude to all my former colleagues in the Natural Sciences Sector, and elsewhere in the Organization, for their kindness, as well as my admiration for their competence, industry and honesty. I feel great joy whenever I hear good news about UNESCO and its science programmes – the return of the United States, a synchrotron radiation source in the Middle East, etc.– and recall some of the colleagues, leading these new ventures, who were then junior, brilliant and full of promise

Great things
Adnan Badran[2]

Adnan Badran sent the following message in 2005 to show his support for the Sixty Years of Science at UNESCO 1945–2005 *project and briefly evoke a few memories.*

GREAT things were done between 1990 and 1998 in UNESCO's Natural Sciences Sector in basic, applied sciences and environmental sciences. It was the era of capacity-building in science, the era of UNESCO's biotechnology and genetic engineering programmes, the era of science ethics, the era of consolidation and coordination of the four great pillars of the Sector's environmental programmes – the Intergovernmental Oceanographic Commission (IOC), the Man and the Biosphere Programme (MAB), the International Hydrological

---

2 Adnan Badran served as Assistant Director-General for Natural Sciences at UNESCO from 1990 to 1993, a post he occupied thereafter by interim until 1996, while serving as UNESCO's Deputy Director-General (1993–98). See Annex 3: 'Heads of Natural Sciences at UNESCO'.

Programme (IHP) and the International Geological Correlation Programme (IGCP) – and the new coastal zones project. It was the era of the launching of the *World Science Report* and of affiliating the Academy of Sciences for the Developing World (TWAS) to UNESCO, the strengthening of cooperation with scientific NGOs such as the International Council for Science (ICSU), the World Conservation Union (IUCN) and TWAS. It was also the era of cooperation with the International Atomic Energy Agency (IAEA), the World Bank, the Global Environment Facility (GEF), and the United Nations Industrial Development Organization (UNIDO). Finally, it was the era of strengthening the Field Science Offices and of decentralization and regional programmes.

22 February 2005

In the service of Member States

Maurizio Iaccarino[3]

WHEN I decided to apply for the post of Assistant Director-General for Natural Sciences, I knew very little about UNESCO and the UN system. I thought that the job was to coordinate and promote scientific activities at a global level and was unaware of the necessity to work to implement the decisions of the Member States. As a consequence, when applying for the job, I mentioned several Nobel laureates as a source of reference letters and did not actively seek political support. Although with some difficulties, I obtained the job.

After joining UNESCO, I soon began to realize the extent of my ignorance of the system. I was much helped by the Director-General, Federico Mayor, who held long meetings of the General Directorate and briefed us in a detailed way on the issues at stake, on what was needed, and why it was important. On at least a couple of occasions, I raised my eyebrows to show my perplexity and he started explaining again what he wished to obtain. Very early after my arrival he asked me to write to all Science Academies of the world, introduce myself as a newly arrived Assistant Director-General and use this opportunity to describe UNESCO and its programmes. He also asked me to visit 'all the countries of Latin America' (I visited only seven). I think these initiatives were useful for UNESCO, but they also showed me what was needed.

I soon learned that our programmes should be at the service of the Member States. I did not know enough, and therefore I chose to consult my collaborators

---

3   Maurizio Iaccarino served as Assistant Director-General for Natural Sciences at UNESCO from May 1996 to February 2000. See Annex 3: 'Heads of Natural Sciences at UNESCO'.

on many issues, facts and even details. They were knowledgeable and helpful, and I learned a lot. I do believe, however, that in some cases my ignorance and my unorthodox views were useful because they showed the possibility of analysing issues from a different perspective. But I also realized that the UN bureaucracy was not always conducive to achieving results in an efficient way. Because of this, I tried to separate the issues: first describe what should be achieved and only later discuss how to obtain it. When my collaborators pointed out to me the difficulties, my answer was that bureaucracy should be at the service of the programmes, and not vice versa. Quite often I failed, but I still think that it was useful to define why we wanted to achieve something and how it should be done.

During my stay, several activities were intended to support small scientific initiatives, generally very valid, but more suitable for a granting agency than for an intergovernmental organization. In most cases, I already knew and appreciated the scientists involved, or I met them and realized that they were proposing very good programmes. These initiatives were useful for UNESCO because they promoted contacts and collaboration with the best scientific experts in the world. However, it was often difficult to justify them with the Member States and to find the financial resources during a period of decreasing budgets. At the same time, the amount of money reserved for these initiatives was very small as compared to the requests, and it was very difficult to justify refusals, since a coherent plan of action was missing. During my stay, it became necessary to cut this type of activities due to lack of resources and this caused deep frustrations in my collaborators. Since I was new to the system I was able to cope with this problem with less anxiety.

At a different level, I found another problem. UNESCO had been 'created for the purpose of advancing, through the educational, scientific and cultural relations of the peoples of the world, the objectives of international peace and of the common welfare of mankind'.[4] However, the concept of Science has substantially changed in these sixty years. Applied science has become more sophisticated, and the activities of several UN agencies deal much more with science than in the past. As a consequence, it frequently happened during my stay that, when I proposed to discuss the role of UNESCO as related to a specific topic, the delegates objected that I was entering the field of another UN agency, such as the World Health Organization (WHO), the Food and Agriculture Organization (FAO), the United Nations Environment Programme (UNEP), and so on. Unfortunately, it is rarely appreciated that a true scientific culture is at the roots of basic science, which is needed to generate technological applications.

When thinking of the major activities organized in UNESCO during the period of my stay, I consider myself lucky because I was involved in several

---

4   The Constitution of UNESCO, 16 November 1945.

activities of global or regional interest: among them, the World Solar Summit (Harare, Zimbabwe, September 1996), the Pan-African Conference on Sustainable Integrated Coastal Management (Maputo, Mozambique, July 1998) and the World Conference on Science (Budapest, Hungary, June 1999). These three events showed me how difficult it is to organize collaboration within the UN system: although UNESCO proposed collaboration, the response was often disappointing, sometimes very poor. Apparently the UN agencies wish to work on a programme only when it is assigned only to them. Within UNESCO, I found difficulties in organizing collaboration with other Sectors: while it was easy to discuss at the Assistant Director-General level, it was difficult for the Programme Officers to work without reporting to their own Assistant Director-General.

I found the organization of the World Solar Summit well advanced, but there still was interesting work to do. I was happy to see that the programme included components relating to education, science, technology, environment, social sciences and culture, and that it was necessary to cooperate with other UN agencies, such as UNIDO and UNEP. The governmental delegations were very interested in the results of the Summit, and this served to demonstrate to me that the UN system works at its best when it indicates the needs of the civil society to the political counterparts at a global level. I was delighted to meet some heads of state (I had a long conversation with the then-president of Pakistan, Mr Leghari, a physicist, with whom I discussed soil microbiology).

My contribution to the organization of the Pan-African Conference on Sustainable Integrated Coastal Management was crucial at the very early stages of the planning of that conference. When my collaborators asked me to approve the organization of a scientific meeting on the subject, I refused and expressed the opinion that UNESCO should only be involved in something more ambitious and with the involvement of the Member States. Eventually, all Sectors of UNESCO and the Intergovernmental Oceanographic Commission took part in the organization of the Conference in close collaboration with UNEP. All African ministers of environment participated and therefore the Conference had not only an intellectual input (the experts identified by UNESCO) but also an endorsement of the documents by the African Member States. Because of this conference, more attention was given to the intergovernmental dimensions of coastal management, and this led to the organization of another conference to discuss the problems of the Mediterranean Sea (Genoa, Italy, May 1998).

The World Conference on Science was organized to discuss several issues, as for example: the relationship between basic and applied science; the role of science for economic development in industrialized and in developing countries; science education; and science and ethics. There was a need to discuss these topics at a global level, in the context of different cultures. They dealt with the social aspects of natural sciences and therefore required an interaction of natural and social scientists. But a very important aspect of these discussions was to listen to the ideas of the politicians and discuss with them what was needed. It is therefore clear why UNESCO was the most appropriate place to hold these discussions. The Sectors of Education, Natural and Social Sciences and Culture had competent Programme Officers who knew the best world-experts in their fields. At the same time, many delegations of the Member States to UNESCO had competent people with whom to discuss the issues at stake.

When the Secretariat of UNESCO proposed to organize this conference, the Executive Board decided that it should be organized in full partnership with the International Council for Science (ICSU). Through this partnership it was possible to have the official opinion of the International Scientific Unions belonging to ICSU. The participation of ICSU in the process was essential and very productive, although many ICSU scientists conceived of the Conference as a purely scientific event and did not appreciate that the aim was to propose a document to be approved by politicians.

The discussions were initiated in UNESCO through the International Scientific Advisory Board, whose members were prestigious (including eight Nobel laureates) and had a role in the management of science (including ministers and ex-ministers). The ideas discussed in this forum led to a provisional document that was later discussed and modified in a smaller body. This document, combined with the reports issued by many preparatory meetings organized all over the world with the participation of scientists and politicians, made it possible to prepare a draft declaration. This was sent to ICSU for consultation and amended accordingly. In order to have the view of the politicians, the declaration was then sent to the national delegations and to the UNESCO National Commissions of the Member States. In this way, the final draft brought to the Conference was prepared after full consultation with scientists (through ICSU) and politicians.

The Conference saw the participation of delegates from 155 countries (including almost eighty ministers of science), and the documents were approved after thorough discussions, which led to many new amendments. The Member States approved the documents in the following General Conference of UNESCO. In doing so, they committed themselves to implementing different activities, listed in a table attached to the documents. I consider that one important conclusion of the World Conference on Science is that developing countries should be helped through the transfer of knowledge. In this respect, I was lucky to have the International Centre of Theoretical Physics and the Third World Academy of Sciences under my authority. And – in my discussions with delegations of the Member States – I cited them as an example to follow.

The discussions at the Conference and the conclusions contained in the documents have influenced many activities throughout the world. The

Secretariat of UNESCO was charged to ask the Member States to report on the implementation of their commitments and to make a compilation to be presented to the General Conference. This report is intended to show if Member States have implemented their commitments or if they wish to change them. At the same time it will be an opportunity for Member Sates to discuss how best to use science for the benefit of humankind and what can be done at the global level.

I consider myself quite fortunate to have had the chance of leading the Natural Sciences Sector of UNESCO. Although I worked with passion and enthusiasm, I know that my contribution to its activities was small. I inherited an efficient organization from my predecessor and I am glad to see that new impetus is being poured into the Sector.

UNESCO AND ICSU: SIXTY YEARS OF COOPERATION[5]

INTRODUCTION

THE International Council for Science (ICSU, formerly International Council of Scientific Unions) has been engaged since 1931 in promoting international scientific cooperation for the benefit of society. Since the establishment of UNESCO in 1945, it has been the Organization's major international non-governmental partner in the field of science. The first indication of the potential of such cooperation was the instruction of the UNESCO Preparatory Commission to its Executive Committee 'to consult with ICSU on methods of collaboration to strengthen the programmes of both bodies in the areas of their common concern'. Two more recent indications have been the World Conference on Science (WCS), organized jointly, and the explicit recognition by the 2003 General Conference of UNESCO that ICSU is a key partner in many of the activities not only of the Natural Sciences Sector, but also the Education, Social and Human Sciences and Communication and Information Sectors. This partnership has evolved over the last sixty years, and there are presently many opportunities to develop additional synergetic joint action, based on the strategic priorities of both organizations, as well as the outcome of the WCS.

In 1986, ICSU, under the pen of its Executive Secretary, F. G. W. Baker, published a booklet entitled *ICSU-UNESCO: Forty Years of Cooperation*, in which the reader may find comprehensive information on the evolution of UNESCO-ICSU cooperation during the period 1946–1986 (Baker, 1986). Hence, while this chapter touches upon a few highlights of cooperation between the two organizations in the forty years following the Second World War, it will focus on new avenues of cooperation during the last two decades.

The special symbiotic link between ICSU and UNESCO was very well described by a former president of ICSU, Harrison Brown:

---

5 Compiled from material provided by previous and current members of the ICSU Secretariat.

The evolving working relationships between intergovernmental organizations, such as UNESCO and WMO, and a responsible non-governmental international organization, such as ICSU, could well turn out to be a development of critical importance in the scheme of international organizations. The symbiotic relationships which have developed thus far between ICSU on the one hand and members of the United Nations family on the other may result in one of the more important technical-social-political interventions of our time (Brown, 1976).

Thirty years later, these symbiotic relationships continue to develop, both with old partners, notably UNESCO, and new ones.

INTERNATIONAL RESEARCH PROGRAMMES

As ICSU and its International Scientific Unions began to recover from the effects of the Second World War, closer contacts with UNESCO developed for a series of programmes, among them ICSU's International Biological Programme (IBP, 1964–74), which was conceived as a global study of the biological basis of productivity and human adaptability, but which also included studies of conservation and the use and management of natural resources. UNESCO played an important part by diffusing information (especially to governments and government scientists), by making space available in UNESCO House for meetings, and by providing contracts for specific parts of the Programme. A former UNESCO Assistant Director-General for Science called IBP one of the parents of the Man and the Biosphere Programme (MAB); and the scientific director of IBP wrote: 'One obvious cause of the success of some IBP projects was the long-lasting interest shown and the continuous support afforded by UNESCO at the international level' (Worthington, 1975). Most of the fifty-five specialists who put forward proposals for the initial MAB programme in 1969 were leading scientists in IBP and later became key players in the programme's evolution. In some fields of study, the transition was so smooth that it is difficult to see where IBP stopped and MAB began. It is interesting to note that one result of the work of the IBP Section on the Conservation of Terrestrial Ecosystems and Communities was the creation (in 1965) of the Azraq Desert National Park in Jordan, a reserve similar in concept to that later used in the creation of other national natural reserves and of the MAB Biosphere Reserves. In September 1981, UNESCO organized with ICSU the 'Ecology in Practice: Establishing a Scientific Basis for Land Management' Conference, to take stock of the results achieved in ten years of research within MAB.

Only a few years after the launching of MAB, two more international scientific cooperation programmes were launched and hosted by UNESCO, with ICSU and its relevant International Scientific Unions playing significant roles therein: the International Hydrological Programme (IHP), and the International Geological Correlation Programme (IGCP). The IHP had been preceded by the International Hydrological Decade (IHD), the idea for which developed from a proposal put forward at the General Assembly of the International Union of Geodesy and Geophysics (IUGG) of ICSU, in 1960, for an international focus for hydrology to help resolve the problems of how to provide access to water for the growing global population, agriculture and industry. The idea was expanded at a meeting of the IUGG International Association of Scientific Hydrology (IAHS) in Athens (Greece) in 1961. A Committee of Experts organized by UNESCO in 1962 took up the proposal and prepared a work plan and programme that was submitted to and adopted by the General Conference of UNESCO, at its twelfth session. Its importance was recognized in 1974 by the decision to transform the Decade into a Programme, which continues today and in which the IAHS continues to play an active role.

In addition to its inputs into UNESCO's International Hydrological Programme, ICSU has co-sponsored with UNESCO a number of International Conferences on Water, such as those on 'Water and the Environment' (in 1992), 'Hydrology towards the Twenty-First Century: Research and Operational Needs' (1993), and 'Water and Sustainable Development' (1998), and provided inputs into the International Year of Freshwater (2003). In 1994, UNESCO, ICSU and the International Union of Technical Associations and Organizations (UATI) published a book – prepared by the ICSU Committee on Water Research (COWAR) – providing a research agenda for sustainable development of water resources, enumerating the basic conditions that must be satisfied for sustainable water systems.

The International Geological Correlation Programme developed as the result of a joint effort of UNESCO and ICSU's International Union of Geological Sciences (IUGS), following discussions during 1966–69. The scope and objectives were outlined at a UNESCO-sponsored meeting in Budapest (Hungary) in September 1969. Since 1974, the IGCP has worked as a joint programme of UNESCO and the IUGS; it was renamed the International Geoscience Programme in 2003. Several independent evaluations have proven the high scientific standard of IGCP and its great success. Due to the evolution of the programme towards making it more policy relevant, and in addition to the name change, it now has the subtitle 'Geology in the Service of Society'. UNESCO and ICSU have also started to collaborate on the International Year of Planet Earth (2008), building on the successes of IGCP.

In 1979, ICSU co-sponsored the first World Climate Conference, which led to the establishment in 1980 of the World Climate Research Programme

(WCRP) by ICSU, the World Meteorological Organization (WMO), and the Intergovernmental Oceanographic Commission (IOC) of UNESCO. The objective of this ongoing programme is to understand and provide the basis for prediction of the Earth's climate system. It is no exaggeration to assert that the WCRP has been the main mechanism by which the scientific community provides to societies and policy-makers the scientific facts and advice on the crucial issue of climate change. ICSU and UNESCO were also sponsors of the Second World Climate Conference in 1990.

In 1986, ICSU, on its own, created the International Geosphere Biosphere Programme (IGBP). IGBP's objective is to describe and understand the interactive physical, chemical and biological processes that regulate the total Earth system – the unique environment it provides for life, the changes occurring in this system, and how these are influenced by human actions. While UNESCO did not become a formal sponsor of IGBP, numerous cooperative activities between UNESCO (notably MAB, IHP and the IOC) and IGBP have been implemented over the years. The results of IGBP will be essential for addressing several contemporary environmental problems, including those related to freshwater and specific aspects of climate change.

Last but not least was the launching, in 1991, of DIVERSITAS, an international research programme on biodiversity. During its first phase, 1991–98, DIVERSITAS was jointly sponsored by UNESCO-MAB, the International Union of Biological Sciences (IUBS), and ICSU's Scientific Committee on the Problems of the Environment (SCOPE). Following a transition period in 1999–2000, ICSU and the International Union of Microbiological Societies (IUMS) joined the original sponsors with the aim of developing DIVERSITAS as part of the group of global environmental change programmes. To assist in the development of the first phase of the programme, an international forum – 'Biodiversity, Science and Development: towards a new partnership' – was held in September 1994 at UNESCO Headquarters. Since 2002, with ICSU becoming a direct sponsor, the programme has focused on promoting integrative biodiversity science, linking ecological and social science disciplines in an effort to produce new, socially relevant knowledge. Advice has been provided to the Convention on Biological Diversity (CBD), and DIVERSITAS has been invited by the CBD to assess the articles of the Convention from a scientific point of view. An International Biodiversity Observation Year (IBOY) was organized for 2001–2002 to bring together global data sets on the status and trends of biodiversity.

Another recent example of UNESCO and ICSU cooperation in the field of environment is their involvement in the international Millennium Ecosystem Assessment, launched jointly by several UN organizations (including UNESCO and relevant global environmental conventions) and non-governmental organizations (NGOs, notably ICSU). Since 2001, the Assessment has produced new insights into the changes underway in ecosystems and their services. The Assessment's reports have specifically addressed the needs of decision-makers and the public for scientific information concerning the consequences of ecosystem change for human well-being, and the options for responding to such changes. As a follow-up to the Millennium Ecosystem Assessment, ICSU and UNESCO are currently designing a research agenda on the basis of the gaps in knowledge identified in the Assessment.

GLOBAL OBSERVING SYSTEMS (GOS)

ICSU and UNESCO have jointly played a major role in the development of the three Global Observing Systems (GOS): the Global Climate Observing System (GCOS), the Global Ocean Observing System (GOOS) and the Global Terrestrial Observing System (GTOS). In each case, other partners, all intergovernmental organizations, have joined UNESCO and ICSU in guiding and supporting the development of the observing systems. GOOS was the first global environmental observing system to be established (in 1991), with sponsorship by the IOC (UNESCO), WMO, the United Nations Environment Programme (UNEP) and ICSU. The objective of GOOS is to enable the state of the ocean to be described, its changing conditions to be forecast, and its effects on climate change to be predicted. In 1992, the second such system, GCOS, was established, as a joint initiative of ICSU, WMO, the IOC of UNESCO, and UNEP. The objectives of the Global Climate Observing System are to provide the observational basis needed to observe, document and understand the present state of the world's climate and, as much as possible, to develop a predictive capacity for its evolution. In 1996, the third system, the Global Terrestrial Observing System (GTOS), was created. It is co-sponsored by the Food and Agriculture Organization of the United Nations (FAO), UNESCO, UNEP, WMO and ICSU to provide policy-makers, resource managers and researchers with access to the data and information they need to detect, quantify, locate and warn of changes (especially reductions) in the capacity of terrestrial ecosystems to support sustainable development. As of 1998, an Integrated Global Observing Strategy (IGOS) has brought together all the agencies of the UN system involved in the GOS, along with ICSU, the WCRP, IGBP, the Committee on Earth Observation Satellites (CEOS) of the group of space agencies, and the three Global Observing Systems. In 2003, at the first Earth Observation Summit, it was agreed to develop and implement a Global Earth Observation System of Systems (GEOSS). UNESCO and ICSU, together with the other partners in IGOS, have supported the development of GEOSS, which was formally approved by the Third Observation Summit in 2005.

PART V: OVERVIEWS AND ANALYSES

WORLD CONFERENCES ON THE ENVIRONMENT

ICSU-UNESCO cooperation has also been important in the context of the United Nations Conferences on the Human Environment (Stockholm, Sweden, 1972) and on Environment and Development (Rio de Janeiro, Brazil, 1992), and their respective follow-ups. At the World Summit on Sustainable Development (WSSD, in Johannesburg, South Africa, 2005), ICSU and UNESCO were co-organizers of several sessions of the WSSD Forum on Science, Technology and Innovation for Sustainable Development, with the overall organization provided by the Government of South Africa. UNESCO-ICSU cooperation has been instrumental in giving science and technology its proper role in the outcomes of these world conferences.

UNESCO-ICSU WORLD CONFERENCE ON SCIENCE

The idea for a World Conference on Science was put forward at the first meeting of the UNESCO International Science Advisory Board in 1997. The Conference – with the subtitle 'Science for the Twenty-First Century: A New Commitment' – was co-organized by UNESCO and ICSU in Budapest in 1999. It brought together over 1,800 delegates from 155 countries, twenty-eight intergovernmental organizations and some sixty NGOs. There were three forums: (1) Science: Achievements, Shortcomings and Challenges; (2) Science and Society; and (3) Towards a New Commitment. At the final session of the Conference, the participants unanimously adopted a Declaration and the Science Agenda: these two documents were subsequently endorsed by the ICSU General Assembly, the UNESCO General Conference, and the UN General Assembly. Since 1999, ICSU-UNESCO cooperation has been based on the achievements of the World Conference on Science.

EDUCATION, TRAINING AND CAPACITY-BUILDING

ICSU's collaboration with UNESCO in the field of Science Education began with the science teaching programmes of the individual Scientific Unions but was later brought into focus by the creation of ICSU's Committee on the Teaching of Science (CTS), which organized in cooperation with UNESCO a number of conferences on science and technology education, such as that on 'Science and Technology Education and Human Needs', held in Bangalore, India, in 1985.

In 1987, ICSU and the Academy of Sciences for the Developing World (TWAS formerly the Third World Academy of Sciences) started a programme (joined by UNESCO in 1989) of lectureships in science, to allow for repeated visits by international experts to developing countries. As a follow-up to the UN

Conference on Environment and Development, the scope of the programme was extended to include sustainable development. The UNESCO-ICSU-TWAS Short-Term Fellowship Programme in the basic sciences, which began in 1991, promotes international cooperation by enabling scientists, especially those from developing countries and central and eastern Europe, to carry out short-term studies in well-established scientific centres, and to learn and use techniques not accessible to them in their own country.

ICSU, TWAS and UNESCO (later joined by the United Nations University, UNU) created a Visiting Scientist programme (to replace the previous two), which gives institutions and research groups in the South, especially those with limited outside contacts, the opportunity to establish long-term links with world leaders in science, so as to help to build capacity in the countries of the visiting scientists. A roster of international specialists has been created which facilitates such links. In October 1998, UNESCO organized the World Conference on Higher Education in the Twenty-First Century, in which ICSU cooperated by organizing a 'Round Table on Higher Education and Research: Challenges and Opportunities'.

ICSU created a Committee on Science in Central and Eastern Europe (COMSCEE) in order to help bring assistance to these regions. With the support of UNESCO, it organized two high-level meetings on the 'Role of Science in Rebuilding Central and Eastern Europe', drawing attention to the dangers of lessened support for science in these regions. The Committee on Science and Technology for Developing Countries (COSTED) – which began in 1966 – cooperated closely with UNESCO in the field of capacity-building in science and technology in developing countries. In 1993, COSTED joined forces with the International Biosciences Networks (IBN) of ICSU and UNESCO, in order to support capacity-building in developing countries also in promising new scientific areas, such as biotechnology. The Committee was dissolved in 2002, and ICSU is now establishing four Regional Offices for Africa, the Arab region, Asia and the Pacific, and Latin America and the Caribbean. The ICSU Regional Office for Africa (based in Pretoria, South Africa) started its work in early 2005. These Regional Offices will work closely with the Regional Bureaux of UNESCO.

In September 2002, eleven global educational, scientific and technical organizations – including UNESCO, the UNU and ICSU – meeting on the occasion of the World Summit on Sustainable Development in the Ubuntu area of Johannesburg, prepared the Ubuntu Declaration, identifying four main goals: curriculum development; North–South networking; strategic educational planning and policy-making; and capacity-building in scientific research and learning, emphasizing that education is essential for reaching sustainable development goals. Shortly thereafter, the signatories to the Ubuntu Declaration created an alliance to promote its implementation throughout the world. One of the recommendations of the World Summit in Johannesburg was to launch a UN Decade on Education for Sustainable Development (2005–2014), with UNESCO as its lead international organization. The Decade will provide an important mechanism for ICSU-UNESCO cooperation in the area of education, training and capacity-building. At the same time, all partners in the Ubuntu Alliance will work together in the context of the Decade.

SCIENTIFIC DATA AND INFORMATION

In 1967, ICSU and UNESCO jointly established a Central Committee to study the feasibility of a world scientific and technical information system (UNISIST, 1968–70). The results of this study were submitted to an Intergovernmental Conference in 1971, accepted and later adopted as part of UNESCO's programme. In its follow-up, ICSU invited a number of NGOs in science documentation and information to a meeting in 1972 to discuss the role of such organizations in the implementation of UNISIST. A second UNISIST Conference was held in 1979, after UNESCO had decided to include UNISIST into its General Programme of Information (PGI). Two other Science Service Systems regarding the collection and preservation of the results of scientific research were established by ICSU with the administrative and financial support of UNESCO: the Federation of Astronomical and Geophysical Services (FAGS), established in 1956; and, within the programme of the International Geophysical Year (1957–1958), a series of World Data Centres to collect and store the geophysical data obtained. Both of these systems continue with the support of UNESCO and, when appropriate, of its IOC and other organizations.

In 1992, ICSU established – in cooperation with UNESCO and TWAS – an International Network for the Availability of Scientific Publications (INASP) to enhance worldwide access to scientific information, and to improve its flow within and between countries, especially those with less-developed systems of publication and dissemination. The Network plays an important role in collecting and making available scientific publications and information to the scientific communities in developing countries. In order to provide information and advice concerning new methods of publishing, a First Conference on Electronic Publishing in Science was organized jointly by UNESCO and ICSU in February 1996. A second Conference was held in February 2001, which had a broader coverage and a greater emphasis on biology and medicine.

In March 2003, ICSU, in partnership with UNESCO and the ICSU Committee on Data for Science and Technology (CODATA), organized a workshop at UNESCO on Science and the Information Society that produced an Agenda for Action. This was adopted by ICSU and its main partners – UNESCO, the European Organization for Nuclear Research (CERN) and TWAS – and was used in negotiating the final documents that were later agreed by the World Summit on the Information Society (WSIS) held in Geneva (Switzerland) in 2003. The majority of the specific actions identified by the international scientific community have now been endorsed by governments, with follow-up discussions held at the second World Summit event, held in Tunis (Tunisia) in 2005.

FREE CIRCULATION AND COLLABORATION OF SCIENTISTS, AND ETHICS

The problem of 'how to assist the free flow of scientists coming and going across national boundaries' was one of the topics raised by the UNESCO Preparatory Commission. In 1963, the ICSU General Assembly created a Standing Committee on the Free Circulation of Scientists to find solutions to problems relating to the free passage of scientists and free collaboration among scientists. The problems were varied: some concerned scientists who disapproved of their government's policies; others the refusal to grant individuals exit or entry visas (often linked to the Cold War); still others, the non-recognition by a nation of the territory in which a scientist worked. There has been an exchange of information and assistance between ICSU and UNESCO on these issues for many years, which has often resolved such problems.

UNESCO's World Commission on the Ethics of Scientific Knowledge and Technology (COMEST) and the former ICSU Standing Committee on Responsibility and Ethics in Science (SCRES) have interacted on issues and principles concerning ethics and the responsibilities of scientists, and particularly in encouraging young scientists to adopt such principles.

CONCLUSION

Overall, the sixty years of close cooperation between ICSU and UNESCO have contributed greatly to the advancement of science in general, and to international scientific cooperation for the benefit of society in particular, and have allowed both organizations to significantly enhance their institutional efficiency. ICSU continues to expand its cooperation with UNESCO, as attested to by the many statements at UNESCO's General Conferences and ICSU General Assemblies, as well as in a wide range of meetings and publications. ICSU-UNESCO cooperation is unique in many respects. It covers a broad range of scientific areas and a number of different modalities. The relationship has important intellectual, financial and (to some extent) political dimensions.

In 1945, when cooperation between UNESCO and ICSU began, ICSU consisted of seven International Scientific Unions, thirty-nine National Scientific

Organizations, five Joint Commissions and a Committee on Science and Social Relations. In 2005, ICSU had no fewer than thirty International Scientific Unions, 104 National Members and eighteen Interdisciplinary Bodies.

Today's societies are faced with increasingly complex issues and problems. Efforts to address these must be based more than ever on sound science. It is also becoming more and more important to capitalize on hybrid models of institutional cooperation, involving both governmental and non-governmental institutions, to create synergies and join forces. Strategic partnerships are needed, and collaboration between ICSU and UNESCO should be based on the strategic plans of both organizations as well as innovative visions for collaboration between them. Much more could and should be done.

BIBLIOGRAPHY

Baker, F. W. G. 1986. *Forty Years of Cooperation*. Paris, ICSU–UNESCO.

Brown, H. 1976. *ICSU: Organization and Activities*. Paris/Washington DC, ICSU.

Greenaway, F. 1996. Science International: A History of the International Council of Scientific Unions. Cambridge (UK), Cambridge University Press.

Worthington, E. B. (ed.). 1975. *The Evolution of IBP*. Cambridge (UK), Cambridge University Press.

BOX V. 4.1: MILESTONES IN UNESCO-ICSU COOPERATION

1946 ICSU becomes the very first non-governmental organization to sign an agreement with UNESCO.

1953 ICSU helps develop and participates in the International Advisory Committee on Scientific Research, set up to advise the UNESCO Director-General on UNESCO's scientific programmes.

1957–1958 UNESCO is a key partner in the International Geophysical Year, co-organized by ICSU and WMO.

1964 Launching of the International Biological Programme (IBP), ICSU's first international interdisciplinary research programme. UNESCO cooperates in IBP until its end in 1974. In 1971, UNESCO launches the Man and the Biosphere (MAB) Programme as a follow-up endeavour to IBP.

1964 ICSU establishes a Committee on Water Research (COWAR) to advise, in particular, the International Hydrological Decade launched by UNESCO in that year.

1966 Joint UNESCO-ICSU Feasibility Study of a World Science Information System, to become known as UNISIST.

1972 Launching of the International Geological Correlation Programme (IGCP) as a joint endeavour of UNESCO and the International Union of Geological Sciences, a union member of ICSU.

1979 The International Biosciences Networks (IBN) launched jointly by ICSU and UNESCO.

1981 The IOC of UNESCO joins ICSU and WMO in co-sponsoring the World Climate Research Programme, launched in 1980.

1991 Launching of the Global Ocean Observing System – jointly sponsored by the IOC of UNESCO, WMO, UNEP and ICSU – as the first of three global environmental observing systems; and, later, of the two others, with ICSU and UNESCO sponsorship, together with other partners.

1992 UNESCO and ICSU collaborate in the preparatory process of the UN Conference on Environment and Development (UNCED), and Agenda 21.

1992 The International Network for the Availability of Scientific Publications (INASP) established jointly by ICSU, UNESCO, and TWAS.

1996 ICSU and UNESCO sign a first Framework Agreement as a new mechanism for setting out the mutually agreed priority areas for working together; renewed in 2002.

1999 The 'World Conference on Science: Science for the Twenty-First Century – A New Commitment' is co-organized by UNESCO and ICSU.

2001 Launching of the second phase of DIVERSITAS, an international research programme on biodiversity, with ICSU joining UNESCO and ICSU international scientific union members as a co-sponsor.

2002 ICSU and UNESCO collaborate at the Forum on Science, Technology and Innovation for Sustainable Development, held under the aegis of the Government of South Africa in Johannesburg, in parallel with the World Summit on Sustainable Development.

2003 As part of the preparatory process of the World Summit on the Information Society, ICSU, UNESCO and the ICSU Committee on Data for Science and Technology (CODATA) develop an Agenda for Action on Science in the Information Society.

PARTNERSHIP IN SCIENCE
Cross-cutting issues in UNESCO's natural sciences programmes
Malcolm Hadley and Lotta Nuotio[1]

THE preceding chapters focus mainly on the substance of UNESCO's work relating to the natural sciences. In this chapter, some cross-cutting issues are addressed. An attempt is made to highlight trends and tendencies and to flag some topics for further consideration, with a view to contributing to the planning and development of future activities, within and outside UNESCO. An underlying thread is that the very nature of the Organization – with its broad scope, limited core-funding and multiple constituencies and ambiguities – has entrained many different types of partnership in the past. Identification and pursuit of new opportunities for partnership is a key ingredient for future development.

PROMOTING INTERNATIONAL SCIENTIFIC COOPERATION

When UNESCO was set up in 1946, the international scientific community was partitioned as a result of the Second World War. Probably never before had scientists in different countries been so relatively isolated.[2] For this reason, the conference convened in London in 1945 for establishing UNESCO incorporated in its Constitution the purpose of maintaining and increasing human knowledge by encouraging cooperation among the nations in all branches of intellectual activity, among which science was one of the most universal.

In these circumstances, the fledgling Natural Sciences Section had first to take action on international scientific relations, in particular on helping to restore the unity of the world scientific community (Box V.5.1) Its primary action was to reinforce the structure and intensify the work of ICSU (at the time, the International Council of Scientific Unions, now the International Council for Science). In the same spirit, UNESCO helped set up the Council for International Organizations of Medical Sciences (jointly with the World Health Organization [WHO]) and the Union of International Engineering

---

1 Malcolm Hadley: Member of staff of UNESCO's Division of Ecological Sciences from the early 1970s to 2001.
Lotta Nuotio, journalist at National Radio Broadcasting of Sweden, was research assistant for the *Sixty Years of Science at UNESCO 1945–2005* project.
2 These introductory paragraphs are culled from a document that figures prominently in this chapter, *Science and Technology in UNESCO* (UNESCO, 1964).

> **BOX V.5.1: ON THE INTERNATIONAL RELATIONS OF SCIENCE**
>
> 'I think we ought to be much better informed about the international relations of science than we generally are in the scientific world. We usually carry on doing our job in our own corner of the field, whatever it may be, and we usually feel we have not enough time to take an interest as biologists, for example, in what geodesists or astronomers are doing. But actually, since science is a unity, we ought to take such an interest; and once we start looking across the international boundaries of the sovereign states which exist in the world today, and considering the question of how scientists can cooperate more effectively across these boundaries, we have to do so.'
>
> —**Joseph Needham**, founding Head of the UNESCO Natural Sciences Section, from a lecture to the Oxford University Scientific Club, 1 June 1948 (Needham, 1949).

Organizations. A framework and network was thereby constituted through which scientists and engineers of all countries could interact and cooperate with each other, through specialized non-governmental organizations (NGOs), which were in turn grouped in three major federations intimately linked with UNESCO. At the same time, Field Science Cooperation Offices were established in parts of the world furthest removed from the major centres of scientific activity.[3]

Subsequently a variety of approaches have been used to foster international scientific cooperation (Box V.5.2). This work has included collaborating in the setting up of institutions which have become leading international presences in their fields of activity (see Appendix 1, at the end of this chapter). Among the prestigious bodies established during the first decade of the Organization's own existence were IUCN (now known as the World Conservation Union) and CERN (European Organization for Nuclear Research).

Many other scientific bodies having a more finely focused brief were also set up under the aegis of UNESCO, such as those grouping botanists interested in the flora of the New World tropics (Flora Neotropica Organization) or chemists of industrialized and developing countries collaborating in work that benefits developing countries in health, agriculture, and industry (International Organization for Chemical Sciences in Development).[4] This work has continued to this day, with an increasing emphasis on encouraging networks of cooperation at regional and international levels, in fields ranging from physics education to biotechnological applications.

---

3   Namely, in Cairo (Egypt), Jakarta (Indonesia), Nanjing (China), New Delhi (India), Manila (the Philippines), Montevideo (Uruguay), Rio de Janeiro (Brazil) and Shanghai (China). The Offices at Nanjing, Manila, Rio de Janeiro and Shanghai were subsequently closed.
4   Not all initiatives for developing new bodies of scientific cooperation were successful. Disappointments included the International Institute of the Hylean Amazon, in the late 1940s, where the ratification of the founding diplomatic convention did not prove possible, and the International Computation Centre (set up in 1961 in Rome, Italy), which mutated into the International Bureau for Informatics but eventually lost momentum and was dissolved in the late 1980s.

> **BOX V.5.2: INTERNATIONAL COOPERATION IN SCIENTIFIC RESEARCH: A 1964 LISTING OF UNESCO-DEVISED METHODS**
>
> - Creation and/or support of intergovernmental scientific research organizations;
> - International cooperative research, carried out and financed by UNESCO;
> - Concerted international research under UNESCO's auspices, financed by Member States on a basis of free voluntary association;
> - Programmes of international research coordinated under UNESCO's auspices by a voluntary association of co-opted scientific institutions;
> - Programmes of international research coordinated under UNESCO's aegis by a voluntary association of co-opted scientists;
> - International programme of joint basic research organized and carried out on the initiative and under the direction of international non-governmental scientific unions, with the sponsorship, collaboration and financial support of UNESCO.
>
> *Source*: UNESCO, 1964, p. 6.

Another important means of promoting international scientific cooperation has been through programmes carried out on the initiative of and coordinated by scientific NGOs, with the cooperation and support of UNESCO. Examples include ICSU's International Geophysical Year (IGY) and International Biological Programme. To the extent possible, the Organization's support to initiatives such as these has been sustained and continued. UNESCO's financial support might not be huge, but it can often be for the long haul.

BUILDING CAPACITIES

If nurturing international scientific cooperation was a primary theme of the early days of the natural sciences programmes, then building scientific capacities was to become of major importance from the second half of the 1950s onwards. There were two combining circumstances: the demands from many newly independent countries and other developing countries for institutions and human resources to enable them better to take responsibility for their own socio-economic development; and the availability of significant levels of international funding to put proposals into action.

In responding to expressed needs and emerging resources, the Organization's work on building scientific capacities took shape. It had several components, often intimately linked, including support for the creation or reinforcement of research and training institutions, group training activities of various kinds, provision of individual fellowships and study grants, and the preparation of teaching and learning materials in different technical fields.

One of the early technical assistance projects in the 1950s was cooperating with the Pakistan Meteorological Service in cloud-seeding experiments and in helping set up a geophysical research centre at Quetta. Here, cloud physicist Michael Fournier d'Albe with an instrument maker in Karachi, Pakistan.

INSTITUTION-BUILDING

Contributing to scientific reconstruction was a task of the highest priority in the immediate post-war years. Among the useful work was the purchase of surplus war equipment to form the equivalent of about eighty workshop sets that were sent to the devastated countries and distributed among laboratories there to help replace their apparatus. According to Joseph Needham, in countries such as China and Poland, the equipment was enormously appreciated at the universities (Needham, 1949, p. 25). Subsequently, support for laboratory equipment was set aside under a Science Credits Scheme, under which countries were allocated funds to spend through the ordinary commercial channels. In another initiative, the East-Asia Office of UNESCO took over the work of the UN Relief and Rehabilitation Administration in distributing US$2 million worth of equipment to the engineering colleges of China.

During the 1950s, building up institutional capacity was a key part of the work on arid lands, through projects associated with the evolving programme on hydrology and water resources, such as that on the use of saline waters in irrigation in Tunisia. The 1950s also saw the beginning of a major effort to help countries set up higher institutes of technology and engineering. One of the first initiatives was in the newly independent India, where, in 1951, an institute was opened at Kharagpur (near Calcutta) with the help of foreign aid, including engineers sent by UNESCO (under the United Nations Expanded Programme of Technical Assistance), who helped establish courses in production technology and other fields. In 1955, an agreement was signed to enable the Organization to use a contribution by the Soviet Union to the United Nations Technical Assistance Fund, to assist a second institution at Powai (near Mumbai) (Behrman, 1964). The agreement called for UNESCO to provide the equivalent of US$3.5 million in aid over five years, but actual inputs far exceeded that figure, in addition to the US$10 million invested initially by the Indian government.

In the 1960s, UNESCO took on the role of executing agency for several tens of projects financed by the United Nations Special Fund, aimed at establishing scientific and technical institutes and getting their work started. This included support to institutes for technical and vocational education, to research institutions and to schools of engineering (as can be seen in a sampling of projects underway in the mid-1960s; see Appendix 2).

Through the 1960s and 1970s, institutional capacity-building continued in various natural sciences fields. Projects included those on hydraulic and applied hydrological research in Buenos Aires (Argentina), support to the Centre of Applied Hydrology in Puerto Alegre (Brazil), soil dynamics research at National Autonomous University of Mexico, and the development of regional networks of seismological observatories in South-East Asia and in the Balkan region. In Japan, support was provided to the International Institute of Seismology and Earthquake Engineering, aimed at making Japanese experience on earthquakes and building in earthquake-threatened zones available internationally.

In the early 1980s, the International Centre for Integrated Mountain Development was set up in Kathmandu (Nepal), with UNESCO providing a channel and facilitating funding from the governments of the Federal Republic of Germany and Switzerland, as well as being decisively involved in the initial years of the work of the International Board of Governors. More recent examples have included support to the establishment of the Regional Postgraduate Training School on Tropical Forest Management at the Mont-Amba University campus in Kinshasa (Democratic Republic of the Congo), and the Synchrotron-light for Experimental Science and Applications in the Middle East (SESAME) Centre at Al-Balqa' Applied University in Jordan. Complementing these sorts of activities at the higher education level is the UNESCO Chairs and UNITWIN[5] university networking programme. Since the programme was launched in 1992, some 590 UNESCO Chairs and

---

5   UNITWIN is the abbreviation for the University Twinning and Networking scheme.

UNITWIN networks have been established in over 120 countries in all major fields of knowledge within the Organization's competence.

GROUP TRAINING COURSES

Across the natural sciences programme, one consistent element of capacity-building has been training activities of various kinds. One snapshot from the mid-1960s (Table V.5.1) gives an idea of the range of subjects addressed, as well as the use of mixed funding sources. In providing training for young specialists from developing countries, an initial focus was through a series of postgraduate training courses organized on a regular basis at specialized institutions, principally in developed countries, mainly in Europe (see Appendix 3). There was one such course in 1960. By 1963, there were five, then eight in 1964, and nearly two dozen in 1965. For practically all of these courses, the vast bulk of the funding came from the host nations, with UNESCO's function largely that of 'internationalizer' of training facilities and activities (Baez, 1962).

Table V.5.1: Funds for training of scientists, for 1965–1966, by subjects of study and sources of finance

Subjects of study	Regular budget (US$)	Technical Assistance (US$)	Special Fund (US$)
Training of secondary-school teachers in new methods and techniques of science teaching	$150 000	–	–
Training of university staff in special subjects on postgraduate levels	$200 450	$368 300	–
Training of scientific documentalists	$80 000	$208 000	–
Training of hydrologists	$100 000	$40 000	–
Training of marine scientists	$220 000	$262 600	–
Training of geophysicists and seismic engineers	$50 000	–	$282 800
Training of geologists	$20 000	$45 000	–
Training of soil scientists	$10 000	$15 000	–
Training of ecologists	$22 400	$62 250	–
Totals	**$852 850**	**$1 001 150**	**$282 800**

Source: UNESCO, 1964, p. 29

During the 1970s, an effort was made to develop training courses in the developing countries themselves, thus helping to reinforce local institutional capacities, as well as contributing to the skills and knowledge of individual scientists, technologists and technicians. Since most developing countries lack the resources to cover the costs of medium- and long-term courses organized on a regular basis, the courses were essentially ad hoc in nature, generally short-term (from one week to a few months) and often jointly organized with other institutions. Early examples included: several three-month courses on water resources management and groundwater exploration held in Haifa and Jerusalem (Israel); and, in 1973–75, extended (three- to six-month) regional training courses on tropical ecology held in Nairobi (Kenya), Los Banos (Philippines) and Caracas (Venezuela), organized in cooperation with the United Nations Environment Programme (UNEP). Examples from the 1980s included a series of regional in-service training workshops on protected area management in Africa, held in Korhogo (Côte d'Ivoire), Dakar-St Louis (Senegal), Mweka (Tanzania) and Virunga (Zaire), among other locations. A common approach in much training is to provide an introduction to the concepts and practice of integrated approaches to natural resources management. Such approaches are often missing in the formal education of specialists in such disciplines as agronomy, economics, engineering, forestry, hydrology, and so on.[6]

Through group activities such as these, contributions have been made to the training of scientists and technicians in many technical fields. Thus, within the International Hydrological Decade (1965–1974), for example, more than 1,700 specialists received training in UNESCO-sponsored courses at the postgraduate level, together with about 160 technicians and professionals in courses on specific hydrological subjects. During the International Hydrological Programme, the number of postgraduate trainees in hydrology during the 1980s was estimated at about 400 per year (Kovar and Gilbrich, 1995).[7]

A somewhat different, ongoing training programme is that of the Floating University – international ship-based programmes combining training and research in a cutting-edge field (e.g. poorly known interactive processes in the deep ocean). During 1991–2005, fifteen major training-through-research cruises were conducted in the Mediterranean and Black Seas and in the north-eastern Atlantic. Twelve post-cruise conferences have been held; and a number of other field exercises (including smaller cruises), group and individual training activities, and presentation and publication of the research results, have been carried out. Some 600 scientists and students, from thirty countries, have taken part in these cruises. Impacts have included important discoveries (e.g. of numerous mud volcanoes in the Gulf of Cadiz, giant cold-water carbonate mounds in the north-eastern Atlantic), with a number of the participants going

---

6   One example from the late 1970s and 1980s was a nine-month training course on ecosystem management sponsored by UNEP and UNESCO at the Technical University of Dresden (German Democratic Republic), with project assignments being published in a series of volumes of readings from the course.
7   For other accounts of training in hydrology, see UNESCO, 1975; Gilbrich, 1991, 1994.

Over its sixty years, UNESCO has organized or sponsored several thousand group training courses at national, sub-regional, regional and international levels – some on a regular basis, others ad hoc – covering a wide variety of scientific fields and issues. Pictured here, participants taking part in courses on:

(a) Mechanical engineering at the Engineering Faculty of the West Indies University College at St Augustine in Trinidad and Tobago in 1967 (faculty set up with the assistance of UNESCO and UNDP) (Photo: © UNESCO/Dominique Roger);

(b) Tropical entomology, Abidjan and Taï (Côte d'Ivoire), July 1988 (Photo: © UNESCO/M. Hadley);

(c) Internet connection for biosphere reserves in Czech Republic, Poland and Slovak Republic, at the University of Warsaw (Poland), September 1995 (jointly organized with Global Environment Facility) (Photo: © UNESCO/Han Qunli);

(d) River-basin management at the International Centre for Water Hazard and Risk Management (ICHARM) in Tsukuba, Japan, in 2006 (Photo: © UNESCO/Kenji Saitoh).

on to become recognized leaders in their fields of research. Central to the success of the programme has been the commitment of a closely knit group of scientists from a handful of countries, who have guided the work, from data collection through analyses to the publication of results.[8]

Synergies between group training and international scientific cooperation have also been seen in the work of the International Cell Research Organization (ICRO), founded in 1962 with the aim of assisting in the implementation of UNESCO's programme in cell biology. Over its first 43 years, more than 12,000 young scientists from all over the world have taken part in some 456 advanced experimental training courses held in various fields of cell biology and biotechnology in 80 countries. Teachers are chosen from among the world's top-level specialists, and the courses have brought together senior and junior scientists from around the world, providing a singularly successful mechanism for stimulating international contacts between scientists across linguistic and cultural boundaries. Although diminishing in recent years, UNESCO's support to ICRO for these courses (generally between 15 and 40 per cent of total costs) plays a 'pump-priming' role, opening the possibility of obtaining further support from other international bodies and local and national sources.

Another approach is to provide support for world-renowned experts to respond to invitations from institutions and research groups in developing countries, especially those with limited outside contacts. There, during a stay of at least one month, the visiting scientist interacts closely with faculty and students of the host institution, with the aim of strengthening existing activities and/or helping set up new ones. The visiting scientist will often give lectures and seminars to research students, supervise students, take part in research and discuss future collaborative partnerships. A recently (2005) revamped example is the Visiting Scientist Programme operated as a joint initiative with the International Council for Science (ICSU), the Academy of Sciences for the Developing World (TWAS), and the United Nations University's Institute for Advanced Studies. A similar scheme for mathematics and physics is the Visiting Scholar/Consultant Programme offered by the Abdus Salam International Centre for Theoretical Physics in Trieste, Italy.

INDIVIDUAL FELLOWSHIPS AND STUDY GRANTS

Since its beginning, the Organization has accorded several thousand fellowships and study grants as a contribution to the training of individual scientists. For instance, within the work on arid zones starting in the early 1950s, some fifty-four fellowships were made available up to 1958, to help young scientists complete

---

8   See Part III, Chapter 9, D. C. Krause et al., 'Building blocks for marine sciences', in the present volume.

their training by a period of study at a well-known specialized laboratory abroad or by a study tour of sites where practical projects involving their speciality were underway. Fellowships were awarded under both the Regular and Technical Assistance Programmes, in such fields as animal and plant ecology, geophysics, hydrogeology and hydraulics, medicinal plants, microclimatology and soil science. Subsequently, the setting up of the Major Project on Arid Lands provided additional support to this form of training (initially, ten fellowships per year, later fifteen per year) (UNESCO, 1961). The importance of the fellowship programme is reflected in the significant contribution that grantees subsequently made to their scientific discipline, as well as to the development of science in their countries and at the international level.[9]

More recently, some of the individual science programmes have organized their own schemes for support to the training of individual scientists, in such fields as the basic sciences, molecular and cell biology, biotechnology, marine sciences, and young women in life sciences. Actions at the international level can have an important exemplary effect. An example is within the Man and the Biosphere Programme (MAB), where the MAB Young Scientists Research Award scheme provides grants for on-the-spot training within MAB field projects and biosphere reserves. Since the scheme was launched in 1989, some 218 grants have been awarded to young scientists from over eighty countries.[10] Similar schemes have been launched at the national level, for example in Sweden in the late 1980s and, more recently, in Indonesia, where a MAB Certificate for Young Researchers and Environmental Managers was set up in cooperation with the UNESCO Jakarta Office. The Canadian Biosphere Reserves Association's Student Network (formalized in 1999) builds linkages between students studying biosphere reserves and the biosphere reserve concept, and supports the development of promising environmental professionals. In China, the Ecological Research for Sustaining the Environment project includes a Young Scientist Prize.

TRAINING AND EDUCATIONAL MATERIALS

Many of the natural sciences programmes have incorporated training and capacity-building as long-term components of their work, including support for creating or reinforcing institutions, for group training activities and for individual study grants. An associated challenge is the development of training materials and curricula, as well as teachers' guides. This work dates back to the late 1940s, when a very popular booklet written by J. F. Stephenson provided

---

9   For example, in the field of the marine sciences, UNESCO fellowship recipients included Konstantin Federov (Soviet Union, who became Director of UNESCO's Office of Oceanography), Selim Morcos (Egypt), Aprilani Soegiarto (Indonesia) and Marta Vannucci (Brazil).
10   See: http://www.unesco.org/mab/mys.shtm.

PART V: OVERVIEWS AND ANALYSES

Suggestions for Science Teachers in Devastated Countries (Stephenson, 1948). The highly appreciated Book Coupon scheme for easing the passage of books from hard- to soft-currency areas was conceived in the Natural Sciences Department (Needham, 1949, p. 27), though its subsequent trials and successful implementation required a great deal of hard work on the part of other organs of the Organization (in particular, the Department of Libraries). A number of teachers' guides were produced on such subjects as inventories of apparatus and materials for teaching science, the teaching of general science, the teaching of health science (including nutrition) and a handbook for instructors who train primary-school teachers in the field of nature study (Valderrama, 1995, p. 75). In the mid-1950s, 'Teaching Science with Odds and Ends' was one of several features in the *UNESCO Courier* providing tips for teachers. In 1956, publication of the *UNESCO Source Book for Science Teaching* had an immediate impact; over the following decades, it was reprinted and updated many times and translated into some thirty languages, becoming one of the Organization's best-known publications, with well over a million copies sold (UNESCO, 1973).

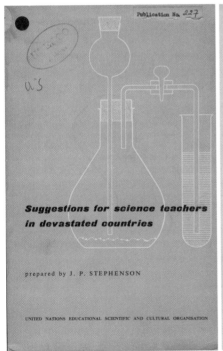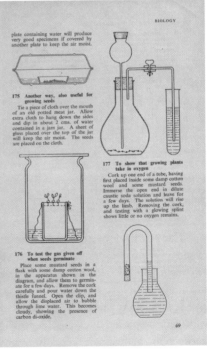

A handful of the 201 suggestions for constructing simple apparatus and conducting experiments in the booklet *Suggestions for Science Teachers in Devastated Countries* (Stephenson, 1948). Ideas were grouped in nine categories: astronomy, weather studies, measurement and properties of matter, heat, light, magnetism, electricity, chemistry, biology.

Thirty science and vocational teachers from Manila secondary schools, participating in a three-day workshop under the guidance of a UNESCO expert from Switzerland on the production of home-made science demonstration apparatus (here, how to convert a light bulb into a laboratory flask), Philippines, 1953.

The work of preparing guidance materials for secondary education had originally been a responsibility of the Natural Sciences Sector, on the principle that scientists should be in charge of the science education process because they know the content.[11] Nevertheless, the decision was subsequently made to shift science and technical education to the Education Sector, with a view to bringing this work closer to the educational and teaching community. Cooperation continued between the natural sciences community and the Education Sector-based work on

---

11   See Part II, Chapter 15, A. Baez, 'Live and learn', in the present volume.

science education, including technical and vocational education and environmental education, but probably not to a level even approaching optimal.

In terms of science education at higher secondary and tertiary levels, guidance materials and teaching aids have continued to be part of the natural sciences programme. Recent examples include toolkits of learning and teaching materials in various renewable energy fields and kits for microscience experiments. In the marine sciences and oceanography, programmes such as that on Training, Education and Mutual Assistance have provided a long-term focus for training materials and activities of various kinds, with an analogous function in hydrology provided by the work of UNESCO's International Hydrological Programme (IHP) on education and training (under the rubric of 'Knowledge, information and technology').

Other materials have been produced and disseminated by the broader UNESCO community. An example is the Asia/Pacific Cultural Centre for UNESCO based in Tokyo. Among its activities is a wide-ranging programme for diffusing materials on disaster management for community empowerment, including education and public awareness packages produced by various organizations, institutions and individuals, with special emphasis on earthquakes, tsunamis and disaster preparedness. This area of disaster preparedness education is one of the key areas being addressed in the UN Decade of Education for Sustainable Development, launched in early 2005.

UNESCO'S OWN SCIENTIFIC COOPERATION PROGRAMMES

In addition to helping create and reinforce institutions promoting science, and contributing to the training of scientists and technologists, UNESCO has used a variety of other approaches to set on foot new kinds of international scientific cooperation. After the aborted plans for the International Institute of the Hylean Amazon in the late 1940s, which had ended in 'utter failure' according to one observer (Florkin, 1956), programmes in the 1950s on the arid zones and the humid tropics had a significant influence on the evolution of UNESCO's approach to international scientific cooperation. Among the lessons learned was that a multidimensional international programme – including very different types of action in research, information exchange, technical assistance and training – can be guided by a simple advisory committee of well-chosen, highly motivated individuals. Scientific advisory committees have continued to constitute a flexible, low-cost means for guiding and shaping programmes in a range of fields.[12]

---

12   Including science popularization, natural resources research, molecular and cell biology, microbial resources, tropical soil biology and fertility, chemistry of natural products, and biosphere reserves. More recently, a scientific advisory committee was adopted as the structure for overseeing and guiding the evolving International Basic Sciences Programme (IBSP).

But these early experiences also suggested that – beyond a certain level – in order to have decisive and lasting effects at the country level, such programmes call for clearer commitment and more formal support from governments. This may be particularly the case where the subject of research is the natural environment and its natural resources, which come under the direct sovereignty of national governments or which require concerted action among countries at the governmental level.

It was within such a context that in the 1960s there emerged what have come to be known as intergovernmental programmes with a geographical dimension – frameworks for scientific cooperation in fields where direct governmental participation and involvement is considered a prerequisite for success, where the vast majority of the research is financed by Member States on the basis of free voluntary association. These fields included oceanography – with its need for large logistic facilities including research vessels and long-distance communications – where the approach was to set up the Intergovernmental Oceanographic Commission (IOC) within the framework of UNESCO.

UNESCO's programmes in the arid zones and humid tropics had underlined the key role of freshwater resources in development. Indeed, by the early 1960s, it had become clear that all parts of the world were facing growing water problems. The response was a major worldwide cooperative study on water, the International Hydrological Decade (IHD). The ambition was comparable to an IGY or an IOC. The method, however, was different. The coordinating mechanism was an international council with a limited number of participating countries elected on a rotating basis by the General Conference, represented by experts and providing guidance to a common programme of research and monitoring designed by consensus. The council thus constituted an intergovernmental steering body that was lighter than a fully fledged commission (such as the IOC), but more representative of world diversity and responsibility than a simple scientific committee. In due course, it was deemed appropriate to explore whether a similar formula could be applied to other global subjects where a need for intergovernmental cooperation seemed apparent. In this way, the Man and the Biosphere Programme was set up along similar structural lines as the IHD.

Within geology – with its well-established surveys and agencies in many countries and a single strong international union – a rather different mechanism was developed. This combined the existing solid cooperation between scientists and services within the International Union of Geological Sciences (IUGS), on the one hand, and intergovernmental support provided by UNESCO, in which developing countries could play an active role, on the other. The result was the setting up of a board of experts appointed jointly by UNESCO and IUGS to lead the programme, with technical assistance from a scientific committee.

Some features of these programmes were discussed by Michel Batisse in the *World Science Report 1993* (Batisse, 1993). He noted that, by the very nature of the issue tackled, the programmes require international cooperation. Both developed and developing countries have an interest in participating, because the programmes contribute to a better knowledge of natural phenomena and resources, and a wider sharing of information. Coordinated by international governing bodies, their execution relies on activities conducted by participating countries. The programmes are problem-oriented, involving interdisciplinary approaches and field activities, and inviting interaction between research workers and decision-makers to allow for the correct formulation of issues and the practical utilization of results. They also include a strong training component, essential for building up national capabilities for participation in international efforts and the achievement of related national goals. Finally, these programmes are implemented with the active participation of other UN agencies and NGOs, which all have statutory representation on the governing body. These so-called intergovernmental programmes have each developed their own identity and constituencies.

But such intergovernmental programmes are not the only approach that UNESCO has adopted for promoting scientific cooperation and encouraging new directions of work. The International Conference – an event familiar to the scientific community – has been an important part of the working methods of the Organization from the beginning (see Appendix 4). Among the landmark scientific conferences was the Conference on Radioisotopes in Scientific Research, held in Paris in September 1957. Conferences and symposia have been an integral part of the long-term, multi-component programmes in the natural sciences, with literally thousands of other scientific gatherings in the form of smaller workshops and seminars. In addition, the convening of major intergovernmental conferences at the ministerial level was a principal tool in the promotion of work on science and technology policies during the 1970s and early 1980s.[13]

COMMUNICATING AND INFORMING

Information exchange is an essential part of any scientific programme. The research results generated within the natural science programmes of UNESCO are primarily published in the open scientific literature by the individual researchers concerned. Building on this primary source of communication, a range of series of documents and publications have been launched for disseminating the results of UNESCO-sponsored activities to scientists taking

---

13   See Part IV, Chapter 2, J. Hillig, 'Helping hands, guiding principles', in the present volume.

The 1957 international scientific Conference on Radioisotopes in Scientific Research (in Paris) brought together some 1,000 scientists from around the world (including several Nobel Prize winners). The Secretary-General of the 1957 Paris Conference was Pierre Auger, Director of the Natural Sciences Department (foreground, right), with Michel Batisse (background, centre) as Secretary.

part in the science programmes of the Organization as well as to those in the broader scientific and technological communities (see Appendix 5). In addition to these, are materials published in out-of-series, self-standing publications, as well as a range of materials produced by individual Field Offices and by collaborating institutions.

RAISING PUBLIC AWARENESS, CONTRIBUTING TO EDUCATION

Efforts to raise public awareness about science figured quite prominently in the early natural sciences programmes. 'Popularization of science' received specific attention at the fifth session of the General Conference held in Florence (Italy) in 1950, and in the same year, four of the forty-one posts in the Natural Sciences Department were attributed to this end. During the 1950s, Natural Sciences staff (including Pierre Auger, Alain Gille, Maurice Goldsmith, Jim Swarbrick and Gerald Wendt) contributed frequently to the *UNESCO Courier*. 'Science Has the Answer' was the somewhat presumptuous title of an occasional feature in the monthly magazine, with question and answer exchanges on such topics as 'What is the cause of tidal waves?', or 'Why is seawater salty and how does it retain its saltiness?'

PART V: OVERVIEWS AND ANALYSES

The work of Science Clubs was actively promoted.[14] At the Mexico City (Mexico) General Conference in 1948, it was decided to initiate discussions each year on a scientific theme of current international importance ('Food and People' was the theme selected for 1949, 'Energy in the Service of Man' for 1950). And eminent scientists from such fields as anthropology, biology, genetics, psychology and sociology took part in the work of a Committee of Experts on Race Problems, whose declaration (released in July 1950) on the 'known facts about human race' generated considerable discussion and debate in public, media and scientific circles.[15]

Science exhibitions were another activity which had a considerable impact, with a first UNESCO 'Exposition Scientifique Internationale' being mounted in Paris in November–December 1946. During the 1950s, millions came to see a series of travelling science exhibitions.[16] Exhibitions remain a prime means of communicating scientific information to the broader public. A notable example from the early 1980s was the poster exhibition 'Ecology in Action', considered by some as the activity having had the single greatest impact in the history of MAB. Recent exhibitions have included: 'Renewable Energies for Sustainable Development' at the World Summit on Sustainable Development in Johannesburg (South Africa) in 2002; 'Celebrating Diversity' at the UNESCO-led United Nations Pavilion of the 2005 World Exposition on 'Nature's Wisdom' held at Aichi (Japan); and 'Experiencing Mathematics', which visited Athens (Greece) and Beijing (China) in 2005.

The Natural Sciences Sector has also used a variety of other means to popularize science. For many years, science journalism and writing featured among the specialities of the Sector's staff.[17] Since 1952, the annual award of the Kalinga Prize for the Popularization of Science has recognized more than sixty scientists and communicators who have used their talents to interpret science and technology for the general public. In addition to the educational and training materials mentioned earlier in this chapter, recent learning materials for the young have included an education kit on desertification and *Discovering the World* booklets (aimed at ages 10+) on such subjects as oceans, biosphere reserves, climate, and the World Heritage sites.

---

14 See, for example, articles in the August 1949 (p. 4) and December 1949 (p. 5) issues of the *Courier*.
15 The text of the declaration was published in the July/August 1950 issue of the *Courier*.
16 See Part I, Chapter 18, A. Gille, 'On the road', in the present volume.
17 Among the science journalists on UNESCO's staff was Dan Behrman (1923–1991), who contributed many articles to such magazines as the *UNESCO Courier*, as well as writing books such as *Science and Technology in Development: a UNESCO approach* (Behrman, 1979), and *Assault on the Largest Unknown: the International Indian Ocean Expedition, 1959–65* (Behrman, 1981).

TAKING ADVANTAGE OF MODERN TECHNOLOGIES

New information and communication technologies (ICTs) have been one of the driving forces in globalization, as well as triggering major changes in how science is done. Among other things, ICTs have allowed large amounts of information to be stored in electronic format and to be made instantaneously accessible through the internet. The evolution of modern electronics has had a revolutionary impact on programmes of scientific cooperation. In oceanography, automated observations, quality control software and electronic communications is leading to the disappearance of the border between archived data and data available in real time. An example is the use of a Global Telecommunications

Encouraging closer links between researcher and decision-maker is a long-standing challenge in much of UNESCO's scientific work. This poster from the 'Ecology in Action' exhibition of the early 1980s addressed the issues of different time-scales in research, management, and policy-making, and ever-evolving land use problems.

System for exchanging ocean data, beginning with temperature versus depth (bathythermograph) data and then expanding to temperature and salinity profiles.[18] In seemingly traditional biological sciences such as taxonomy and systematics, multimedia computer techniques have provided opportunities for handling large data sets. A striking example is the system of biodiversity documentation and species identification developed by networks of specialists

---

18   See Part III, Chapter 8, G. Holland, 'Observing and understanding planet ocean', in the present volume.

PART V: OVERVIEWS AND ANALYSES

associated with the UNESCO-affiliated Expert Centre for Taxonomic Identification (ETI), based at the University of Amsterdam.[19]

Since the mid-1990s, several UNESCO science programmes have taken advantage of the opportunities presented by ICTs and put online much textual and other information. But progress has been spotty, particularly compared to other sectors. In each of the four other programme sectors, there is a unit responsible for the websites of the whole sector and its divisions and programmes. In contrast, in Natural Sciences, the various substantive programmes and divisions each have individual responsibility for website management.[20] Not surprisingly, the websites of some programmes are information-rich and regularly updated, whereas, with others, the opposite is the case. In addition to the science-related websites operated by UNESCO (at Headquarters and in Field Offices and Institutes), a considerable effort has been made to link up with websites developed by collaborating scientific institutions. For example, within the International Geoscience Programme (IGCP), the first websites began to appear in the early 1990s, and a recent search brought up over 5,500 websites dedicated to IGCP projects.[21]

Modern information technologies have also revolutionized science publication, with many publications now being accessible electronically. One example is the *Encyclopedia of Life Support Systems* (EOLSS). Since its launching, the online source of information has been built up to comprise an integrated compendium of sixteen encyclopedias, equivalent to 200 hard-copy volumes. Another new medium of information exchange allowed by ICTs is internet discussion forums. Pioneering exemplars include two moderated multilingual discussion forums operated by UNESCO's Environment and Development in Coastal Regions and Small Islands (CSI) Platform.[22]

UNESCO COMMUNITIES

All of UNESCO's activities are undertaken within the framework of the programme and budget approved by its members, that is – as UNESCO is an autonomous international organization at the intergovernmental level – its Member States.

---

19   The Expert Center for Taxonomic Identification is an NGO operationally affiliated with UNESCO, at the University of Amsterdam (the Netherlands). As is not unusual in the history of science and scientific cooperation, its origins and links with UNESCO are serendipitous, and date back to a chance conversation (in 1983) at a coffee machine between a researcher at the University of Amsterdam and a staff member of UNESCO's Natural Sciences Sector. Further background is given in Schalk and Troost, 1999.
20   Although there is a website facility attached to the Executive Office of the Assistant Director-General.
21   See Part III, Chapter 5, S. Turner, 'Rocky road to success', in the present volume.
22   See Part III, Chapter 10, D. Troost and M. Hadley, 'Breaking down barriers, building bridges', in the present volume.

NATIONAL COMMISSIONS FOR UNESCO AND ASSOCIATED BODIES

The National Commissions are advisory, liaison, information and executive bodies established by Member States, pursuant to the Constitution, in order to associate their major institutions in education, science, culture and communication with the Organization's work. UNESCO is the only specialized agency of the United Nations to have a system of this kind, and the impact of the National Commissions can be considerable. Scientists figure prominently in the commissions in many countries, while a fair number of commissions have a special sub-commission responsible for natural sciences. Another option adopted by some countries is to request one or several of its leading scientific institutions (e.g. academy of sciences, national scientific research councils) to take responsibility for articulating with the natural sciences programme of UNESCO.

Also at the national level, the aim of encouraging international understanding and links between UN organizations and civil society finds expression and support from such bodies as UNESCO Clubs, Centres and Associations, of which there are (in 2005) more than 3,600, in eighty-nine Member States (UNESCO, 2005). Bodies such as these may in turn be active in projects and programmes related to the natural sciences, more particularly those with a significant education and public-awareness focus. A recent example concerns beach monitoring in the Caribbean and the 'Sandwatch' initiative undertaken jointly with the Education Sector's Associated Schools Project Network.

GENERAL CONFERENCE

The General Conference is the Organization's supreme authority, which decides policy, votes the budget, and elects the members of the Executive Board and the Director-General. Since its first session, membership has grown steadily, making the Organization more representative of that one world which the founders had set out to serve. As of mid-2005, there were 191 Member States and six Associate States.

Originally the General Conference met yearly and was invited to different cities by the Member States. In addition to Paris, it was convened in Mexico City, Beirut (Lebanon), Florence, Montevideo (Uruguay) and New Delhi (India). In 1952, a decision was made to review UNESCO's programmes every other year. With the completion of permanent headquarters in Paris in 1958, however, it became convenient to hold most General Conferences there.

General Conferences are now generally held over a three-week period in the October of odd-dated years. The natural sciences programmes are debated in one of the programme commissions of the General Conference, and scientists active in UNESCO's programmes at the national level are often present in many of the national delegations taking part in those discussions. However, it is important to realize that, in general, the heads of delegations to the General Conference are not part of the scientific community, but rather usually come from ministries of education or foreign and/or development affairs. This can result in difficulties for some programmes and bodies.

EXECUTIVE BOARD

In assuring the overall management of UNESCO, the Executive Board prepares the work of the General Conference and sees that its decisions are properly carried out. The functions and responsibilities of the Executive Board – which meets twice a year (currently, in April and September) – are derived primarily from the Constitution and from rules or directives laid down by the General Conference. The Board comprises Member States (at present, fifty-eight in number) drawn from the different electoral regions recognized by the Organization, with four-year terms of office.

This contrasts with the early years, when Board members were elected as individual personalities. As Richard Hoggart (Assistant Director-General for Social Sciences, Human Sciences and Culture, 1970–75) once explained, the theory was that the Board

> should help provide a counterweight, on behalf of the Constitution itself, to the one-directional pressures of national politics. Hence, the original notion that members of the Board (should) have three simultaneous roles: to represent their States or regions; to be eminent intelligences in their own right; and to serve the Constitution (as embodied in the General Conference) in itself ... Back in the beginning, it was possible to argue that the Board would eventually become a sort of world intellectual, scientific and cultural Senate. This proved almost as idealistic a notion as that of a world government (Hoggart, 1978).

That said, this purported triple role of Board members (and their six-year terms of office) meant that individual 'intelligences' could, and indeed did, have a considerable long-term impact on UNESCO's work. An example is Brazilian chemist Paulo E. de Berrêdo Carneiro (1902–1982), who was a member of the Executive Board on five different occasions from 1946 to 1982 and was considered to be 'one of the most outstanding speakers ever to be heard in the fora of UNESCO' (Conil-Lacoste, 1994, p. 84).[23] Paulo Carneiro is credited with the idea of the Hylean Amazon project and oversaw the work of the more

---

23   Paulo Carneiro had said, 'It is UNESCO's vocation to be a perpetual question mark' (Conil-Lacoste, 1994, p. 55).

than 500 scholars and researchers who contributed to the *Scientific and Cultural History of Mankind.*[24]

FIELD OFFICES AND INSTITUTES

As noted above, the setting up of Field Science Cooperation Offices was one of the very first actions in UNESCO's natural sciences programmes, with a first batch of offices in Cairo (Egypt), Jakarta, Nanjing (China), New Delhi (India), Manila (the Philippines), Montevideo (Uruguay), Rio de Janeiro (Brazil) and Shanghai (China). Over the years, some of these offices were closed down, such as those in Nanjing, Manila, Rio de Janeiro and Shanghai. Others have changed in their functions (e.g. New Delhi), so that in mid-2005 there were Regional Offices for Science and Technology (ROSTs) in Cairo, Jakarta, Montevideo, Nairobi (Kenya) and Venice (Italy).

In the field of education, there are a set of specialized UNESCO institutes and centres. Apart from the Abdus Salam International Centre for Theoretical Physics in Trieste, such has not been the case in the natural sciences, at least until recently. In 2003, the UNESCO-IHE Institute of Water Education was formally set up in Delft (the Netherlands, with the Director a staff member of UNESCO), following an agreement with the Dutch Government. A number of other regional and international centres have also become affiliated with the International Hydrological Programme.[25]

RELATIONS WITH OTHER INTERNATIONAL ORGANIZATIONS AND INITIATIVES

As reflected in the many tens of acronyms dotted throughout this volume, numerous are the organizations with which UNESCO cooperates in the natural sciences. Relationships have taken diverse forms, including multiple mutually supportive links with the major scientific NGOs, such as ICSU, and joint activities over several decades, such as that with IUGS in the International Geoscience Programme (IGCP).

In terms of trends, one feature of the changing international scene over the last decade and a half has been the launching of a whole series of new initiatives related to science and technology, environment and development, which have required inputs

---

24 This ambitious project comprised 12,000 pages, in 7 volumes; see Part I, Chapter 19, P. Petitjean, 'The ultimate odyssey', in the present volume.
25 Examples of regional centres affiliated with the IHP are those on: water in the humid tropics of Latin America and the Caribbean (Panama City, Panama), training and water studies of arid and semi-arid zones (Cairo), humid tropics hydrology and water resources of South-East Asia and the Pacific (Kuala Lumpur, Malaysia), and urban water management (Tehran, Islamic Republic of Iran). International centres include those on qanats and historic hydraulic structures (Yazd, Islamic Republic of Iran), erosion and sedimentation (Beijing, China), groundwater resources assessment (Utrecht, the Netherlands), and urban drainage (Belgrade, Serbia and Montenegro).

and collaboration from the UNESCO side. Examples include new international conventions associated with the Rio 'Earth Summit' (in such fields as biodiversity, climate change and desertification), wide-ranging reviews (such as the Millennium Ecosystem Assessment) and multimedia global observing systems (such as those contributing to the Global Earth Observation System of Systems). The so-called transaction costs of participating in such initiatives are not negligible. A challenge for UNESCO is finding an optimal balance between contributing usefully to these initiatives and maintaining its own distinctive activities in the natural sciences field.

FUNDS AND BUDGETS

The core of UNESCO's budget is that financed by its Member States through assessed contributions adopted by biennial sessions of the governing body of the Organization, the General Conference. Thus the approved Regular Programme and Budget for the two-year period 2004–2005 was approved by the thirty-second session of the General Conference, which met in Paris in October 2003. There are also so-called extrabudgetary resources, made available for defined activities and projects by various funding sources operating at the national, regional and international levels, both private and public.

REGULAR PROGRAMME BUDGET

Over the last half-century, the total Regular Programme budget of UNESCO has increased thirty-fold, from US$ 21.6 million in 1955–1956 to US$610 million in 2004–2005 (Figure V.5.1). During the same period, the budget for the Natural Sciences increased from US$2.1 million to US$58.2 million.

In terms of the Regular Programme funds available for the various programme sectors, since the early days of the Organization, the Education

Figure V.5.1: Comparative table of Regular Programme budgets for the Natural Sciences, all programmes, and all of UNESCO, 1955–2005 (in millions of US dollars)

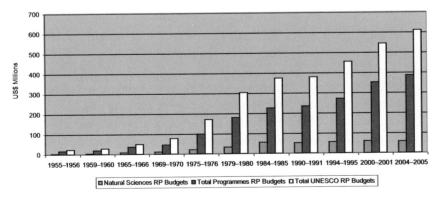

Sector has customarily received the highest allocations, and that pre-eminence has continued to this day. During the fifty-year period starting from 1955, the budget for the Natural Sciences has oscillated between 13 and 25 per cent of the total funds available for programme activities (Figure V.5.2).

Figure V.5.2: Natural Sciences Sector budget as a percentage of the budget of all programme sectors under the Regular Programme

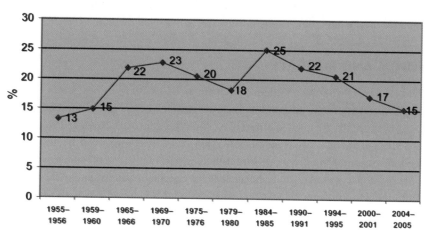

One period of expansion was the early 1960s, as part of the process associated with the UN Conference on the Application of Science and Technology for the Benefit of the Less Developed Areas, held in Geneva (Switzerland) in February 1963. Prior to that Conference, in October 1962, Director-General René Maheu had reflected:

> The study of the main trends in scientific research ... and the ten-year plan adopted by the General Conference enable us to establish a scheme of priorities to meet the major needs of our times and to frame an international scientific policy valid for the Organization and its Member States alike and backed up by the international structures required for international action (Maheu, 1962).

One of UNESCO's aims, he added, should be 'to give the natural sciences the kind of impetus that it has recently given, and rightly given, to education.'

Subsequently, in March 1963, after the Geneva Conference had taken place, the Executive Board approved in principle the proposal that 'scientific questions be accorded an importance in UNESCO's programme similar to that given to education'. The result was a substantial increase – both in relative and absolute values – in the allocations for science and technology in the Organization's

Regular Programme and Budget (Table V.5.2). An absolute peak was apparently reached in the mid-1980s, when the Natural Sciences represented 25 per cent of the Organization's budget for programme activities (Figure V.5.2). The proportion has since declined to its present level of about 15 per cent.

Within the changing total financial resources for the Natural Sciences Sector, internal allocations have followed the rising and falling status of the various

Table V.5.2: Regular Programme budgets for 'Science and technology' and for 'Education', for three consecutive biennia in the early 1960s

Years	Total UNESCO budget for biennium (US$)	Total implementation of the programme (US$)	Sciences and technology (US$) (% of total programme)	Education (US$) (% of total programme)
1961–1962	$32 513 228	$23 704 029	$3 432 224 (16.7%)	$6 589 675 (32.7%)
1963–1964	$39 000 000	$29 580 672	$4 775 084 (16.7%)	$8 057 720 (34.7%)
1965–1966	$46 800 000	$33 950 000	$7 455 135 (22%)	$11 150 000 (32.8%)

Note: Figures in the two final columns do not include certain programme activities such as statistics, information and international exchanges, parts of which are accounted for in the budgets of the Departments of Social Sciences and Mass Communication and the International Exchange Service.
Source: UNESCO, 1964, p. 22.

Figure V.5.3: Environmental funding in the Natural Sciences Sector of UNESCO Regular Programme budgets, 1973–2005 (in millions of US dollars)

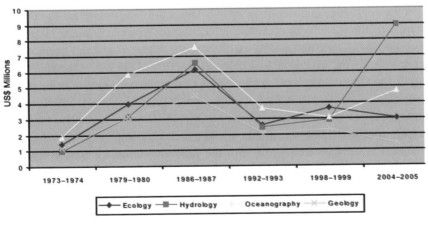

Sources: Biennium Regular budgets: 17 C/5 approved for 1973–1974, 20 C/5 approved for 1979–1980, 23 C/5 approved for 1986–1987, 26 C/5 approved for 1992–1993, 29 C/5 approved for 1998–1999, and 32 C/5 approved for 2004–2005.

programmes and groups of programmes, such as the increase of resources to work on science policy in the 1960s and 1970s and subsequent decline from the mid-1980s onwards. The various environmental science programmes received expanding budgets during the 1970s, with more recent increases (both in absolute and proportional terms) accorded to work on freshwater resources and oceanography (Figure V.5.3), with absolute priority within the Natural Sciences being accorded since 2002 to the theme of 'freshwater and associated ecosystems'.

EXTRABUDGETARY FUNDING

In addition to the resources made available through the Regular Programme, substantial levels of so-called extrabudgetary funds have been generated for projects related to the natural sciences, with the Organization taking on the role of executing agency for these projects and funds. The two main sources are through multilateral funding bodies of various kinds[26] and bilateral funds-in-trust arrangements. By their very nature, many of these funds have been linked to technical assistance for building scientific capacities in particular countries, regions or professional fields. Increasingly, extrabudgetary funds have focused on building scientific partnerships of various kinds, and many of these 'partnerships for development' can be considered as an extension of Regular Programme activities.

UNESCO's Regular Programme has always included some technical assistance activities; but, given the modest level of its budget, these were necessarily few and low key. More substantive actions had to await the availability of greater funding. This was to come with the creation, by the United Nations, of the Expanded Programme of Technical Assistance (EPTA), following the call by US President Harry Truman (in his January 1949 State of the Union Address) that the advanced countries of the world should combine to give aid to raise the standard of living in poor countries. UNESCO took part in discussions at Lake Success (New York State, USA) with officials of the specialized agencies, on proposals for the Technical Assistance Programme. This was subsequently set up by the General Assembly, with UNESCO's first expert mission being that of Norwegian mathematician Karl Borch to Iran in 1950 (Valderrama, 1995, p. 67). But resources were limited and focused on a small number of activities, including expert missions, supply of equipment and the allocation of a few fellowships.

Larger-scale projects came with the introduction of the Special Fund in 1958, which provided the means for the setting up of scientific and technical education institutions in more than a score of countries in the early 1960s (with one 1964 listing of projects indicating a total Special Fund contribution of

---

26   For instance, the UN Development Programme (UNDP) and UN Environment Programme (UNEP).

US$26.2 million to projects in twenty-three countries). Additional support was provided to schools and faculties of engineering in countries such as Algeria, Chile, Colombia, Kenya, Pakistan, Syrian Arab Republic, Turkey and Venezuela, as well as at the regional University of the West Indies.

In 1965, the merging of the Technical Assistance Programme and the Special Fund gave birth to the United Nations Development Programme (UNDP), which rapidly became the main source of funding for operational programmes for the next decade and a half. As an ensemble, projects were catholic in focus and scope, many linked to capacity-building. The early 1970s saw the creation of the United Nations Environment Programme (UNEP) and the associated Environment Fund. Over more than a decade, a whole series of activities were carried out on a joint basis by UNEP and UNESCO, with projects ranging from state of knowledge reviews (in such fields as tropical forest ecosystems and tropical grazing lands) to planning meetings, support to field projects and education and training activities of various kinds. In terms of UNESCO's budget, these funds were labelled extrabudgetary resources. But substantively, the vast majority represented an extremely valuable and valued additional resource for Regular Programme activities.

Things were to change in the 1980s, with disbursements from the Environment Fund being shifted primarily to UNEP's own activities, many related to proposals from environment ministries in developing counties. Some joint UNEP-UNESCO activities did however continue, for instance in environmental education and through activities financed using UNEP's 'soft currency' resources. In a similar way, UNDP funds entrusted to UNESCO declined in the 1980s, as UNDP increasingly took on the role of executing agency for its projects, and as countries' needs changed.

Offsetting to some extent the decline in UN funding during the last two decades of the twentieth century was the support made available through funds-in-trust arrangements by certain countries (including Belgium, the Federal Republic of Germany, Italy, Japan, the Republic of Korea, Libya, the Netherlands, and the Nordic countries), as well as from regional funding sources such as the World Islamic Call Society and the InterAmerican Development Bank. Such funds-in-trust support has enabled innovative work to be carried out in a whole range of fields, from harmful algal blooms to reinforcing ecological research in China, from the exploration of deep water sources in the northern Sahara to setting up research capacity for cell and molecular biology, from ethnobotany and the sustainable use of plant resources to regional cooperation in the basic sciences and research and training in the marine sciences. In addition to support for specific projects, funds-in-trust have provided a significant resource for certain overall Natural Sciences initiatives, such as the special IOC Trust Fund set up by the Intergovernmental Oceanographic Commission.

Figure V.5.4: Natural Sciences Sector and UNESCO total extrabudgetary funds, 1981–2003 (in millions of US dollars)

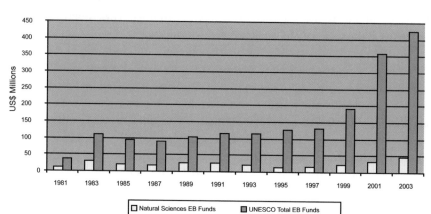

Figure V.5.5: Natural Sciences Sector extrabudgetary funds as percentage of total UNESCO extrabudgetary funds, 1981–2003

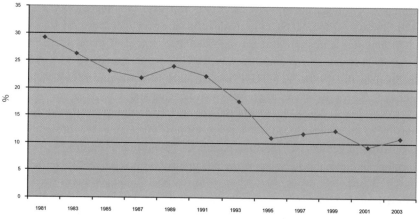

Sources: Budgetary Status Reports on Extrabudgetary Operational Projects, Analysis by Sector, as at the end of 1981, 1983, 1985, 1987, 1989, 1991, 1993, 1995, 1997, 1999, 2001 and 2003

More recently, private sources of funding have contributed to specific Natural Sciences Sector activities. Examples include: the Mondialogo partnership (with DaimlerChrysler), supporting engineering awards for reducing poverty and promoting sustainable development in developing countries; the joint work with the L'Oréal company to promote the access of women to science and technology, including annual awards for women in science;[27] and cooperation with Agfa Gevaert to provide chemistry teachers with a high-standard educational aid, adaptable to multi-language needs and in various media forms.

In sum, extrabudgetary funding has been an important yet variable source of support for Natural Science activities (Figures V.5.4 and V.5.5). Over the years, there have been changes in the relative weight of the different sources of such funding. Overall, the last quarter-century has seen a proportional decline in the percentage of the extrabudgetary resources received by the Organization that are attributed for the natural sciences. One reason for this proportional decline is a series of decisions by some major funding partners to concentrate their cooperation with UNESCO on basic education. Another is donor dissatisfaction with the Organization's capacity to disburse in a timely way the amounts allocated for individual projects.

THE NOTION OF CATALYTIC, PUMP-PRIMING RESOURCES

'UNESCO is not a funding agency' is a phrase quickly learned by successive generations of staff members. Its own budget – voted by its members – is limited, and can be compared to that of a medium-sized university in a medium-sized industrialized country. Given the wide range of the Organization's fields of activity, it is essential to use scarce resources in an optimal way, particularly in seeking to mobilize the additional resources and support that are required for a project or programme to be carried out satisfactorily. Illustration is provided by a handful of examples from the natural sciences programmes.

At the fifth session of the General Conference held in Florence in 1950, the Director-General was authorized to assist the creation of regional research centres and laboratories in fields where the efforts of a single country would be inadequate.[28] This resolution sought to establish the important principle that UNESCO should help stimulate and plan such centres, but should not build or operate them. Rather, the Organization's help was needed in making plans and in translating them into reality. Informal discussions led to agreement that one such centre should bring together the countries of Western Europe for cooperative research on nuclear energy.

Preliminary plans for the European centre were developed by the UNESCO Secretariat, with cooperation from outside scholars and specialists. The Director-General invited all of the European Member States to take part. An interim council of interested countries was set up to study the issues involved in establishing an international laboratory and to draft a convention to that end. A scientific conference was held to survey the general situation in nuclear research

---

27 See Part IV, Chapter 6, R. Clair, 'Tapping at the glass ceiling', in the present volume.
28 This example is distilled from Laves and Thomson, 1957, pp. 194–96.

that might most usefully be tackled by an international organization. Geneva was selected as the site, and plans for building and equipment drawn up. In July 1953, the convention was signed to establish the European Organization for Nuclear Research (CERN)[29] and promptly ratified by the twelve founding governments.

And the money for all this? Three governments – Belgium, France and Italy – provided some US$500,000 for the preliminary studies. The full membership later pledged US$28 million for construction, equipment and operation over the first seven years. The Ford Foundation provided a grant of US$400,000 to cover the research expenses at the centre by scientists from countries not members of the Organization. For its part, UNESCO's budgetary contribution over three years was about US$25,000.

Thus, the creation of CERN provides an early example of how the Organization can make an effective – even decisive – contribution, with its modest financial resources being made available at the early shape-making stage of a project. Two other factors were central to the success of CERN: the resolve and converging will of a regional group of industrialized countries having shared interests and a critical pool of talent in a new field (nuclear physics); and the fact that none of these governments working alone could readily meet the cost of building and operating the experimental facility, in this case a giant accelerator.

A second example is the International Tsunami Information Center (ITIC), established by the IOC in 1965. Located in Hawaii (USA), ITIC maintains and develops relationships with scientific research and academic organizations, civil defence agencies, and the general public, in order to carry out its mission to mitigate the hazards associated with tsunamis by improving tsunami preparedness for all Pacific Ocean nations. UNESCO's contribution to the overall system is through the Intergovernmental Coordination Group for the Tsunami Warning System in the Pacific, which is a subsidiary body of the IOC, formed in 1968. The main purpose of the group is to assure that tsunami watches, warning and advisory bulletins are disseminated throughout the Pacific to Member States, in accordance with procedures outlined in the Communication Plan for the Tsunami Warning System. The present annual financial support is US$30,000, most of which is provided to ITIC. In other words, the principal contribution of UNESCO's IOC to a major regional system of disaster warning has been through its role in helping to set up the system, and through its long-term involvement in providing an international framework for cooperation among countries, agencies and scientific personnel. Most of the financial resources required for building and

---

29 See Part I, Chapter 8, P. Petitjean, 'Cool heads in the Cold War', in the present volume. CERN's provisional name was the Conseil Européen pour la Recherche Nucléaire (European Council for Nuclear Research), and it kept the acronym.

maintaining the physical elements and providing human resources are provided by the participating Member States.

An evaluation of the MAB Programme conducted in 1981 cited, as one indication of its effectiveness, 'the important multiplying effect – from tenfold to a hundredfold – of relatively modest amounts of funds invested (by UNESCO) in the MAB Programme, in attracting additional support for the Programme from national, bilateral and international resources' (UNESCO, 1981). As a nationally based, international programme of scientific cooperation, the primary commitment and source of funding within MAB has been at the national level. Some countries have set aside special funds for MAB research, particularly in its first two decades. In most cases, however, support for MAB field projects and work in biosphere reserves comes from the normal operating budgets of participating institutions and national grant-giving agencies. In a number of countries, particularly in developing regions, outside support and endorsement has proved of strategic importance, in some cases permitting MAB groups to secure additional national support for research. In other cases, it has provided a useful framework at the national level to bring together specialists from different disciplines in the natural and social sciences to examine the possibilities for collaborative action. Generally, such outside funding has been channelled on a bilateral basis, or through a collaborating international organization.

NATURAL SCIENCES STAFF

Once approved by the General Conference, the Regular Programme and Budget is implemented by an international secretariat appointed by the Director-General, based in UNESCO Headquarters and in a network of Field Offices and specialized institutes and centres. In addition to the Office of the Director-General, sectors responsible for administration and for external relations and cooperation, and what are called central services, there are currently five substantive programme sectors – Education, Natural Sciences, Social and Human Sciences, Culture, and Communication and Information.

REGULAR PROGRAMME

As with the other programme sectors, the professional staff responsible for the planning and implementation of activities in the Natural Sciences Sector are appointed in large part as a function of their expertise in a particular area, as well as their experience in such fields as international scientific cooperation. Changes in the way that staff resources are indicated and grouped in various planning and administrative documents make comparison over time somewhat hazardous. This said, from an initial staff of twelve in 1946 and fourteen in 1947, the number of staff associated with Natural Sciences increased progressively

Figure V.5.6: Natural Sciences Sector staff, compared to total UNESCO posts, 1946–2000

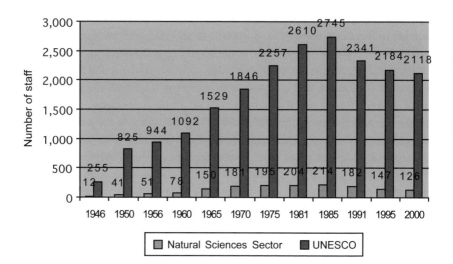

Sources: Information on Natural Sciences Sector staff taken from the Lists of Secretariat Members (1945–85), List of Staff (1991) and UNESCO Telephone Directory (1995, 2000); on UNESCO posts from: 1946: Approved programme and budget; 1950 and 1956: the Chronology of UNESCO 1945–87; 1960: 10 C/5 approved; 1965: 13 C/5 approved; 1970: 15 C/5 approved; 1975: 18 C/5 approved; 1981: 21 C/5 approved; 1985: 22 C/5 approved; 1991: 25 C/5 approved; 1995: 25 C/5 approved; 2000: 30 C/5 approved. Note: Data compiled and analysed by Lotta Nuotio (November 2004).

Figure V.5.7: Personnel of the Natural Sciences Sector, 1946–2000

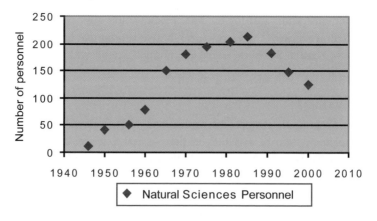

Sources: Lists of Secretariat Members (1945–85), List of Staff (1991), UNESCO Telephone Directory (1995, 2000).

over the next forty years, reaching a peak of 214 in 1985 and declining from that time to 126 in 2000 (Figures V.5.6 and V.5.7).[30] Another trend suggested by Figure V.5.6 is that the Natural Sciences Sector has suffered disproportionately from staff reductions since the mid-1980s. In 1985, Natural Sciences represented 8.1 per cent of UNESCO staff (218 posts out of a total of 2,745), compared to 5.7 per cent in the year 2000 (126 out of 2,118).

In 1947, out of a total staff of fourteen, there was one staff member for medical science and another for agricultural science, reflecting that in the first few years the science mandate of UNESCO within the UN system covered technical areas that have since been incorporated by sister agencies (in these two cases, WHO for medical science, and the Food and Agriculture Organization [FAO] for agricultural science). There were four staff associated with Field Science Cooperation Offices: two for Latin America and two for the Middle East.

In terms of the disciplinary backgrounds of the technical personnel of the Natural Sciences Section of this period, in a 1948 lecture, Joseph Needham (1949, p. 24) provided the following listing of staff 'competences': agriculture (1), anthropology (1), astronomy (2), biochemistry (3), botany (2), chemistry (2), aeronautical engineering (1), civil engineering (2), geography (2), mathematics (1), metallurgy (1), neurology (1), parasitology (1), pathology (1), physics (3), and radiology (2). Needham commented that it was 'necessary to have a wide range of competences like this if you are going to operate what is in effect a world liaison centre for the natural sciences'.

In 1970, the total staff of the sector numbered 181, with six staff in the Office of the Assistant Director-General and twenty-one in the Administrative Unit. At this time, there were three departments. First, the Department of Science Policy and Promotion of Basic Sciences included divisions dealing with Science Policy (14 posts), Scientific Documentation and Information (6) and International Cooperation in Scientific Research (11). Second, the Department of Science Teaching and Technological Education and Research included divisions concerned with Technological Education and Research (36 posts, including units for the promotion of engineering science and research, engineering education and training of technicians), Science Teaching (20) and Agricultural Education and Science (10). Third, the Department of Environmental Sciences and Natural Resources Research comprised the Division of Studies and Research Relating to Natural Resources (19 posts), the Office of Hydrology (14) and the Office of Oceanography (16). In contrast with the situation in 1965, staff dealing with extrabudgetary projects were no longer in a separate operational division with

---

30 For both Figure V.5.6 and Figure V.5.7, different sources were used in compiling numbers of staff, including periodically issued lists of Secretariat members and the internal telephone directory. Thus the basic data are not strictly comparable, and the figures should be understood as indicative only.

regional sub-units, but were incorporated in the various substantive divisions of the sector (Box V.5.3).

BOX V.5.3: TECHNICAL ASSISTANCE PROJECTS: BACKSTOPPING OPTIONS

As early as 1950, UNESCO set up a department for technical assistance, which had full control over all extrabudgetary projects and fellowships. After that, following various structural reforms, the implementation of extrabudgetary projects was entrusted to individual sectors 'and then to the programme units and Field Offices concerned, so as to establish permanent interaction between what was being said and what was being done' (UNESCO, 1997, p. 225).

In respect to the natural sciences, during at least two periods (in the mid-1960s and early 1980s), there existed discrete units or divisions responsible for the planning and implementation of technical assistance projects financed through extrabudgetary resources. These discrete operational units were staffed by technical personnel with background and experience in fields such as mechanical engineering or hydraulics (i.e. not by generalized project managers).

For most of the period under review, technical assistance and other extrabudgetary projects came under the purview of the individual divisions of the Natural Sciences Sector. At times, this work was entrusted to one or several individuals with particular experience and expertise in running such projects, and on occasions there have been specific units identified within particular divisions for this work. But the trend since the mid-1980s has been that individual programme specialists handle both Regular Programme and extrabudgetary activities, and that extrabudgetary funds have come to be perceived as a reinforcement and sometimes invaluable ingredient of Regular Programme activities.

In 1975, staff resources totalled 195 posts. Attached to the Office of the Assistant Director-General (13 posts) and the Administrative Unit (21), there appeared a publications unit for the sector (2), a science cooperation bureau for Europe (3) and a unit on 'Science in the Seventies' (5). By this time, the intermediary departmental structures (between sector and divisions) had disappeared. The main changes in divisional structures related to work in the environmental sciences, with discrete divisions for the Ecological Sciences (including MAB, 17 posts), Earth Sciences (including IGCP, 10) and Water Science (including IHP, 12). The previous Office of Oceanography had been split into the Intergovernmental Oceanographic Commission (IOC, 15 posts) and the Division of Marine Sciences (11).

Six years later, in 1981, the structure of the sector remained largely unchanged, with a 5 per cent increase in number of personnel to 204. The main changes included the first mention of a Coordination and Evaluation Unit (1 post) attached to the Office of the Assistant Director-General and the reinstitution of a Division of Operational Programmes (32), with four regional sections. Similarly, the situation in 1985 showed only minor changes. Total numbers peaked, at 214. Within the Division of Technological Research and Higher Education, there was an increase in staff working on energy issues (9 posts).

By 1991, overall post numbers had started to decline (to 182). The setting up of a Bureau for Coordination of Environmental Programmes (6 posts) reflected the Director-General's resolve that UNESCO should make an important and distinctive contribution to the preparations for the UN Conference on Environment and Development in Rio de Janeiro (June 1992). In the list of staff for the Basic Sciences Division, discrete units were identified for mathematics and computing (2 posts), physics (2), natural products and chemical sciences (2), bioscience and microbiology networks (2), molecular, cell biology and biotechnology (2) and basic medical sciences (1). In 2000, post numbers had been reduced to 126. In qualitative terms, there is a long-standing gender imbalance in professional staff in the natural sciences programmes, particularly at higher managerial levels (Box V.5.4).

> **BOX V.5.4: ON GENDER AND THE NATURAL SCIENCES 'GLASS CEILING'**
>
> The gender representation in the higher management of the Natural Sciences Department/Sector might be considered regrettable by many, exceptionable by some. In contrast to the other programme sectors, over a sixty-year period, no woman has been appointed to the posts of Assistant Director-General for Natural Sciences, Deputy Assistant Director-General, or director of a Natural Sciences Division.[1] The list of directors of Field Science Offices includes but one woman.[2] Breaking the 'glass ceiling' in the Natural Sciences Sector, as in the broader scientific world, remains a challenge.[3]
>
> ---
> 1   Mireille Jardin was acting director of the Division of Ecological Sciences from August 2003 to June 2004. Several women have headed (and are currently heading) units or sections (i.e. structures at the sub-divisional level) within the Natural Sciences Sector. But to the authors' best knowledge, there have been no women in higher management posts in the Sector.
>
> 2   Marine biologist Marta Vannucci was director of the Regional Office for Science and Technology for South and Central Asia (ROSTSCA) in New Delhi (India) in the early 1980s.
>
> 3   The term 'glass ceiling' was coined in 1986 by the *Wall Street Journal* to describe the apparent barriers that prevent women from reaching the top of the corporate hierarchy.

OTHER HUMAN RESOURCES

If the core personnel involved in the planning and implementation of the natural sciences programme are financed through the biennial provisions of the Regular Programme and Budget, other mechanisms have contributed significantly to the deployment of scientific and technical staff by UNESCO, including experts financed through extrabudgetary resources and specialists seconded by national institutions. And perhaps most important of all is the voluntary, unpaid contribution of untold thousands of individual scientists to the programmes and projects of the Organization.

Extrabudgetary posts

From the late 1950s, UNESCO took on the role of executing agency for several hundred projects to develop science and technology in Member States. Their implementation entailed the deployment of many high-level specialists, generally on fixed-term (e.g. two-year) assignments financed by such sources as the UN Expanded Programme of Technical Assistance (EPTA), the UN Special Fund, UNDP, and funds-in-trust arrangements. An indication of the importance of this deployment is given in a comparison of the sources of funding of UNESCO's scientific and technical staff in 1963–1964. During this period, some 475 experts were deployed by UNESCO within Technical Assistance and Special Fund projects, ten times the number of specialists financed through the Regular Programme (Table V.5.3).

Table V.5.3: Established posts in the Department of Natural Sciences in 1963–1964

	Headquarters staff	Headquarters staff	Field staff	Field staff
	Regular Programme	UN Special Fund	UNESCO Officials	Technical Assistance and Special Fund experts
Staff with scientific and/or technical training at university (or equivalent) level	41	15	8	460
Administrators with university or equivalent training	3	9	–	–
Staff in the General Service category	34	18	16	–
Partial totals (Grand total)	78	42	24	460 (604)

Source: UNESCO, 1964, p. 42.

The Associate Expert scheme

The Associate Expert scheme is among the mechanisms that provide an opportunity for young professionals to contribute to international technical cooperation. Young postgraduates from several of the countries funding the scheme (presently sixteen in number) have made a distinctive contribution to the science and technology programmes of the Organization, in fields such as agroforestry, coral reef monitoring, mineralogy, ecological economics, forest biology, hydrology, geography, soil survey and evaluation. Several have subsequently joined the staff of UNESCO and other international organizations.

Secondments

Another important source of technical expertise has been the detachment of scientists by national institutions to assist the Secretariat in various activities. Generally, the arrangement has entailed the continued payment of salaries by the scientist's employer, with a position being kept open when he or she returns home after a fixed-term (generally one to three years) detachment. On its part, the Organization has customarily provided a nominal ($1) contractual contribution or top-up support to cover additional living expenses.

A good example would be the successive secondments to the MAB Secretariat in the 1970s by the US National Park Service and the US Forest Service, which proved decisive in the development of the biosphere reserve concept. In the 1980s, two senior French scientists seconded to the Secretariat were instrumental in developing the comparative studies on 'Land–inland water ecotones' and on 'Rural landscape dynamics in temperate zones'. Another example, from the 1990s, was the Director and Project Coordinator for the IUBS-MAB collaborative study on Tropical Soil Biology and Fertility (TSBF), based in the UNESCO Regional Office in Nairobi and funded directly by the Rockefeller Foundation and the UK Natural Environment Research Council. Secondments such as these provide fresh inputs and new perspectives on the challenges faced by the programmes of the Organization.

In oceanography and the marine sciences, the special status of the IOC has facilitated secondment of specialists to the Secretariat. In the early 1970s, a staff member was seconded from each of FAO, WMO and the Intergovernmental Maritime Consultative Organization (IMCO) to facilitate cooperation among UN institutions, and in 1974 about one-quarter of IOC staff salaries and operational funds were provided by members of the Inter-Secretariat Committee on Scientific Problems Related to Oceanography. Another example, which stretches over many years and continues to this day, is the secondment to the IOC Secretariat of specialists by the National Oceanographic and Atmospheric Administration (NOAA) of the United States, in such fields as fisheries biology, physical oceanography and ocean observations for climate.

Contributions of individual scientists

Like many international scientific organizations, UNESCO is indebted to the many thousands of individual scientists who have contributed their know-how and expertise to programmes in a number of fields. In some cases (particularly for operational projects linked to capacity-building), scientists and other technical staff may be recruited as paid consultants and experts. More often, individual scientists have generously made their skills and knowledge available free-of-charge, with UNESCO's contribution generally that of an economy-class air ticket and support to living expenses, and providing outlets for communication. One example is the 600-odd authors from 42 countries who wrote the 7,600 pages of the 30 volumes of the Arid Zone Research Series in the 1950s and 1960s, who neither asked for nor received the slightest remuneration for their endeavours (Batisse, 2005).

GOVERNANCE OF THE NATURAL SCIENCES PROGRAMMES

At the level of the overall programme in the natural sciences, the General Conference, at its seventh session (in 1952), authorized the Director-General to establish an International Advisory Committee on Scientific Research, which was duly set up in 1954. Its statutory functions were to advise the Director-General in two main areas: first, the fields and methods of action which the Natural Sciences Department should start to plan for, based on the changing requirements of the world scientific community and of the scientific and technological policies of Member States; and second, the best methods of carrying out the approved programme of the Department and action planned in light of recommendations of the Specialized Committees of the Department.

The Committee comprised seventeen members, of whom fourteen were scientists of different nationalities representing their countries' top research organizations. The other three were representatives of ICSU, the Council for International Organizations of Medical Sciences (CIOMS) and the Union of International Engineering Organizations (UIEO). In addition, the UN system and (on occasion) other intergovernmental organizations were invited to send representatives to each session of the Committee. Individual specialists were also invited to several sessions. The Committee met on a near-annual basis for a decade from 1954, with its ninth and final session in Ottawa (Canada) in June 1963.

As the work of the Committee was phased out, with ICSU being invited by the Executive Board to act as a permanent advisory body to UNESCO on the Natural Science Programme (Baker, 1986), the various programmes and projects in the natural sciences field were to use a variety of mechanisms for ensuring technical assessment and quality control, including advisory groups of individual specialists, scientific boards, inter-institutional coordinating committees and intergovernmental structures of various kinds, each of which has strengths and weaknesses.

In terms of international scientific cooperation, it may be that it is through the so-called intergovernmental programmes that UNESCO has contributed most distinctively to the store of experience worldwide. Many organizations have used advisory groups, scientific boards and inter-institutional coordinating committees. But relatively few are the scientific programmes with their own intergovernmental structures, such as those of IHP and MAB – intergovernmental in that they each have their own governmental bodies within the overall intergovernmental framework of the originating institution (in this case UNESCO), including national committees, an international policy-shaping body and a programme-servicing secretariat. It may be of interest to attempt an assessment of the strengths and weaknesses of the mechanisms that have been used to facilitate planning, implementation and quality control of the intergovernmental programmes, as a springboard for future work.

National Committees provide the mechanisms for orchestrating national contributions to the programme. Optimally, they bring together representatives from a range of disciplines and governmental and non-governmental institutions. They should be capable of mobilizing the human and financial resources needed for meaningful national participation in the programme, including participation in collaborative ventures at bilateral, subregional, regional and interregional levels.

In some countries, these conditions have been met more or less satisfactorily. In others, National Committees often lack resources, both human and financial. In certain countries, National Committees may be an obstacle rather than a facilitator to the participation in the programme of scientists in that country. Some Committees have become ossified, beset by inertia and marginalized within the national scientific community and/or policy-making machinery.

Guiding and supervising the intergovernmental programme is the responsibility of the International Coordinating Council. Functions include reviewing progress, recommending research projects to countries, making proposals on the organization of regional or international cooperation, assessing priorities among projects, coordinating international cooperation of Member States, liaising with other international scientific programmes, and consulting with international NGOs on scientific or technical questions. The Council reports via the Director-General on its activities to each ordinary session of the General Conference.

The Council is composed of a representative group of the Member States. At each ordinary session of the General Conference, normally held every two years, half the members of the Council are elected (members can be re-elected). As noted in the Council's own statutes, the individuals who represent the Member States on the Council are generally experts in the technical fields covered by the programme and may be active in implementing programme activities at national, regional and international levels. They are designated by their countries. The Council normally meets once every two years, usually at UNESCO's Headquarters. Although each Member State has only one vote, it can send as many experts or advisors as it wishes to the Council sessions. In addition, other Member States which are not members of the Council can send representatives as observers, as can other UN institutions and the several NGOs which may act as advisory bodies to the Council.

Several factors appear to militate against the Council, in its present form, discharging its functions in an optimal way. Selection of proposed members within each regional group may be determined as much by conditions affecting the diplomatic process as by the extent of a country's participation in the programme. This means that some Council members have no real stake in the programme. In addition, the Council is intergovernmental, with the costs of national representation being borne by governments, and for many countries it is not easy to finance the participation of technical personnel. Representation may be solely through diplomats posted in Paris.

ON COOPERATION AMONG DISCIPLINES AND SECTORS

In much of science, the cross-fertilization of ideas and techniques between disciplines is a principal motor of progress. Generally, cooperation and cross-fertilization takes place between closely related disciplines, or within the broad families of what might be called the 'natural sciences' and the 'social sciences'. Within the natural sciences programmes, there are many examples of successful collaboration and cooperation among specialists with backgrounds in different disciplines: among geoscientists, medical doctors and veterinarians in IGCP Project 454 on Medical Geology; among soil biologists, pedologists and agronomists in the IUBS-MAB collaborative project on Tropical Soil Biology and Fertility (TSBF); between analytical chemists and ethnobotanists in the work on the chemistry of natural products in East and South-East Asia.

Several programmes and projects have contributed to a better understanding of what elements might be important in the planning and conduct of research that cuts across disciplines. More importantly, they have served one of UNESCO's key roles as a 'laboratories of ideas', allowing the testing of concepts that might otherwise remain just that ... theories untested by practice. An example is the MAB Programme, which – with its focus on complex environmental and land use problems and involvement of researchers from both the natural and social sciences – has provided a testing ground for interdisciplinary approaches to problems in a variety of contexts.

A 1986 assessment suggested a number of steps that could be taken to help ensure such research attains its objectives (di Castri and Hadley, 1986), including: identifying key problem to be addressed, distilling the precise scientific target from the social demands, selecting the relevant disciplines and researchers, choosing the area and scale of study, adapting to a moving target, obtaining the involvement of all the 'actors', developing a continuum of actions from basic and applied research through to training and information diffusion, and early and explicit definition of the criteria for evaluation (the 'rules of the game').

If complete success has proved elusive, there are sufficient examples to show that cross-disciplinary (natural and social science) approaches to resource and environmental problems can work, given the right mix of ingredients.[31] But while interdisciplinary approaches have shown their worth, they founder easily. Obstacles include the personalities of individual scientists, differences in scientific method, and administrative, institutional, political and financial bottlenecks (Table V.5.4).

An issue related to cooperation across disciplines is cooperation across *sectors*, where 'sector' has two senses. First, there are the different economic

Table V.5.4: 'Interdisciplinary research for land use planning' at a glance

Rationale for an interdisciplinary approach involving natural and social sciences	• As an operational and social response to the complexity of today's environmental and land use problems.
Common failures and shortcomings	• Scientifically, results often mediocre, not published; tendency for researchers to become marginalized within scientific community. • Socially, primary emphasis on local situations not sufficiently placed within context of national and international driving forces; danger of local communities becoming more marginalized within national society.
Obstacles encountered	• Difficulty of distilling clear scientific target from confused, poorly articulated and often contradictory demands. • Lack of procedures for periodic, unbiased evaluation of programmes and projects – a motor for institutional innovation. • Intrinsic resistance to change of sectoral institutions, from academic to administrative, national to international. • The effort, the risk involved in breaking new ground beyond and away from traditional science, safe careers, sources of research funding, and lack of incentives for individual scientists to even try.
Challenges	• Increase timescales (historical context) and space scales (from site to regional and international connections) within which research planned and undertaken. • Encourage fluidity and flexibility in institutional capacities to respond to surprise and change. • Develop the theoretical basis for studying man–environment interactions.

Source: di Castri and Hadley, 1986.

---

31 The issue of interdisciplinarity within MAB has been addressed in such articles as: (a) di Castri, 1976; (b) Zube, 1980; (c) Whyte, 1982 ; (d) Whyte, 1984; (e) Spooner, 1984; (f) di Castri and Hadley, 1986; and (g) Price, 1990.

sectors (agriculture, manufacturing, education, etc.). An example of a project that sought to develop mutually supportive links between economic sectors is the interagency (FAO, UNESCO, WMO) agricultural biometeorology project of the 1960s, to promote development and field testing of methodological guidelines for agroclimatological surveys leading to better matching of crops with soils and climate conditions.[32]

Second, there are the various programme sectors within UNESCO, which have evolved in name and number over the years. Currently (2006), they number five: Education, Natural Sciences, Social and Human Sciences, Culture, and Communication and Information. What has not changed much is their vertical structure – each is headed by an Assistant Director-General, subtending a handful of individual structural units each with its own director and staff. Over six decades, programmes have tended to be planned, budgeted and implemented within the hierarchical vertical framework of the different sectors (Box V.5.5), which has presented a recurrent challenge to the Organization in terms of options for addressing issues that cut across the concerns of two or more sectors.

Many are the projects that have been conceived as cross-sectoral or even house-wide operations. Success has been mixed. Thus, in the late 1940s, the Hylean Amazon project was selected as one of four all-UNESCO major initiatives. But activities remained largely in the Natural Sciences, apart from some ethnological work. Similarly, in the late 1950s, the Major Project on Arid

---

BOX V.5.5: UNESCO'S HIERARCHICAL STRUCTURE: REFLECTIONS OF AN INTERNAL WORKING GROUP IN 1970

'The UNESCO Secretariat was conceived at the beginning on the basis of a classically modelled hierarchical structure, such as that found in national administrations. The strongly demarcated system of grades gives an almost military feel to this system. One might ask whether a system of this kind is suited to the tasks of a secretariat like that of UNESCO. In any event, experience would seem to suggest that, over the years, UNESCO has tended to conserve the drawbacks of such a system (rigidity, excessive importance of grades, paralysing fear of sanctions, discouragement of initiative, adhesion to the letter rather than the spirit of regulations, encouragement of flattery towards superiors, etc.) rather than the advantages (clear lines of responsibility, efficiency in communications and execution of orders, etc.)

At the same time, it is clear that the classical hierarchical system theoretically in place does not function in practice and that it is increasingly undermined by 'back room' practices and parallel structures, which lead to confusion about roles and responsibilities.'

—Extract from 'Rapport du groupe de travail sur les lignes d'autorité et les niveaux de responsabilité' (UNESCO, 1970; in French only, translated by the authors).

---

32   See Part III, Chapter 2, M. Hadley, 'Nature to the fore', in the present volume.

Lands was ranked as one of three house-wide projects, and should in principle have involved the other sectors of UNESCO (such as Education, Communication or Social Sciences), in addition to the Natural Sciences. But, as related by Michel Batisse, the intention to develop intersectoral action remained largely symbolic, apart from the interest of anthropologist Alfred Metraux (a staff member of what was then known as the Department of Social Sciences) for studies on nomadism (Batisse, 2005, p. 42). In short, issues of professional territoriality and personal job security may represent important obstacles to cooperation across and among disciplines, sectors and organizations, in UNESCO as elsewhere (Box V.5.6).

BOX V.5.6: ON TERRITORIALITY AND JOB SECURITY

'Why is it so difficult to get agreement on co-ordination? I came to the conclusion that the international bureaucrat has most of the bad features of a national bureaucrat, but he has them in a greater order of magnitude. The international bureaucrat has only one thing he can call his own: that is his program. If he admits that somebody in another agency can do some of the things he is doing, then part of his shield is broken. So he will just not agree to co-ordinate.'

—Jim Harrison, Assistant Director-General for Natural Sciences (January 1973 to March 1976); from 'James Harrison: A Brighter Side of UNESCO' (Government of Canada).

A range of issues of intersectoral concern have been flagged in many chapters of this volume, including the use of scientific information in educational and public awareness activities, ethical dimensions of scientific research, and protection of the world's natural and cultural heritage. Several positive examples of effective cooperation in addressing cross-sectoral issues have been described, including the experimental creation of semi-autonomous units outside the vertically structured programme sectors, such as that created in the early 1990s for the servicing of the World Heritage Convention.[33]

Another approach has been the setting up of intersectoral initiatives, such as the four projects launched by the Director-General as an immediate follow-up to the UN Conference on Environment and Development in Rio (UNCED) in 1992, dealing (respectively) with information challenges, adaptation of training programmes and institutional functioning, studying and testing diverse approaches to sustainable development, and biological diversity and its functional significance and dynamics in time and space (UNESCO, 1992). More recent cross-cutting projects – entailing the provision of funds within the budgets of two or more sectors – have included those on Local and Indigenous Knowledge Systems

---

33   See Part III, Chapter 13, B. von Droste, 'A gift from the past to the future', in the present volume.

(LINKS) and on Small Islands Voice. Another mechanism is that of units or focal points responsible for promoting and coordinating house-wide cooperation in particular fields, such as inputs to the planning, convening and follow-up of UNCED, and of the UN Conferences on Sustainable Development of Small Island Developing States (held in Barbados in 1994 and Mauritius in 2005).[34]

More generally, experience in fostering intersectoral and interdisciplinary cooperation suggests that it is often much more readily achieved within the real-world concerns of field projects and working laboratories than in the rarefied confines of centralized bureaucracy (i.e. the UNESCO Secretariat). This is reflected in the widely held view among staff that intersectoral cooperation does not represent a major issue or difficulty in UNESCO's various Field Offices. Staff resources are generally tightly stretched. Staff attached to one sector are often called upon to undertake or contribute to activities headed by staff with other sectoral affiliations. And several Field Offices will use a programme related to one sector as an initial building block for initiating activities that are of interest to a particular country or community; from that starting point, components relating to other programmes and sectors are added. Over time, this leads to activities that relate to several disciplinary fields and sectoral interests. According to a discussion paper prepared by two directors of Regional Field Offices, 'the reason that multidisciplinarity works at field level is that invariably the combination of discipline elements and their respective valences, are fundamentally dependent on the specific situation' (Hill and Moore, 2005).

For instance, the Jakarta Office started its activities in the Siberut Biosphere Reserve (situated off the north-western coast of Sumatra, Indonesia) in 1998 with a small-scale research project on gender-related resource management.[35] In the first year, the project covered two villages, with just three field workers. Over the following six years, the field team expanded to more than fifty, representing a broad variety of stakeholders that included indigenous communities, national park staff, local and national NGOs, and government institutions. The team comprised people with a variety of disciplinary backgrounds, expertise and skills in conservation, education, community development and agroforestry. UNESCO inputs came from three sectors (Natural Sciences, Social and Human Sciences, and Education), with recent component projects including an archaeological survey and work on village regulations for forest management.

---

34   In addition to officially designated coordinating units and focal points, there presently exists within the Natural Sciences Sector an informal listing of staff members responsible for keeping abreast of contacts with other sectors, institutions and programmes and of activities in particular countries, regions and technical fields. List available from the Executive Office of the Assistant Director-General for Natural Sciences.
35   Accounts of intersectoral activities at Siberut are given in the Annual Reports of the UNESCO-Jakarta Office, such as those for 2002 (p. 13) and for 2003 (pp. 7–8). Or see the overview of the work at Siberut at http://www.unesco.or.id/activities/science/env_sci/sitsup_env/85.php.

Other experience has suggested that science and technology – particularly in the emerging economies in such regions as East and South-East Asia – has become increasingly multidisciplinary and 'contextualized' in the last couple of decades, rather than driven by questions that arise out of disciplinary discourses as such. This trend has been described by the Director of the Jakarta Office, Stephen Hill, in terms of the dissolution of boundaries between disciplines and the increasing presence of complex webs of personal relationships embodying transfers of knowledge across institutional boundaries (Hill, 1995; Hill and Turpin, 2003; Hill, 2004). Disciplines appear to have increasingly less relevance in driving research fields; the majority of research is now published across disciplinary boundaries. The driving dynamic at the forefront of new scientific knowledge today is what Hill describes as a 'multi-type complexity'. Not only is more than one scientific 'discipline' involved in finding solutions to problems, but so too are different types of knowledge, both explicit and tacit. And – in line with these changes in the way that science is done and applied – the research supported and sponsored by the natural sciences programmes of the Organization has also changed, with an increasing emphasis on research that cuts across disciplines and specialities.

There is also evidence at the programme level that, over time and with the increasing availability of resources, relatively focused activities in a particular field can blossom in many directions, taking on an increasingly intersectoral and interdisciplinary character. An example is the evolution of work on freshwater resources. From the launching of the International Hydrological Decade in 1965 to its transformation into the IHP in the mid-1970s, this work remained technically focused, largely corresponding to the research and training agenda of those scientists (the hydrologists) dealing with the study of the occurrence and distribution of water within the natural environment. This concern remains. But for the last decade and a half, many additional perspectives have become prominent, reflected in the focus of IHP-V (1996–2001) on stimulating stronger interrelation between scientific research, application and education, and the concern of IHP-VI (2002–2007) with 'Water interactions: systems at risk and social challenges'.

Related changes in emphasis have included: increasing focus on biological dimensions of freshwater ecosystems (including biological diversity of freshwaters, and strengthening of work on ecohydrology and land/water ecotones); increased attention to the historical dimensions of water use over time (including lessons to be drawn from traditional water harvesting systems such as qanats and the use of archival resources for climate history research); the relations of different cultures and civilizations to water; increased attention to issues linking water and security (including the launch of the IHP PCCP [From Potential Conflict to Cooperation Potential] initiative); increasing focus on addressing issues linking water and poverty and the Millennium Development Goals (such as the urban water development project led by the Institute for Water Education in Delft, with major European Union funding); the reinforcement of cooperation between regional and international organizations dealing with freshwater resources (as in the World Water Assessment Programme [WWAP]); the decision of the UN General Assembly to proclaim the period from 2005 to 2015 as the International Decade for Action 'Water for Life'; and the designation by the UN of 'Water and Culture' as the theme for World Water Day 2006, with UNESCO requested to take the lead-agency role.

POLITICAL DIMENSIONS

UNESCO is a specialized agency of the United Nations, an intergovernmental body, and as such subject to pressures from the governments of its Member States and any number of blocs and groupings. All Member States have accepted the idealistic constitution of the Organization, with its references to the disinterested exchange of knowledge, internationally, and that is a process that has not always been popular with governments. This is the paradox and ambiguity of UNESCO and the roots of the tension that is inseparable from its work.

Throughout its sixty years, there have been explicitly political debates about the Organization's work in its governing organs – and within its connected intellectual communities and civil society associations. One observer, Richard Hoggart, has commented that the regrettable thing about such debates was not that they had a political inspiration – UNESCO is and always has been political and cannot avoid being so – but rather that very often these debates were in themselves distorted and conducted on false premises (Hoggart, 1978, p. 80).

During the Cold War years especially, UNESCO's programmes were used extensively to promote the political agendas of particular countries and groups of countries, and indeed the usefulness of the Organization in this regard was central to the Soviet Union becoming a member in 1954 (Gaiduk, 2005). The natural sciences programmes were no exception. Examples include the attempts of both the United States and the Soviet Union to use the Organization to assert leadership within the marine sciences, a domain that was both strategically important and crucial for understanding the source of the world's food supply (Hamblin, 2005). In the hydrological field, Batisse (2005) has described how several sessions of the Coordinating Council of the International Hydrological Decade (IHD, 1965–1974) were enlivened by attempts by the Soviet Union and its allies to introduce 'politically motivated' issues into the debate, for example, on the effects of defoliants on freshwaters. But that same initiative (IHD) also provides an excellent example of how the Organization was able to bring together, in a shared cooperative endeavour, scientists from countries of very different political affiliations.

In terms of substance, work in the natural sciences field has been relatively free of political overtones compared with activities related to culture, human rights, the free flow of information, and education. This said, some within the scientific community have commented on the pervasiveness of political considerations in the Organization's scientific programmes, at least in the intergovernmental programme with which they were most familiar.[36] But perhaps most of those with extensive experience in promoting international scientific cooperation in the period since the Second World War would argue that the natural sciences programmes have been relatively 'politics-free', and that (in the main) the Organization has succeeded in preventing its science and technology programmes becoming political footballs.

Political pressures and favours have sometimes come to bear in the process of appointments of staff, and this is inevitably and understandably the case for appointments at higher management levels. But again, the natural sciences programmes have probably been less subject to 'inward parachuting' of unqualified or inappropriate personnel than some other parts of the Organization.

To the extent that UNESCO's programmes enter into policy domains (and considering that there is a single word for 'policy' and 'politics' in some languages), it is scarcely surprising that political dimensions do surface. In some countries, this may impinge on the relations of part of civil society with central government, as in the debates in the United States over biosphere reserves and World Heritage sites. But, more generally, some would argue that if science is indeed to provide information useful for policy and management purposes, then inevitably political dimensions and considerations will come into play. And so much to the good.

If the natural sciences programmes have indeed been relatively free of political dimensions, then one basic reason may be that the bedrock of these programmes is research scientists. Many – perhaps most – members of the 'scientific tribe' are intellectual monogamists, whose closest contacts tend to be members of the shared scientific discipline, or rather sub-sub-discipline, to which the individual researcher belongs. Nationality and political affiliation count for comparatively little in determining such contacts and relationships.

If an intrinsic characteristic of the natural sciences is its universality, this does not mean that there is a single way of looking at the world among scientists in a particular field. UNESCO has sometimes been able to use its own universality in helping to bring about understanding and cooperation among scientists of different schools, through promoting inter-calibration and harmonization of different methodologies and terminologies. One of the best-known examples is

---

36   'Politics permeate all MAB debates', according to two scientists seconded to the UNESCO-MAB Secretariat in the mid-1980s (Dyer and Holland, 1988).

the process associated with the *FAO/UNESCO Soil Map of the World* (UNESCO, 1974). A seventeen-year long project (starting in 1961), it entailed plotting the world's soils on eighteen detailed maps (at a scale of 1:5,000,000), using a standard terminology and system of classification. Most important, the project had to reconcile significant differences between the systems of representation used by the American, French and Russian schools of soil science. The project was initiated and nurtured by Victor Kovda, the renowned Russian soil scientist and Director of the Natural Sciences Department (1959– 1964), who was highly respected by the international scientific community and who had the charisma and political finesse to bring together scientists from East and West during one of the tensest periods of the Cold War. He was also able to conclude an agreement with a sister agency of the United Nations (FAO) to cooperate over a long-term period on a subject of shared interest and concern. Among other advantages, this enabled funding to be continued – admittedly at times at 'hibernating' levels – in spite of the vagaries of individual institutional budgets.

In line with the universality of science, UNESCO's natural sciences programmes have tended to be firmly international (i.e. interregional) in their concept and substance, even though a regional approach may often be taken in promoting linkages and cooperative activities among scientific communities in different countries. Even then, the regional groupings are often not the geopolitical groupings of the United Nations,[37] or even the groupings of countries recognized officially within UNESCO, with a view to the execution of regional activities in which the representative character of states is an important factor.[38] Rather, the regional groupings in the Organization's natural sciences programmes have often been biogeographical and not geopolitical. Thus, the Major Project on Arid Lands of the late 1950s spanned the arid and semi-arid areas of the globe, covering vast tracts of land in more than fifty countries and territories.[39] Taking part in the Indian Ocean Expedition were scientists and research vessels from countries bordering the Indian Ocean (such as India, Indonesia, Pakistan and Thailand), as well as other countries with strong maritime and oceanographic

---

37   As reflected for example in the UN Economic Commissions for such regions as Asia and the Pacific (ESCAP), Europe (EEC) and Latin America and the Caribbean (ECLAC).
38   Within UNESCO, the five regional groupings are Africa, Arab States, Asia and the Pacific, Europe, and Latin America and the Caribbean. The listing of countries within each regional grouping is given in the successive versions of the 'Manual of the General Conference' and of a collection of 'Basic Texts'.
39   Within the Major Project, the arid and semi-arid areas of the world were grouped in five great provinces: North African–Eurasian, North American, South African, Australian and South American. Continuing work on arid lands brings together scientists and planners from countries with very different socio-economic and geopolitical associations. For example, a workshop on traditional knowledge and modern technology for the sustainable management of dryland ecosystems, held in the Republic of Kalmykia (Russian Federation) in June 2004, brought together scientists, conservation experts and government officials from ten countries: Azerbaijan, Egypt, India, Iran (Islamic Republic of), Jordan, Mongolia, Morocco, Russian Federation, Sudan and Turkmenistan.

traditions. Hydrological studies of large water basins have very often brought together scientists from countries of different political affiliations, such as studies on the Danube and Volga basins. Some of the regional and subregional networks of biosphere reserves have brought together scientists and managers from countries of different economic and political groupings. Examples here are the subregional networks of biosphere reserves in East Asia (members are China, Japan, Democratic People's Republic of Korea, Republic of Korea, Mongolia, Russian Federation) and in the East Atlantic–Macaronesian region (Cape Verde, Mauritania, Morocco, the Azores and Madeira archipelagos of Portugal, Senegal, and the Canary Islands of Spain). And within the Earth sciences, almost every one of the 500 projects undertaken since the launch of the IGCP (now the International Geoscience Programme) in the early 1970s has involved geoscientists from a wide range of countries in different geopolitical and socio-economic regions.[40]

In the basic and engineering sciences too, activities have cut across political boundaries and groupings. Witness the projects and alumni of the Abdus Salam International Centre for Theoretical Physics, or the countries taking part in the work of the Synchrotron-light for Experimental Science and Applications in the Middle East (SESAME) Centre.[41]

Scientific specialists, members of the Secretariat and those involved in policy and politics may also have very different perceptions on priorities and practicalities. An example from the early 1950s relates to a listing of priorities on the establishment of international institutions, based on a compromise between the urgency of the scientific need and the efficiency factor. Following the recommendations of a group of experts in 1949, a joint committee of representatives of UN specialized agencies adopted, in December 1950, a revised list of priorities as a basis for action by UNESCO. Top priority was accorded to setting up an International Computation Centre. At the very bottom of the list – fourteenth out of fourteen – figured a Regional Laboratory for the Physics of High Energy Particles.

Yet it was the proposal that received the lowest expert ranking which received the political support that led to the creation of the European Organization for Nuclear Research (CERN), a jewel in the list of institutions that UNESCO helped catalyse. Of course there were other factors that contributed to the failure of one (the fact that computer developments were outside the direct

---

40   Thus, IGCP Project 471 (2002–06) on 'Evolution of Western Gondwana during the Late Palaeozoic' includes the participation of scientists from some eighteen countries: Argentina, Bolivia, Brazil, Democratic Republic of Congo, Czech Republic, Iran, Madagascar, Mozambique, Namibia, Paraguay, Peru, South Africa, Spain, United States, Uruguay, Venezuela, Zambia, Zimbabwe.
41   The SESAME Council is currently (in mid-2005) made up of the following Founding Members: Bahrain, Egypt, Iran, Israel, Jordan, Pakistan, Palestine, Turkey and the United Arab Emirates.

control of governments) and the success of the other (as noted by one member of the Natural Sciences Sub-Commission of the General Conference, Marcel Florkin, the Natural Sciences Department was headed at the time, by a happy coincidence, by an eminent nuclear physicist [Florkin, 1956]). But the point remains valid that ranking by experts is not necessarily reflected in governmental priorities and actions.

A final political dimension concerns the perception in different countries of the Organization's role in promoting international scientific cooperation. Again an example is provided by geologist Jim Harrison, Assistant Director-General for Natural Sciences from 1973 to 1976:

> One of the things I discovered, when I came back in 1976, was that, while UNESCO may not be important to the USA or even to Canada, it is enormously important to Third World countries. I fear this is not recognized well enough by the industrialized world, who see it too much from their own individual point of view. One minor example in the Canadian program of Man and the Biosphere: we organized a conference on 'The Child in the City'. The Sick Children's Hospital in Toronto had a project going on this and had put in a big chunk of money, but they couldn't get the people they wanted from developing countries because they had no link. However, once the Canadian Commission for UNESCO got UNESCO itself to sponsor the meeting, several people from the Philippines and Indonesia were able to come simply because it was MAB and UNESCO. The powers-that-be in industrialized countries don't seem to realize this (Government of Canada).

BIBLIOGRAPHY

Baez, A. V. 1962. *The Work of UNESCO in the Field of Higher Scientific and Technological Education.* Document NS/ST/1962/5. Paris, UNESCO.

Baker, F. W. G. 1986. *ICSU-UNESCO: Forty Years of Cooperation.* Paris, ICSU, p. 7.

Batisse, M. 1993. Intergovernmental cooperation. In: UNESCO (ed.), *World Science Report 1993,* pp. 152–60. Paris, UNESCO.

——. 2005. *The UNESCO Water Adventure: from desert to water.* Paris, UNESCO.

Behrman, D. 1964. The India of the engineer. In: D. Behrman, *Web of Progress: UNESCO at work in science and technology.* Paris, UNESCO, pp. 11–27.

——. 1979. *Science and Technology in Development: a UNESCO approach.* Paris, UNESCO, 104 pp.

——. 1981. *Assault on the Largest Unknown: the International Indian Ocean Expedition, 1959–65.* Paris, UNESCO, 96 pp.

Conil-Lacoste, M. 1994. *The Story of a Grand Design, UNESCO 1946–1993: people, events and achievements.* Paris, UNESCO.

di Castri, F. 1976. International, interdisciplinary research in ecology: some problems of organization and execution. The case of the Man and the Biosphere (MAB) Programme. *Human Ecology*, Vol. 4, No. 3, pp. 235–46.

di Castri, F. and Hadley, M. 1986. Enhancing the credibility of ecology: is interdisciplinary research for land use planning useful? *GeoJournal*, Vol. 13, No. 4, pp. 299–325.

di Castri, F., Baker, F. W. G. and Hadley, M. (eds). 1984. *Ecology in Practice.* Vol. 2: *The Social Response.* Dublin, Tycooly International Publishing Company, and Paris, UNESCO.

Dyer, M. I. and Holland, M. M. 1988. UNESCO's Man and the Biosphere Program. Roundtable. *BioScience*, Vol. 38, No. 9, pp. 635–41.

Florkin, M. 1956. Ten years of science at UNESCO. *Impact of science on society*, Vol. 7, No. 3, pp. 121–46.

Gaiduk, I. 2005. The Soviet Union and UNESCO during the Cold War. Paper presented at International Symposium 'Sixty Years of UNESCO's History' (Paris, 16–18 November 2005). http://portal.unesco.org/en/ev.php-URL_ID=30447&URL_DO=DO_TOPIC&URL_SECTION=201.html

Gilbrich, W. H. (ed.). 1991. *Twenty-five Years of UNESCO's Programme in Hydrological Education under IHD/IHP.* Technical Documents in Hydrology. Paris, UNESCO.

——. (ed.). 1994. *Hydrological Education during the Fourth IHP Phase (1990–1995).* Technical Documents in Hydrology. Paris, UNESCO.

Government of Canada. Department of Foreign Affairs and International Trade website. 'James Harrison: a brighter side of UNESCO – Science', from 'Canada and the United Nations'. http://www.dfait-maeci.gc.ca/ciw-cdm/caun/Harrison-en.asp (accessed 15 April 2006).

Hamblin, J. D. 2005. The politics of international cooperation in science. Paper presented at International Symposium 'Sixty Years of UNESCO's History' (Paris, 16–18 November 2005). http://portal.unesco.org/en/ev.php-URL_ID=30435&URL_DO=DO_TOPIC&URL_SECTION=201.html

Hill, S. 1995. Confronting the eleven myths of research commercialization. *Nature and Resources*, Vol. 31, No. 4, pp. 2–15.

——. 2004. Laying foundations for innovation bridges in developing countries, or 'Creating a world of butterflies, spiders and bees'. Keynote speech to International Conference on 'Regional Innovation Systems and Science and Technology Policies in Emerging Economies: Experiences from China and the World'. Zhongshan University, Guangzhou (China), 19–21 April 2004.

Hill, S. and Moore, H. 2005. UNESCO Science Sector futures: a view from the field. Discussion paper prepared for Natural Sciences Sector Retreat, 29–31 August 2005. Document dated 11 July 2005.

Hill, S. and Turpin, T. 2003. Supporting key technologies in the APEC region: a new framework for cooperation in human resource capacity-building. Discussion paper for the APEC Industrial Science and Technology Working Group, Seoul, STEPI, August 2003.

Hoggart, R. 1978. *An Idea and its Servants: UNESCO from within.* London, Chatto & Windus, pp. 102–03.

Kovar, P. and Gilbrich, W. H. 1995. *Postgraduate Education in Hydrology: a state-of-the-art report.* Technical Documents in Hydrology. Paris, UNESCO.

Laves, W. H. C. and Thomson, C. A. 1957. *UNESCO: Purpose, Progress, Prospects.* Bloomington, Indiana University Press.

Maheu, R. 1962. UNESCO 1960–1962: Developments and prospects. *UNESCO Chronicle*, Vol. 8, No. 10 (October 1962), pp. 358.

Needham, J. 1949. *Science and International Relations.* Fiftieth Robert Boyle Lecture. Delivered before the Oxford University Scientific Club. 1 June 1948. Oxford, UK, Blackwell Scientific Publications.

Price, M. F. 1990. Humankind in the biosphere: the evolution of international interdisciplinary research. *Global Environmental Change* (December), pp. 3–13.

Schalk, P. H. and Troost, D. G. 1999. Computer tools for accessing biodiversity information. *Nature and Resources*, Vol. 35, No. 3, pp. 31–8.

Spooner, B. 1984. The MAB approach: problems, clarifications and a proposal. In: di Castri et al., 1984, pp. 324–39.

Stephenson, J. P. 1948. *Suggestions for Science Teachers in Devastated Countries.* Paris, UNESCO, 88 pp.

UNESCO. 1961. Arid Zone Fellowship Programme. *Arid Zone* newsletter, No. 12, June 1961, pp. 10–14.

——. 1964. *Science and Technology in UNESCO.* Document UNESCO/NS/ROU/43 (WS/1263.141 [NS]), dated 15 January 1964. Paris, UNESCO.

——. 1970. Rapport du groupe de travail sur les lignes d'autorité et les niveaux de responsabilité. UNESCO Document ADG/ADM/3085/0408, dated 4 August 1970 (French only, translated by the authors).

——. 1973. *New UNESCO Source Book for Science Teaching.* Paris, UNESCO.

——. 1974. *FAO/UNESCO Soil Map of the World*, 1:5,000,000, 10 vols. Paris, UNESCO.

——. 1975. Unesco's activities in promoting education in water resources. *Nature and Resources*, Vol. 11, No. 2.

——. 1981. *General Recommendation of the International Coordinating Council for MAB, meeting for its seventh session on the occasion of the tenth anniversary of the MAB Programme (Paris, 30 September–2 October 1981).* MAB Report Series No. 53, p. 34.

——. 1992. Report of the Director-General on the Follow-up to the United Nations Conference on Environment and Development (UNCED). UNESCO Document 140/EX/10, dated 18 September 1992.

——. 1997. *UNESCO, Fifty Years for Education.* Paris, UNESCO.

——. 2005. UNESCO Flash Info, No. 128-2005 ('Director-General Opens World Conference for the UNESCO Clubs Movement'), dated 18 July 2005.

Valderrama, F. 1995. *A History of UNESCO.* Paris, UNESCO.

Whyte, A. V. 1982. The integration of natural and social sciences in the MAB Programme. *International Social Science Journal*, Vol. 93, pp. 411–26.

——. 1984. Integration of natural and social sciences in environmental research: a case study of the MAB Programme. In: di Castri et al., 1984, pp. 298–323.

Zube, E. H. (ed.). 1980. *Social Sciences, Interdisciplinary Research and the US Man and the Biosphere Program.* Washington DC, US MAB Program.

APPENDIX 1

PROMOTING INTERNATIONAL SCIENTIFIC COOPERATION

HELPING OTHERS GET STARTED: SOME EXAMPLES OF UNESCO'S ACTION

- IUCN: World Conservation Union (1948)
- CIOMS: Council for International Organizations of Medical Sciences (1949)
- IAU: International Association of Universities (1950)
- UIEO: Union of International Engineering Organizations (1950)
- UATI: International Union of Technical Associations and Organizations (1951)
- CERN: European Organization for Nuclear Research (1952)
- ICLA: International Committee on Laboratory Animals (1956)
- Charles Darwin Foundation for the Galapagos Islands (1958)
- IBRO: International Brain Research Organization (1960)
- ICPE: International Commission on Physics Education (1960)
- ICRO: International Cell Research Organization (1962)
- Flora Neotropica Organization (1962)
- CLAF: Latin American Centre for Physics (1962)
- WFEO: World Federation of Engineering Organisations (1968)
- ICASE: International Council of Associations for Science Education (1973)
- FACS: Federation of Asian Chemical Societies (1978)
- ICPAM: International Centre for Pure and Applied Mathematics (1978)
- ANSTI: African Network of Scientific and Technological Institutions (1980)
- ICCS: International Centre for Chemical Studies (1980)
- ASPEN: Asian Physics Education Network (1981)
- IOCD: International Organization for Chemical Sciences in Development (1981)
- ICIMOD: International Centre for Integrated Mountain Development (1982)

- INISTE: International Network for Information in Science and Technology Education (1985)
- INSULA: International Scientific Council for Island Development (1989)
- MCBN: Molecular and Cell Biology Network (1991)
- INASP: International Network for the Availability of Scientific Publications (1992)
- World Foundation for AIDS Research and Prevention (1993)
- STEMARN: Science and Technology Management Arab Regional Network (1994)
- ICET: International Council for Engineering and Technology (1994)
- Tropical Soil Biology and Fertility (TSBF) Institute of the International Centre for Tropical Agriculture (CIAT) (2001)
- INWES: International Network of Women Engineers and Scientists (2001)
- WAYS: World Academy of Young Scientists (2003)
- SESAME: Synchrotron-light for Experimental Science and Applications in the Middle East (2003)

APPENDIX 2

RESEARCH AND TRAINING IN SCIENCE AND TECHNOLOGY
CREATING AND REINFORCING INSTITUTIONAL CAPACITIES: SOME EXAMPLES OF UNESCO'S ACTIONS IN THE MID-1960s

- Algeria: Faculty of Civil and Electrical Engineers, University of Algiers. 1963–68. US$1.24 million.
- Argentina: National Institute of Petroleum. 1963–68. $1.27 m.
- Cambodia: National School for Public Works, Building and Mining. 1964–68. $0.77 m.
- Chile: Faculty of Engineering, University of Concepción. 1961–65. $1.05 m.
- Colombia: Industrial University of Santander, Bucaramanga. 1962–66. $1.50 m.
- Congo (now Democratic Republic of Congo): National Institute of Building and Public Works, Léopoldville (now Kinshasa). 1963–69. $1.42 m.
- Ecuador: National Polytechnic School, Quito. 1961–66. $1.32 m.
- Greece: Training of technical teachers for vocational, industrial schools, Athens. 1963–66. $0.89 m.
- India: Power Engineering Organization (Bhopal and Bangalore). 1960–65. $1.99 m.
- Iran: Tehran Polytechnic, Tehran. 1960–65. $1.56 m.
- Iraq: Technical Training Institute, Baghdad. 1961–66. $0.91 m.
- Kenya: Kenya Polytechnic, Nairobi. 1963–69. $1.42 m.
- Laos: Technical and vocational teacher training. Vientiane. 1961–66. $0.62 m.
- Lebanon: Technical Training Institute, Beirut. 1961–65. $0.71 m..
- Libya: College of Advanced Technology, Tripoli. 1961–68. $1.22 m.
- Malta: Malta Polytechnic. 1961–66. $0.61 m.
- Mexico: National Centre for Technical Teacher Training, Mexico City. 1962–67. $0.90 m.
- Morocco: Engineering School, Rabat. 1960–65. $0.74 m.
- Pakistan: Training of engineers and other technical personnel in West Pakistan. 1961–65. $2.27 m.

- Pakistan (East Pakistan, present-day Bangladesh): Chittagong Polytechnic, Chittagong. 1963–67. $0.91 m.
- Peru: School of Technology, National Engineering University, Lima. 1963–68. $1.15 m.
- Saudi Arabia: Higher Institute of Technology, Riyadh. 1962–67. $1.04 m.
- Syrian Arab Republic: Technological Institute, Damascus. 1963–67. $1.10 m.
- Thailand: Thonburi Technical Institute, Thonburi. 1962–67. $1.14 m.
- Turkey: Middle East Technical University. 1960–64. $1.85 m.
- Uganda: Kampala Technical Institute, Kampala. 1962–68. $1.28 m.
- United Arab Republic (now Egypt): Mansoura Institute of Higher Education, Mansoura. 1963–69. $1.70 m.
- Venezuela: National Polytechnic Institute, Barquisimeto. 1963–68. $1.32 m.
- West Indies (regional): Faculty of Engineering, University of West Indies. 1961–66. $0.97 m.

Source: Based on UNESCO, 1964, pp. 30–32.

Note: For each institution, information is provided on the total extrabudgetary funding for the entire duration of the project.

PART V: OVERVIEWS AND ANALYSES

APPENDIX 3

UNESCO-SPONSORED INTERNATIONAL POSTGRADUATE TRAINING COURSES IN SCIENCE AND TECHNOLOGY
SOME EXAMPLES OF ANNUAL COURSES AT SPECIALIZED INSTITUTIONS, 1960s–1970s

Basic sciences
- Research Techniques in Chemistry. Sydney, Australia.
- Molecular Physical Chemistry. Leuven, Belgium.
- Biophysics and Chemistry of Natural Products. Rio de Janeiro, Brazil.
- Probability Theory and Mathematical Statistics. Budapest, Hungary.
- Basic and Applied Research in Microbiology. Osaka, Japan.
- Experimental Physics. Uppsala, Sweden.
- Computer Sciences. London, United Kingdom.

Technology and engineering sciences
- Metallurgy. Buenos Aires, Argentina.
- Chemical Engineering. Karlsruhe, Federal Republic of Germany.
- Heat and Mass Transfer. Bangalore, India.
- Theoretical Mechanics. Udine, Italy.
- Petroleum Technology. Bucharest, Romania.

Earth sciences
- Prospection, Exploration and Mining. Leoben, Austria.
- Quaternary Geology. Ghent, Belgium.
- Environmental Geology. São Paulo, Brazil.
- Geochemical Prospection Methods. Prague/Bratislava, Czechoslovakia.
- Engineering Geology. Paris, France.
- Geothermal Energy. Fukuoka, Japan.

Hydrology
- Hydrological Data for Water Resources Planning. Prague, Czechoslovakia.
- Hydrological Methods for Developing Water Resources Management. Budapest, Hungary.
- Surface and Sub-Subsurface Hydrology. Roorkee, India.
- General Hydrology. Padova, Italy.

- Hydraulic Engineering. Delft, the Netherlands.
- General and Applied Hydrology. Madrid, Spain.

Ecology
- Limnology. Vienna, Austria.
- Pedology and Soil Cartography. Ghent, Belgium.
- Study and Management of the Natural Environment. Paris/ Montpellier/Toulouse, France.
- Integrated Surveys. Enschede, the Netherlands.
- Soil Sciences and Plant Biology. Granada/ Seville, Spain.
- Natural Resources Research and Land Evaluation. Sheffield, United Kingdom.

Marine sciences
- Marine Biology. Puerto Deseado, Argentina.
- Marine Biology. Copenhagen, Denmark.
- Marine Chemistry. Barcelona, Spain.
- Mangrove Productivity and Ecology. Phuket, Thailand.
- Marine Sciences. Duke University, United States.

APPENDIX 4

INTERNATIONAL SCIENTIFIC CONFERENCES AND SYMPOSIA ORGANIZED OR SPONSORED BY UNESCO

A sampling of subjects and issues

- Science Abstracting. Paris, France. 1949.
- Protection of Nature. Lake Success, New York, USA. 1949.
- High-Altitude Biology. Lima, Peru. 1949.
- Hydrology of the Arid Zone. Ankara, Turkey. 1952.
- Wind and Solar Energy. New Delhi, India. 1954.
- Study of Tropical Vegetation. Kandy, Sri Lanka. 1956.
- Physical Oceanography. Tokyo, Japan. 1956.
- Radioisotopes in Scientific Research. Paris, France. 1957.
- Salinity Problems in the Arid Zones. Tehran, Iran. 1958.
- Plant–Water Relationships in Arid and Semi-Arid Conditions. Madrid, Spain. 1959.
- Termites in the Humid Tropics. New Delhi, India. 1960.
- Changes of Climate. Rome, Italy. 1961.
- Environmental Physiology and Psychology in Arid Conditions. Lucknow, India. 1962.
- Scientific Knowledge of Tropical Parasites. Singapore. 1962.
- Kuroshio. Tokyo, Japan. 1963.
- Arid Lands of Latin America. Buenos Aires, Argentina. 1963.
- Man's Place in the Island Ecosystem. Honolulu, Hawaii, USA. 1963.
- Ecological Research in Humid Tropics Vegetation. Kuching, Malaysia. 1963.
- Problems of the Savannah/Tropical Forest Boundary. Caracas, Venezuela. 1964.
- Scientific Problems of the Humid Tropical Zone Deltas and Their Implications. Dacca, East Pakistan (now Dhaka, Bangladesh). 1964.
- Oceanography and Fisheries Resources of the Tropical Atlantic. Abidjan, Côte d'Ivoire. 1966.
- Methods in Agroclimatology. Reading, UK. 1966.
- Gondwana Stratigraphy and Palaeontology. Mar del Plata, Argentina. 1967.
- Brain Research and Human Behaviour. Paris, France. 1968.

- Trends in the Teaching and Training of Engineers. Paris, France. 1968.
- Investigations and Resources of the Caribbean Sea and Adjacent Regions. Curaçao, Netherlands Antilles. 1968.
- Use of Analog and Digital Computers in Hydrology. Tucson, Arizona, USA. 1968.
- Productivity of Forest Ecosystems. Brussels, Belgium. 1969.
- Environmental Changes and the Origin of *Homo sapiens*. Paris, France. 1969.
- Plant Response to Climatic Factors. Uppsala, Sweden. 1970.
- Geology and Genesis of Precambrian Iron-Manganese Formations and Ore Deposits. Kiev, Ukraine, USSR. 1970.
- Culture and Science. Paris, France. 1971.
- Young Scientists: Population and the Environmental Crisis. Paris, France. 1972.
- Human Implications of Scientific Advance. Paris, France. 1973.
- Biochemistry of Estuarine Sediments. Melreux, Belgium. 1976.
- Education and Training of Engineers and Technicians. New Delhi, India. 1976.
- Biology and Ethics: Problems and Positive Results of Scientific Research in Genetics. Madrid, Spain. 1977.
- Strategies and Policies for Informatics. Torremolinos, Spain. 1978.
- Environmental Management and Economic Growth in the Smaller Caribbean Islands. Barbados. 1979.
- Ophiolite. Nicosia, Cyprus. 1979.
- Mangrove Environment. Kuala Lumpur, Malaysia. 1980.
- Ecology in Practice: Establishing a Scientific Basis for Land Management. Paris, France. 1981.
- Coastal Lagoons. Bordeaux, France. 1981.
- Scientific Forecasting and Human Needs. Trends, Methods and Message. Tbilisi, Georgia, USSR. 1981.
- State of Biology in Africa. Accra, Ghana. 1981.
- International Biosphere Reserve Congress. Minsk, Belarus, USSR. 1983.
- Hydrology of Large Flatlands. Olavarria, Argentina. 1983.
- 100 Years Development of Krakatau and its Surroundings. Jakarta, Indonesia. 1983.
- Science and Society. Bangalore, India. 1984.

PART V: OVERVIEWS AND ANALYSES

- Next Twenty Years in Plasma Physics. ICTP Twentieth Anniversary Symposium. Trieste, Italy. 1984.
- Arid Lands: Today and Tomorrow. Arizona, USA. 1985.
- Integrated Global Monitoring of the State of the Biosphere. Tashkent, Uzbekistan, USSR. 1985.
- Tin and Tungsten Development. Chiang Mai, Thailand. 1985.
- Sustainable Development and Management of Small Islands. Puerto Rico. 1986.
- Mineral Deposit Modelling. Paris, France. 1986.
- Hydrology and the Scientific Bases of Water Resources Management. Geneva, Switzerland. 1987.
- Microbiology and Biotechnologies. Hong Kong. 1988.
- Biotechnology on the Threshold of the Twenty-First Century. Kiev, Ukraine, USSR. 1989.
- New Prospects for Biological Research in the Arab Countries. Amman, Jordan. 1989.
- Second World Climate Conference. Geneva, Switzerland. 1990.
- Tropical Forests, People and Food: Biocultural Implications and Applications to Development. Paris, France. 1991.
- Environmentally Sound Socio-Economic Development in the Humid Tropics. Manaus, Brazil. 1992.
- Science and Technology in Africa. Nairobi, Kenya. 1994.
- Measuring and Monitoring Forest Biological Diversity. Washington DC, USA. 1995.
- Electronic Publishing in Science. Paris, France. 1996.
- Sustainable Integrated Coastal Management. Maputo, Mozambique. 1998.
- World Conference on Science. Budapest, Hungary. 1999.
- Biodiversity and Society. New York, USA. 2001.
- Integrated Global Observing System (IGOS) – Geohazards. Frascati, Italy. 2002.
- Role of Science in the Information Society. Geneva, Switzerland. 2003.
- Geoparks. Beijing, China. 2004.
- Conserving Cultural and Biological Diversity: The Role of Sacred Natural Sites and Cultural Landscapes. Tokyo, Japan. 2005.
- Albert Einstein's Century. Paris, France. 2005.
- Physics and Sustainable Development. Durban, South Africa. 2005.

APPENDIX 5

UNESCO NATURAL SCIENCE PUBLICATIONS AND DOCUMENTS

A DIGEST OF SERIES AND COLLECTIONS

Periodicals/magazines

- Impact of Science on Society. 1950–92 (168 issues)
- Nature and Resources. 1965–99

Technical papers/documents/booklets

- Inventories of Apparatus and Materials for Teaching Science. 1950–57 (8)
- IOC Information Documents Series. 1961–present (1,212 to date)
- Science Policy Studies and Documents. 1965–94 (74)
- Technical Papers in Marine Science. 1965–94 (67)
- IOC Technical Series. 1965–present (68 to date)
- Annual Summary of Information on Natural Disasters. 1969–79 (relating to disasters in years 1966–75) (10)
- Technical Papers in Hydrology. 1970–85 (27)
- Man and the Biosphere (MAB) Report Series. 1971–present (71 to date)
- Technical Documents in Hydrology. 1972–present (193 to date, 118 not numbered, 75 numbered)
- Geological Correlation. 1973–present (33 to date)
- Studies in Engineering Education. 1974–89 (13)
- IOC Workshop Reports. 1974–present (198 to date)
- MAB Technical Notes. 1975–86 (18)
- IOC Manuals and Guides. 1975–present (45 to date)
- GESAMP Reports and Studies. 1975–present (74 to date)
- Reports in Marine Science. 1977–96 (69)
- Natural Hazards. 1980–84 (4)
- Studies in Mathematics Education. 1980–92 (9)
- IOC Training Course Reports. 1980–present (77 to date, mainly published only in electronic format since 2002)
- IOC Reports of Governing and Major Subsidiary Bodies. 1984–present (106 to date)

- IOC Reports of Meetings of Experts and Equivalent Bodies. 1984–present (203 to date)
- MAB Digests. 1989–98 (19)
- IHP Humid Tropics Programme Series. 1990–2001 (15)
- Environment and Development Briefs. 1991–94 (7)
- People and Plants Working Papers. 1993–2003 (11)
- IOC Annual Reports. 1994–present (11 to date)
- South–South Working Papers. 1995–present (35 to date)
- Coastal Region and Small Island Papers. 1997–present (19 to date)
- CSI Info. 1997–present (15 to date)
- MAB Drylands Series. 2000–present (4 to date)
- MAB Biosphere Reserves Technical Notes. 2003–present (1 to date)
- IHP Groundwater Series. 2002–present (7 to date)
- IHP PC–CP (From Potential Conflict to Cooperation Potential) Series. 2003–present (31 to date)
- IHP Water and Ethics Series. 2004–present (12 to date)

Book series

- Arid Zone Research Series. 1952–66 (30)
- Humid Tropics Research Series. 1958–66 (6)
- Natural Resources Research Series. 1964–83 (20)
- New Trends in Mathematics Teaching. 1966–79 (4)
- New Trends in Biology Teaching. 1967–87 (5)
- New Trends in Chemistry Teaching. 1967–91 (6)
- New Trends in Physics Teaching. 1968–84 (4)
- Monographs on Oceanographic Methodology. 1966–present (11 to date)
- Earth Sciences Series. 1969–2000 (20)
- Studies and Reports in Hydrology. 1969–98 (59)
- Ecology and Conservation Series. 1970–73 (6)
- New Trends in Integrated Science Teaching. 1971–90 (6)
- Studies in Engineering Education. 1974–89 (13)
- Innovations in Science and Technology Education (1986–2003) (7)
- Man and the Biosphere Series. 1989–2002 (28)
- International Hydrology Series. 1993–present (12 to date)
- World Science Report. 1993–98 (3)
- IOC Ocean Forum Series. 1994–present (5 to date)
- Coastal Management Sourcebooks. 1996–present (3 to date)

- Encyclopedia of Life Support Systems (EOLSS). 2002–present (Virtual online, equivalent to 200 volumes)
- Renewable Energies Series. 2003–present (4 to date)
- LINKS Knowledges of Nature. 2005–present (1 to date)

Newsletters/information bulletins

- Monthly Bulletin of Scientific Documentation and Terminology (originally Monthly Circular). 1952–60
- Arid Zone Newsletter. 1958–64 (26)
- IBRO Bulletin. 1962–68 (28)
- Tsunami Newsletter. 1968–present (87 to date)
- International Marine Science (IMS) Newsletter. 1973–96 (76)
- MIRCEN News. 1980–91 (13)
- Geology for Sustainable Development. 1982–98 (10)
- InfoMAB. 1984–96 (24)
- IHP Information. 1985–94 (39)
- IOC Harmful Algae News. 1992–present (27 to date, plus special issues)
- IHP Waterway. 1994–99 (19)
- GOOS News. 1994–2001 (11)
- IOC Window (Western Indian Ocean Waters) Newsletter. 1994–present (45 to date)
- Biosphere Reserve Bulletin. 1995–present (13 to date)
- People and Plants Handbook. 1996–2002 (8)
- A World of Science. 2002–present (12 to date)

Notes: For each series of publications and documents, an indication is given in parentheses of the number of issues/volumes in that series. For ongoing series, 'to date' refers to mid-2005. Some of the series have been placed somewhat arbitrarily, since they include both short technical documents and bulky multi-authored books (e.g. the Earth Sciences series ranges from a 44-page Explanatory Note to the Tectonic Map of the Carpathian-Balkan System [1975] to a 1,173-page volume on Gondwana Stratigraphy [1969]).

OVERVIEW
The Natural Sciences Sector, 2005
Walter Erdelen, Assistant Director-General for Natural Sciences[42]

INTRODUCTION

> 'It is time to breathe new life ... into the intergovernmental organs of the United Nations'—UN Secretary-General Kofi Annan (United Nations, 2005, p. 40)

WE have come a long way in this book so far: we've seen the changes in the Natural Sciences Sector, from the beginnings of UNESCO in the mid-1940s to up until quite recently. Now, we need to assess where we are, and – based on what we have learned – where we should be going. For this reason, this chapter describes the current (as of January 2006) situation in the Sector in terms of mission, structure, programme and budget, while the following section – Part VI: 'Looking Ahead' – addresses global forces which will shape our programmes, and key areas where additional concentration will need to be given, and suggests a new vision and mission to guide us as we go forward to meet the challenges ahead.

At the time of writing, UNESCO – as well as the whole UN system – is experiencing a massive reform process. As Secretary-General Kofi Annan expressed in his report 'In Larger Freedom: Towards Security, Development and Human Rights for All', written to assess our progress five years after the United Nations Millennium Declaration:

> We also need agile and effective regional and global intergovernmental institutions to mobilize and coordinate collective action. As the world's only universal body with a mandate to address security, development and human rights issues, the United Nations bears a special burden. As globalization shrinks distances around the globe and these issues become increasingly interconnected, the comparative advantages of the United Nations become ever

---

42 Walter R. Erdelen became Assistant Director-General for Natural Sciences in 2001. See Annex 3: Heads of Natural Sciences at UNESCO.

more evident. So too, however, do some of its real weaknesses. From overhauling basic management practices and building a more transparent, efficient and effective United Nations system to revamping our major intergovernmental institutions so that they reflect today's world and advance priorities set forth in the present report, we must reshape the Organization in ways not previously imagined and with a boldness and speed not previously shown (United Nations, 2005, p. 6).

Subsequently – with the formulation of the eight Millennium Development Goals (MDGs) and their associated eighteen targets – new commitments have been made globally towards improving the situation on our planet, with a particular focus on poverty. One must view UNESCO's current activities against this background, but one must also see in this the potential for the Natural Sciences Sector to reshape itself both in terms of structure and programmatic content.

The year 2005, in addition to being UNESCO's sixtieth anniversary, was a year during which the General Conference convened and the normal programming cycle was followed, but it was also one in which a new Medium-Term Strategy (2008–2013) was to be set in motion. In other words, this chapter was written right in the middle of substantive decisions the Organization was taking concerning programmatic planning for the years ahead. This included the definition of priorities for the different programme sectors. Before we can address where the Sector should go, we need a snapshot of where it is currently, since this is our point of departure. The following pages will provide this overview.

NATURAL SCIENCES SECTOR: MISSION AND STRUCTURE

MISSION AND VISION OF THE SECTOR

The Organization's current Medium-Term Strategy (2002–2007) is formulated around a single unifying theme: UNESCO's contribution to peace and human development in an era of globalization, through education, the sciences, culture and communication. Within the framework of this objective, the overall vision for UNESCO's Natural Sciences Sector is to ensure that the creativity of science is used for the benefit of society by providing world leadership in expertise and international cooperation in natural and environmental sciences and engineering, and thereby to contribute to the safety and well-being of people throughout the world and to the economic well-being of nations.

Based on its overall vision, the Sector must make strategic choices and full use of its comparative advantages, combining both its intellectual and operational mandates while thinking globally and acting locally. Its niche is defined by its specific role as an honest promoter and broker of science and of 'science without frontiers' at all levels (local, regional, global), and by its intersectoral potential and the societal and political relevance of its programmes.

The mission of the Sector is to further the advancement and sharing of scientific knowledge and to promote the application of this knowledge and its understanding to the pursuit of sustainable development. In performing its functions, the Sector contributes to understanding the Earth systems and preserving their diversity. It is called upon to foster the role of science in peace processes and conflict resolution. It should work with the social and human sciences as an advocate for ethics in science and technology and for the contribution of science to preserving human rights, while also promoting the peaceful use of science and technology. Its programmes should integrate a social contract for science and attempt to improve the image of science, aiming to attract young people to scientific studies. In the pursuit of its development mission, the Sector strives to promote equitable access to science and technical knowledge, as well as to the benefits of science, and contributes to the upgrading of scientific knowledge and its use in developing countries, including related capacity-building activities, both human and institutional.

An increasingly important challenge for the Sector is to link science and society at large. In the years to come, its role will be to further promote productive linkages between scientists and decision-makers in government and in the private and public sectors. The Sector's programmes are called upon to support the creation and delivery of scientific knowledge for local, national and international policy-making and problem-solving.

In working towards the attainment of its objectives, the Sector will endeavour to be a forum for excellence in science. Its programmes should develop innovative approaches to science education, capacity-building and science policies. They should rely on new forms of partnerships and new ways of cooperating, while constantly promoting collaboration among scientists and facilitating the exchange and transfer of scientific and technical knowledge.

STRUCTURE, STAFFING AND FIELD REPRESENTATION

Structure

The Natural Sciences Sector of UNESCO is headed by the Assistant Director-General for Natural Sciences. The Executive Office attached to the Assistant Director-General pools programme coordination and evaluation, administration and information services. The programme services are currently organized as shown in Figure V.6.1.

Staffing

As of January 2006, there are 1,879 posts at UNESCO funded by the Regular Programme budget, according to the Programme and Budget for 2006–2007

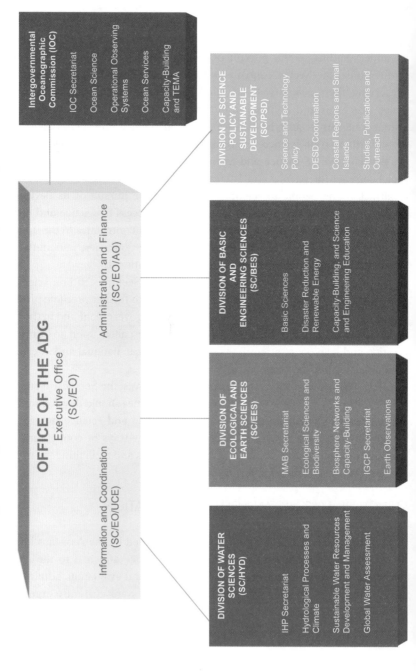

Figure V.6.1: Organizational chart of the Natural Sciences Sector, as of January 2006

(33 C/5). These fall into two main categories: Professional (928 posts) and General Services (951 posts). A total of 157 regular posts are established under the Regular Programme budget in natural sciences at UNESCO, 101 of which are professional posts, including eight National Professional Officers (NPO) in the field (see Figure V.6.2). Forty-four professional science staff (including NPOs) are posted in UNESCO's regional and cluster offices. These figures do not include staff working at the Abdus Salam International Centre for Theoretical Physics (ICTP, in Trieste, Italy), the UNESCO-IHE Institute for Water Education (Delft, the Netherlands) or other UNESCO-affiliated institutes.

At Headquarters, there is an even ratio between professional and support staff. For the breakdown of professional staff at the different seniority ranks between Headquarters and the field, see Figure V.6.2. In addition to its regular staff, the Natural Sciences Sector employs a number of consultants on a short-term basis. UNESCO's Natural Sciences programme is implemented by staff at UNESCO Headquarters in Paris, the five Regional Bureaux for Science and other Field Offices.

Figure V.6.2: Distribution of professional posts between Headquarters and the field in the Natural Sciences Sector (excluding directors of the five Regional Bureaux for Science)

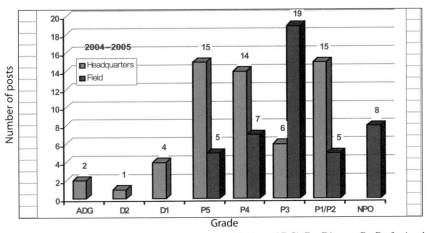

Notes: ADG = Assistant Director-General (IOC is also headed by an ADG), D = Director, P = Professional staff, NPO = National Professional Officer.

PROGRAMME ACTIVITIES, FLAGSHIP PROGRAMMES, BUDGET AND FIELD REPRESENTATION

PROGRAMME ACTIVITIES

At present, there are five major units covering programme activities within the Sector, namely: the Division of Water Sciences, the Division of Ecological and Earth Sciences, the Intergovernmental Oceanographic Commission, the

Division of Basic and Engineering Sciences, and the Division of Science Policy and Sustainable Development. These are briefly described below.

Water sciences

Water is an issue of strategic importance with strong environmental, social and economic implications. The Division of Water Sciences is the focal point for all activities related to freshwater, with 'freshwater and associated ecosystems' currently being the principal priority of the Natural Sciences Sector. The Division's main objective is to enhance the understanding of hydrological processes occurring in nature and society in order to develop sustainable and integrated approaches to water resources management. Activities focus on capacity-building related to the assessment, development and operation of water systems. They are based on the fundamental principle that freshwater is as essential to sustainable development as it is to life, and that water – beyond its geophysical, chemical and biological function in the hydrological cycle – has social, economic and environmental values that are interlinked and mutually supportive.

BOX V.6.1: UNESCO-IHE INSTITUTE FOR WATER EDUCATION

The need for a more integrated approach to water and environmental resources management calls for professionals with a high degree of specialization, as well as generalists equipped to lead and manage multidisciplinary efforts, individuals and organizations in the water and environment sectors worldwide.

The International Institute for Infrastructural, Hydraulic and Environmental Engineering (IHE-Delft) was established in the Netherlands in 1957. UNESCO's General Conference in 2001 gave the green light for IHE-Delft to become an integral part of UNESCO. It was renamed the UNESCO-IHE Institute for Water Education. Its mandate includes strengthening and mobilizing the global education and knowledge base for integrating water resources management and contributing to meeting the water-related capacity-building needs of developing countries and countries in transition.

Through the use of the internet, satellite technology and other advanced ICTs, UNESCO-IHE is consolidating collaboration in the water and environment sectors worldwide. The UNESCO-IHE global alumni community comprises over 13,000 mid-career, senior professionals and decision-makers active in water, environment and infrastructure communities and networks, representing over 120 countries worldwide.

Core activities of IHE include Masters of Science and Ph.D. programmes on water, the environment and physical infrastructure, which instil a problem-solving approach. Tailor-made short training courses are organized on demand. Knowledge is transferred through lectures, workshops, role-playing, videos, study tours, lectures and printed media. Other activities of the Institute include institutional capacity-building, including provision of expertise on institutional or organizational reform, such as training needs assessment, decentralization strategies, and support for strengthening in-house training centres of sectoral ministries or other authorities.

The Division provides the Secretariat for the International Hydrological Programme (IHP), UNESCO's long-standing intergovernmental scientific cooperative programme in hydrology and water resources. IHP is a vehicle through which Member States can upgrade knowledge of the water cycle and thereby increase their capacity to better manage and develop their water resources. The Division also houses the World Water Assessment Programme (WWAP), involving twenty-four United Nations agencies and Convention Secretariats.

The Division has three sections. The *Hydrological Processes and Climate* section aims at the understanding of hydrological and biogeochemical processes connected with water transfer into, through and out of, drainage basins impacted by anthropogenic pressures and climate variability, and (at the interface between societal and environmental needs) promoting sustainable water resources management and sound policy-making. *Sustainable Water Resources Development and Management* focuses on the development and operation of water systems within the framework of integrated water resources management; enhancement of the understanding of how society interacts with water systems at different scales (national, regional and international); and scientific input to water-related policies, education, conflict resolution and other measures. Finally, *Global Water Assessment* covers the coordination of activities of the WWAP – which coordinates the efforts of all UN agencies concerned with freshwater issues – and publishes the triennial *World Water Development Report*.

Ecological and Earth sciences
In December 2004, it was decided to merge the Division of Earth Sciences with the Division of Ecological Sciences to create the new Division of Ecological and Earth Sciences. The new Division contributes to advancing scientific knowledge about the environment and to applying innovative policy approaches to ecology and biodiversity science.

The Division houses the Secretariat of the Man and the Biosphere (MAB) Programme, the major goal of which (as currently endorsed by its Intergovernmental Coordinating Council) is to develop the basis for improving the relationship between people and their environment. MAB's operational tool is the World Network of Biosphere Reserves (WNBR) – comprising 482 sites in 102 countries. Biosphere reserves are living laboratories for testing and demonstrating the sustainable use of natural resources; as such, they represent a concrete means of contributing to the meeting of many of the MDGs, and to the implementation of the UN Decade of Education for Sustainable Development (DESD, 2005–2014).

The Division of Ecological and Earth Sciences manages initiatives to build human and institutional capacity through interdisciplinary training of young scientists and decision-makers in the fields of ecology, conservation biology and sustainable development. It is the focal point for biodiversity issues at UNESCO. Its section on Ecological Sciences and Biodiversity coordinates research and capacity-building activities of the MAB Programme focusing on different ecosystems. Studies of novel ecological theory, emerging ecosystems, global and climate change, land degradation and rehabilitation, linkages between biological and cultural diversity, and the key role of freshwater resources for ecosystem functioning have become the main research issues in recent years. Its section on Biosphere Networks and Capacity-Building deals with the evaluation of new nominations for the WNBR, support for the regional WNBR networks, the periodic review mechanisms and legal aspects of the functioning of biosphere reserves, and the fostering of cooperative activities among Member States and individual biosphere reserves.

The Division contributes to the understanding of the Earth's complex system (on a broad spectrum of scales in space and time) and to wise management practices and sustainable development of the Earth's crust and natural resources. (UNESCO is the only UN organization dealing with interdisciplinary research training, education and capacity-building in geology.) It also provides the Secretariat of the International Geoscience Programme (IGCP), which is implemented in cooperation with the International Union of Geological Sciences (IUGS) and constitutes (through UNESCO) the UN-wide core component for fundamental and applied Earth sciences research, as well as global monitoring, geoscience and space education, and the popularization of Earth sciences through support for the Global Network of National Geoparks.

The Earth Observation Section (EOS) develops cooperation with space agencies and space-related institutions, organizations and UN specialized agencies, initiates and coordinates space-related activities within the Organization, and represents UNESCO in space-related events. Through cooperation in the Integrated Global Observing Strategy Partnership (IGOS) and the Global Earth Observation System of Systems (GEOSS) Implementation Plan, the better use of Earth observation data for the planning of sustainable development is promoted, and capacity-building in the use of *in situ* and space data is enhanced. It also develops capacity-building activities related to the use of Earth observation data from space and *in situ*. As an innovative approach to science education, space-related subjects are being promoted and enhanced in schools and universities, particularly in developing countries, in the framework of the Space Education Programme (SEP). The use of remote sensing for the management of biosphere reserves and World Heritage sites is enhanced through the activities of the Open Initiative with the space agencies. International Cooperation and Capacity-building in Earth Sciences and Space Activities include: the IGCP, development of mechanisms for global environmental monitoring within IGOS, and the Space Education Programme and coordination of outer-space activities.

The Intergovernmental Oceanographic Commission (IOC)
The IOC promotes international cooperation and coordinates programmes in research, services and capacity-building, in order to learn more about the nature and resources of the ocean and coastal areas. Through the application of this knowledge, the Commission aims to improve management practices and the decision-making process of its Member States, foster sustainable development, and protect the marine environment. In addition, the Commission strives to further develop oceans governance, which necessitates strengthening the institutional capacity of Member States in marine scientific research and of ocean management. The vision guiding the development of a Global Ocean Observing System (GOOS) is one of a world where the information needed by governments, industry, science and the public to deal with marine-related issues, including environmental issues and the effects of the ocean upon climate, is supported by a unified global network.

With the advances in oceanography – from a science dealing mostly with local processes, to one that is also studying ocean basin and global processes – researchers and a wide spectrum of users depend critically on the availability of an international exchange system to provide data and information from all available sources. The IOC's Ocean Sciences programmes address scientific uncertainties for the management of the marine environment and climate change, catalysing and coordinating international oceanographic research, and communicating the results of these investigations to the Member States of the IOC, the United Nations and the general public.

The IOC's Operational Observing Systems programme aims at: monitoring and forecasting capabilities within the Global Ocean (GOOS) and Global Climate Observing Systems (GCOS) needed for the management and sustainable development of the open and coastal ocean; integrating and distributing oceanic observations; and generating analyses, forecasts and other useful products. Capacity-building (CB) and teaching, education and mutual assistance (TEMA) deal with the capacity-building process (CB-TEMA) to link the IOC's programme activities to existing and planned national and regional programmes, namely institutional capacity-building, training, education, and mutual assistance in marine sciences. CB-TEMA is central to the overall IOC strategy. The International Oceanographic Data and Information Exchange (IODE), Ocean Mapping, and Tsunami programmes are all part of the ocean services development and the strengthening of a global mechanism to ensure full and open access to ocean data and information for all.

Basic and engineering sciences
Capacity-building in basic and engineering sciences as a prerequisite for the advancement, transfer, sharing and dissemination of knowledge, is the priority

in the Division of Basic and Engineering Sciences. Activities compliment other issue-driven scientific and environmental programmes in the natural sciences. In the basic sciences, activities relate to tertiary-level education and to research in mathematics, physics, chemistry, biology, biotechnology and basic medical sciences. In the field of engineering and technology, activities are designed to strengthen human resources in developing countries in engineering education and to promote selected technological applications, including the use of renewable sources of energy and natural disaster prevention and awareness building. The Division is the focal point for the science education programme and for the follow-up to the World Conference on Science (Budapest, Hungary, 1999).

Within the Basic Sciences Sections, activities in mathematics, physics and chemistry focus on: capacity-building in tertiary-level education and research, lifelong learning and the training of scientists and teachers from developing countries to promote endogenous science, and activities carried out in cooperation with ICTP and other organizations (mentioned above).

The life sciences focus on strengthening national and regional capacities in the various fields of biology (including marine and agricultural biology, biochemistry and biophysics, immunology, neurobiology, genetics and molecular biology of micro-organisms, plants, animals and humans), as well as in all biotechnological applications of these fields. Support of tertiary-level education and research is of special interest, as are exchange of knowledge, South–South and North–South scientific networking, and training of scientists and teachers from developing countries. Activities are carried out in cooperation with numerous partner organizations.

The Disaster Reduction and Renewable Energy section includes: promoting cooperative education and training programmes in renewable energy; fostering activities to reduce urban vulnerability to man-made and natural disasters, including the study of natural hazards and disaster-risk reduction, and promotion of the assessment of natural hazards including earthquakes, landslides, volcanic eruptions and hydrological risks; and fostering measures for disaster prevention and preparedness. These functions are undertaken in the framework of the UN International Strategy for Disaster Reduction.

In Capacity-Building and Science and Engineering Education, the overall focus is on human and institutional capacity-building, with specific activities designed to strengthen human resources in developing countries in engineering education, curricula, learning and teaching materials and distance learning, and to promote technological innovation and applications useful to achieving the MDGs.

Science policy and sustainable development

In conformity with the mandate of UNESCO in the field of science and in response to the demand of Member States and the recommendations of the

World Conference on Science (Budapest, 1999), the mission of the Division of Science Policy and Sustainable Development (SC/PSD) is to contribute to the advancement of science, to promote a new contract between science and society, and to advocate the adoption of science and technology (S&T) policies that imply: consistent and long-term support for S&T, strengthening the human resource base, establishment of scientific institutions, improvement and upgrading of science education, integration of science into the national culture, development of infrastructures, and promotion of technology and innovation capacities at the national, regional and global levels; to improve and strengthen the delivery of UNESCO's programmes on sustainable development, including an enhanced contribution to the DESD; to assume overall responsibility for the Organization's follow-up to the Mauritius Programme of Action for the Sustainable Development of Small Island Developing States (SIDS); and, last but not least, to use scientific and technological resources to contribute towards the achievement of the MDGs.

In order to implement the above, the Division is structured around four themes, as outlined below.

Science and Technology Policy comprises advisory services to Member States in formulating their national science, technology and innovation policies, strategies and programmes, formulation of methodologies and guidelines for policy formulation, capacity-building in policy analysis, evaluation of S&T programmes and the promotion of university–industry partnership, and science system management.

DESD Coordination coordinates the Natural Sciences Sector's inputs towards the formulation and implementation of an overall strategy for the Decade, under the leadership of the Education Sector. A major contribution of the Sector is the *Encyclopedia of Life Support Systems* (EOLSS), the largest body of knowledge for sustainable development ever compiled.

Coastal Regions and Small Islands fosters the sustainable development of Small Island States by contributing to the implementation of the Mauritius Declaration and Strategy through an intersectoral initiative that provides a global platform for environmentally sound, socially equitable, culturally respectful and economically viable development in coastal regions and small islands. This is a successful pilot for intersectorality within UNESCO, and it plays a leadership role in two cross-cutting projects: Small Islands Voice, an interregional initiative focusing on small islands in the Caribbean, Indian Ocean and Pacific regions; and Local and Indigenous Knowledge Systems (LINKS: see Box VI.1.1), a project to strengthen dialogue between indigenous and scientific knowledge holders to reinforce biodiversity conservation and enhance equity in resource governance.

Studies, Publications and Outreach works on major science policy issues such as ethics of science and technology, valorization and strengthening of local and

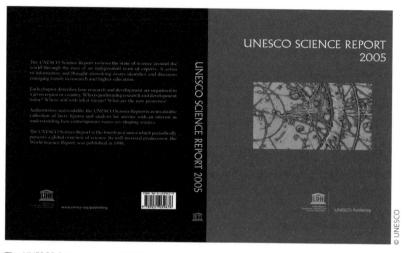

The *UNESCO Science Report, 2005* (274 pp.) analyses the current state of science and technology around the globe, through the eyes of an international team of experts.

indigenous knowledge and culture, science and society, science legislation, S&T indicators, gender issues in science, and the publication of these studies in a series of thematic reports, as well as the conduct of comparative studies and analysis of the status of science in the world, and publishing the *UNESCO Science Report*.

The Division also leads the overall efforts of the sector in favour of the development of S&T in Africa, in cooperation with the African Union (AU) and the New Partnership for Africa's Development (NEPAD). This major effort is based on supporting the formulation and implementation of an international initiative adopted by the G8, with a view to marshalling S&T for development in Africa, through the support for and creation of centres of excellence in S&T, as well as the revitalization of higher education and research systems on the continent.

FLAGSHIP PROGRAMMES FOR THE 2006–2007 BIENNIUM

The purpose of flagship programmes is to focus extra attention on certain issues, which are key to achieving certain overall goals set by UNESCO's governing bodies. Therefore, establishing a 'flagship' on an issue within the programme and budget creates a structure and mechanism through which issue advocates may interact. Such programmes are alliances of organizations willing to work together to further the goals highlighted by the issue given flagship status. They can be instrumental in helping countries and donors alike to address the specific flagship issue as they develop and fund country-level plans to achieve specific goals. In the current biennium, there are three natural science flagship programmes: Tsunami Early Warning Systems, International Basic Sciences Programme (IBSP), and Natural Disaster Reduction.

PART V: OVERVIEWS AND ANALYSES

Tsunami Early Warning Systems
The establishment of a global tsunami warning system developed within the United Nations' international strategy and coordinated by the IOC of UNESCO will benefit from the forty-year experience of the Tsunami Warning System in the Pacific, and will be implemented in close collaboration with the present International Coordination Group for the Tsunami Warning System in the Pacific of the IOC, in cooperation with the World Meteorological Organization (WMO) under their Joint Commission for Oceanography and Marine Meteorology. The International Strategy for Disaster Reduction will be called upon to ensure the synergy between disaster reduction activities and those in the socio-economic and humanitarian fields.

International Basic Sciences Programme (IBSP)
IBSP will focus on major region-specific actions involving a network of national, regional and international centres of excellence or benchmark centres in the basic sciences. It will concentrate on building national capacities for basic research, training and science education, through international and regional cooperation in development-oriented areas of national priority. Relying on the services of existing or newly created centres of excellence, IBSP will foster excellence in other national, regional and international institutions and involve them in the centres' activity, in line with the needs of Member States and international partners. IBSP will also focus on transfer and sharing of scientific information and excellence in science through North–South and South–South cooperation.

The quarterly newsletter *A World of Science* has been published by UNESCO's Natural Sciences Sector since 2002. Here a selection of six issues.

Natural Disaster Reduction: knowledge, education and information
Building a culture of resilient communities requires active and knowledgeable citizens and informed decision-makers. The enhancement and use of scientific and indigenous knowledge for protecting people, habitat, livelihoods and cultural heritage from natural hazards will be pursued through this flagship activity.

UNESCO's activities will help to encourage formal and non-formal education to become an instrument for disaster prevention and will also promote information and public awareness on measures to reduce risks from natural hazards.

BUDGET

The Regular Programme budget in the Programme and Budget for 2006–2007 (33 C/5) baseline scenario for the natural sciences amounts to about US$23 million, of which US$700,000 is allocated to cross-cutting projects (see Figure V.6.3). Extrabudgetary resources contribute an estimated further US$182 million to the overall programme budget, including for ICTP and IHE.

Since 'water and associated ecosystems' continues to be the principal priority for the Natural Sciences Sector, the programme in this area has been allocated almost US$9 million in regular funds and is expected to obtain about US$4 million in extrabudgetary funds. Some 34 per cent of the Natural Sciences Sector Regular Programme budget is decentralized to the Field Offices (see Figure V.6.4). As with the structure of the network of the Field Offices in general, science is presented in the field at the national, cluster and regional levels.

Figure V.6.3: Regular Programme resources (in US dollars)

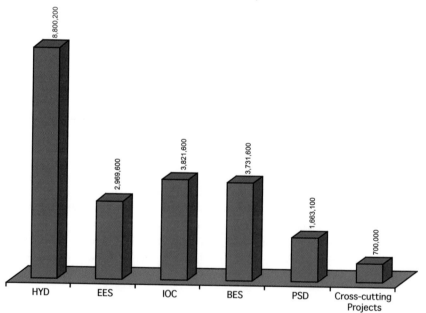

Notes: Distribution of Regular Programme resources for activities during the 2006–2007 biennium, as finalized in October 2005 during the thirty-third General Conference. Extrabudgetary sources are expected to be roughly twice these amounts. HYD = Water Sciences; EES = Ecological and Earth Sciences; IOC = Intergovernmental Oceanographic Commission; BES = Basic and Engineering Sciences; PSD = Science Policy and Sustainable Development; Cross-cutting = intersectoral.

Figure V.6.4: Decentralization of Regular Programme funds in the Natural Sciences Sector during the 2006–2007 biennium

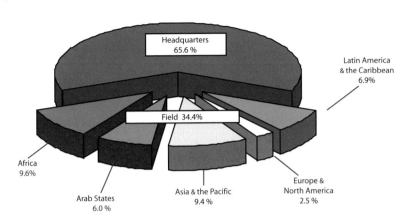

Decentralization is one of the key parameters of the reform process. A new programme management cycle and process has allowed a more pronounced involvement of Field Offices in the elaboration of various components of the programme and budget, in particular through the preparation of an integrated strategic plan by each Field Office. Field Offices are clearly the central pivot of the Organization's action in Member States.

NATURAL SCIENCES IN THE FIELD

The structure of the natural sciences at UNESCO includes both representation in the field and at Headquarters. UNESCO has a total of fifty-two regional, national and cluster Field Offices around the world, including liaison offices in Geneva (Switzerland) and New York (USA) (see Figure V.6.5). These include five Regional Bureaux for Science: in Cairo (Egypt), Jakarta (Indonesia), Montevideo (Uruguay), Nairobi (Kenya) and Venice (Italy).

The Cairo Office, in addition to its role as the Regional Bureau for Science in the Arab States, serves the needs of Arab States in science and technology, is the Cluster Office for Egypt, Sudan and Yemen, and addresses the needs of these countries in the areas of education, culture, and communication and information. The Office places great emphasis on promoting intellectual cooperation among Arab Member States and in contributing to their development through a number of major interdisciplinary programmes and activities in UNESCO's fields of competence in coordination with National Commissions, sister UN organizations, and development and donor organizations.

As a Regional Bureau for Asia with particular responsibility for science and technology, within an Organization committed to increasing intersectoral activities, the Jakarta Office – in consultation with National Commissions and

Member States of the region – builds educational, cultural, communication and gender perspectives onto the Natural Sciences Sector-based programmes and networks already established. The Jakarta Office is also the Cluster Office for Brunei Darussalam (which joined the Organization in 2005), Indonesia, Malaysia, the Philippines and Timor-Leste.

The mandate of the Montevideo Regional Bureau is to reinforce multilateral, technical cooperation in science and technology in Latin America and the Caribbean, as part of the regional strategy to stimulate sustainable development and a culture of peace and tolerance in all countries of the region. This Office is the Cluster Office for Argentina, Brazil, Chile, Paraguay and Uruguay; it houses the Offices of the World Heritage Centre for Latin America and the Caribbean, and the Regional Adviser in Informatics and Telematics in the framework of the Information Society Division.

The Nairobi Office provides services to African Member States in its cluster region (Burundi, Kenya, Rwanda and Uganda) in all of UNESCO's fields of competence, but especially in science and technology, and coordinates the Organization's activities in sub-Saharan Africa in these two areas. In all cases, the Office's concern is to encourage and support initiatives in Member States that contribute to achieving the objectives of UNESCO's programme and to disseminating the Organization's ideals.

UNESCO's Office in Venice promotes and facilitates capacity-building activities and cooperation in science, technology and culture between the countries of Central and Eastern Europe, and in recent years has had a particular focus on the countries of south-eastern Europe. As the Regional Bureau for Science in Europe, it has an important role in facilitating the application of knowledge and the mobilization of skills for the solution or prevention of problems facing Member States at the regional level.

ACKNOWLEDGEMENTS

This chapter has greatly benefited from the contributions of L. Anathea Brooks, Pilar Chiang-Joo, and B. Djaffar Moussa-Elkadhum.

BIBLIOGRAPHY

United Nations. 2005. *In Larger Freedom: Towards Development, Security and Human Rights for All*. Report of the Secretary-General. United Nations General Assembly, fifty-ninth session. UN A/59/2005 (21 March 2005). http://daccessdds.un.org/doc/UNDOC/GEN/N05/270/78/PDF/N0527078.pdf?OpenElement

UNESCO OFFICES WORLDWIDE.   Map pages 588–589

I. Regional Bureaux for Science: ○
Cairo, Egypt
Jakarta, Indonesia
Montevideo, Uruguay
Nairobi, Kenya
Venice, Italy

II. Cluster Offices: ●
Accra, Ghana
Addis-Ababa, Ethiopia
Almaty, Kazakhstan
Apia, Samoa
Bamako, Mali
Beijing, China
Beirut, Lebanon
Bucharest, Romania
Dakar, Senegal
Dar es Salaam, United Republic of Tanzania
Dhaka, Bangladesh
Doha, Qatar
Hanoi, Vietnam
Harare, Zimbabwe
Havana, Cuba
Kabul, Afghanistan
Kingston, Jamaica
Libreville, Gabon
Moscow, Russia
New Delhi, India
Phnom Penh, Cambodia
Quito, Ecuador
Rabat, Morocco
San José, Costa Rica
Santiago, Chile
Tehran, Islamic Republic of Iran
Windhoek, Namibia
Yaoundé, Cameroon

III. National Offices: □
Abuja, Nigeria
Amman, Jordan
Bangkok, Thailand
Brasilia, Brazil
Brazzaville, Congo
Bujumbura, Burundi
Guatemala City, Guatemala
Iraq (located in Amman, Jordan)
Islamabad, Pakistan
Kathmandu, Nepal
Kinshasa, Democratic Republic of Congo
Lima, Peru
Maputo, Mozambique
Mexico City, Mexico
Port-au-Prince, Haiti
Ramallah, Palestine
Tashkent, Uzbekistan

IV. Liaison Offices: ☆
Geneva, Switzerland
New York, USA

V. Affiliated Institutions with SC staff: △
ICTP, Trieste, Italy
TWAS, Trieste, Italy
IHE, Delft, The Netherlands

MAP OF UNESCO OFFICES

PART V: OVERVIEWS AND ANALYSES

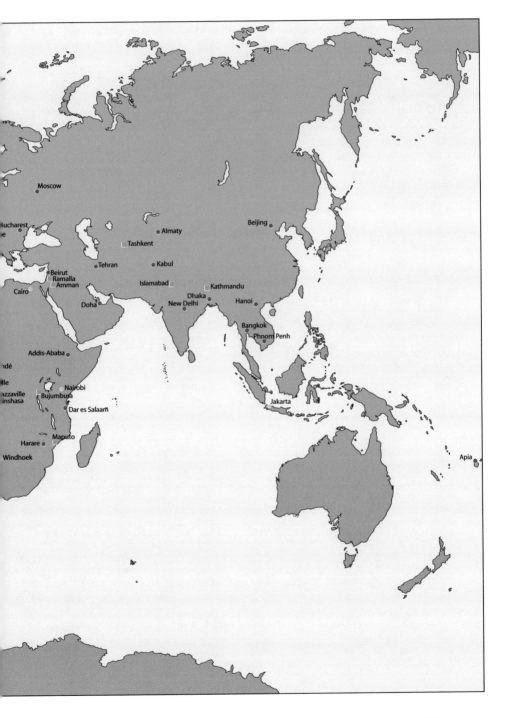

PART VI
LOOKING AHEAD

PART VI: LOOKING AHEAD

THOUGHTS ON THE NATURAL SCIENCES AT UNESCO
Beyond 2005

Walter Erdelen, Assistant Director-General for Natural Sciences

INTRODUCTION

WRITING about the future of the natural sciences at UNESCO necessarily entails writing about one of a number of possible futures. The thoughts presented here include current and most recent trends, based on the work of our executive organs and the Natural Sciences Sector, and their extrapolation into the future; they provide a scenario we may wish to realize at some stage. Our ideas and possible futures at UNESCO are not developed in a void, but rather with respect to major global trends and changes and in alliance with the United Nations and the scientific communities. Virtually all these ideas could be subsumed under the rubric *ensuring global quality of life and alleviating poverty and hunger in this world*. Our generation could put an end to extreme poverty, 'but only if we grasp the historic opportunity in front of us' (Sachs, 2005).

UNESCO is an intergovernmental organization, with a very limited regular budget and limited staff resources in its programme sectors. These limitations, however, should not be seen as obstacles but as opportunities. Indeed, UNESCO's history has shown the enormous opportunities it has had in its past and the potential it has for the future. As written in its Medium-Term Strategy 2002–2007, the functions of the Organization are to work as a laboratory of ideas, a standard-setter, a clearinghouse, a capacity-builder in its Member States, and as a catalyst for international cooperation.

In my view, UNESCO's diverse set-up and its multiple mandates, as reflected in its programme sectors, offer not only challenges but also opportunities to address the most important of today's issues and problems. New domains entered by the natural sciences, emerging ethical questions and issues, the ever-growing relevance of the social and human sciences, the key importance of global quality education at all levels, the culture-related frameworks in which many if not all of our problems need to be seen, and, last but not least, the new tools provided by modern information and communications technologies (ICTs) to improve living conditions and communication – all these are at the heart of what UNESCO does.

SHAPING THE EARTH'S FUTURE: MAJOR GLOBAL TRENDS, FORESIGHT, AND SCIENCE IN THE TWENTY-FIRST CENTURY

MAJOR GLOBAL TRENDS AND FORESIGHT

Let us begin with a look at major global trends relating particularly to demography, natural resources and environment, globalization, social change and governance. Demographic trends relate to population size, and to geographical, age and other distribution factors. The current world population of 6.5 billion people is expected to increase to 7.2 billion by 2015 (average growth rate of 1.1 per cent) (UNDP, 2005). Of this, the population of developing countries is expected to increase from 5 billion to 5.9 billion (average annual growth rate, 1.5 per cent), and that of developed (OECD) countries to remain around 1.2 billion. Urban populations will increase from 42 per cent to 49 per cent of this total in developing countries, and from 76 per cent to 79 per cent in developed countries. By 2007, more than half of the human population will – for the first time in history – live in urban environments. It has been theorized that, due to intrinsic factors, the world population should stabilize at 10–12 billion by 2050, following a 'global demographic transition'.

Consumption may be stabilizing in developed countries, due to constant and declining populations and greater efficiency in resource use and recycling. Consumption patterns are, however, changing dramatically in developing countries such as China and India – leading again to questions of the sustainability of development and the population/natural resources 'Limits to Growth' debate of the 1970s and 1980s among the Club of Rome and others (see Meadows et al., 2004), where the role of new technology was seen increasingly as a vital and dynamic factor in the process of change. Environmental changes are also more evident now than in the 1970s – at the macro scale with climate change, global warming and sea-level rise, and in the physical, natural and social changes we see worldwide in terms of population increase, urbanization, the expansion of agricultural and settled areas, and the loss of natural areas and of biodiversity. Water and fossil fuel energy resources are set to become critical in the coming decades. These changes, and the increasingly rapid pace of technological advances, are creating the stage for future transformation relating to climate, changing rainfall patterns and water use.

These changes will continue to have impacts on natural resource use and the environment, as will greater population numbers and increasing consumption. Interest in the role of technological change on resource use, availability, depletion and associated issues – such as climate change – has developed significantly since the 1970s. Technology is seen now as a crucial driver of globalization, particularly through ICTs and the knowledge and information societies and economies. Technology will continue to drive future change. This is illustrated in the growth of interest in forecasting, foresight, and innovation systems, which is matched by the increasing concern for developing countries on the other side of the knowledge divide, with limited capacity in science and engineering. The role of technological change is illustrated in such areas as voice-operated internet protocol (VOIP) – a 'disruptive technology', which could displace conventional telephony in the next few years. VOIP also illustrates that most recent change is technology-based (in a fusion of science, engineering and technology), rather than the increasingly outdated 'linear' model of science-led innovation. Other areas of technological change, including ICTs, biotechnology, materials technology, nanotechnology and control systems technology (for example, robotics and remote systems technology), are discussed below. VOIP and other technological change will lead to further social change and issues relating to governance, at the national and international level, as well as to international relations, especially between the North and South.

MAJOR TRENDS IN SCIENCE, ENGINEERING AND TECHNOLOGY

Major trends in science, engineering and technology (SET) depend on various complex and interrelated factors. These include, particularly, the role of SET in shaping society while, at the same time, being shaped by society. Most research and development (R&D) in science and engineering takes place in Western industrialized societies in response to the needs, conditions and opportunities of these societies. Technological innovation, however, takes place on a global scale and is indeed a significant factor in the process of globalization. This represents a mismatch of profound importance regarding the sharing of knowledge and capacity-building in SET. From the science perspective, it is apparent that there are decreasing numbers of young people going into science subjects at the secondary level, and into science and engineering at the tertiary level, around the world. This could have adverse consequences for the developed countries, and also for developing countries as net brain drain.

Social change – and social and economic development – are driven and shaped significantly by technological change, which in turn is shaped by innovation and change in science and engineering. Past waves of technological innovation and change include water and steam power, iron and steel, electricity, the internal combustion engine, the chemical and electronics industries, and information technology. In this process, existing knowledge disciplines are transformed and new disciplines arise, further stimulating innovation, leading to increasingly knowledge-based societies and economies. Innovation also hastens technological obsolescence in a process of 'creative destruction', as waves of innovation occur over decreasing timescales. The overall trend is that the time between discovery and application in science and technology continues to decline, with the transformation of existing disciplines into new areas, leading to an increase in innovation. Which leads to a pressing question: Can our educational systems and their contents at all levels keep pace with these rapid developments?

Current waves of technological change and innovation focus mainly on such 'platform technologies' as ICTs, biotechnology, materials technology and nanotechnology. Broader social and economic change includes such issues as: increasing urbanization and the growth of mega-cities; the need for efficient water, sanitation, energy and transportation systems; and the development of alternative energy sources. The change to increasingly knowledge-based production also requires the development of organizational and management forms for SET to cope with increasingly complex systems. This has important implications in terms of SET policy and planning, human resource, education and training, and equity issues – including the need to promote the entry and participation of women in SET, and associated gender issues.

Developments in nanotechnology, for example, have actual and potential applications in the chemical and pharmaceutical industries in disease diagnosis, drug development and delivery, agricultural production and food processing, water remediation and control of air pollution, energy storage, production and conversion. Continued ICT miniaturization, the development of information-technology-enabled devices and their widespread integration into products will lead to increasing embedded intelligence and wireless connectivity. Biotechnology-based production will be used to make an increasing range of chemicals, although medical applications of biotechnology will continue to be costly. Materials technology will continue to provide an increasingly diverse range of products.

Technological change will have profound effects on industrial production, benefiting economies of smaller-scale production, as will the trend towards customization – as 'just-in-time' production is increasingly geared to individual orders, requiring less mass-production, smarter machines and people. R&D expenditures will increase, as will the need to maintain and upgrade skill levels by continuous training and retraining. The need for energy efficiency in industry and transportation will also continue to increase, based mainly on fossil fuels – with a particular need for clean coal technologies, while the use of renewable and alternative sources of energy increases. The need for air- and water-pollution management technology will also continue to increase.

The demographic changes mentioned above will have a major impact on developing countries. The need for increasing knowledge- and skills-based labour forces at all levels emphasizes the importance of continuous education and training, and the greater inclusion of previously under-represented groups, especially women, in science and engineering and associated areas of management. Greater electronic networking in organization and management will also lead to forms of inter-organizational cooperation and more complex organizational forms, especially in the area of smaller-scale production.

In many scientific fields, it is now relatively meaningless to distinguish between basic and applied research – in fact, a continuum – as the basic research problem itself may be set within industrial application parameters, such as in biotechnology and electronics. What are increasingly important are flexibility, interdisciplinarity, and the capacity to continue learning on the job, rather than to leave university with all the knowledge needed for a technical or scientific career. A workforce with these attributes will better contribute to national innovation systems, by capturing the precise knowledge needed for a specific application from constantly changing networks of knowledge (Hill, 2005).

Most recent foresight research and mega-trend analysis (FOREN, 2001; UN Millennium Project, 2005) has focused on the delineation of trends in SET and implications for social and economic development, policy and planning in developed, industrialized countries, although the results of such studies have equal importance for developing countries. The transfer of foresight research and mega-trend analysis into SET policy, planning and management for developing countries presents a major opportunity as well as a challenge for UNESCO. Developments across a range of platform technologies, particularly in nanotechnology, new materials, ICTs and biotechnology, are stimulating innovation within and between these fields through synergy and crossover. This will lead to many new applications in the chemical and pharmaceutical industries and products, medicine, agriculture, food processing, water- and air-pollution control and remediation, energy production, conversion and storage, increasing embedded intelligence and wireless connectivity. As with other waves of transformative technologies, new applications, approaches and products will change patterns of resource use, leading to a possible increase in the North–South knowledge divide, especially for those countries with weaker systems of science and technology. There is a clear need and role for UNESCO in the development of SET policies, national and regional centres of excellence and networks as a bridge to such innovative technologies.

SCIENCE IN THE TWENTY-FIRST CENTURY

Science in the new millennium is the science of a world of complexity. Simple structures interact to create new phenomena and assemble themselves into devices; large complicated structures can be designed atom by atom for desired characteristics. In general, with new tools and new understanding – and a developing convergence of the disciplines of physics, chemistry, materials science and biology – we build on our twentieth-century successes and begin to ask and to solve questions that were, until recently, in the realm of science fiction.

The term 'ignorance' has a different meaning in this context than in its daily use. Scientific ignorance is lack of knowledge but also of awareness of the role of science for society. From an epistemological point of view, it would be interesting to look more deeply into the relation between ignorance in this sense and scientific illiteracy (see, e.g., Marshall et al., 2003). In other words, what does science literacy mean in the context of the development processes outlined here?

At the Conference on Knowledge and Society, held at the World Summit on the Information Society in Geneva (Switzerland) in December 2003, it was stressed that the building of knowledge societies must be inclusive (see UNESCO, 2005*a*). This means that everyone, without distinction, has the right to build his or her knowledge, so that all are empowered to create, receive, share and utilize information and knowledge freely for their benefit – whether for economic betterment, social recreation, cultural expression and enjoyment, or civic participation. More specifically, the conference recognized scientific research and discovery (and their technological applications) as the driving forces behind the creation of knowledge societies. Yet, science's impact on society is, in turn, dependent not only on scientific findings but also on the ways in which society shapes science – through national policies for science and technology, perceptions, and through the institutional mechanisms for organizing research programmes and activities, for understanding the implications of scientific findings, and for promoting their application (UNESCO, 2003*b*).

This is why we must strive towards *scientific literacy*: in a society that is becoming increasingly driven by science and technology, scientific literacy is not a luxury; rather, it is a necessity (Trefil, 2003). Science education thus should be at the forefront of UNESCO's efforts – and has been for many years now. The World Conference on Science (WCS), held in Budapest, Hungary, in 1999, provided us with a clear vision and guidance on how to implement science education (UNESCO, 2000). This was followed by the March 2005 Venice (Italy) meeting, 'Harnessing Science for Society: further partnerships', which reviewed progress at the level of regions and noted that 'every child is a potential scientist; every woman, who is the first and primary educator of every child, needs herself to be well educated in science; [and that] there is an immense capacity in modern media to assist with education at all levels' (UNESCO, 2005*b*).

An increasingly scientifically oriented society needs *science popularization* in the widest sense – to promote an improved understanding of science and adequately orient public perceptions and attitudes about science and its applications. We may speak of a new social contract for science, because 'Society cannot be "for science" unless science is "for society"; but equally, science cannot serve society unless society is prepared to give its full support to the scientific endeavour' (Erdelen and Moore, 2005). In this context, I note that the role of research and education in promoting good citizenship through consciousness-raising has been identified as requiring special attention. ICTs have accelerated synergy among different science specialities. Subsequent to the formulation of integrated approaches to scientific research (but also thanks to the intervention of ICTs) the knowledge base is increasingly interdisciplinary – as witness fields such as 'biodiversity informatics', 'bioinformatics' or nanobiotechnology.

Changes in paradigms – or the 'new interdisciplinarities' – are evinced also by new terms such as 'biocultural diversity', 'environmental and ecological economics', and 'bioethics'. These 'interface disciplines' are a result of a knowledge explosion in genetics, molecular biology, and the biotechnologies. Interdisciplinarity is a necessity when addressing complex environmental issues, which in many cases are of a transboundary nature. This is well illustrated by concepts such as that of the river continuum, and by realities such as transboundary protected areas, which reflect efforts to protect extensive and continuous natural areas. Interestingly, interdisciplinarity has not only been instrumental in the development of new technologies; its practice has also led to relocation of traditional discipline boundaries. 'Disciplinary mobility' has become a buzzword for scientists and institutions alike, to address current complex issues and problems in different fields (see Hill, 2005).

Paradigm shifts have been made possible by: looking at the basic and applied sciences from a continuum, rather than a discrete, perspective; exploring the interlinkages of modern science and local indigenous knowledge systems; trying to combat ignorance by valuing diversity, both of a biological and a cultural nature; and by the dynamic networking of institutions at national, subregional, regional and global levels. It is UNESCO's position that scientific disciplines should increase their contact not only with one another but also with other modes and traditions of inquiry, especially in terms of the ethical, social and environmental implications of scientific and technological developments. To quote our Director-General, speaking at the (above-mentioned) Knowledge and Society Conference,

> Scientific advance and technical innovation are closely associated with the emergence of new capabilities. This is not new. From the very beginning of the age of scientific discoveries, people have devised tools for improving health, raising productivity and facilitating learning and communication. What is new are the pace of change, the range of its impact and the unprecedented character of some of the challenges and opportunities being generated. Today's science is marked by digital, genetic and molecular breakthroughs

that are pushing far beyond yesterday's frontiers of knowledge. These breakthroughs are creating new possibilities for improving health and nutrition, expanding knowledge, eradicating poverty and stimulating economic growth (UNESCO, 2003*b*).

So, what are the challenges for science in the twenty-first century? Let me mention a few, based on current debates on the subject: the emergence of a global civil society, in terms of multinational businesses and global dynamics of stock markets; the 'evolution' of ICTs; and the recognition of global issues and problems, such as global concern about ethical issues, climate change, and the North/South debate. More specifically, environmental challenges in the twenty-first century include, among other topics: enormous recent impacts on the atmosphere; health risks from ground-level ozone, smog and fine particulates; overexploitation of surface freshwater resources as well as aquifers; increasing threats to biodiversity; increasing exploitation and depletion of wild fish stocks; worsening land and soil degradation, in particular in developing countries; the degradation and fragmentation of many remaining forest ecosystems; increase in frequency and intensity of natural disasters; and climate change and its impacts at all levels.

The four-year Millennium Ecosystem Assessment (MA, 2001–2005) was designed to address these issues. Formally launched by UN Secretary-General Kofi Annan, in his report *We the Peoples* (Annan, 2000), the MA was carried out by more than 1,350 scientists from all regions of the world and representing the main current scientific disciplines in the fields of natural and social sciences, from ecology to anthropology. Its aim was to assess the status of ecosystem services worldwide, based on a series of sub-global assessments, at different scales. The ensuing results were used to advise policy-makers on how to deal with ecosystems in the context of landscape planning and management in the years and decades to come, so as to best meet the Millennium Development Goals (MDGs). Some forty sub-global assessments were carried out, including in UNESCO's biosphere reserves. Results were organized according to condition and trends of ecosystem services, scenarios, and policy-responses. The main multilateral environmental conventions (e.g. the Convention on Biological Diversity, the UN Framework Convention on Climate Change, the Ramsar Convention on Wetlands) provided an enabling environment for the Assessment. As a co-sponsor, UNESCO backed up the work of the Assessment's main sponsors – the United Nations Environment Programme (UNEP), the United Nations Development Programme (UNDP), the World Bank and the Global Environment Facility. UNESCO and the International Council for Science (ICSU) plan to design a research agenda for filling the gaps in knowledge as identified by the Millennium Ecosystem Assessment.

In addition to environmental issues, social transformations continue to occur with increasing frequency and intensity; institutional arrangements adapt to these changes with difficulty; some cultures are on the verge of extinction; and overall cultural diversity, including linguistic diversity, is being eroded and lost (see Wurm, 2001). These are all challenges for the new millennium, and their effects are often cumulative, making the global situation and its local manifestations even more difficult to deal with.

What to do, then? At UNESCO, based on the needs expressed by Member States, we of course have a duty to endeavour to respond to these issues through our programmes. I believe that our main asset is that we – quite successfully – foster and catalyse delivery of on-the-ground science and services and associated capacity-building, and that governments find in the Organization the necessary and appropriate forum to discuss science and other issues related to intercultural dialogue, and to identify the principles that should guide how social transformations take place.

The recommendations of the World Conference on Science and the Venice meeting remain valid. What is missing, however, is a true partnership for implementing a global science agenda, as called for in the WCS (see the 'Science Agenda – Framework for Action', in UNESCO, 2000). Implementation of a global science agenda should be based on existing science networks around the world and on the centres of excellence involved in implementing our respective programmes in various fields of the basic and environmental sciences, engineering and science education. This subject was further addressed at the 'Ministerial Round Table on the Basic Sciences: the Science Lever for Development', held during UNESCO's General Conference in 2005 (see UNESCO, 2005c).

SCIENCE IN THE UN SYSTEM – IMPACTS FOR THE FUTURE OF SCIENCE AT UNESCO

The future of UNESCO's science programmes will be affected not only by major global trends, the changing culture of science, and changes within UNESCO itself, but by changes at the overall UN-system level as well. The recent Millennium Project task force reports – and, in particular, that of Task Force 10 on science, technology, and innovation (UN Millennium Project, 2005) – have indicated new directions. Moreover, with the increasing awareness in the UN of the relevance of science and technology for development in general and for achieving the MDGs in particular, a stronger focus on science and science-related activities may be expected in the near future. This process will certainly also include strengthening of the social sciences which, in my view, have not received the necessary attention in contexts such as sustainable development.

OVERVIEW OF SCIENCE AND TECHNOLOGY ACTIVITIES IN THE UN SYSTEM

Science and technology (S&T) related activities are distributed widely within the UN system and reflect the different mandates and financial resources of its numerous bodies. Activities include: high-level scientific advice; environmental assessment and monitoring; norms and standard setting; research and development; technical assistance and technology transfer; management of large intergovernmental programmes, education and research training; in-country capacity-building; S&T diplomacy; and 'flagship reporting', such as, UNEP's Global Environmental Outlook and UNDP's Human Development Reports.

Major efforts are devoted to water management and land-use planning, biodiversity conservation, monitoring climate change, promoting renewable energy and providing 'carbon financing'. Many UN programmes are technical assistance in nature, or form part of large UN-wide or international global monitoring consortia. A small number of organizations conducts research in-house, related to their specific mandates. Advisory activities take place through the Commission on Science and Technology for Development (CSTD); the Commission on Sustainable Development (CSD); and the UN Conference on Trade and Development (UNCTAD), which carries out S&T policy activities, particularly in the area of technology transfer.

UNEP's science-related priorities are the production of global environmental assessments and information on the state of the planet's natural resources, particularly for international environmental policy-making. It coordinates numerous environmental conventions, including their scientific advisory groups. UNDP and the World Bank Group are among the largest multilateral sources of funding for economic and social development, together with the Global Environment Facility (GEF), which is supported by UNEP, UNDP and the World Bank. UNDP's water programmes deal primarily with water governance, development of waste management systems and providing access to municipal water supplies and sanitation. Its 'biodiversity for development' portfolio involves mainstreaming commitments under the Convention of Biological Diversity (CBD) into national governance frameworks and supporting community efforts to conserve biodiversity. UNDP is active in land management programmes (especially drought-related ones) and supports extensive energy-related activities through the GEF in energy efficiency, renewable energy and sustainable transportation. The World Bank funds major programmes in water resources management, biodiversity (also through the GEF), energy, rural land resources management and forestry. In the area of climate change, the Bank supports extensive 'carbon finance' initiatives.

Aside from UNESCO, six UN specialized agencies support S&T activities directly related to their more narrow mandates. The Food and Agriculture

Organization (FAO) has programmes in food and nutrition, agriculture, fisheries, forestry and sustainable development. FAO and the International Atomic Energy Agency (IAEA) have a joint programme to use nuclear techniques and related biotechnologies for sustainable food security. The World Meteorological Organization (WMO) facilitates international cooperation in establishing networks of stations for meteorological and hydrological observations, and promotes the rapid exchange of meteorological information, including education and training programmes. The International Maritime Organization (IMO) conducts programmes to prevent marine pollution from ships. The United Nations Industrial Development Organization (UNIDO) has the mission to promote industrial development, focusing on the least developed countries, particularly in Africa, on agro-industries and on small and medium-size enterprises. Its water management and energy-related services concentrate on introducing environmentally sound technologies to prevent discharges of industrial effluents and to provide renewable energy resources for the rural poor. The World Health Organization (WHO) targets its water and sanitation activities on drinking water quality and hygiene, water-related diseases and wastewater use.

The United Nations Human Settlements Programme (UN-HABITAT) runs major programmes for addressing the urban water crisis in Africa and Asia through improved water demand management, while the United Nations Children's Fund (UNICEF) manages a Water, Environment and Sanitation Programme, which supports government programmes on a global scale to provide a minimum level of water supply and sanitation for those most in need. Lastly, the United Nations University (UNU) operates several research training centres and programmes, as well as a network of cooperating institutions, many of which address natural resources management, climate change, global water problems, biodiversity, and S&T innovation.

OPPORTUNITIES FOR UNESCO

This brief overview of science and technology in the UN system shows that almost no activity is taking place outside UNESCO, including its major partners ISCU and the Academy of Sciences for the Developing World (TWAS), in the following areas:

- programmes in general dealing with research and education in the basic and engineering sciences;
- building basic-science and engineering research capacity at the university level;
- programmes dealing with research and education in ecological, Earth, water and ocean sciences;

- regional and country-specific science policy advice, including how to improve S&T advisory mechanisms in governments and how to integrate research and science education in national planning;
- programmes at all levels aimed at increasing public understanding of science;
- networks for centres of basic sciences research in developing countries; and
- support of activities linking science and culture.

Opportunities for UNESCO to enhance its scientific profile within the UN might include:

- expanding the portfolio in the basic sciences and engineering, ecological and Earth sciences, and water sciences programmes to emphasize building science education and research capacities from the primary school through to the tertiary level;
- creating 'flagship' programmes linking science, culture, and education for knowledge promotion and greater public understanding of science;
- expanding current efforts to link research centres of excellence in key disciplines for developing countries, with or even beyond the International Basic Sciences Programme (IBSP), and taking into account regional activities such as the New Partnership for Africa's Development (NEPAD); and
- expanding the portfolio for science policy to include advising a broader range of Member States on the integration of research and education in national development planning.

KEY SUBJECT AREAS FOR UNESCO

In this chapter, a few subject areas and modus operandi which most likely will be of importance in the years to come for the Sector's activities are described in greater detail. Some of them, such as sustainable development, have had a long history in the Organization. Others, such as capacity-building, have gained new meaning and importance in the UN and especially for UNESCO.

Sustainable development is a notion that was already anticipated by the Organization before the famous Brundtland Report was published (WCED, 1987), taken up further at the UN Conference on Environment and Development (Rio de Janeiro, Brazil, 1992), and most recently at the World Summit on Sustainable Development (WSSD), held in 2002 in Johannesburg

(South Africa). Sustainable development continues to be an issue that needs more attention and focus within the Organization and in partnership with other organizations both inside and outside the UN system. Here, as a basis for further discussions, we introduce a new approach to sustainable development. *Capacity-building* is just beginning to become a more quantifiable and results-based element of our programmes, though it has been dealt with since UNESCO's early days.

SUSTAINABLE DEVELOPMENT

Sustainable development is the unifying thread for the Natural Sciences Sector in the 2006–2007 Programme and Budget. All activities are subdivided into two main programme areas: 'science, environment, and sustainable development' and 'capacity-building in science and technology for sustainable development.' The latter has a sub-programme area called 'science policy and sustainable development'. This is the division that is coordinating the Sector's activities in the area, including its contribution to the UN Decade of Education for Sustainable Development (DESD).

A new approach to sustainable development
Following the concept that sustainable development rests upon four pillars, the three 'classical' pillars are the economic, the social, and the environmental. The fourth, favoured by UNESCO, is the cultural pillar (this seemingly new dimension was already discussed within UNESCO some ten years ago [di Castri, 1995]). Alternatively, one could maintain that culture and science, as well as education (UNESCO's own main areas of concern and operation), represent cross-cutting elements of the sustainable development pillars, in that (for example) to mainstream education, science and culture into development will reduce inequalities and their associated social costs, while at the same time guaranteeing environmental sustainability. Furthermore, there may be reason to question the notion of balance or equilibrium – even though this has been and still is commonly seen as the basis of sustainable development. If 'sustainable development' is considered as identical to best practice for assessing development and coping with complex problems, it is possible that development largely 'benefits from not being "perfectly" balanced' (Mawhinney, 2002).

The second element in the notion of sustainable development is the time factor – that is, the intergenerational aspect, as in the classical definition used in the Brundtland Report (WCED, 1987): 'Sustainable development is development that meets the needs of the present without compromising the ability of future generations to meet their own needs.' Governments do not always manage to connect the short-term priorities of development with the longer-term goal of sustainability. However, this gap has started to close, with the adoption by the

United Nations General Assembly of the Millennium Declaration and the related development goals.

As an organization composed of 191 Member States, the first questions we must address are: how do Member States perceive sustainable development; what are their priorities; are there commonalities at larger geographical scales such as subregional and regional levels; and how do these relate to UNESCO's mandate, capabilities, capacities and programmes? Answers to these questions will define the levels at which issues should be analysed and sustainable development implemented. One end of a potential continuum would be if we were to find shared priorities (or homogeneity) that could then be translated into UNESCO programmes or activities at regional levels. At the other end of the spectrum, there could be such fundamental differences that implementation of sustainable development might need to be addressed at individual national levels, or on the basis of small clusters of Member States only. This is the 'space factor' of sustainable development, which together with the 'time factor' (mentioned above) illustrates the complexity of the scales of problem assessment and intervention in implementing sustainable development.

Activities should be outcome-related with clearly identified quantitative targets to be achieved over well-defined timelines. A results-based approach should be used. Programme implementation should include rigorous monitoring and adaptive management of activities. Evaluations should be included from the planning or design process. This is precisely what the Sector has initiated, starting with a pilot project, 'Sustainable Development – A Pacific Islands Perspective', undertaken in cooperation with the Apia (Samoa) Office, the Coastal Regions and Small Islands (CSI) platform, and the Centre for Environment and Sustainable Development at the University of the South Pacific. Results of this pilot project will be adopted and modified for application in other regions, especially the Caribbean Region.

The UN Decade of Education for Sustainable Development (DESD), 2005–2014
The challenge of the Natural Sciences Sector is to keep the clear centrality of science in the agenda of the DESD, now that international focus seems to have shifted to the cultural and educational aspects of sustainable development. It is essential for joint activities to be developed between the Natural Sciences and Education sectors, primarily relating to science and education policies and the relationship between scientific research (through the IBSP, among others) and institutions of higher education in developing countries.

We must keep in mind that education is a process, and that it takes place at all levels (both formal and informal), at all ages of life, and everywhere we are. Education contributes to our fundamental ethics and basic behavioural patterns, and it therefore must begin early in life. We must make a special effort in education in order to implement sustainable development. In addition to 'packaging' sustainable development, UNESCO should contribute to incorporating sustainable development topics into university curricula. Moreover, additional UNESCO Chairs in this area may be another important contribution to strengthening efforts towards implementing sustainable development. I would envision a new meta-structure to share innovations in sustainable development between nations in a region, then between regions, as each nation and region learns from others.

CAPACITY-BUILDING AND NETWORKING

THE NEED FOR CAPACITY-BUILDING

One long-term focus of the Natural Sciences Sector has been human and institutional capacity-building and networking, in the basic and engineering sciences as well as in the various environmental sciences programmes of the Organization. UNESCO's experience with these various dimensions of capacity-building is both long-standing and wide-ranging, although it has gone by other names in the past. What is perhaps lacking is an overall appreciation and stocktaking of these various activities and their impact, which would act as a springboard for the design and implementation of future directions of work. Also needed are methodologies and procedures for demonstrating the effectiveness of capacity-building initiatives, in quantitative and qualitative terms, and for providing a basis for monitoring progress. Developing and testing such methodologies will be a primary challenge, entailing assessment of experience at various (national, regional, and international) levels.

Classically, capacity-building in the Natural Sciences Sector includes a whole series of actions, often interlinked:

- creating or reinforcing research and training institutions in particular disciplinary or technical areas in developing countries;
- support for short-term training courses of various kinds in developing regions to contribute to the skills and aptitudes of participants, often organized on a regional or subregional basis;
- support for postgraduate training courses organized in specialized institutions (in both developing and developed regions) and designed principally for students and in-service personnel from developing countries;
- using modern ICTs for innovative distance-learning programmes and for promoting exchanges of experience between practitioners in different parts of the world;
- award of individual study grants and fellowships, through competitive training schemes with independent assessment procedures; and

- elaboration and testing of learning and teaching manuals and materials of various kinds.

The role of S&T for achieving the MDGs is undisputed. Capacity-building is perhaps the most important element in this process. This should include reinforcing national strategies, involving society at large, and building strong regional and international scientific communities. As we recognize the global importance of capacity-building in S&T, science is experiencing a major crisis, in essence due to poor communication between science and society (see ICSU, 2005a). The 'science–society context' – already underscored in the World Conference on Science in Budapest in 1999 – has become ever more problematic (ICSU, 2005b). The culture gap between science and society has widened; we may even speak of a science–society divide.

In an editorial in the journal *Science,* Kofi Annan stated, 'No nation can afford to be without its own independent S&T capacity' (Annan, 2004). This editorial marked the launch of the InterAcademy Council's (IAC) report on global S&T capacity-building (IAC, 2004), which recommends that every nation develop an S&T strategy reflecting local priorities and that developing nations commit 1–1.5 per cent of their GDP to S&T capacity-building.

To achieve the needed global S&T capacities would require building new capacities among the relevant organizations. A concerted approach is needed. Our relevant partners would be (among others) IAC, ICSU, TWAS, and the UNU. Increased and more effective cooperation should include mutual reinforcement through the field networks of these organizations. For instance, we foresee closer cooperation of UNESCO's Regional Bureaux for Science with those of ICSU (for ICSU-UNESCO cooperation, see ICSU, 2005c). Such schemes of mutually reinforced capacities in the field could benefit even more from the different types of expertise and experience available in these different organizations. For instance, UNESCO could be a partner in all its fields of competence. In fact, we see beyond the important linkage between capacity-building in S&T and education, to potential linkages with the social sciences, local cultural contexts and the use of modern ICTs. In other words, capacity-building is at the heart of not only what UNESCO should and indeed does do in science, but what it does in all its areas of competence.

Initiatives, however, should not come from international organizations alone. Regional, subregional and local institutions would be vital in the context of building the relevant capacities. For instance, the African Academy of Sciences created a new capacity-building initiative on 'Strengthening Education and Research in Basic Sciences in Africa'. UNESCO's involvement, following the recommendation contained in the UK's Report of the Commission for Africa (Commission for Africa, 2005) and its lead role in the NEPAD S&T Cluster, will be channelled into this new initiative, which could make a continent-wide difference in the basic sciences. It would also fit nicely into the overall framework of UNESCO's IBSP, launched in 2004.

Capacity-building – from new concepts to new action
The newest conception of capacity-building is as a systemic approach to strengthen collective network capacities or systems of actors interactively linked to innovation (e.g. Chataway et al., 2005; Oyelaran-Oyeyinka, 2005). In-country and between-country partnerships should be viewed in this context of systems conceptualization (Oyelaran-Oyeyinka, 2005). These new approaches – at the heart of which are well-coordinated and optimized partnership arrangements – go far beyond the traditional views of capacity-building as being mainly the building of infrastructure or the training of scientists or S&T users.

In these new conceptions capacity-building includes individuals, organizations, institutions and systems of innovation. Furthermore, it encompasses support for skills and activities, use of S&T knowledge and products, development of means to improve S&T knowledge, and issues related to management and governance of R&D facilities (Chataway et al., 2005). Chataway et al. have identified three key ingredients for capacity-building programmes: understanding of the local context; the right mix of short-, medium- and long-term interventions; and the encouragement of systems of innovation.

To be successful, capacity-building efforts need to be based on quantifiable approaches, monitoring, and well-defined targets over given timelines. Appropriate indicators for measuring S&T capacity are particularly needed for developing countries. These indicators would allow measurement of progress (or lack thereof) towards goals outlined in the planning process for each capacity-building initiative. In sum, UNESCO's capacity-building activities need a fresh look in terms of the latest conceptual development in the field, and in view of the increasing importance of results-based programming in the Organization.

NETWORKS AND NETWORKING

Capacity-building and networking are closely related. In a sense, networks build the framework for all processes related to capacity-building. Networks are themselves dynamic entities, through both the elements of networks (nodes, hubs) and the interactions (connections and connectivity) among these elements. The study of networks has developed into a new branch of science, known as *network science* (see, e.g., Barabasi, 2002, 2005). Barabasi (2002) has even referred to the next scientific revolution as the new science of networks.

Many complex systems in both nature and society can indeed be described using the new concepts of network theory and network science – from the living cell to the World Wide Web. More specifically: in ecosystems, interactions among species can be mapped to show food web structure; and in social systems, networks can describe how individuals interact in a given organization. Large infrastructures such as power grids and transportation networks may also be analysed using network science tools.

For UNESCO as an international organization, networks of collaboration – their structure and dynamics – are of particular importance (see, e.g., Guimera et al., 2005). Network science provides novel perspectives for issues as varied as the spread of the HIV/AIDS virus, and the interconnectedness among staff in organizations. Network science could provide important insights for optimizing our work with scientific networks and for analysing our network of Field Offices and other institutions, which could contribute to our presence in the field. For instance, institutions such as the Abdus Salam International Centre for Theoretical Physics (ICTP, in Trieste, Italy) and the UNESCO-IHE Institute for Water Education (IHE, in Delft, the Netherlands) can be seen as 'our' hubs, linking their respective networks to UNESCO as their home organization.

Finally, the success of any capacity-building measure relies on building up or building upon relevant collaborative networks. This should be seen in a North–South as well as South–South context. In fact, our approaches to the 'use' of networks are largely based on a passive view, which uses existing structures rather than creating new and dynamic ones. Going a step further, how would we design networks to accomplish certain tasks? For instance, what would an 'optimized network' linking institutions for scientific exchange and training (with, for example, development-related goals) look like – as given, for example, in the Third World Network of Scientific Organizations (TWNSO, 2003), which lists 525 institutions working in the fields of mathematics, engineering and technology, physics, chemistry, and the biological, medical, environmental, agricultural, and Earth sciences?

Certainly we should not aim at connecting everything. As an organization with clear objectives to be achieved in particular areas and programmes, we need to be selective. Indeed, as a first step, we should look into existing networks and structures and examine their usefulness for our purposes. Second, we need to identify our partners and contacts for the respective networks – in other words, our future hubs in the systems – which allow us to communicate to as many links as we wish, in the shortest time possible. The IBSP provides a good example of making use of existing networks in the basic sciences, optimizing their function, or even designing new elements for them.

NEW MODUS OPERANDI FOR THE FUTURE OF THE SECTOR

Let us now consider two important aspects of our future work – intersectorality, and our capacities in the field. *Intersectorality* is among the recurring issues that – despite numerous different efforts – have never been fully and successfully implemented on a larger scale. The Headquarters and Field Office 'divide' has occasioned many discussions, in particular in view of the ongoing decentralization process (a major component of the Director-General's overall reform process). Clearly, we need to evaluate our programme delivery in the field – and in priority regions such as Africa. In other words, we need to reflect more about the *geography of our action*.

BEYOND SECTORALITY: TOWARDS INTERSECTORALITY AND INTERDIVISIONALITY

Both the basic sciences per se and the multiple contexts of their application can no longer do without transdisciplinary approaches. These, of course, cannot replace classical single-discipline approaches, but rather are complements needed to address many of the complex issues and problems facing today's world. In my view, sustainable development and addressing the MDGs are not only among the most important, but also, currently, *the* most relevant such examples. 'Transdisciplinarity' is the cooperation among academic disciplines to address and solve real-world problems.[1] It represents a new way to address complex societal issues, based on a multi-stakeholder and practice-oriented approach. In a scientific context, it may be seen as a tool to bring science and society closer together. Transdisciplinarity at UNESCO means first of all transcending disciplines as conducted *within* each of its five programme sectors. In other words, we need to examine inter-unit or interdivisional cooperation within each of the sectors but also – at a higher level of complexity – transdisciplinarity from an intersectoral point of view. The sectoral set-up (in particular, as regards programmes) certainly has its advantages, as do single-discipline approaches; however, Member States continue to request more intersectoral elements in UNESCO's programmes. Therefore, all that transcends classical disciplines needs to be given more serious thought by the Organization in the future.

The new millennium has brought increasing efforts to include greater intersectorality into the biennial planning process. What started as a 'Transdisciplinary Project' and 'Transverse Activities' (in the Programme and Budget for 2000–2001 [30 C/5]), evolved in the following biennium – based on a decision of the General Conference in 2001 – into the two cross-cutting themes 'Eradication of poverty' (coordinated by Social and Human Sciences) and 'Information and communication technologies for knowledge societies' (coordinated by Communication and Information). The Natural Sciences Sector

---

1   See, e.g., Somerville and Rapport, 2000; Thompson Klein et al., 2001. For a discussion of the terms 'transdisciplinarity', 'interdisciplinarity' and 'multidisciplinarity' see, e.g., Brown, 2002; Thompson Klein, 2000.

contributed five initiatives to the poverty theme and two to the communication theme. One of the most successful of the projects under the two cross-cutting themes has been the Local and Indigenous Knowledge Systems (LINKS) project (see Box VI.1.1). As stated in the Programme and Budget for 2004–2005 (32 C/5), 'the LINKS project promotes local knowledge, values and world views as tools to shape and achieve poverty eradication and environmental sustainability. It seeks to empower rural and indigenous communities through local and indigenous knowledge systems'. LINKS was the first project to be integrated into the Regular Programme activities, in the Programme and Budget for 2006–2007 (33 C/5).

In a sense, the Natural Sciences Sector has taken a lead in intersectorality. For the first time,[2] the Sector realized a new and truly intersectoral approach, based on so-called joint Main Lines of Action – major programme components to be carried out jointly with other programme sectors. The first two were with

BOX VI.1.1: THE FUTURE OF LINKS

The rapid increase of international interest in indigenous knowledge is cause for concern as well as encouragement. To develop a full appreciation of the complex and dynamic synergies (at once social, ecological and spiritual) between traditional knowledge, indigenous peoples and their environments demands a long-term commitment and cooperation with local communities and a full range of interdisciplinary expertise.

Yet today, a bewildering array of actors flies the 'indigenous knowledge' banner. Can the rigour and integrity of work in this field be maintained when so many have 'jumped on the bandwagon'? Might not an increasingly cursory, superficial and perhaps even romanticized treatment of traditional knowledge and indigenous peoples ultimately undermine the aspirations of local communities for equity, resource security and the right to determine their own future?

In this rapidly expanding domain, there is a need for some reliable reference points. The LINKS project, by consolidating and expanding a growing network of leaders in the local knowledge field (both scientific and traditional knowledge holders), and in cooperation with other UN agencies, programmes and conventions, will work towards elaborating appropriate benchmarks and indicators to ensure a rigorous and respectful inclusion of local and indigenous knowledge in sustainable development efforts, and also to ensure its continuing vitality and transmission within local indigenous communities.

Through these efforts a network of indigenous knowledge centres and experts will be constituted. This will offer the possibility to build the capacities of local communities, state resource managers and decision-makers, to accommodate indigenous knowledge, values and worldviews in sustainable development, biodiversity management, and education for sustainable development.

---

2   In the Programme and Budget for 2004–2005 (32 C/5).

the Education and Culture Sectors, respectively. These have continued into the 2006–2007 biennium, as 'Promoting education and capacity-building in science and technology' and 'Enhancing the linkages between biological and cultural diversity'. For additional intersectoral activities planned for the 2006–2007 biennium, see Table VI.1.1.

Intersectoral cooperation should not just be a concept, but a daily practice. Given the great work pressure on all professional staff in all programme sectors, it should be kept simple and offer added value towards attainment of programme objectives without additional bureaucracy or complex procedures.

SCIENCE CAPACITY IN THE FIELD

The Regional Bureaux for Science are our hubs for coordinating programme implementation in the field (see Figure V.6.5: Map of UNESCO Offices Worldwide). They work together with other Field Offices and (except for Venice) function also as cluster offices. The Regional Bureaux play the important roles of moderators and coordinators during the programming and work plan cycles. It has long been clear that science capacity in the Field Offices is not sufficient. The future of our programme delivery therefore depends on our ability to raise capacities in the field. Since internal reinforcement of the current forty-four science specialists seems rather unlikely, in the near future even our cluster offices may not all have science staff. As expressed by the Director-General, in his response to the general debate of the 171st session of the Executive Board:

> decentralization encompasses not only field units, institutes and National Commissions but also many relays that our Organization possesses in civil society. One illustration of this growing tendency is shown by the desire of several Member States to establish Category II centres aligned with UNESCO's programmes and priorities (Matsuura, 2005).

UNESCO's science presence and capacity for action around the world can be enhanced by: further establishment of Category I and II centres and institutes, particularly in the water centres with their global distribution; twinning with other organizations with a field presence (ICSU, TWAS, etc.); extension of the UNESCO Chairs network; intensified cooperation with existing networks (e.g. the InterAcademy Panel on International Issues [IAP], TWNSO and the African Academies Network); and increased cooperation with our National Commissions.

Institutes and Centres
Creativity and targeted approaches will be required in order to develop the necessary science field presence for better, more intense programme

Table VI.1.1: Intersectoral initiatives planned for the biennium 2006–2007, by programme sector

Major Programme Subject/Area	I: Education Sector	II: Natural Sciences Sector	III: Social and Human Sciences Sector	IV: Culture Sector	V: Communication and Information Sector
Biodiversity – cultural diversity		x		x (including WHC)	
Global Initiative on HIV/AIDS and education	x	x	x	x	x
E-learning and ICTs in education	x (ED + IITE)				x
Access to scientific and technical information and ICTs		x (including ICTP)	x		x
United Nations Decade of Education for Sustainable Development (DESD)	x	x (including IHE)		x	x
Small Island Developing States (SIDS)	x	x	x	x (WHC)	
Promoting World Heritage values in education policies and practices	x			x (WHC)	
Science and technology education	x	x			
Disaster prevention and preparedness, including tsunami warning system	x	x		x	x
Languages and Multilingualism	x	x	x	x	x

Source: UNESCO Document 33 C/5.
Notes: IITE = UNESCO Institute for Information Technologies in Education; WHC = World Heritage Centre; IHE = UNESCO-IHE Institute for Water Education; ICTP = Abdus Salam International Centre for Theoretical Physics.

implementation. Very promising signs have emerged, especially in 2005. A number of water- and other science-related centres are currently being established by Member States in close cooperation with the Natural Sciences Sector. The Sector has in fact taken a lead in using centres to increase its capacities and programme delivery in the field.

It is timely to think about further strategizing our approach to establishing future centres. As highlighted in the recent strategy document for institutes and centres (submitted to the Executive Board at its 171st session and approved by the General Conference in October 2005): 'If new institutes/centres are established, they should also aim to strengthen a horizontal division of labour among the institutes/centres'.[3] The institutes and centres are embedded in networks of experts and institutions of relevance to their fields of expertise. This hub-like function within networks needs permanent monitoring with regard to the Organization's programme priorities. To date, these networks have not been carefully mapped out. Nor has there been a planning process in place which would take into account the need to optimize communication and cooperation among governmental and non-governmental institutions.

In order to have the relevant global capacities in place, the Natural Sciences Sector, in cooperation with Member States, is currently emphasizing the development of more regional centres, in particular in the fields of indigenous knowledge and biotechnology. For the latter, the thirty-third General Conference in 2005 approved a proposal for the establishment of a regional centre for biotechnology training and education in India. It is only recently that biotechnology has been seen as a potentially very powerful tool to contribute to development, particularly in developing countries (see, for example, UN Millennium Project, 2005). As outlined in the relevant document submitted to our Executive Board,[4] this regional centre will not only have clearly identified key objectives and modalities, but we expect a significant regional and international impact from the centre (see Box VI.1.2).

Water-related centres
The demand for the establishment and continuation of water-related centres under the auspices of UNESCO is on an upswing due to several factors: the increasing currency of water issues at a local, regional and global scale; the realization that building regional know-how is essential to solving water problems; the recognition that the knowledge base on specific topics can be best handled by specific centres; and the recognition that the International Hydrological

---

3   UNESCO Document 171 EX/18.
4   UNESCO Document 171 EX/9.

> **BOX VI.1.2: REGIONAL AND INTERNATIONAL IMPACT OF THE PROPOSED REGIONAL CENTRE FOR BIOTECHNOLOGY TRAINING AND EDUCATION IN INDIA, UNDER THE AUSPICES OF UNESCO**
>
> - The proposal views the Regional Centre as 'the focal point for cooperation among Member States of the Asia region' and refers to South-East Asia, South Asia and the South Asian Association for Regional Cooperation (SAARC) countries.
> - Upgrading of national and regional capacity in terms of biotechnological expertise and ensuring effective transfer of appropriate technologies are an important means of securing long-term self-reliance and sustainable development; these remain a priority for many Member States.
> - Scientific exchange will reinforce existing collaboration in the region and promote new partnerships through the development of mutually beneficial research and development programmes.
> - The focus of activities of the Regional Centre will be demand-driven and directed towards problems relating to food supplies, human health, environmental issues, etc., indigenous to the region. This will address priority issues for the region and also promote sustainable exploitation of indigenous biological resources.
> - The Regional Centre will aim towards the establishment of a functional infrastructure within the region for collaborative research, technology transfer and information dissemination. Furthermore, it will foster the development and expansion of biotechnology-related industries in the region.
>
> *Source*: UNESCO Document 171 EX/9.

Programme (IHP) is an effective and proven mechanism for instituting centres with an international or regional scope on water issues.

The overall strategy in setting up water centres was outlined by the Director-General in 2003, at the opening of the UNESCO-IHE Institute for Water Education:

> Looking towards the future, let me emphasize that recently UNESCO has been pursuing a very deliberate policy of building up a network of regional centres of excellence which will spearhead research, education and capacity-building in freshwater throughout the world ... I view the creation of such a network as exemplary of the new approaches I have been promoting to reform our Organization ... This coalition of expertise, outreach and professionalism would spearhead a concerted effort by all stakeholders. Together we can make a giant leap forward in both qualitative and quantitative terms,

to serve the capacity-building needs underpinning the international water agenda (UNESCO, 2003a).

In 2004, the Intergovernmental Council of IHP examined the policy of establishing Category II water centres under the auspices of UNESCO and adopted the following proposal:

> The expanding network of international and regional water centres under the auspices of UNESCO should be increasingly used to execute the relevant parts of the IHP plan in areas of their geographical and scientific competence. The 33 C/5 should therefore explicitly recognize the existence and role of the UNESCO water centres for the implementation of the 2006–2007 programme (IHP, 2004).

The IHP Bureau recommended in 2005 that the expanding network of international and regional water centres under the auspices of UNESCO should be increasingly used to execute the relevant parts of the IHP plan in areas of their geographical and scientific competence. The UNESCO-IHE Institute for Water Education, in its unique status of being legally part of UNESCO, should be responsible for the execution of the educational, training and transfer-of-knowledge components of IHP, in full cooperation with other educational elements of the UNESCO water family, including the existing courses, the international educational and training networks, and the UNESCO Water Chairs.

THE NATURAL SCIENCES AT UNESCO

In the following section, I present thoughts on the possible future directions of the Natural Sciences Sector's current divisions. These ideas were developed from discussions with the directors and other professional staff in the divisions concerned, as a basis for reflection on possible futures. Of course, it is the UNESCO Member States who shall determine the actual course of priorities and actions the Sector will undertake.

THE WATER PROGRAMMES

UNESCO has gained a fairly strong position within the UN system when it comes to water issues. With the three strong pillars, IHP, UNESCO-IHE and the UNESCO-led World Water Assessment Programme (WWAP), the Organization clearly leads UN efforts not only in the area of water sciences and education, but in policy-relevant assessment as well. A convergence is expected in the near future among these three pillars.

IHP is expected to continue to have a wide-ranging focus on issues both at the global and the local scales. Water in the hydrological cycle acts as the essential 'bloodstream' for all terrestrial and coastal systems. Water provides environmental services, and at the same time is a vital resource, both in the built and the natural environment. At a global scale, UNESCO's water programmes should thus see their role as promoter of the study, observation and quantification of uncertain global impacts that arise from the continuing expansion of human populations and infrastructure. At the local scale, IHP's role is more complex and some priorities

BOX VI.1.3: CATEGORY II CENTRES

Fully operational existing Centres
- CATHALAC – Water Center for the Humid Tropics of Latin America and the Caribbean (Panama City, Panama)
- HTC Kuala Lumpur – Regional Humid Tropics Hydrology and Water Resources Centre for Southeast Asia and the Pacific (Kuala Lumpur, Malaysia)
- ICQHHS – International Center on Qanats and Historic Hydraulic Structures (Yazd, Islamic Republic of Iran)
- IRTCES – International Research and Training Center on Erosion and Sedimentation (Beijing, China)
- IRTCUD – International Research and Training Centre on Urban Drainage (Belgrade, Serbia)
- RCTWS – Regional Center for Training and Water Studies of Arid and Semi-Arid Zones (Cairo, Egypt)
- RCUWM – Regional Center on Urban Water Management (Tehran, Islamic Republic of Iran)

Potential future Centres
Four Centres were approved by the thirty-third session of the General Conference of UNESCO:
- CAZALAC – Water Center for Arid and Semi-Arid Regions of Latin America and the Caribbean (La Serena, Chile)
- European Regional Centre for Ecohydrology (Lodz, Poland)
- IHP-HELP Centre for Water Law, Policy and Science (Dundee, United Kingdom)
- ICHARM – International Centre for Water Hazard and Risk Management (Tsukuba, Japan)

Three further Centres will be considered by the Executive Board at a later stage:
- IGRAC – International Groundwater Resources Assessment Centre (Utrecht, the Netherlands)
- Regional Centre for the Management of Shared Groundwater Resources (Tripoli, Libyan Arab Jamahiriya)
- Regional Center on Urban Water Management for Latin America and the Caribbean (Bogotá, Colombia)

need to be set. Water is fundamental for societal needs in terms of ensuring environmental sustainability (which is one of the MDGs); with the widespread recognition of this fact, a huge number of worldwide actions are being mounted, resulting in a tangled 'web' of actions. IHP should consider placing itself at the apex of what may be called a 'water-for-life web of actions', in the sphere of science and policy. IHP is expected to ensure that the important function of the ecohydrological 'services' in the planet's life support system are addressed in key areas – especially: in water for food security, in assuring the vital ingredient for human survival and in alleviation of water-health-related problems. At the catchment scale, IHP's role in water-environment management – such as on erosion impact, dilution of waste, and in translating land use activities into ecological sustainability involving secure habitats for aquatic biota – cannot be forgotten. In addition, at this scale (also known as the *landscape level*), water is perceived as more of a finite resource, and so its wise sharing between water-dependent activities – and neighbouring political units – are challenges that IHP must not ignore.

Taking into account the previous achievements and historical strategic evolution of the phases of IHP (in which societal aspects of hydrology increased in importance over time), the designers of the future phases of IHP have advised that it would not be appropriate to adopt any 'step change' in the direction of the IHP. Rather a 'continuity with change' approach is recommended. A possible framework would ensure continuity with the high-quality efforts of IHP's previous phases, while expanding the arena in which UNESCO would assist and support water-related actions across the UN system, as well as with other international agencies, governments, Member States and stakeholders. For example, there is a need to widen UNESCO's scope in several specific areas, such as socioe-conomics, health, groundwater, governance and ecohydrology. To best express the transition from the current (sixth) phase to subsequent ones, and its continuity, we can speak of three threads that are likely to run through all water-related activities of the Organization – system interdependencies, systems under stress and societal responses.

These three threads would perhaps be best encompassed by a programme on 'Water Dependencies: Systems under Stress and Societal Responses'. In order to be effective, UNESCO's water activities should continue to be based on the three main pillars of hydrological research: water resources management, education and capacity-building. They should continue to focus on strengthening scientific knowledge to provide new directions for science and research to develop tools to solve challenges posed by the adverse effects of global changes. The Four Themes might be:

- Theme I: Global Change, Watersheds and Aquifers
- Theme II: Governance and Socio-Economics

- Theme III: Ecohydrology and Environmental Sustainability
- Theme IV: Water Quality, Human Health and Food Security

It must be remembered that education, training and technology transfer remain among IHP's most important roles. IHP capacity to provide education and training has improved during IHP V (1996–2001) and VI (2002–07) through the establishment of several regional and international centres and of the UNESCO-IHE Institute for Water Education. It is to be hoped that the role and capacity of IHP to develop water education opportunities will be strengthened. UNESCO needs to increase its capacity by: (a) creating a real network among the existing centres, Chairs and related institutes, and (b) by setting up an efficient coordination mechanism for this network, one which allows these centres to take responsibility for implementation of activities, and which facilitates the identification of gaps, so as to orient decisions of Member States in the establishment of new centres. This mechanism should be set up using the new capacity existing at the UNESCO-IHE.

This approach also proposes that UNESCO's water programmes provide an improved focus on the interdependency of water and society – including on water governance, economics, culture and ethics – considering especially the emerging problems posed by the international scientific community and recent UN summits. IHP, UNESCO-IHE and WWAP are expected to create the groundwork for developing valuable contributions to the achievement of the MDGs and to the UN International Decade for Action 'Water for Life' (2005–2015).

THE ECOLOGICAL SCIENCES

What the ecological sciences at UNESCO should do in the future closely resembles the twenty-first century 'Vision' of the Ecological Society of America (Palmer et al., 2004), namely that the three areas of focus should be: informed decision-making, promotion of anticipatory research, and influence on cultural changes in the style and working methods of ecology practitioners (particularly in terms of how they relate with policy- and decision-making communities).

We should convene, in cooperation with other relevant UNESCO or UN units, one or more dialogues to review changes in the perceptions of human-environment relations that have taken place since the 1970s. Such dialogues may reveal a variety of ethical, scientific and operational perspectives that could feed into efforts to elaborate the new MAB research agenda. MAB has a history of more than thirty years as a research programme. It should make a systematic effort to contribute towards the MDGs (particularly MDG 7, on environmental sustainability) and the Convention of Biological Diversity (CBD) target to minimize biodiversity loss by 2010. MAB contributions could include: developing and testing indicators to track and illustrate hitherto undetected patterns and trends in environmental sustainability and biodiversity loss; synthesizing results of relevant MAB research implemented at national and local levels over the last three decades; holding scientific seminars and workshops to review MAB's contributions deriving from a particular ecosystem or theme focus; and launching well-designed, short-term (two to three years) studies and projects to make specific contributions towards MDGs and the CBD 2010 target.

MAB is encouraged to make a determined effort to contribute to UNESCO-wide planning and programming priorities – particularly the 'water and associated ecosystems' priority of the Natural Sciences Sector – and to explore interlinkages between cultural diversity and biodiversity. Sustainable development capacity-building opportunities under MAB should be significantly increased and expanded. The MAB Young Scientists Award scheme should be a prime target for scaling up and fundraising. The MAB International Coordinating Council might consider options for linking specific numbers of awards each year for selected priority research – such as studies focusing on contributions to MDGs or measuring ways and means of minimizing biodiversity loss – for the period 2005–2010. The European Union and the Government of Belgium support the development of the ERAIFT[5] School in the post-conflict Democratic Republic of the Congo, linking it to the ENEF[6] School in Gabon; these are positive developments and should at all costs be sustained at least until 2010. ERAIFT, or programmes based upon it, might fruitfully be expanded to other forested regions of the tropics. A new series of research projects linking the implications of land/resource-use development options to carbon emissions – and their consequences for ecosystem management, sustainable use of biodiversity, economic growth and human development – should be designed and launched. In addition, Transboundary Biosphere Reserves, which can contribute to peace and cooperation through joint development planning, should be promoted wherever possible.

The Ecological and Earth Sciences Division and MAB should interact with all Natural Sciences Sector and UNESCO efforts to ensure that the best conservation sciences knowledge and practice are taken into consideration in programmes and projects with implications for the sustainable use of biodiversity. In particular, collaboration with the World Heritage Centre and the World Conservation Union (IUCN) is necessary to ensure that UNESCO Member States and Parties to the Convention do not confuse World Heritage sites and biosphere reserves, but use the two instruments to optimally combine protection and sustainable use options for biodiversity at national and subregional levels.

---

5   Ecole régionale post-universitaire d'aménagement et de gestion intégrés des forêts tropicales.
6   Ecole Nationale des Eaux et Forêts.

More and more data and experience point to increasing risks associated with limiting conservation of globally significant biodiversity to legally protected areas only, as currently practiced in the identification and nomination of World Heritage sites.

In the current DESD, the Ecological and Earth Sciences Division and MAB should emphasize educational interventions to improve future options for ecosystem management and sustainable use of biodiversity. Field ecology, natural history and biodiversity awareness provide a knowledge cluster that needs to be taken to rural and remote-area schools, often situated in or around biosphere reserves. Where sustainable tourism generates income and employment opportunities, education of youth on fauna, flora, natural history, ecology and behaviour of animals and plants, and regional cultural history and geography could enhance their ability to interpret their environment and heritage for visitors. Rural-area educational content and methods could emphasize practical and field-based learning about students' environment, ecology and biodiversity. Taxonomic knowledge is a diminishing area of expertise in countries and regions containing the most unique biodiversity; an ability of rural youth to contribute to taxonomic knowledge (as parataxonomists) could increase both interest and knowledge in areas where people live closest to rich biodiversity. UNITWIN/UNESCO[7] Chairs could be created to help address this lack of taxonomic expertise. Education for sustainable development could also make useful interventions at the level of tertiary, non-formal and professional education, particularly for practitioners of public administration and international relations in environment and biodiversity conservation. Opportunities for such interventions should be explored in full consultation with MAB National Committees, UNESCO National Commissions and the Permanent Delegations of Member States to UNESCO.

THE EARTH SCIENCES

The International Geoscience Programme (IGCP) is one of UNESCO's science programmes with a tradition going back over thirty years. Its history shows it has made constant adaptations to changing orientations in priority themes within the geoscience community. In the future, the IGCP anticipates continuing the tradition of serving as a catalyst for international cooperation, emphasizing the advancement and sharing of knowledge between developed and developing nations, promoting interdisciplinary dialogue and networking. But the IGCP is now coming to a turning point regarding its traditional basic geosciences themes. After the IGCP reform (which will bring structural changes in management and functioning), the programme most likely will concentrate mainly on projects

---

7 UNITWIN is the abbreviation for the university twinning and networking scheme.

grouped around geoscience research themes with high relevance to the needs of society. They would be limited in number and address specific, well-defined niches, where Earth science expertise contributes to sustainable development. In addition, some IGCP projects should cover cutting-edge fundamental geological and geophysical research with a regional or worldwide scope. Earth System Science – which takes into account the biotic and abiotic nature–human interaction – will be a priority. This includes formal cooperation, which would take place with biological, ecological, hydrological, oceanographic, conservation and environmental monitoring and assessment programmes.

A major initiative of the Earth science community is to demonstrate the importance of geoscience in the development process. This is foreseen through the UN General Assembly's declaration of 2008 as the International Year of Planet Earth. Preparations for the Year will start in 2006, and activities will culminate at the quadrennial International Geological Congress, to take place in Norway in 2008.

Assistance to national Geoparks initiatives could also be provided in the future at the request of Member States, providing a platform of cooperation and exchange between experts and practitioners in geological heritage matters, under the umbrella of UNESCO. UNESCO Geoparks may become the driving force behind the increasing success of national Geoparks networks. UNESCO's support helps networks to attract new members and gives them publicity and visibility. UNESCO Geoparks may, in the future, become complementary to the WH and MAB programmes. The Global Geoparks Network – or a future 'UNESCO Geoparks Programme' – would then be the only international undertaking focused on enhancing the value of geological heritage sites, while integrating the preservation aspect into a strategy for sustainable regional development.

The Earth sciences community needs to strengthen its participation in the Global Earth Observation System of Systems (GEOSS) by working towards the development of harmonized priorities regarding the creation of an integrated Earth system observation mechanism. UNESCO should take the lead role in developing the disaster prevention networks and observation systems, especially in the fields of geohazards and tsunamis. The Geological Applications of Remote Sensing (GARS) programme might gradually evolve from a research and capacity-building programme into a worldwide observation system for the solid geosciences, covering geology, geophysics and geodesy. GARS would thus become the fourth worldwide observing system – joining the Global Ocean Observing System (GOOS), the Global Climate Observing Systems (GCOS) and the Global Terrestrial Observing System (GTOS).

Cooperation with the space agencies for the monitoring of biosphere reserves and World Heritage sites will continue. It is foreseen that, in the future, the partnership between UNESCO, the space agencies, space-related research institutions and universities (as well as the associated governments sponsoring these activities) will be further reinforced through new partners. While continuing with a project-by-project approach, we also foresee using an ecosystem approach. The concept is to use, for example, space technologies to better understand the current status of tropical forests worldwide that fall under some sort of 'UNESCO protection', and then assess where conservation measures are, or are not, having a positive effect. This type of analysis will assist UNESCO in setting up priorities to work jointly with the associated Member States to improve their related conservation efforts.

Activities in space science, including geological studies of other planets, will be highlighted in celebration of the International Year of Astronomy in 2009. This Year should prove another excellent opportunity to communicate the excitement of science to young people, in conjunction with ongoing space-related educational programmes, such as Bringing Space to Schools and Universities, the Space Volunteer Programme and the UN-affiliated Regional Centres for Space Science and Technology Education.

THE INTERGOVERNMENTAL OCEANOGRAPHIC COMMISSION OF UNESCO

The Intergovernmental Oceanographic Commission of UNESCO (IOC) will face the challenge of promoting and supporting a major institutional change at the national and international level to develop a whole new series of public services based on ocean data and information. The Indian Ocean tsunami of December 2004 demonstrated the importance of ocean services that are able to mitigate the effects of natural disasters, and prevent much loss of life and property. More than fifteen years ago, the IOC endorsed the development of GOOS as a multi-purpose system, serving many needs beyond those of climate studies. The fully operational GOOS will contribute to our understanding of physical/chemical/biological cycles on a basin-wide scale, and provide the information to develop an ecosystem-based approach to the management of oceans and coasts. In recent years, GOOS has become the major programme of the IOC. Focusing first on the physics of the ocean, the climate component of GOOS has enabled a level of *in-situ* observation of the global ocean never before achieved. The observing systems comprising GOOS, developed initially in support of research, have evolved to become permanent instrumental networks that can and do provide useful services to a wide range of users. In many areas, organizations such as FAO and WMO already operate services such as warning of locusts outbreaks, and early warnings for floods, tropical storms, hurricanes, storm surges and tsunamis. However the real challenge is to promote an institutional change. What is needed is to build the organizations capable of exploiting this new wealth of data, while maintaining those services for the public good.

For example, GOOS and the development of operational oceanographic applications require the support of new institutions at national, regional and global levels. The new Joint Technical Commission for Oceanography and Marine Meteorology – which reports simultaneously to WMO and the IOC – represents a giant step forward in our ability to implement an integrated system. As we move to monitor the quantity and quality of surface- and groundwater, to assess the total output of organic matter and fertilizers reaching the coastal zone or the changing chemistry of the atmosphere, we will see new mechanisms joining the IOC, FAO, WMO, IMO and UNEP. In the coming years, the IOC will continue to support and cooperate with other organizations in the effort to develop the Earth System Science agenda.

The 1992 'Earth Summit' in Rio de Janeiro succeeded in focusing world attention on the threats to the environment posed by the intensive use of the natural resources and natural systems of the planet. For the first time, the long-term risks to the life-support system of the Earth were made clear, and world leaders agreed to take action. Yet, with the exception of the elimination of the emissions to the atmosphere of Freon and other man-made gases which threaten the ozone layer, all the other factors and trends responsible for global and climate change remain unaltered or have been even further exacerbated. Part of the difficulty lies in the limited knowledge available on how the life-support system works, how resilient it is to perturbations, and the inherent uncertainty of the final outcome of the interaction of many natural processes that are inadvertently being modified as a result of development. In addition, the net effects of measures adopted to reverse negative trends in the global environment – such as the accumulation of greenhouse gases in the atmosphere – will be extremely difficult to measure. These incremental planet-wide changes will take years to detect and ascertain scientifically and will be difficult to differentiate from natural variability. For example, determining the efficacy of the very costly measures agreed on to mitigate climate change will pose a substantial technological challenge.

Fulfilling the promise of sustainable development will ultimately require a better understanding of how the planet works as an integrated system, which will in turn require extensive research involving sustained and high-quality atmospheric, oceanic and terrestrial observations. Because of the scale at which the observations will be needed, this will require the deployment of instrumental systems across national borders, over many jurisdictional spaces. The full sharing of data between nations is essential to this effort – including making data available in time to be of operational use. Achievement of the full benefits of the new observation systems for sustainable development will require the strengthening of the scientific and technical capabilities of countries (especially in the developing world) to use this information.

Today, national institutions can tap the hourly and daily flow of data and information originating in oceanic and atmospheric networks of instruments managed by the IOC and WMO. These and other international organizations are training human resources and providing access to the new technologies. But much more is needed as the products synthesizing environmental information become more sophisticated, requiring non-trivial skills to handle and interpret them correctly. This will continue to be one of the priority tasks for the IOC in the coming years, which will require the full application of the new strategy for capacity-building adopted in 2005.

For the oceans, the UN Convention of the Law of the Sea (UNCLOS) provides a general framework which, complemented by the new fisheries and environmental agreements adopted in the last decade, is slowly testing and building up an integrated system of governance for oceans and coasts. The ability of the international community to provide appropriate stewardship for oceans and coasts depends critically on the success of these efforts.

The ongoing effort to eliminate the uncertainties of global change is increasing the availability of valuable environmental data and information, a large fraction of which, however, has not been fully utilized. This information is slowly starting to be used to reduce the risks inherent to the operation of many industrial systems, making them more efficient in their consumption of energy and use of resources. However, if this information could be incorporated in a more systematic way by the private sector in their everyday operations, it would have an enormous impact on sustainable development and economies. These are not contradictory goals; rather, they are complementary.

For the IOC, as focal point of the UN for ocean science and ocean services, this translates into a specific challenge, as outlined in the biennial report of the IOC to the thirty-first General Conference of UNESCO:

> The long-term challenge for IOC is to define a global framework in which the development of GOOS as a single, permanent, global, public- oriented service, can be achieved, with the active contribution of different segments of the society, including the private sector. This requires demonstration of the economic benefits of a common shared strategy between the public and private sectors, the identification of the public and private services that can be derived and/or shared through a common observing platform and the appropriate segmentation of public and private products and users. Achieving this new vision will require the development, negotiation and adoption of international norms and agreements, especially in the area of data and information exchange and sharing.[8]

---

8   UNESCO Document 31 C/REP.10 (11 September 2001).

PART VI: LOOKING AHEAD

THE BASIC AND ENGINEERING SCIENCES

Although the basic sciences offer opportunities to meet societal needs, many developing countries find themselves excluded in one way or another from the endeavour to create – and, consequently, to fully benefit from – scientific knowledge. At a time when the welfare of emerging knowledge societies and the future of humankind have become – more than ever before – dependent on an equitable production, distribution and use of knowledge, the divide in the basic sciences cannot but deepen the divide in technology, agriculture, health care, ICTs, science education and, finally, between the North and South. Hence, in line with the unique mandate of UNESCO for the basic sciences, the Organization should strive to promote worldwide action and regional cooperation in the basic sciences at a scale that will ensure that science becomes a truly shared asset, benefiting all.

UNESCO's programmes in the basic sciences respond to the expectations of Member States, and they should embrace the promotion of national capacity in the basic sciences and of excellence of basic research in areas of national priority. Science capacity-building should emphasize the renewal, expansion and diversification of education in the basic sciences for all – focusing on the knowledge and skills necessary for the creation of highly qualified specialists and responsible citizens who can participate in the society of the future. This strategy should particularly emphasize sustained assistance to improve the quality of science education in the developing and least developed countries, to foster the use of ICTs, as well as to create and develop a world-class university in each developing country, which can attract and nurture young talents.

International and regional cooperation in basic research should be promoted in order to ensure high scientific standards in the science services sought by Member States. From now on, this should be achieved primarily through the development of services of a wide range of networks and centres of excellence in the basic sciences, consolidated into the framework of the International Basic Sciences Programme (IBSP), launched in 2004 by UNESCO as part of the follow-up to the World Conference on Science (WCS). The IBSP will also seek to form new partnerships with non-governmental and intergovernmental scientific organizations in order to pool intellectual and material resources for the attainment of the MDGs.

A Ministerial Round Table on the Basic Sciences, the 'Science Lever for Development', was convened by UNESCO in conjunction with the thirty-third session of the General Conference in October 2005, in order to provide a forum for exchange of views and political debate between high-level policy- and decision-makers on the challenges to be taken up by the basic sciences in their service to society, and on actions to be taken by governments and the scientific community so that adequate capacity in the basic sciences can be built (see

UNESCO, 2005a). The Round Table gathered participants from 125 countries, and focused on four main themes: challenges in the twenty-first century, national and regional cooperation priorities, capacity-building in developing countries, and science policy and the role of the basic sciences for governmental decision-making. A Final Communiqué (posted on the UNESCO website) was approved and is to be circulated to the Secretary-General of the United Nations, the UNESCO National Commissions, ministers of science (and those with science or science education in their portfolio), and the heads of our partner organizations in the basic sciences.

The Final Communiqué called upon UNESCO to, *inter alia*, promote the basic sciences and science education to attain a science culture as precursor to a knowledge-based society; to work for youth and for gender parity; to emphasize capacity-building in educational curricula; to strengthen UNESCO Chairs and centres of excellence in the basic sciences, and foster networking between these; to promote the mobility of teachers and researchers in science and technology, in particular from developing countries; to promote the training of scientists from developing countries to help them to negotiate with donors and other development partners; to promote equitable access to scientific information and literature for scientists and researchers, particularly from developing countries; and to foster partnerships and coordination across the UN system, and with other international organizations. This is a diverse set of mandates, and taken together they pose many challenges. The IBSP represents the soundest way towards their achievement.

Biotechnology
The United Nations General Assembly, at its fifty-eighth session,[9] recognized

> the vital role of new and emerging technologies in raising the productivity and competitiveness of nations and the need, *inter alia*, for capacity-building, measures promoting the transfer and diffusion of technologies to developing countries, and the promotion of private sector activities and public awareness of science and technology (Global Biotechnology Forum, 2004).

This was the objective behind the Global Biotechnology Forum (GBF) – organized by UNIDO and the Government of Chile (in Concepción, Chile, March 2004) – which examined the potential of biotechnology to, among other things, meet the needs of the poor and foster creation of bio-industry, especially

---

9   UN General Assembly Resolution 58/200.

for the developing world (ibid.). The GBF also recognized the potential role of biotechnology in fostering socio-economic growth and in contributing to attainment of the MDGs. As a follow-up, the Inter-Agency Cooperation Network on Biotechnology (UN-Biotech) was initiated. UNESCO is a member of this network, which offers a unique opportunity to develop a coordinated approach to the development and promotion of biotechnology in various high-priority applications worldwide.

Biotechnology remains at the forefront of many agendas, in both developing and industrialized countries, as a tool to promote development and address problems in public health, food production and environmental management. NEPAD has put biotechnology high on its agenda to assist in the development of African countries (Nkuhlu, 2005); and, in keeping with this, the UN Economic Commission for Africa (UNECA) is also represented on the UN-Biotech. UNECA is actively seeking to establish a project on UN Inter-Agency Partnerships on Biotechnology for Africa's Development (UN-Biotech/Africa). This project's major objective would be to review the biotechnology-related activities of the UN bodies for Africa's development in order to identify priority areas for potential joint action. Consultations have already been held with representatives of UNECA in this regard, and an active follow-up is envisaged. Despite great global effort and investment, many developing countries still lack adequate trained human resources, appropriate infrastructure, and an enabling environment for development of science. Capacity-building needs to continue in new and creative ways to meet the human resource needs of Member States in the field of biotechnology.

UNESCO's global action in biotechnology has up to now been developed with, and implemented through, the Microbial Resources Centres network (MIRCENs) and the Biotechnology Action Council (BAC), which includes the Biotechnology Education and Training Centres (BETCENs). These centres of excellence in the regions have proved to be a major and effective tool in promoting scientific exchange and collaboration and should be strengthened further.

In light of the new developments and advances in biotechnology, and the overriding need for additional capacity-building in this domain, UNESCO could envisage focusing its support for training activities and project development on a number of these. In this context, there is much scope for revitalizing the currently established UNESCO Chairs in diverse fields of biotechnology and biological sciences. A more focused approach to establishment of such Chairs in the future is necessary to take into account needs, priorities and current scientific advances. For example, a Chair in nanotechnology could be envisaged. The possibility of increased activities also applies to the UNESCO-affiliated centres which are under development or already established. One such example is the Regional Centre for Biotechnology Training and Education in India, mentioned above (see Box VI.1.2). This initiative may serve as a model for similar developments in the future. Such initiatives are not region-limited, and in the long-term a wider impact through intra-regional and indeed international collaboration and exchange can be envisaged. It is evident that biotechnology, bioinformatics and nanotechnology are likely to be the dominant technologies of the future and that an interdisciplinary approach is imperative for their effective application to solving problems.

Engineering

From the founding of UNESCO to the 1980s, the focus of activity in the Natural Sciences Sector was on core areas of the sciences, engineering and technology, with particular emphasis on human and institutional capacity-building. Since the 1980s, the focus of activity has shifted towards the environmental sciences, with the engineering sciences and technology programme now facing particular human and financial resource constraints. At the same time, many Member States – developed and developing countries alike – are expressing increasing concern regarding the decline in student enrolments and growing weakness in human and institutional capacity in engineering. This, and continued brain drain, will have a particular impact on social and economic development in developing countries.

Moreover, engineering and technology are vital for: addressing basic human needs; poverty reduction; promoting secure and sustainable development; emergency and disaster prevention, preparedness, response and reconstruction; bridging the 'knowledge divide'; and promoting intercultural dialogue and cooperation. As noted and supported by many Member States in the Decision of the Executive Board in April 2005, 'Capacity-building in basic and applied sciences, engineering and technology is a critical aspect of reducing poverty and establishing sustainable economic and social development in developing countries.'[10] The overall focus of the engineering sciences and technology at UNESCO should be on human and institutional capacity-building, enhancing the application of engineering to sustainable social and economic development, poverty eradication and other MDGs and related priorities.

UNESCO has a unique mandate in the engineering sciences and a comparative advantage in networking and international cooperation, and needs to develop and build upon past experience and respond to present needs, opportunities and calls to promote human and institutional capacity-building in engineering. Specific programme activities need to be developed in the following areas:

---

10   UNESCO Document 171 EX/54.

- strengthening of engineering education, training, research and professional development;
- exchange and development of standards, quality assurance and accreditation;
- sharing and development of curricula, learning and teaching materials and methods;
- promoting distance and interactive learning (including virtual universities and libraries);
- international cooperation in the development of engineering ethics and codes of practice;
- promoting advocacy and the public understanding of engineering and technology;
- sharing and development of indicators, information, publications and communication systems in engineering and technology;
- understanding and addressing women and gender issues in engineering (including entry and participation);
- application of engineering to specific issues – including poverty reduction, sustainable development, and disaster reduction, response and recovery;
- international cooperation in the development of engineering and technology policy, planning and management to promote the above.

A variety of types of delivery mechanism or 'modalities' of activity are required, and partnerships with governments, professional organizations, NGOs, and the private sector should be considered essential. The use of ICTs will facilitate and enhance activities in such areas as distance and open learning, virtual meetings and conferences, multi-media information, training materials and electronic networking.

SCIENCE AND TECHNOLOGY POLICY AND SUSTAINABLE DEVELOPMENT

UNESCO programmes on science and technology policy have witnessed a remarkable development. In response to the outcome of the World Summit on Sustainable Development (WSSD) – which recognized S&T policies as fundamental tools for attaining sustainable development, along with the Decade of Education for Sustainable Development (DESD) – UNESCO responded by focusing its programme on providing assistance to developing countries to integrate sustainable development priorities into their national policies on science, technology and innovation. The programme should have a strong emphasis on the needs of African Member States, within the framework of NEPAD, and on Small Island Developing States. In addition, UNESCO's programme should include studies on the contribution of S&T towards the achievement of the MDGs. UNESCO advocates that science policies take into account gender equality and improve women's access to science education and support networks of women scientists and engineers.

Through this programme, UNESCO should work on improving governance of national and regional S&T systems and deliberate on emerging science policy and ethical issues related to making science and technology more participatory. UNESCO will promote cooperation among universities and industries through national and regional partnerships (UNISPAR), as well as promoting virtual networks of laboratories and universities. The participatory process should be encouraged, with more involvement of forums of parliamentary science committees, scientists, the private and public sector, representatives of the media and members of civil society. In cooperation with the UNESCO Institute for Statistics (UIS), production of policy-relevant S&T gender-aggregated indicators should be strengthened at the international level.

UNESCO's role might cover the following domains:

- governance of S&T and its implications, through the support of participatory policy reviews and policy formulation, resulting in improved management of S&T effort at the national level;
- promoting cooperative networks with and among parliamentarians on issues of legislation;
- promoting public debate on S&T issues and public participation in science policy options and raising public understanding of science;
- human resource development and capacity-building in the formulation, implementation, and review of S&T policies and plans; and
- studies on the ethical and social implications of scientific and technological developments.

In charting its future, the Natural Sciences Sector will give special priority to the African region. An important element in the Sector's strategy for future cooperation in Africa is the promotion of collaboration among UN agencies, given the importance attached to this issue and the numerous resolutions adopted by the General Assembly of the United Nations. UNESCO will cooperate with the African Union (AU) and NEPAD in implementing the 'Plan of Action on Science and Technology', adopted by the Second African Ministerial Meeting on Science and Technology (Dakar, Senegal, September 2005). The plan contains flagship programmes and specific policy issues and stresses the need for building the continent's capacities to harness, apply and develop S&T in order to eradicate poverty, fight diseases, stem environmental degradation, and improve economic competitiveness. We are very proud that the Ministerial Meeting adopted a resolution on the establishment of an AU/NEPAD/UNESCO High-Level Working Group to prepare a comprehensive implementation plan for establishing and funding centres of excellence in accordance with the Commission for Africa Recommendations. UNESCO, in close cooperation with the other UN agencies in the Science and Technology Cluster, will focus on consolidating existing centres of excellence, establishing new centres while, at the same time, strengthening the university system to solidly ground capacity-building in the education of future researchers. The initiative embodies the vision and commitment of all African governments for peace and development.

In addition to its focus on Africa, UNESCO should further efforts to address the sustainable development needs of vulnerable and marginalized groups that are often sidelined in global efforts to achieve the MDGs. While the MDGs have been designed to rally development efforts, recent reviews suggest that groups particularly in need of assistance are being left by the wayside. Small-island communities and indigenous peoples provide two prime examples. As recognized in the Mauritius Declaration, Small Island Developing States (SIDS) remain a 'special case' for sustainable development. They remain vulnerable to global and regional socio-economic, cultural and ecological forces; and many figure on the list of the least developed countries. Indigenous peoples are disproportionately represented among the world's most disadvantaged and marginalized populations. SIDS are sidelined by the quantitative MDG approach because statistics from populous continental countries completely overwhelm data that monitor small island progress (or lack thereof) towards designated targets. Indigenous peoples suffer the same fate, as the absence of disaggregated statistics at the national level masks their specific conditions and development aspirations. In short, MDGs, due to their universality and simplicity of expression, fail to provide for development approaches tailored to the realities and priorities of small-island and indigenous communities.

To ensure that the MDGs fulfil their fundamental goal of reducing global disparities, the Coastal Regions and Small Islands (CSI) platform has combined forces with the LINKS project (see Box VI.1.1), which has evolved from a cross-cutting project to a regular component of the Natural Sciences programme. This joint CSI-LINKS platform will help ensure that sustainable development policy remains strongly anchored in on-the-ground implementation. CSI and LINKS will focus on reducing disparities, through a human-rights-based approach to the MDGs. To this end, the unit will continue to serve as the house-wide focal point for Mauritius Strategy[11] implementation, while also augmenting international efforts to address indigenous concerns, voiced with renewed vigour through the UN Permanent Forum on Indigenous Issues (established in 2002). The WCS underlined the interrelationship and interaction between scientific

---

11   The thirty-page 'Mauritius Strategy for the Further Implementation of the Programme of Action for the Sustainable Development of SIDS'.

and indigenous knowledge systems as issues of major importance with respect to natural resource access, conservation and benefits sharing. Recognition of the critical importance of women's knowledge and of the need to ensure intergenerational transmission brings both gender and youth perspectives to sustainable development action. This integrated approach that acknowledges the importance of building capacities – whether in small islands or indigenous communities – based upon local ecological and sociocultural systems, provides fertile ground for contributions to the DESD.

TOWARDS OVERARCHING CONCLUSIONS

The future of the Natural Sciences Sector will be shaped by developments within science itself and how these are integrated into the Sector's programmes and related to the overall mandate and mission of the Organization as defined in its Constitution. The future relationship between science and society will determine whether the full potential of S&T will be used to improve living conditions in the developing world. For instance, the current agricultural assessment (International Assessment of Agricultural Science and Technology for Development, or IAASTD) will examine various possibilities concerning the use (or not) of genetically modified organisms (GMOs) to improve agricultural production. The arguments behind this are at least partially ethical in nature. In other words, this and other contexts highlight the growing importance of ethical considerations in debates on scientific issues. This is also reflected in UNESCO's principal priority in the Social and Human Sciences Sector: the ethics of science and technology, with emphasis on bioethics.

Without a doubt, the Sector's future will critically depend on significant improvement of the budget and staffing situations. Whereas for the regular budget only slim possibilities exist, staff reinforcement is still possible. The future of the budget lies with extrabudgetary funding – for which professional fundraising skills (a capacity largely non-existent in our programme sectors) are of the essence. In short, we need to look in a holistic fashion into the capacities of the Organization – both at Headquarters and in the 'extended' Field (including our centres and institutes, and those of our partners) – to be clear what we can meaningfully deliver in terms of programmes and activities.

In parallel, the science programmes certainly are in need of better visibility. The water programmes and possibly the MAB Programme are the best known among Member States. The success of many of the Sector's publications – such as the *World Water Development Report* and the *Encyclopedia of Life Support Systems*, to cite but two – are not sufficient to raise awareness of the Sector's programmes. Closer cooperation with the Bureau of Public Information could lead to a 'PR Strategy' for making the Sector's activities and products better known. This strategy should go far beyond mere publishing and include work with the media and the press, the holding of exhibitions and other attractive events, as well as regular informational meetings and other exchange with our Permanent Delegations and the National Commissions.

The more important intersectorality becomes in the Organization, the less important permanent organizational and administrative structures within the Sector (and other sectors) will be. Structure would essentially facilitate management and break down into increasingly cooperating and interdependent units. One is tempted to speculate whether this might (at some stage) mean dissolving the classical programme sectors – or at least some of the units within them – in favour of more task-related and/or functional groups. These could be highly dynamic, depending on UNESCO's priorities, and be fully in line with the new views of a long-term staff policy in the Organization.

Last, but certainly not least, the future of science at UNESCO will also depend upon currently ongoing overall changes and reform processes in the Organization, for instance the forward-looking review of the programmes in the sciences in both UNESCO and in the UN system as a whole.[12] How is the UN going to respond to future needs to invest in S&T, to transfer technologies and to build indigenous capacities in science in the least developed countries? The crisis that science is undergoing today may be detrimental to its ability and (increasingly necessary) efforts to ensure quality of life for present and future generations.

Let me conclude with a quotation from Albert Einstein: 'One thing I have learned in a long life: that all our science, measured against reality, is primitive and childlike – and yet it is the most precious thing we have'.

ACKNOWLEDGEMENTS

This chapter could not have been written without the contributions and expertise of our colleagues working in the Sector, both at Headquarters and in the Field. In addition to colleagues 'from within', other colleagues and friends helped us by providing their views and sharing their experience. My special thanks to Mohammed Abdulrazzak, Patricio Bernal, Mario Bertero, L. Anathea Brooks, Pilar Chiang-Joo, Ehrlich Desa, Mustafa El Tayeb, Christine Galitzine, Gisbert Glaser, Jorge Grandi, Mohamed Hassan, Stephen Hill, Lucy Hoareau, Natarajan Ishwaran, Tony Marjoram, Joseph Massaquoi, Robert Missotten, Howard Moore, B. Djaffar Moussa-Elkadhum, Douglas Nakashima, Maciej Nalecz, Yoslan Nur, Folarin Osotimehin, Jane Robertson-Vernhes, Thomas Rosswall, Badaoui Rouhban, András Szöllösi-Nagy and Dirk G. Troost.

---

12 See the outcomes of the 2005 World Summit (UNESCO, 2005) and related UN documents.

BIBLIOGRAPHY

Annan, K. 2000. *We the Peoples: The Role of the United Nations in the Twenty-First Century.* New York, United Nations.

——. 2004. Science for All Nations. *Science*, Vol. 303, p. 925.

Barabasi, A.-L. 2002. *Linked: The New Science of Networks.* Cambridge, UK, Perseus.

——. 2005. Network theory – the emergence of the creative enterprise. *Science*, Vol. 308, pp. 639–41.

Brown, E. N. 2002. Interdisciplinary research: a student's perspective. *Chemical Education Today*, Vol. 79, pp. 13–14.

Chataway, J., Smith, J. and Wield, D. 2005. Partnerships for building science and technology capacity in Africa: Canadian and UK experience. Paper prepared for the conference 'Africa-Canada-UK Exploration: Building Science and Technology Capacity with African Partners', 30 January to 1 February 2005, Canada House, London.

Commission for Africa. 2005. *Our Common Interest: Report of the Commission for Africa.* London, Penguin.

di Castri, F. 1995. The Chair of Sustainable Development. *Nature and Resources*, Vol. 31, No. 3, pp. 2–7.

Erdelen, W. and Moore, H. 2005. 'Science yes, but what kind?'. *UNESCO today* (Magazine of the German Commission for UNESCO), No. 2/2005, Special Edition: Sixtieth anniversary of the founding of UNESCO. Bonn, Germany, German Commission for UNESCO, pp. 2–6.

FOREN. 2001. FOREN Workpackage 5. Final Report on Deriving Policy Actions from Issue Specific Foresight Results: How foresight actions can be used to support the long-term competitive position of local SME systems, by Carlotta Ca'Zorzi and Michele Capriati, June 2001. [The FOREN network is a Thematic Network of the Fifth RTD (Research, Technological Development and Demonstration) Framework Programme of the European Commission.] http://foren.jrc.es/Docs/WP5%20ROME%20version%20finale.pdf.

Global Biotechnology Forum. 2004. Final Statement. Global Biotechnology Forum, Concepción, Chile, 2–5 March 2004. Convened under the auspices of the United Nations Industrial Development Organization (UNIDO) and the Government of Chile. http://www.unido.org/file-storage/download/?file%5fid=21564

Guimera, R., Uzzi, B., Spiro, J. and Amaral, L. A. N. 2005. Team assembly mechanisms determine collaboration network structure and team performance. *Science*, Vol. 308, pp. 697–702.

Hill, S. 2005. Empowering innovation in the Third World: the case for promoting a new global localism. Plenary address given at the First Asia-Pacific Congress and Exhibition on Innovation Management, Bangkok, Thailand, 21 September 2005.

IAC (InterAcademy Council). 2004. *Inventing a Better Future: A Strategy for Building Worldwide Capacities in Science and Technology.* Amsterdam, InterAcademy Council.

IHP (International Hydrological Programme). 2004. IHP Proposals to the Programme and Budget for 2006–2007 (33 C/5). Sixteenth session of the IHP Intergovernmental Council (Paris, 20–24 September 2004). IHP/IC-XVI/8 (13 July 2004). http://unesdoc.unesco.org/images/0013/001356/135612E.pdf

CSU (International Council for Science). 2005a. Priority Area Assessment on Capacity Building in Science. Draft Report. Paris, ICSU.
——. 2005b. Science and Society: Rights and Responsibilities. ICSU Strategic Review. Paris, ICSU.
——. 2005c. Strengthening International Science for the Benefit of Society: a strategic plan for the international council for science, 2006–2011. Paris, ICSU.
Marshall, S. P., Scheppler, J. A. and Palmisano, M. L. (eds) 2003. *Science Literacy for the Twenty-First Century*. Amherst (Mass.), New York, Prometheus Books.
Matsuura, K. 2005. Address by the Director-General of the United Nations Educational, Scientific and Cultural Organization (UNESCO) to the 171st Session of the Executive Board, on occasion of the reply to the general debate on Items 3, 4, 5 and 20. UNESCO, 20 April 2005.
Mawhinney, M. 2002. *Sustainable Development: Understanding the Green Debates*. Oxford, Blackwell Science.
Meadows, D., Randers, J. and Meadows, D. 2004. *The Limits to Growth: the thirty-year update*. White River Junction, Vermont, Chelsea Green Publishing Company.
National Research Council. 2002. *Knowledge and Diplomacy: Science Advice in the United Nations System*. Washington, The National Academies Press.
Nkuhlu, W. L. 2005. *The Journey So Far*. The New Partnership for Africa's Development (NEPAD) Secretariat. http://www.nepad.org/2005/files/documents/journey.pdf
Oyelaran-Oyeyinka, B. 2005. Partnerships for building science and technology capacity in Africa: African experience. Paper prepared for the conference 'Africa-Canada-UK Exploration: Building Science and Technology Capacity with African Partners', 30 January to 1 February 2005, Canada House, London.
Palmer, M., Bernhardt, E., Chornesky, E., Collins, S., Dobson, A., Duke, C., Gold, B., Jacobson, R., Kingsland, S., Kranz, R., Mappin, M., Martinez, M., Micheli, F., Morse, J., Pace, M., Pascual, M., Palumbi, S., Reichman, O. J., Townsend, A. and Turner, M. 2004. *Ecology for a Crowded Planet: ecological science and sustainability for a crowded planet*. Twenty-First Century Vision and Action Plan for the Ecological Society of America. http://www.esa.org/ecovisions
Sachs, J. 2005. *The End of Poverty: how we can make it happen in our lifetime*. London, Penguin.
Somerville, M. A. and Rapport, D. J. (eds) 2000. *Transdisciplinarity: Recreating Integrated Knowledge*. Oxford, EOLSS Publishers.
Trefil, J. 2003. Two Modest Proposals Concerning Scientific Literacy. In: S. P. Marshall, J. A. Scheppler and M. L. Palmisano (eds), *Science Literacy for the Twenty-First Century*. Amherst (Mass.), New York, Prometheus Books, pp. 150–56.
Thompson Klein, J. 2000. Voices of Royaumont. In: M. A. Somerville and D. J. Rapport (eds), *Transdisciplinarity: Recreating Integrated Knowledge*. Oxford, EOLSS Publishers, pp. 3–13.
Thompson Klein, J., Grossenbacher-Mansuy, W., Haeberli, R., Bill, A., Scholz, R. W. and Welti, M. (eds) 2001. *Transdisciplinarity: Joint Problem Solving Among Science, Technology, and Society*. Basel, Switzerland, Birkhaeuser Verlag.
TWNSO (Third World Network of Scientific Organizations). 2003. Profiles of Institutions for Scientific Exchange and Training in the South. Trieste, Italy, TWNSO.
UNDP. 2005. *Human Development Report*. http://hdr.undp.org

UNESCO. 2000. *World Conference on Science. Science for the Twenty-First Century: A New Commitment*. Paris, UNESCO.
——. 2003a. Meeting the Educational and Capacity-Building Challenges of the International Water and Sustainable Development Agendas: Towards 2015. Keynote Address by Mr Koïchiro Matsuura, Director-General, at the UNESCO meeting on strategies, actions and coalitions in water education and capacity-building; Delft, the Netherlands, 17 July 2003. DG/2003/106. http://unesdoc.unesco.org/images/0013/001307/130767e.pdf
——. 2003b. *Science in the Information Society*. UNESCO Publications for the World Summit on the Information Society. Paris, UNESCO.
——. 2005a. Towards knowledge societies. *UNESCO World Report*. Paris, UNESCO.
——. 2005b. Harnessing Science for Society: further partnerships. Report on International UNESCO/ICSU/TWAS symposium on WCS follow-up. Venice, Italy, 2–5 March 2005.
——. 2005c. *The Basic Sciences: The Science Lever for Development*. Report on the Ministerial Round Table held during the Thirty-Third Session of the General Conference, October 2005. (In preparation.)
United Nations. 2005. Implementation of decisions from the 2005 World Summit Outcome for action by the Secretary-General. Report of the Secretary-General. United Nations Document A/60/430, dated 25 October 2005.
UN Millennium Project. 2005. *Innovation: Applying Knowledge in Development*. Task Force on Science, Technology, and Innovation. London, Earthscan.
WCED (World Commission on Environment and Development). 1987. *Our Common Future*. Oxford, UK, Oxford University Press.
Wurm, S. A. (ed.). 2001. *Atlas of the World's Languages in Danger of Disappearing* (2nd edn). Paris, UNESCO.

ANNEXES

ANNEX 1

ACRONYMS

Included in this list are acronyms appearing in the volume independent of full title.

AAAS	American Association for the Advancement of Science
ABN	African Biosciences Network
ACC	United Nations Administrative Committee on Coordination
ADG/SC	Assistant Director-General for Natural Sciences
AEC	Atomic Energy Commission
ANAIC	Asian Network for Analytical and Inorganic Chemistry in Asia
ANBS	Asian Network of Biological Sciences
ANSTI	African Network of Scientific and Technical Institutions
AScW	(British) Association of Scientific Workers
ASPEN	Asian Physics Education Network
BAAS	British Association for the Advancement of Science
BAC	Biotechnology Action Council
BETCENs	Biotechnology Education and Training Centres
CAME	Conference of Allied Ministers of Education
CAST	(Regional) Conference on the Applications of Science and Technology to Development
CASTAFRICA	Conference of African Ministers Responsible for the Application of Science and Technology to Development
CASTALAC	Conference of Ministers Responsible for the Application of Science and Technology to Development in Latin America and the Caribbean
CASTALA	Conference on the Application of Science and Technology to the Development of Latin America
CASTARAB	Conference of Arab Ministers Responsible for the Application of Science and Technology to Development
CASTASIA	Conference on the Application of Science and Technology to the Development of Asia
CCCO	Committee on Climate Change and the Ocean
CERN	European Organization for Nuclear Research (provisionally, European Council for Nuclear Research)
CEOS	Committee on Earth Observation Satellites
CIOMS	Council for International Organizations of Medical Sciences
CLAF	Latin American Physics Centre
CLAM	Latin American Centre for Mathematics
CNRS	French National Centre for Scientific Research (Centre National de la Recherche Scientifique)

COMAR	Major UNESCO Interregional Project on Research and Training leading to the Integrated Management of Coastal Systems (often referred to as Coastal Marine Programme/Project)
COMEST	World Commission on the Ethics of Scientific Knowledge and Technology
COR/ENV	Bureau for Coordination of Environmental Programmes
COSPAR	Scientific Committee on Space Research
COSTED	Committee on Science and Technology in Developing Countries
CSD	United Nations Commission on Sustainable Development
CSI	Environment and Development in Coastal Regions and Small Islands platform
DESD	Decade of Education for Sustainable Development
DOEM	Designated Officials on Environmental Matters
ECB	Environment Coordination Board
ECOSOC	United Nations Economic and Social Council
EMBC	European Molecular Biology Conference
EMBL	European Molecular Biology Laboratory
EMBO	European Molecular Biology Organization
EOLSS	Encyclopedia of Life Support Systems
EPD	Environment Population and Development (formerly Environment and Population Education for Human Development Project)
ERAIFT	Regional Post-Graduate Training School on Integrated Management of Tropical Forests
ESCAP	Economic and Social Commission for Asia and the Pacific
ESRO	European Space Research Organization
FAO	Food and Agriculture Organization of the United Nations
FEBS	Federation of European Biochemical Societies
FRIEND	Flow Regimes from International Experimental and Network Data
GARP	Global Atmospheric Research Programme
GARS	Geological Applications of Remote Sensing
GEBCO	General Bathymetric Chart of the Oceans
GEOSS	Global Earth Observing System of Systems
GESAMP	Joint Group of Experts on the Scientific Aspects of Marine Environmental Protection
GIAM	Global Impact of Applied Microbiology
GIS	Geographic Information System
GOOS	Global Ocean Observing System
GTOS	Global Terrestrial Observing System
HELP	Hydrology for the Environment, Life and Policy

ACSD	Inter-Agency Committee on Sustainable Development
ACOMS	International Advisory Committee on Marine Sciences
IAEA	International Atomic Energy Agency
IAHS	International Association of Hydrological Sciences
IAPO	International Association of Physical Oceanography
IASH	International Association of Scientific Hydrology
IBC	International Bioethics Commission
IBE	International Bureau of Education
IBI	Intergovernmental Bureau of Informatics
IBN	International Biosciences Networks
IBP	International Biological Programme
IBRO	International Brain Research Organization
IBSP	International Basic Sciences Programme
ICAM	Integrated Coastal Area Management
ICC	International Computation Centre
ICCS	International Centre for Chemical Studies
ICES	International Council for the Exploration of the Sea
ICHS	International Congresses of the History of Science
ICIC	International Commission for Intellectual Co-operation
ICOMOS	International Council on Monuments and Sites
ICPAM	International Centre for Pure and Applied Mathematics
ICRO	International Cell Research Organization
ICSOPRU	International Comparative Study on the Managemant, Productivity and Effectiveness of Research Teams and Institutions
ICSPRO	Inter-Secretariat Committee on Scientific Programs Relating to Oceanography
ICSU	International Council for Science (formerly International Council of Scientific Unions)
ICTs	Information and communication technologies
ICTP	Abdus Salam International Centre for Theoretical Physics
IDNDR	UN International Decade for Natural Disaster Reduction (1990–1999)
IEEP	International Environmental Education Programme
IFAP	Information for All Programme
IGBP	International Geosphere Biosphere Programme
IGC	International Geological Congress
IGCP	International Geoscience Programme (formerly International Geological Correlation Programme)
IGOS	Integrated Global Observing Strategy
IGOSS	Integrated Global Ocean Services System
IGY	International Geophysical Year (eighteen-month period 1957–1958)
IHD	International Hydrological Decade (1965–1974)
IHE	UNESCO-IHE Institute for Water Education
IHO	International Hydrographic Organization
IHP	International Hydrological Programme

IIHA	International Institute of the Hylean Amazon
IIIC	International Institute for International Cooperation
IIOE	International Indian Ocean Expedition
IIP	Intergovernmental Informatics Programme
IITAP	International Institute of Theoretical and Applied Physics
ILO	International Labour Organization
ILSI	International Life Sciences Institute
IMO	International Maritime Organization
IMS	International Marine Science Newsletter
IMU	International Mathematical Union
INCE	International Network for Chemical Education
INFN	(Italian) National Institute of Nuclear Physics (Istituto Nazionale di Fisica Nucleare)
INIS	International Nuclear Information System
INISTE	International Network for Information in Science and Technology Education
IOC	Intergovernmental Oceanographic Commission of UNESCO
IOC/MRI	Office of the IOC and Marine-Science Related Issues
IOCD	International Organization for Chemical Sciences in Development
IODE	International Oceanographic Data Exchange
IRC	International Research Laboratory
ISES	International Solar Energy Society
ISWA	International Science Writers Association
ITIC	International Tsunami Information Center
ITSU	The Tsunami Warning System in the Pacific
IUBS	International Union of Biological Sciences
IUCN	World Conservation Union (formerly International Union for the Conservation of Nature and Natural Resources)
IUGG	International Union of Geodesy and Geophysics
IUGS	International Union of Geological Sciences
IUHPS	International Union of History and Philosophy of Science
IUHS	International Union of History of Science
IUMS	International Union of Microbiological Societies
IUPAC	International Union of Pure and Applied Chemistry
IUPAC-CTC	(IUPAC) Committee on Teaching of Chemistry
IUPAP	International Union of Pure and Applied Physics
IUPN	International Union for the Protection of Nature
JPOTS	Joint Panel on Oceanographic Tables and Standards
LACDOS	Latin American Conference for the Development and Organization of Science
LEPOR	Long-Term and Expanded Program of Oceanic Research
LINKS	Local and Indigenous Knowledge Systems
MAB	Man and the Biosphere Programme

ANNEXES

MCBN	Molecular and Cell Biology Network
MDGs	Millennium Development Goals
MEDNET	Mediterranean Network on Science and Technology of Advanced Polymer Based Materials
MINESPOL	Conference of Ministers Responsible for Science and Technology Policies in the European and North American Region
MIRCENs	Microbial Resources Centres network
MOST	Management of Social Transformations Programme
NATO	North Atlantic Treaty Organization
NEPAD	New Partnership for Africa's Development
NGO	Non-governmental organization
NSF	National Science Foundation (USA)
OCE	UNESCO Division of Marine Sciences
OECD	Organisation for Economic Co-operation and Development
OIC	Organization of Intellectual Cooperation
OST	United Nations Office for Science and Technology
PAC	Physics Action Council
PGI	General Informatics Programme
PICMA	Programme for the International Cooperation in Mathematics and its Applications
POEM	Physical Oceanography of the Eastern Mediterranean
RELAB	Latin American Network of Biological Sciences
ROSTA	Regional Office for Science and Technology for Africa
ROSTLAC	Regional Office for Science and Technology for Latin American and the Caribbean
ROSTs	Regional Offices for Science and Technology
SAC	Scientific Advisory Committee
SBPC	Brazilian Society for Scientific Progress
SCOR	Scientific Committee on Oceanic Research
SEAMS	South-East Asian Mathematical Society
SEP	UNESCO Space Education Programme
SESAME	International Centre for Synchrotron-light for Experimental Science and Applications in the Middle East
SET	Science, engineering and technology
SIDA	Swedish International Development Authority
SIDS	Small Island Developing States
SIMDAS	Sustainable Integrated Management and Development of Arid and Semi-Arid Regions of Southern Africa
SIV	Small Islands Voice

SPINES   Science and technology Polocies Information Exchange System
STA   Scientific and technical activities
STEPAN   Science and Technology Policy Asian Network
STET   Scientific and technological education and training
STID   Scientific and technological information and documentation
STS   Scientific and technical services

TTR   Training-through-Research Programme
TWAS   The Academy of Sciences for the Developing World (formerly Third World Academy of Sciences)
TWNSO   Third World Network of Scientific Organizations

UATI   International Union of Technical Associations and Organizations
UFSO   UNESCO Field Science Offices
UN   United Nations
UNACAST   United Nations Advisory Committee on the Applications of Science and Technology
UNAEC   United Nations Atomic Energy Commission
UNCAST   United Nations Conference on the Application of Science and Technology (for the benefit of less developed countries)
UNCED   United Nations Conference on Environment and Development
UNDP   United Nations Development Programme
UNECA   United Nations Economic Commission for Africa
UNEP   United Nations Environment Programme
UNESCO   United Nations Educational, Scientific and Cultural Organization
UNEVOC   International Project on Technical and Vocational Education
UNIDO   United Nations Industrial Development Organization
UNSCCUR   United Nations Scientific Conference on the Conservation and Utilization of Resources
UNISIST   United Nations Information System for Science and Technology
UNISPAR   University-Industry-Science Partnership
UNITWIN   University Twinning and Networking scheme
UNU   United Nations University

WCRP   World Climate Research Programme
WFCC   World Federation for Culture Collections
WFEO   World Federation of Engineering Organisations
WFSW   World Federation of Scientific Workers
WHC   World Heritage Centre
WHO   World Health Organization
WMO   World Meteorological Organization
WNBR   World Network of Biosphere Reserves
WSIS   World Summit on the Information Society
WSP   World Solar Programme (1996–2005)

WSSD	World Summit on Sustainable Development
WWAP	World Water Assessment Programme
WWF	World Wide Fund for Nature (formerly World Wildlife Fund)
WYP	World Year of Physics (2005)

ANNEX 2

UNESCO SCIENCE MILESTONES, 1945-2005
Gail Archibald

This is a brief overview of major events and achievements of UNESCO in science, including intergovernmental and non-governmental organizations, centres and networks set up by UNESCO or with its assistance.

1945	'In these days when we are all wondering, perhaps apprehensively, what the scientists will do to us next, it is important that they should be linked closely with the humanities and should feel that they have a responsibility to mankind for the result of their labours'. -- Ellen Wilkinson, British Minister of Education at Conference to create UNESCO, Institution of Civil Engineers, London (UK), 1-16 November.
1946	UNESCO Preparatory Commission: Julian Huxley, zoologist and author, assumes post of Secretary-General in February. In June, International Computing Centre (ICC) proposed by French and American delegations at UNESCO's Preparatory Commission. Julian Huxley elected Director-General for two years, at UNESCO's first General Conference, 6 December. Joseph Needham, biochemist, appointed Head of Natural Sciences Section, December. First UNESCO agreement with a non-governmental organization (NGO) signed, with the International Council of Scientific Unions (ICSU), 16 December.
1947	Creation of three regional Field Science Cooperation Offices: Cairo (Egypt), Rio de Janeiro (Brazil), Nanjing and Shanghai (China). Hylean Amazon project (ecology of tropical forests) is one of four large-scale projects adopted at the Executive Board, Paris, April. International Union of the History of Science set up in Lausanne (Switzerland), October. 'Protection of Nature' is made a UNESCO General Conference priority in Natural Sciences Section, November-December.

ANNEXES

1948   Creation of fourth regional Field Science Cooperation Office, in New Delhi (India).

Publication: Suggestions for Science Teachers in Devastated Countries.

Pierre Auger appointed Director of Natural Sciences Section (renamed Natural Sciences Department in July).

UNESCO organizes International Conference on High Altitude Research Stations, Interlaken, Switzerland, August–September.

International Union for the Protection of Nature (IUPN) founded, at Fontainebleau (France) Conference, held at instigation of Julian Huxley and French Government, October.

1949   Council for International Organizations of Medical Sciences (CIOMS) created by UNESCO and the World Health Organization (WHO) to carry out programmes in Biological Sciences.

Parallel with the United Nations Scientific Conference on the Conservation and Use of Natural Resources (UNSCCUR), UNESCO and IUPN organize the Technical Conference on the Protection of Nature, at Lake Success, New York (USA), 22–29 August.

1950   Kalinga Prize created to reward activities related to the popularization of science, technology and general research in the improvement of public welfare, the enrichment of the cultural heritage of nations, and solutions to problems of humanity; first awarded in 1952.

Proposition to create a nuclear research laboratory – eventually to become CERN (in 1954) – made by member of the US delegation to General Conference, Florence (Italy).

First issue of UNESCO's quarterly journal *Impact of Science on Society*, in English and French.

First of the Travelling Science Exhibitions organized, in Latin America, to present important scientific discoveries, and their applications, to the public. These popular exhibitions continued for fifteen years, worldwide.

'Commission for Scientific and Cultural History of Mankind' created, in collaboration with ICSU and the International Council of Philosophical and Humanistic Studies, by the UNESCO General Conference, Florence, May–June.

General Conference decides to set up an interim International Arid Zone Research Council for promoting research and development; remains in existence until 1964.

1951   Union of International Technical Associations (UITA) created, under UNESCO's auspices.

International Mathematical Union (IMU), originally founded in 1936, is re-established through UNESCO resolution.

	First session of Advisory Committee on Arid Zone Research, Algiers (Algeria).
1952	Statistics Division established, in the Social Science Department.
	First World Hydrology Colloquium on arid zone hydrology, Ankara (Turkey).
1953	International Advisory Committee on Scientific Research, co-piloted with ICSU, set up to advise UNESCO's Director-General on the Organization's scientific programmes, December.
1954	European Organization for Nuclear Research (CERN) becomes operational, near Geneva (Switzerland), 29 September.
1956	Publication: *UNESCO Source Book for Science Teaching*, which becomes a 'best-seller' for the Organization.
	First meeting of International Advisory Committee on Marine Sciences, Lima (Peru), 22–24 October.
	Scientific Research on Arid Lands is one of three Major Projects for 1957–1958, launched at General Conference, New Delhi (India), November–December.
	Launch of Humid Tropics Research Programme, through preparatory meeting of specialists in Kandy (Sri Lanka), and associated symposium on the Study of Tropical Vegetation.
1957	Creation of network of International Centres for Advanced Training in Journalism.
	First meeting of the International Advisory Committee for Humid Tropics Research, Manaus (Brazil), 29–31 July.
1957–1958	International Geophysical Year (July 1957 to December 1958).
1959	Victor Kovda appointed Director of the Department of Natural Sciences.
	UNESCO instrumental in setting up the Charles Darwin Foundation for the Galapagos, Ecuador.
1960	International Brain Research Organization (IBRO) founded, through collaborative effort of UNESCO and CIOMS, to encourage scientific discussion and training of neuroscientists.
	Intergovernmental Conference on oceanographic research, convened by UNESCO; recommends creation of the Intergovernmental Oceanographic Commission (IOC) of UNESCO, Copenhagen (Denmark), 11–16 July.
	Reorganization of natural resources and environmental sciences programmes, under the aegis of the newly created Division of Studies and Research relating to Natural Resources.
1961	Report by Pierre Auger, *Current Trends in Scientific Research*.

ANNEXES

Ratification of the convention of International Computation Centre (ICC), which becomes operational (Rome, Italy).

Initiation of FAO/UNESCO Soil Map of the World project, joint venture with the Food and Agriculture Organization of the United Nations.

UNESCO represented at panel of eminent physicists at International Atomic Energy Agency (IAEA) in Vienna (Austria) to discuss international centre for theoretical physics, as proposed by Abdus Salam (21–22 March).

Executive Board decision constitutes first statement by governing body of UNESCO on the importance of international cooperation in the field of water resources and the inclusion of this domain in the Organization's programmes and budget, November.

1962 UNESCO launches the International Cell Research Organization (ICRO) to promote research and knowledge of cell biology.

Latin American Centre for Mathematics (CLAM) created in Buenos Aires (Argentina).

Eighteen Latin American countries sign the agreement for establishment of Latin American Physics Centre (CLAF).

1963 Publication: First volume of the *History of Mankind: Cultural and Scientific Development* (in English).

UNESCO's Research Organization Unit becomes the Science Policy Unit.

UN Office for Science and Technology (OST) set up in New York.

Division of Science Teaching launches regional pilot project in physics in Latin America.

Publication: Results of survey of African natural resources (in English and French); first in the Natural Resources Research series.

United Nations 'Conference on the Application of Science and Technology for the Benefit of the Less Developed Areas' held, with organizational support of UNESCO, Geneva, February.

1964 Alexey Matveyev becomes UNESCO's first Assistant Director-General for Natural Sciences.

Publication: Geological Map of Africa (scale: 1:5,000,000), by UNESCO in collaboration with the Association of African Geological Surveys, first in a series of thematic maps.

International Centre of Theoretical Physics (ICTP, in Trieste, Italy) inaugurated; Abdus Salam appointed Director, 5 October.

First UNESCO Intergovernmental Conference on Assessment and Mitigation of Earthquake Risk, Paris.

1965 First regional Conference on the Applications of Science and Technology to Development (CAST), in Santiago de Chile (Chile), organized by UNESCO

Regional Office for Science and Technology in Africa (ROSTA), Nairobi (Kenya), established by UNESCO

First issue of long-running Science Policy Studies and Documents series.

Research programme launched on microbiology, in collaboration with the International Union of Microbiological Societies (IUMS) and ICRO.

First session of Advisory Committee on Natural Resources Research, Paris.

First issue of *Nature and Resources*, a quarterly magazine on international environmental research, natural resources and nature conservation.

1965–1974 International Hydrological Decade, to improve water management and training of specialists needed to develop water resources.

1967 First UNESCO project concerning women in science, in Chile.

Four regional pilot projects to improve science teaching in secondary schools launched – covering physics in Latin America, chemistry in Asia, biology in Africa and mathematics in Arab States.

1967–68 First volumes of series of New Trends in the teaching of basic sciences published.

1968 The UNESCO Science Prize created, for an outstanding contribution to the technological development of a developing country or region.

The 'Conference on the Application of Science and Technology to the Development of Asia' (CASTASIA), New Delhi (India).

UNESCO participates in creation of World Federation of Engineering Organisations (WFEO) for national and regional engineering institutions and associations.

Intergovernmental 'Conference of Experts on the Scientific Basis for Rational Use and Conservation of the Biosphere' (the 'Biosphere Conference'), Paris, 4–13 September, which leads to the creation of the Man and the Biosphere Programme (MAB) in 1971.

1969 Office of Statistics (responsible for scientific statistics) becomes the Division of Science Statistics.

Regional pilot project in mathematics set up in the Arab States.

1970 The 'Conference of Ministers Responsible for Science Policy in European Member States' (MINESPOL) considers general thrust of European scientific research, Paris.

UNESCO and ICRO create the World Federation for Culture Collections (WFCC), for the listing and conservation of microbial strains of importance for medicine, agriculture and industry.

Publication: Climate Atlas of Europe, by UNESCO and World Meteorological Organization (WMO), first in a series of regional climate maps.

ANNEXES

1971	ICC becomes the Intergovernmental Bureau of Informatics (IBI).
	UNESCO Regional Bureau for European Scientific Cooperation established in Paris.
	First meeting of the International Coordinating Council of the Man and the Biosphere (MAB) Programme, October.
1972	South-East Asian Mathematical Society (SEAMS) created.
	'Convention concerning the Protection of the World Cultural and Natural Heritage' adopted by UNESCO's General Conference, 16 November.
	United Nations Conference on the Human Environment (in Stockholm, Sweden) leads to creation of United Nations Environmental Programme (UNEP).
1973	International conference 'The Sun in the Service of Mankind' at UNESCO, on how to respond to the energy crisis; launches World Solar Summit Process.
	Working agreement is signed delineating the responsibilities of UNESCO and WMO in the field of hydrology, and providing mechanisms for cooperation, including joint international conferences.
1974	Adoption by UNESCO of 'Recommendation: The civic and ethical aspect of scientific research'.
	The final statement of the Brezhnev/Nixon summit meeting (Moscow, USSR) supports UNESCO's biosphere reserves.
	'Conference of Ministers Responsible for the Application of Science and Technology to Development in Africa' (CASTAFRICA) held in Dakar (Senegal).
1975	First World Conference on Women, in Mexico, precedes United Nations Women's Decade (1976–1985).
	World Network of Microbiological Resources Centres (MIRCENs) set up by UNESCO to provide a stable basis for cataloguing and conservation of microbe strains.
	First Intergovernmental Coordinating Council of the International Hydrological Programme (IHP).
1976	Intergovernmental Conference on Assessment and Mitigation of Earthquake Risk, Paris.
	The 'Conference of Ministers of Arab States Responsible for the Application of Science and Technology to Development in Arab States' (CASTARAB) held in Rabat (Morocco).
	First thirty-seven biosphere reserves of the Man and the Biosphere Programme (MAB), which reconcile the conservation of ecosystems, biological diversity and the development of local populations.

African Mathematical Union created.

Publication: *Geological World Atlas* (scale: 1:10,000,000) completed, after twelve years of effort, compiling geological research carried out worldwide.

1977 Publication: *The Scientific Enterprise, Today and Tomorrow*.

UNESCO's Carlos J. Finlay Prize created, at initiative of the Cuban Government, for contributions to research and development in the field of microbiology and its applications, including immunology, molecular biology and genetics; first awarded in 1980.

Creation of the South-East Asian Network for the Chemistry of Natural Products, with support from the Government of Japan.

1978 Intergovernmental Conference on Strategies and Policies for Informatics, Torremolinos (Spain).

'Recommendation concerning the International Standardization of Statistics on Science and Technology' adopted by General Conference, Paris.

MINESPOL II held in Belgrade (Yugoslavia).

International Centre for Pure and Applied Mathematics (ICPAM) set up in Nice (France), to increase manpower in mathematics and provide assistance to national institutions of developing countries.

UNESCO contributes to creation of Federation of Asian Chemical Societies.

Word Heritage Convention operational, with first twelve sites in World Heritage List.

Publication: FAO/UNESCO Soil Map of the World (scale: 1:5,000,000) completed, accompanied by 1,600 pages in 10 volumes.

1978–79 UNESCO/UNEP/FAO state-of-knowledge reports on tropical forest ecosystems and tropical grazing land ecosystems.

1979 Launching of South and Central Asian Medicinal and Aromatic Plants Network (SCAMAP) for research, training and sharing of information.

1980 African Network of Scientific and Technological Institutions (ANSTI) established by UNESCO and UNDP, January.

International Biosciences Networks (IBN) set up as a partnership between ICSU and UNESCO in Latin America, Asia, Africa and the Arab Region.

International Centre for Chemical Studies (ICCS) established.

1981 Pluridisciplinary scientific conference 'Ecology in Practice' marks tenth anniversary of the MAB Programme, in Paris. Accompanied by prototype of thirty-six-poster exhibit 'Ecology in Action', translated into a score of languages.

Asian Physics Education Network (ASPEN) established within UNESCO programme of science education.

ANNEXES

International Organization for Chemical Sciences in Development (IOCD) set up.

Symposium on the State of Biology in Africa, in Ghana, April.

Second phase of the International Hydrological Programme (1981–83) launched in Paris at International Hydrological Conference, in cooperation with WMO.

1982 CASTASIA II held in Manila (the Philippines).

Interregional Project on Research and Training Leading to the Integrated Development of Coastal Systems (COMAR) extended to cover the study and sustainable use of mangroves, estuaries, lagoons and coral reefs.

1983 Asian Network for Analytical and Inorganic Chemistry in Asia (ANAIC), established by UNESCO and the Federation of Asian Chemical Societies (FACS), begins its activity in building up regional research capacities.

First International Biosphere Reserves Congress, organized jointly with UNEP in cooperation with FAO and IUCN, Minsk (Belarus, USSR).

International Centre for Integrated Mountain Development (ICIMOD) opened in Kathmandu (Nepal).

1984 Publication: *World Directory of National Science and Technology Policy-Making Bodies.*

CASTARAB II held in the Sudan.

The IOC starts work on the World Ocean Circulation Experiment (WOCE) and study of the Tropical Ocean and Global Atmosphere (TOGA).

Project on the interpretation of remote-sensing imagery for use in oceanography launched under auspices of the Marine Science Training and Education Programme (TREDMAR).

Action Plan for Biosphere Reserves adopted by MAB Council.

1985 World Conference of the International Women's Year, Nairobi.

Intergovernmental Informatics Programme (IIP) launched, to improve access to information technology in countries not possessing such technology.

International Network for Information in Science and Technology Education (INISTE) established.

First issue of *IHP Information* newsletter (called *Waterway* as of October 1994), which disseminates information, free of charge, on the International Hydrological Programme.

1986 *Asia Physics News*, first all-Asia bulletin for research and teaching of physics, is created, with UNESCO assistance.

1987 CASTAFRICA II held in Arusha (United Republic of Tanzania).

UNESCO's Javed Husain Prize for Young Scientists established (with funds donated by Javed Husain), to reward pure or applied research by a young scientist in the natural or social sciences, technology, medicine or agriculture. First EuroMAB meeting in Berchtesgaden (Federal Republic of Germany).

1988 First meeting of International Scientific Council for Science and Technology Policy Development, Paris.

Regional Bureau for European Scientific Cooperation relocated (from Paris) to Venice, Italy (eventually renamed the Regional Office for Science and Technology for Europe, ROSTE).

1989 Launch of MAB Young Scientist Award scheme, with the awarding of ten fellowships.

1989 Establishment of the UNESCO Programme on Human Genome, for training of researchers and sharing of knowledge.

1990 UNESCO/Biotechnology Action Council (BAC) Programme set up, to promote biotechnologies in developing countries.

UNESCO creates the Molecular and Cell Biology Network (MCBN), which becomes an independent NGO in 2002.

Second World Climate Conference, associating the IOC and IHP, adopts ministerial declaration on unprecedented climate-change rate over next century due to greenhouse gases in atmosphere, Geneva, October–November.

1990–1993 International project on the model university foundation courses for undergraduate studies in mathematical, physical, chemical and biological sciences carried out through four regional pilot projects in Africa and Asia.

1991 Inaugural award of UNESCO's Sultan Qaboos Prize for Environmental Preservation, to Instituto de Ecologia AC of Mexico.

Creation of the Global Ocean Observing System (GOOS) at UNESCO's IOC Assembly, in cooperation with WMO, UNEP and ICSU.

Programme on the Promotion of Marine Sciences (PROMAR) establishes the Expert Centre for Taxonomic Identification (ETI) in Amsterdam.

First international 'Training-through-Research' (TTR) cruise, in the Mediterranean and Black Seas.

Biosphere Reserve Integrated Monitoring (BRIM) proposed at EuroMAB meeting in Strasbourg (France).

The International Union of Biological Sciences (IUBS), Scientific Committee on Problems of the Environment (SCOPE) and UNESCO are original partners in DIVERSITAS, set up to catalyse and catalogue knowledge about biodiversity.

1992	Establishment of a regional office of ICPAM in Chile: the International Centre for Mathematics and Computer Science in Latin America.
	French scientific community appoints UNESCO as depository of new data on the analysis of the human genome obtained by the French National Centre for Scientific Research (CNRS) and the Institut Généthon, October.
	Establishment of the World Heritage Centre as an integrated secretariat of both the natural and cultural heritage components of the World Heritage Convention.
	United Nations Conference on Environment and Development (the 'Earth Summit'), Rio de Janeiro (Brazil); with Agenda 21 of special interest to UNESCO.
	Launching of the Mediterranean Network on Science and Technology of Advanced Polymer Based Materials (MEDNET), November.
	South–South cooperation programme in the humid tropics launched through conference in Manaus (Brazil).
1993	Creation of UNESCO's International Bioethics Committee (IBC).
	Establishment of regional office of ICPAM in China: International Centre for Mathematics and Computer Science.
	UNISPAR (University-Industry-Science partnership) launched by UNESCO, in collaboration with the Union of International Technical Associations, World Federation of Engineering Associations, and United Nations Industrial Development Organization (UNIDO).
	Silver Jubilee of the 1968 Biosphere Conference.
	Impact of Science on Society ceases publication.
1994	First issue of *World Science Report*, describing major trends in exact and natural sciences worldwide (other issues follow, in 1996 and 1998).
	UNESCO's scientific programs in the field of physics strengthened with the appointment of Physics Action Council (PAC).
	UNESCO – in collaboration with Agfa-Gevaert and the International Union of Pure and Applied Chemistry (IUPAC) – produces first 'DIDAC' chemistry-teaching materials, including posters and booklets, advanced-level books and CD-ROMs.
	UN Global Conference on the Sustainable Development of Small Island Developing States, Bridgetown (Barbados).
	Two inaugural meetings in China of East-Asian Biosphere Reserve Network.
1995	Fourth World Conference on Women, Beijing (China).
	Creation of the UNESCO/MCBN International Institute for Cell and Molecular Biology in Warsaw (Poland).

	Establishment of biennial UNESCO/Institut Pasteur Medal.
	International Conference on Biosphere Reserves convened by UNESCO and hosted by the Spanish authorities in Seville (Spain), March.
1996	First six-year UNESCO/ICSU Framework Agreement on cooperation signed.
	Trace Element Institute for UNESCO established, Lyon (France).
	Environmental Development in Coastal Regions and in Small Islands (CSI) platform launched, January.
1996–2005	World Solar Programme.
1997	First World Water Forum, Marrakech (Morocco).
	Universal Declaration on the Human Genome and Human Rights adopted by UNESCO's General Conference, November.
	UNESCO supports ICTP and the University of Wisconsin (USA) in launch of five-year Programme for the International Cooperation in Mathematics and its Applications (PICMA).
1998	Intergovermental Bioethics Committee of UNESCO (IGBC) established.
	World Commission on the Ethics of Scientific Knowledge and Technology (COMEST) set up by UNESCO.
1999	'World Conference on Science (WCS) for the Twenty-First Century: A New Commitment' convened jointly by UNESCO and ICSU, Budapest (Hungary), June–July.
2000	World Mathematical Year.
	IOC/World Bank Working Group on Coral Bleaching and Local Ecological Responses initiated, September.
	'Seville+5' review meeting in Pamplona (Spain), including task force on transboundary biosphere reserves.
	Second World Water Forum (the Hague) announces the UN system-wide World Water Assessment Programme (WWAP), March.
2001	'Prince of Asturias Award for Concord' awarded to the World Network of Biosphere Reserves.
	International Conference on Freshwater and the ensuing Ministerial Declaration, Bonn (Germany).
	UNESCO Institute for Statistics relocates to Montreal (Canada).
	Launch of UNEP-UNESCO GRASP initiative, aimed at mobilizing support for conservation of the great apes.
2002	Intersectoral project on Local and Indigenous Knowledge Systems (LINKS) for sustainable development launched.

	World Summit on Sustainable Development (WSSD), ten years after UNCED, Johannesburg (South Africa).
2002–2007	Sixth phase of the International Hydrological Programme (IHP), based on the principal that freshwater is as essential to sustainable development as it is to life.
2003	UNESCO one of the lead agencies in the International Year of Freshwater.

Entry into operation of the new UNESCO-IHE Institute for Water Education, Delft (the Netherlands), March.

Third World Water Forum, Kyoto (Japan), March.

Ground-breaking ceremony of the International Centre for Synchrotron-light for Experimental Science and Applications in the Middle East (SESAME), Jordan, January.

UNESCO General Conference adopts resolution on Sustainable Development of Small Island Developing States for further implementation and review of the Barbados Programme of Action.

Study on Geohazards, carried out by the British Geological Survey, European Space Agency and UNESCO.

Publication: Environmental Education Kit on Combating Desertification, comprising teacher's guide, book of case studies, cartoon and poster.

2004	Creation of UNESCO's International Basic Sciences Programme (IBSP).
2005	International Conference at UNESCO House (Paris) on Biodiversity, Science and Governance, organized by Institut française de biodiversité.

World Year of Physics; UNESCO participates in its international programme.

Publication: *UNESCO Science Report 2005* (fourth in previously named World Science Report series).

World Conference on Disaster Reduction, Kobe (Japan), 18–22 January.

UNESCO contributes to the International Decade for Action 'Water for Life' (2005–2015), proclaimed by the United Nations.

International Meeting to Review Implementation of the Programme of Action on the Sustainable Development of Small Island Developing States, Port Louis (Mauritius), 8-12 January.

ANNEX 3

HEADS OF NATURAL SCIENCES AT UNESCO[1]
Gail Archibald

Walter R. Erdelen, Germany, Assistant Director-General for Natural Sciences, since March 2001.

Gisbert Glaser, Germany, Assistant Director-General for Natural Sciences a.i.,[2] March 2000 to February 2001

Maurizio Iaccarino, Italy, Assistant Director-General for Natural Sciences, May 1996 to March 2000

Adnan Badran, Jordan, Assistant Director-General for Natural Sciences, January 1990 to April 1996

Sorin Dumitrescu, Romania, Assistant Director-General for Natural Sciences a.i., July 1988 to December 1989

Abdul-Razzak Kaddoura, Syrian Arab Republic, Assistant Director-General for Natural Sciences, April 1976 to November 1988

James M. Harrison, Canada, Assistant Director-General for Natural Sciences, January 1973 to March 1976

Adriano Buzzati-Traverso, Italy, Assistant Director-General for Natural Sciences, September 1969 to December 1972

Malcolm Adiseshiah, India, Assistant Director-General for Natural Sciences a.i., April to September 1969

Alexey Matveyev, Soviet Union, Assistant Director-General for Natural Sciences, October 1964 to March 1969

Victor A. Kovda, Soviet Union, Director of Natural Sciences Department, January 1959 to December 1964

Pierre V. Auger, France, Director of Natural Sciences Section/Department, April 1948 to December 1958

Joseph Needham, United Kingdom, Head of Natural Sciences Section, December 1946 to April 1948

---

1 1946–1948: Head of Natural Sciences Section; July 1948–1964: Director of Natural Sciences Department; September 1964 onwards: Assistant Director-General for Natural Sciences.
2 a.i. = *ad interim*.

WALTER R. ERDELEN
ASSISTANT DIRECTOR-GENERAL FOR NATURAL SCIENCES
MARCH 2001–

BORN IN ANSBACH, GERMANY, in 1951, Walter R. Erdelen obtained a B.Sc. in Zoology, Botany, Genetics and Chemistry in 1973; in 1977, a M.Sc. in Zoology, Botany, Genetics and Chemistry; and, in 1983, a Ph.D. in Ecology and Zoology from the University of Munich (Germany); followed by Habilitation in Biogeography, in 1993, from the Saarland University (Germany).

From 1981 to 1988, Mr Erdelen was a Lecturer and Researcher at the Department of Zoology, University of Munich, and a Research Associate at the Zoological Museum in Munich, where he became a Senior Lecturer and Researcher and then Associate Professor at the Institute of Biogeography at Saarland University, Saarbrücken. In 1995, he was appointed Professor of Ecology and Biogeography at the Department of Zoology, Institute for Animal Ecology and Tropical Biology and Director of the Ecological Field Station, University of Würzburg (Germany). In 1997, he left Germany for invaluable experience in Asia to become a visiting Professor at the Department of Biology, Institute of Technology, Bandung (Indonesia), a post he occupied until his nomination as Assistant Director-General for Natural Sciences, UNESCO, in 2001.

Since 1975, Mr Erdelen has taught at universities and public schools in Germany and abroad in the fields of environmental sciences, conservation biology, ecology, systematics and evolutionary biology. His teaching has covered basic and applied sciences at pre- and postgraduate levels, benefiting from accumulated research experience in these fields obtained over twenty-five years, particularly in the tropics. His experience covers training needs assessment at foreign universities; planning and implementation of university education programs; and planning, organization and realization of national and international research programs and conferences. He has fulfilled a number of advisory and evaluation tasks for national and international organizations, especially in relation to ecological research and conservation programs in developing countries. Many years of working experience in the African and South and South-East Asian regions have given him a good understanding of these regions.

Author of over seventy scientific papers and reviews published in international journals, Walter Erdelen is also the editor of three books on tropical ecosystems, landscape management in Sri Lanka, and sustainable use of reptiles in Indonesia. He is also the translator and editor of a book on the natural history of New Zealand and of the *Environmental Monitoring Handbook for Tsetse Control Operations*.

His professional affiliations include: the World Conservation Union (IUCN) and, in the United States, the Society for Conservation Biology, the Ecological Society of America, American Society of Naturalists, and the Society of Systematic Biology.

GISBERT GLASER
ASSISTANT DIRECTOR-GENERAL FOR NATURAL SCIENCES a.i.
MARCH 2000 TO FEBRUARY 2001

BORN IN GERMANY, in 1939, Gisbert Glaser obtained a doctorate in Geography from the University of Heidelberg (Germany) in 1965. He then undertook a university career with teaching and postdoctoral research tasks (Heidelberg, 1965 to 1971), including a year of field research in Brazil.

Mr Glaser joined UNESCO in October 1971 as a Programme Specialist in the Division of Natural Resources Research. He joined the new Division of Ecological Sciences in 1974. His main responsibility in the Man and the Biosphere Programme (MAB) during its first decade was to coordinate activities in island and mountain regions, and in arid and semi-arid lands. In 1982, he was entrusted with managing a large portfolio of extrabudgetary projects, as well as interagency activities. In 1990, he was appointed deputy Coordinator, and subsequently (following the 1992 UN Conference on Environment and Development) Coordinator for Environmental Programmes.

Entrusted with new intersectoral coordination responsibilities (for sustainable development programme elements in the Natural Sciences, Education, and Social and Human Sciences Sectors), Mr Glaser became a member of the Directorate General and was promoted to Assistant Director-General in January 1998. Before retiring in March 2001, he served as acting Assistant Director-General for Natural Sciences at UNESCO from 1 March 2000 to 28 February 2001. He then joined the International Council for Science (ICSU) as a senior advisor.

Gisbert Glaser has published over forty papers in refereed journals and edited scientific books on his research in geography, MAB research work, and on science for sustainable development issues. He is also the author of a large number of reports published by UNESCO and ICSU (after 2001) on international environmental sciences domains, and member, and sometimes director, of the editorial team/board of numerous UNESCO and several other publications.

MAURIZIO IACCARINO
ASSISTANT DIRECTOR-GENERAL FOR NATURAL SCIENCES
MAY 1996 TO FEBRUARY 2000

BORN IN ROME, ITALY, in 1938, Maurizio Iaccarino obtained an M.D. degree in Medicine and Surgery, University of Naples (Italy) in 1962. From 1966 to 1968, he was a postdoctoral fellow at Stanford University (USA).

In 1963, he became a staff scientist at the Institute of Genetics and Biophysics (IGB), Naples (founded by a previous Head of Natural Siences, Adriano Buzzati-Traverso). Mr Iaccarino headed IGB's Molecular Biology Department from 1976 to 1979, was Director of Research from 1980 to 1984, and became Director of the Institute (1985–93). He was also Professor of Biological Chemistry (1971), member of a Ph.D. programme in Molecular and Cellular Genetics at the University of Naples (1983–96), and Professor of Microbiology from 1986 onwards. In 1994–95, he was Director of the Institute of Molecular Genetics, Alghero, Sardinia (Italy). He became Assistant Director-General for Natural Sciences, UNESCO, in May 1996.

Mr Iaccarino is the recipient of several prizes, including from the Accademia dei Lincei (1985), the Academy of Medical Sciences of the Consorzio Interuniversitario di Biotechnologia (1996), and the Accademia Nazionale delle Scienze (Italy, 1997). In 1997, he received an Honoris Causa degree in Biological Sciences from the University of Tuscia (Italy).

Maurizio Iaccarino is co-author of some ninety publications in international scientific journals and of more than 100 scientific reports, and he has edited several books. His research interests deal with protein structure and characterization, DNA methylation, genetics and molecular biology of Rhizobium-legume symbiosis. He has published several papers on 'science and society' issues.

He is a member of the European Molecular Biology Organization (EMBO) and the Academy of Sciences for the Developing World (TWAS), as well as several other scientific societies, and has also been a member of the Board of Trustees of the Human Frontier Science Programme.

ADNAN BADRAN
ASSISTANT DIRECTOR-GENERAL FOR NATURAL SCIENCES
JANUARY 1990 TO APRIL 1996

BORN IN JARASH, JORDAN, in December 1935, Adnan Badran earned his B.Sc. in Physiology from Oklahoma State University (USA), and his M.Sc. and Ph.D. in Biology from Michigan State University (USA). Between 1960 and 1966, he conducted research at Michigan State University and in Boston (USA).

Mr Badran returned to Jordan in 1966, where he taught and conducted research at the University of Jordan, becoming Professor and founding Dean of its Faculty of Science (1971–76). Between 1976 and 1988, he was the founding President of Yarmouk University; founder of the Science and Engineering Campus at JUST University for Science and Technology, Jordan; and founding Secretary-General of the Higher Council for Science and Technology in Jordan (1986–87). In early 1989, he became Jordan's Minister of Agriculture and, in 1989, Minister of Education.

Mr Badran was Assistant Director-General for Natural Sciences from 1990 to 1995, and continued in the post *ad interim* to April 1996. In addition, he was Deputy Director-General a.i. from February 1993 to June 1995, continuing in this post until 1998.

Adnan Badran is the author of fifteen books on science and the environment, and more than seventy scientific papers in the field of plant phenolics and oxidation enzymes, as well as various articles on higher education, and science and technology policy. His professional affiliations include: Secretary-General, Vice-President and fellow of the Academy of Sciences for the Developing World (TWAS) since 1991; member of the American Association for the Advancement of Science (AAAS) since 1993; fellow of the Comité des Hautes Institutions Scientifiques et Culturelles of the European Academy of Arts, Sciences and Humanities since 1997; Aspen executive fellow since 1985; and member of the Arab Thought Forum and World Affairs Council since 1980. He is currently President of Philadelphia University in Jordan, and President of the Arab Academy of Sciences (AAS) in Beirut (Lebanon).

SORIN DUMITRESCU
ASSISTANT DIRECTOR-GENERAL FOR NATURAL SCIENCES a.i.
JULY 1988 TO DECEMBER 1989

BORN IN ROMANIA IN 1928, Sorin Dumitrescu graduated from the Civil Engineering University in Bucharest (Romania) in 1950 (diploma in Hydraulic Engineering). He obtained his M.Sc. (St Petersburg, USSR, 1963) and Ph.D. (Bucharest, 1969) in Hydrology. From 1950 to 1959, he was hydrologist and then Chief Hydrologist at Romania's Hydrological and Meteorological Service, and then Director of the Water Research Institute in Bucharest until 1969.

From 1964 onwards, Mr Dumitrescu participated in all major meetings related to the planning and implementation of the International Hydrological Decade (IHD). He was vice chairman of the IHD Coordinating Council, 1967–69. Director of the Water Sciences Division, UNESCO, and Secretary of the IHP (1969–88), he became Deputy Assistant Director-General for Natural Sciences and focal point on environmental matters (1984–88). He was acting Assistant Director-General for Natural Sciences at UNESCO from July 1988 to December 1989, with the grade of Assistant Director-General (1988–90).

Sorin Dumitrescu was awarded the International Association of Hydrological Sciences (IAHS) International Hydrology Prize in 1988. His publications include various papers on different aspects of hydrology, a monograph on Romania's water resources (co-author) and on the hydrology of Romanian rivers (co-author), and a comprehensive study on river flow fluctuations (Ph.D. thesis).

ABDUL-RAZZAK KADDOURA
ASSISTANT DIRECTOR-GENERAL FOR NATURAL SCIENCES
APRIL 1976 TO NOVEMBER 1988

BORN IN SYRIA in 1928, Abdul-Razzak Kaddoura began his studies at Brussels University (Belgium), and obtained his Ph.D. in Nuclear Physics from Bristol University (UK) in 1961. He was Professor of Physics and then Director of the Engineering Science Faculty of Damascus University (Syrian Arab Republic) until 1968, when he became the university's Vice President (1968–69). He was then invited to carry out research at the nuclear physics laboratory of Oxford University (UK). In 1971, he returned to Damascus University.

Mr Kaddoura is the author of several books written for Damascus University, and of articles on high energy nuclear physics published in scientific reviews in Syria, Italy, the United Kingdom and elsewhere.

Member of UNESCO'S International Commission for the Development of Education (1971–72), from 1974 onwards Abdul-Razzak Kaddoura was Rector of Damascus University, a post he occupied until being nominated Assistant Director-General for Natural Sciences, UNESCO, in April 1976.

JAMES M. HARRISON
ASSISTANT DIRECTOR-GENERAL FOR NATURAL SCIENCES
JANUARY 1973 TO MARCH 1976

BORN IN CANADA in 1915, James Harrison obtained his B.Sc. degree at the University of Manitoba (Canada) in 1935 and, after transferring to Queen's University (Canada), earned his M.A. in 1941 and Ph.D. in 1943.

His early field work for the Geological Survey of Canada (GSC) produced authoritative reports on the mineral-bearing regions of the Canadian Shield and provided him with knowledge and experience that would stand him in good stead as he rose, during his seventeen-year tenure, through the ranks of the GSC to become Director-General in 1956. Under his guidance, the organization enjoyed one of its most successful periods – more than doubling its staff, greatly increasing its budget and (in 1959) moving to new headquarters and decentralizing across the country. A variety of new challenges were met, including mapping the Canadian Arctic, studying the huge continental shelves and slopes, and increasing research and innovative applications in the newly developing fields of geochemistry and geophysics, to name but a few.

Mr Harrison was also one of the founding fathers of the Union of Geological Sciences (IUGS) and was elected its first President in 1961. He was then Under-Secretary of State in the Canadian Department of Energy, Mines and Resources, to become, in 1972, its Deputy Minister. From 1966 to 1968, he was also President of the International Council of Scientific Unions (ICSU). He became Assistant Director-General for Natural Sciences at UNESCO in January 1973. After retirement, he served as Chairman of ICSU's Committee on Nuclear Waste Disposal. James Harrison passed away in 1990.

MALCOLM ADISESHIAH
ASSISTANT DIRECTOR-GENERAL FOR NATURAL SCIENCES a.i.
APRIL 1969 TO SEPTEMBER 1969

MALCOLM ADISESHIAH was born in 1910, in Vellore, India. He had eight years of university education at Madras (India), the London School of Economics, and Cambridge University (UK), where he obtained his Ph.D. Mr Adiseshiah taught economics for ten years in Calcutta (India) and Madras before joining UNESCO in 1949 to head what was at first a modest Department of Technical Assistance, but which soon grew thanks to his pioneering work of launching and developing technical assistance projects in the developing world.

Mr Adiseshiah served as acting Assistant Director-General for Natural Sciences at UNESCO from April to September 1969. He had been nominated Deputy Director-General in 1963, a post he occupied until his retirement from UNESCO in 1970. An outspoken proponent for a more lucid understanding of education as a vital factor for socio-economic development, and a leading educator, he was the author of many books and taught at several Indian universities before and after his work for UNESCO. Malcolm Adiseshiah passed away in 1994.

ADRIANO BUZZATI-TRAVERSO
ASSISTANT DIRECTOR-GENERAL FOR NATURAL SCIENCES
SEPTEMBER 1969 TO DECEMBER 1972

BORN IN ITALY in 1913, Adriano Buzzati-Traverso did his undergraduate and postgraduate studies in the United States. His research, on population genetics, focused on the Drosophila fruit fly. After completing his Ph.D., in 1937 he joined the Faculty of Zoology in Pavia, Italy. He left this post in 1944, to avoid serving the Fascist Government of the Italian Social Republic. In late 1945, he was nominated Professor of Zoology at the University of Milan (Italy), where he was one of the first to teach Genetics. Between 1953 and 1959, he was Professor of Biology, University of California, Scripps Institution of Oceanography, La Jolla (USA).

Founder (1961) and Director of the International Institute of Genetics and Biophysics in Naples (Italy), Mr Buzzati-Traverso introduced and developed molecular genetics and molecular biology in Italy, which, in turn, encouraged the return of Italian researchers working abroad and attracted foreign ones.

His professional affiliations included: President of the Association of Radiobiologists of the EURATOM countries, member of the Executive Council of the International Cell Research Organization (ICRO), President of Pugwash-

Italy, member of the Club of Rome, founding member of EMBO, senior scientific adviser of UNEP (1973–80) and President of the Societé Européenne de Culture.

Mr Buzzati-Traverso was the author of several scientific books and articles, including, with Luigi Luca Cavalli-Sforza, *La teoria dell'Urto e le Unità Biologiche Elementari* (1948), and the UNESCO report *The Scientific Enterprise, Today and Tomorrow* (1977). From 1968 onwards he directed the scientific publication *Sapere*.

Adriano Buzzati-Traverso passed away in 1983. The institute he founded is now the Institute of Genetics and Biophysics 'Adriano Buzzati-Traverso'.

ALEXEY MATVEYEV
ASSISTANT DIRECTOR-GENERAL FOR NATURAL SCIENCES
OCTOBER 1964 TO MARCH 1969

A LEXEY MATVEYEV was the first Assistant Director-General for Natural Sciences. Born in Moscow, USSR, in 1922, he studied at the Moscow State University, where he was awarded his doctoral degree in 1959 and became Professor of Theoretical Physics. He is the author of numerous scientific papers and two books: *Electrodynamics and the Theory of Relativity*, and *Quantum Mechanics and Structure of Atoms*.

The appointment of an Assistant Director-General for Natural Sciences followed the reorganization of UNESCO's Natural Sciences Department into two new departments, one concerned with the advancement of science and the other with the application of science to development.

VICTOR A. KOVDA
DIRECTOR OF THE DEPARTMENT OF NATURAL SCIENCES
JANUARY 1959 TO DECEMBER 1964

B ORN IN VLADIKAVKAZ, RUSSIA, in 1904, Victor Kovda graduated from the Agricultural Institute in Krasnodar (USSR) in 1927. In 1931, he organized a laboratory of saline soils in the V. V. Dokuchaev Soil Institute, which he directed until 1959. He was a professor at Moscow State University (USSR) from 1939 to 1941, and from 1941 to 1942 directed the Institute of Botany and Soil Science of the Uzbek branch of the Academy of Sciences. He was Professor and Chairman of the sub-department of soil science of the biology and soil department of Moscow State University from 1953 onward. As Director of the Natural Sciences Department at UNESCO from January 1959 to December 1964, Mr Kovda played a decisive role in the development of UNESCO's environmental programmes and was the initiator and director of the *FAO/UNESCO Soil Map of the World*, an international project of the Food and Agriculture Organization of the United Nations and UNESCO. He became President of the International Society of Soil Sciences in 1968.

After leaving UNESCO, Mr Kovda continued his scientific career in the USSR as Professor at the University Lomonossov, Moscow, and as a corresponding member of the USSR Academy of Sciences, where he developed the Institute of Soils Sciences at Pushino. He continued to actively participate in the International Council of Scientific Unions (ICSU) and was President of the Scientific Committee for Environmental Problems (SCOPE).

Mr Kovda was the author of several publications, including on the development of new lands, construction of irrigation systems and reclamation of solonetz and saline soils. His main studies dealt with the soils of the USSR, China and Egypt. He was awarded the State Prize of the USSR (1951, 1953), the V. V. Dokuchaev Gold Medal (1967) and the Silver Medal of the French Association of Soil Scientists (1971). Victor Kovda passed away in 1991.

PIERRE V. AUGER
DIRECTOR OF NATURAL SCIENCES SECTION/DEPARTMENT
APRIL 1948 TO DECEMBER 1958

BORN IN PARIS in 1899, Pierre Auger attended the Ecole Normale Supérieure, originally to study biology but quickly showing an interest in atomic physics. From 1922 to 1942, he was at the Sorbonne, where he earned the Doctor of Science degree, became Assistant Professor, and then Professor of Physics (1937), devoting his professional life to experimental physics in the fields of atomic (photoelectric effect), nuclear (slow neutrons) and cosmic ray physics (atmospheric air showers). Between 1939 and 1941, while Head of the documentation service of the French National Centre for Scientific Research (CNRS), he joined the Free French Forces and also participated in the creation of a French/British/Canadian group on atomic energy research at the Universities of Chicago (USA) and Montreal (Canada), before joining operational groups in London (UK) in 1944.

In 1945, Mr Auger became Director of Higher Education in the French Ministry of National Education, where he participated in the creation of the French Atomic Energy Commission. He was a member of UNESCO's Executive Board from 1946 until 1948, when he was appointed Director of the Natural Sciences Section (soon after renamed the Natural Sciences Department), playing a decisive role in the creation of the European Organization for Nuclear Research (CERN) in 1954.

After leaving UNESCO in 1959, and among many other activities, he was Director of the Cosmic Physics Service at the CNRS (1959–62), and Director-General of the European Space Research Organization (ESRO, 1962–67), of which he was also for a time Chairman (1964–67). He was the author of many scientific works on X-rays, neutrons and cosmic rays. Pierre Auger passed away in Paris in 1993.

In 2002, the Pierre Auger Cosmic Ray Observatory, in western Argentina, began studies of the universe's highest energy particles, with unprecedented collecting power and experimental controls.

JOSEPH NEEDHAM
HEAD OF NATURAL SCIENCES SECTION
DECEMBER 1946 TO APRIL 1948

BORN IN LONDON (UK) in 1900, Joseph Needham studied biochemistry at Cambridge University (UK), where in 1924 he obtained his Ph.D. and in 1933 was appointed Sir William Dun Reader in that subject. From 1942 to 1946, he was Head of the British Scientific Mission in China and Counsellor at the British Embassy, Chungking, to establish liaisons between Chinese and Western scientists. Between 1943 and 1945, he was very active in promoting an 'International Science Co-operation Service'. Ultimately his international lobbying and diplomacy helped to ensure the incorporation of natural sciences in UNESCO's mandate. After his term as first Head of UNESCO's Natural Sciences Section, he resumed his work at Cambridge University.

Mr Needham was the author of numerous scientific and philosophical works, including *Chemical Embryology* (1931), *The Levellers and the English Revolution* (1939), and the sixteen-volume *Science and Civilization in China* (Vol. 1, 1954). A fervent member of the Church of England, he was also a Marxist, loyal to the Chinese Communist regime until the double shock of the Cultural Revolution and the 1989 massacre of Tiananmen Square. Joseph Needham passed away in 1995.

ANNEX 4

A TRIBUTE TO TWO OF OUR OWN: MICHEL BATISSE AND YVAN DE HEMPTINNE

MICHEL BATISSE (1923–2004)

MICHEL BATISSE was born on 3 April 1923 in Châteauroux in the Indre department of central France. After his studies in Paris as an engineer and physicist, he was appointed in 1951 as UNESCO's science liaison officer for the Middle East, based in the Organization's Bureau for Science Cooperation in Egypt.

He returned to UNESCO Headquarters in 1957, where he took responsibility for the so-called Major Project on Arid Lands. In the early 1960s, Batisse's lifelong interest in water resources led him to take a key role in shaping and launching the International Hydrological Decade (1965–1974), which in turn gave rise to the International Hydrological Programme.

In 1968, Batisse was secretary-general of the pioneering intergovernmental conference of experts on the 'scientific basis for rational use and conservation of the resources of the biosphere' – the 'Biosphere Conference' – which resulted in the Man and the Biosphere Programme (MAB) and which also helped pave the way for the UN Conference on the Human Environment held in Stockholm, Sweden, in 1972. Within the framework of MAB, Batisse saw the need for setting up areas where conservation of biological diversity would be combined with the satisfaction of the basic needs of local people and associated with ecological research and training. This idea gave birth to the 'biosphere reserves' – which now number 482 sites in 102 countries, and which can rightly be considered as constituting a lasting testimony to the imagination, perspicacity and tenacity of a man who combined global vision with an acute sense of what was politically possible and practically achievable.

During the 1970s, Batisse was given broad responsibilities for promotion and coordination of UNESCO's programmes relating to environmental sciences and natural resources – where the geographic anchoring of the subject calls for cooperation at the intergovernmental level. Notable among projects was the launching of the International Geological Correlation Programme (IGCP), a joint initiative of UNESCO and the International Union of Geological Sciences of ICSU (International Council for Science). Batisse played a crucial role in the preparation and negotiation of the Convention concerning the Protection of the World Cultural and Natural Heritage (1972), where he made sure that the natural heritage would be introduced and given its proper place.

After his retirement in 1984 as Assistant Director-General, Batisse was invited by successive directors-general of UNESCO to remain as honorary scientific adviser. He also became a senior scientific adviser to the Executive Director of the United Nations Environment Programme (UNEP) and to the World Bank. At the same time, UNEP and the French Government asked him to take responsibility for the Mediterranean Blue Plan. In 1989, he was invited to serve on the Board of Directors of Conservation International. He served as special adviser on biodiversity in the preparations for the 1992 UN Conference on Environment and Development (the 'Earth Summit', in Rio de Janeiro, Brazil) and played an active role in many other conservation initiatives. Among these, he was special adviser to UNESCO's Advisory Committee for Biosphere Reserves and participated incisively in the committee's work from its establishment in 1992.

During his long career, Batisse authored or co-authored over one hundred scientific papers devoted to the environment and natural resources. Among honours and distinctions, he was awarded the John C. Phillips Medal for Nature Conservation (1988), UNEP's Global 500 Award (1988) and the UNEP Sasakawa Environment Prize (2000). Other decorations included the medal of Officer of the French Legion of Honour, and UNESCO's Avicenna Silver Medal and Einstein Gold Medal.

Michel Batisse was above all a creator and a constructor. He was a man of vision who also understood the practicalities of making things work. His was an open door to colleagues from many cultural, disciplinary and institutional backgrounds, who came to seek his advice and counsel. He was an individual who truly marked the history and development of UNESCO, as well as the broader international community of which UNESCO forms a part.[1]

---

1   Adapted from UNESCO's MAB website, 2004.

YVAN DE HEMPTINNE (1924–2002)

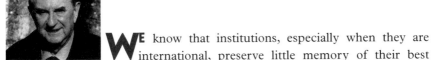

WE know that institutions, especially when they are international, preserve little memory of their best servants. And yet UNESCO's history, stretching over more than a half-century, is studded with a small number of personalities whose human qualities, vision and achievements should not be forgotten – and who could set the example for those who follow. There is little question that Yvan de Hemptinne is among these persons.

Born in a French-speaking family in Flanders, Yvan studied Chemical Engineering during the Second World War. His courage and convictions drove him into active resistance against the German occupation of Belgium and France. When the war was over, he resumed his studies in chemistry at the University of Ghent, after which he managed to combine a job in industry with both research and teaching. But it proved impossible to place his children, as he wished, in French-speaking schools in Flanders. These unacceptable conditions were the main reason why he applied for employment at UNESCO, which he joined in 1954.

At the pleasure of the King of Belgium, Yvan was recalled to Belgium in 1957 to serve as secretary-general of the prime minister's National Commission for Science. This first-hand experience with 'science policy' in a medium-sized country emerging from war would determine the future thrust of his career. Following this leave without pay from UNESCO, Yvan returned to UNESCO, where he soon became Director of the Science Policy Division, a position that he held until retirement in 1985.

Most countries have a ministry of science (or research), just as they have ministries of education or culture. For UNESCO, an intergovernmental body, it is thus just as basic a premise to ensure progress and help Member States in terms of science policy as it is in the fields of educational or cultural policy. There are rich countries that have more science than science policy, to be sure. For the less wealthy, however, and especially the developing countries, it is vital to help guide the choices to be made in scientific and technical research in relationship to their economic, social and environmental needs, as well as to their physical and human resources.

This is the mission that Yvan set for himself and that he accomplished over more than twenty years – with determination, care, imagination and generosity.

The entire philosophy of his action was based on the notion that the acquisition of science and technology by a country cannot be the result of simple knowledge transfer. Such acquisition must depend on the development of a nation's scientific and technical potential. And the latter must conform to a wilful policy adapted to the country's particular situation. As for UNESCO's role, this means, then (on the one hand), stimulating an international exchange of experiences and methods, and (on the other), undertaking specific actions in support of each country. In every case, this requires assuring the supremacy of the ethical requirements inherent in the Organization's charter.

As UNESCO's history will be written, the name of de Hemptinne should be associated chiefly with the regional Interministerial Conferences on the Application of Science and Technology to Development – the CAST Conferences. Yvan organized these at the Earth's four corners: CASTALA for Latin America in 1965 and 1985, CASTASIA for Asia in 1968 and 1982, CASTAFRICA for Africa in 1974, CASTARAB for the Arab States in 1976 and 1984, MINESPOL for Europe in 1970 and 1978. At the same time, he and his team published fifty-some documents and studies on different aspects of science policy. Yvan's particular concern was the status of those working in the research field, which resulted in a specific Recommendation made to UNESCO's General Conference. He developed a comparative study, too, of the organization and performance of research units, country by country. He also initiated a complex system of documentation related to scientific policy, known as SPINES.

One cannot forget, above and beyond his professional achievements, Yvan's strong personality. Always a bear for work, he had a particularly unswerving attitude in regard to public service. Known also for his courage and integrity, he stood by his beliefs with exceptional energy, sometimes even drawing the very wrath of his Director-General, René Maheu. He could be a fierce defender of UNESCO when it came under fire from other organizations, especially from the United Nations on the occasion of the major scientific conferences held in Geneva (1963) and Vienna (1979). And he rose in defence of his co-workers, when necessary, with the very same energy.

Yvan, with your professional competence, the firmness of what you did, the extent of your commitment, the respect you had for others as well as for yourself, you knew how to put to use the noblest virtues of the international civil service. All those who remember these things thank you.

Michel Batisse

Association of Former UNESCO Staff Members (AFUS)[2]

---

2   Reprinted from the Association of Former UNESCO Staff Members magazine *Link*, No. 83, January/March 2003.

ANNEXES

ANNEX 5

CHRONOLOGY OF UNESCO INTERNATIONAL SCIENCE PRIZES

UNESCO's international science prizes are awarded on either an annual or biennial basis.

The **Kalinga Prize for the Popularization of Science** has been awarded on a yearly basis since 1952. The Kalinga Prize bears the name of its donor, the Kalinga Foundation Trust in India, which created the prize to recognize an outstanding interpretation of science and technology for the general public.

The **UNESCO Science Prize**, created in 1968, is awarded biennially to an individual or group in recognition of their contribution to the technological development of a developing Member State or region through the application of scientific and technological research.

The **Carlos J. Finlay Prize for Microbiology** is made possible by a generous grant from the Government of Cuba. The Prize is named after Dr Carlos Juan Finlay, a prominent nineteenth-century Cuban scientist, whose research established a new branch in biology – vector biology – by proving that insects could be vectors in the aetiology of a series of illnesses, including the devastating Yellow Fever. The prize has been awarded biennially since 1980 in recognition of an outstanding contribution to microbiology.

The **Javed Husain Prize for Young Scientists** was established by UNESCO in 1987 as the result of a generous donation by Professor Javed Husain of India. The purpose of the biennial prize is to accord recognition to outstanding pure and applied research carried out by scientists of no more than 35 years of age.

The **Sultan Qaboos Prize for Environmental Preservation**, created in 1991, is awarded biennially for outstanding contributions to the management or preservation of the environment. The prize is made possible through a generous donation from His Majesty Sultan Qaboos Bin Said Al-Said of Oman.

The biennial UNESCO/Institut Pasteur Medal was created by UNESCO and the Pasteur Institute in 1995 on the occasion of the 100th anniversary of the death of Louis Pasteur. It rewards an outstanding contribution to the development of scientific knowledge that has a beneficial impact on human health.

The L'ORÉAL-UNESCO Awards for Women in Science, created by UNESCO and L'Oréal in 2000, are awarded each year to five outstanding women researchers in science – one from each continent – in life sciences and material sciences. The international juries are chaired by three Nobel Prize winners: Christian de Duve, Pierre-Gilles de Gennes and Günter Blobel.

The Great Man-Made River International Prize for Water Resources in Arid and Semi-Arid Areas, created in 2001, is awarded biennially for remarkable scientific research and scientific studies and discoveries in the field of exploration of groundwater and surface water usage in arid zones subject to drought and desertification and contributing to environmental and human development. It is made possible by a grant from the Government of Libyan Arab Jamahiriya.

In March 2005 the 171st Executive Board called for a thorough review of all prizes awarded by UNESCO.

ANNEX 6

CONTRIBUTORS

Mohammed Shamsul **Alam**: UNESCO (1982–2005), Head of UNESCO Offices in New Delhi, India, and Tehran, Iran (Islamic Republic of); UNESCO representative to Iran (Islamic Republic of); Senior Programme Specialist for Chemistry, Division of Basic and Engineering Sciences.

Jorge **Allende**: Expert, United Nations Development Programme–UNESCO Regional Programme on Postgraduate Training in Biological Sciences; President, RELAB Corporation; Former Chilean Minister of Science and Technology; Member, UNESCO Scientific Board for International Basic Sciences Programme; Director, Institute of Biomedical Sciences at the University of Chile.

Gail **Archibald**: Ph.D. in History. UNESCO since 1981, Division of Basic and Engineering Sciences. Author of *Les Etats-Unis et l'UNESCO 1944–1963* (Paris, Publications de la Sorbonne, 1993).

Angelo **Azzi**: Ph.D.; Scientist. Director, Institute of Biochemistry and Molecular Biology, University of Berne, Switzerland. Chairman of the UNESCO Global Network for Molecular and Cell Biology (MCBN); Chairman, International Advisory Board of International Institute of Biochemistry and Molecular Biology (IIBMB).

Adnan **Badran**: Assistant Director-General for Natural Sciences at UNESCO from 1990 to 1995, and then by interim until 1996, while serving as UNESCO's Deputy Director-General (1993–98) (see Annex 3).

Albert V. **Baez**: Ph.D.; Physicist. First Director of the UNESCO Science Education Department (1961–67). In 1948, as a graduate student at Stanford University (USA), collaborated with Professor Paul Kirkpatrick in developing the X-ray microscope. Author of, inter alia, *The New College Physics: A Spiral Approach* (W. H. Freeman, 1967) and *Innovation in Science Education Worldwide* (UNESCO, 1976).

Osman **Benchikh**: Doctorate in Physics, Chemistry. UNESCO, Programme Specialist responsible for renewable energies (since 1998), Division of Basic and Engineering Sciences.

Edward **Beresowski**: UNESCO, Chief, Section of Energy Development and Coordination, Division of Technological Research and Higher Education (1979–86).

Renée **Clair**: Doctorate in Physics; Counsellor at French National Commission for UNESCO (1993–97), coordinating quadrilingual book on scientific education of girls. Joined UNESCO in 1997, charged with special project on Women, Science and Technology; helped launch the L'Oréal-UNESCO Prize for Women Scientists; since 2002, Executive Secretary for this prize.

Rene Paul **Cluzel**: Master's in Mathematics and Informatics. UNESCO Programme Specialist in Informatics; administered and coordinated projects on application of ICTs for capacity-building and education, Natural Sciences Sector (1987–89), Sector of Communication, Information (since 1989).

Georges **Cohen**: Executive Secretary, International Cell Research Organization (ICRO), since 1987; Emeritus professor, Institut Pasteur, Paris.

Rita **Colwell**: Director, Microbial Resources Centres (MIRCENs); UNESCO Panellist for the MIRCENs and Biotechnology Education and Training Centres (BETCENs) programmes (1980–98); President of the University of Maryland Biotechnology Institute (1991–98); Director, US National Science Foundation (1998–2004).

Bernd **von Droste zu Hülshoff**: Joined UNESCO in 1971; Director of UNESCO's Division of Ecological Sciences and Secretary of the MAB Programme (1984–92); subsequently built up UNESCO's World Heritage Centre, which he directed from 1992 to 1999 (to retire at level of ADG). Currently UNESCO's Special Advisor for World Heritage.

Franck **Dufour**: UNESCO consultant in the basic sciences (2004–05); Scientific Director, Vaincre la Mucoviscidose, Paris, France.

Sorin **Dumitrescu**: Chairman International Hydrological Decade (IHD) Coordinating Council (1967–69); joined UNESCO in 1969, as Director of the Office of Hydrology (subsequently Division of Water Sciences); Deputy Assistant Director-General for Science, focal point environmental activities (1985–88). Assistant Director-General a.i. for Natural Sciences (1988–89) (see Annex 3).

Walter **Erdelen**: Ph.D. in Ecology and Zoology; appointed Professor of Ecology and Biogeography, University of Würzburg, in 1995. In 1997, became visiting Professor at the Department of Biology, Institute of Technology, Bandung, Indonesia, until nomination as Assistant Director-General for Natural Sciences at UNESCO in 2001 (see Annex 3).

Clarissa **Formosa Gauci**: UNESCO (since 1981), Assistant Programme Specialist, Division of Basic and Engineering Sciences.

Alain **Gille** (1922–2005): Agronomist; joined UNESCO in 1949 to head programmes concerning protection of the environment and the popularization of science. Organized the Travelling Science Exhibitions and prepared a series of twenty-four publications providing an inventory of scientific equipment. Beginning in the 1960s, worked on Africa's natural resources, first at Headquarters then as Director of the Regional Office for Science and Technology (ROSTA) in Nairobi, Kenya. He ended his career coordinating UNESCO's network of Regional Offices for Science and Technology.

Gisbert **Glaser**: Doctorate in Geography. Joined UNESCO in 1971, as Programme Specialist (Division of Natural Resources Research). Member of the new Division of Ecological Sciences in 1974, main responsibility in MAB; appointed Deputy Coordinator (1990), then Coordinator (1993) of Environmental Programmes. Assistant Director-General for Natural Sciences a.i. (2000–2001) (see Annex 3).

Charles **Gottschalk**: UNESCO, Chief of Section, Section of Energy Information Systems, Division of Technological Research and Higher Education (1979–88).

Ray **Griffiths**: Master's degree in Zoology; has had a long international career in the marine sciences, specializing in biological oceanography, fisheries and marine pollution.

Assigned by the Food and Agriculture Organization (FAO) to the Intergovernmental Oceanographic Commission (IOC) of UNESCO (1972–88).

Santiago **Grisolia**: Chairman, UNESCO Scientific Coordinating Committee (SCC) for the Human Genome Programme (1990–2000); Principe de Asturias Award for Scientific and Technical Research, and distinguished Professor, Department of Biochemistry of the Kansas Medical Center. The Santiago Grisolia Chair gives master courses and awards in biomedicine and neuroscience.

Malcolm **Hadley**: Ph.D. in Soil Biology; special interest in tropical ecology and the communication of scientific information. UNESCO, Programme Specialist, Division of Ecological Sciences from the early 1970s to 2001.

Jacob Darwin **Hamblin**: Ph.D. in History of Science; teaches history at Clemson University (USA); author of *Oceanographers and the Cold War: disciples of marine science* (University of Washington Press, 2005) and of *Science in the Early Twentieth Century: an encyclopedia* (ABC-CLIO Press, 2005).

André M. **Hamende**: Associated with the Abdus Salam International Centre for Theoretical Physics since 1964. At his retirement in 1990, he was Senior Administrative and Scientific Information Officer. Author of *A Guide to the Early History of the Abdus Salam International Center for Theoretical Physics 1960–1968* (Consorzio per l'Incremento degli Studi e delle Ricerche dei Dipartimenti di Fisica, University of Trieste, Italy, 2002).

Julia **Hasler**: Ph.D. in Biochemistry. UNESCO, Programme Specialist for the Life Sciences, Division of Basic and Engineering Sciences, since 2003.

Lucy **Hoareau**: M.Sc. in Pathological Sciences. UNESCO, Programme Specialist for the Life Sciences and Biotechnology, Division of Basic and Engineering Sciences, since 1997.

Jürgen **Hillig**: Ph.D. in Economics. UNESCO, Division of Science and Technology Policies, then Coordinator for entire Natural Sciences Sector, 1967–88; Director, Regional Office for Science and Technology, Jakarta, 1989–94; Director, then Assistant Director-General, Division of Decentralization and Field Relations, 1994–97.

Geoffrey **Holland**: Canadian delegate to the Intergovernmental Oceanographic Commission (IOC) of UNESCO and many of its programmes since 1970. IOC Vice Chairman and Chairman (1993–99). Chaired the IGOSS Working Committee, the preparatory and the first intergovernmental meetings for GOOS, and has served on many committees and task teams.

Maurizio **Iaccarino**: Joined the Institute of Genetics and Biophysics in Naples, Italy, as a staff scientist in 1963, to become Director of the Institute from 1985 to 1993. Director of the Institute of Molecular Genetics, Alghero, Sardinia, in 1994–95, and then Assistant Director-General for Natural Sciences at UNESCO (1996–2000) (see Annex 3).

Abdul-Razzak **Kaddoura**: Ph.D. in Nuclear Physics. Professor of Physics and then Director of the Engineering Science Faculty at Damascus University until 1968; Vice President (1968–69), then President of Damascus University (1973–76). Assistant Director-General for Natural Sciences at UNESCO (1976–88) (see Annex 3).

Dale C. **Krause**: Ph.D. in Oceanography. Marine scientist; geology and geophysics, and pelagic ecology. UNESCO 1973–89, becoming Director of the Division of Marine Sciences.

Irving A. **Lerch**: Ph.D. in Biological and Medical Physics. Member of Physics Action Council (PAC) of UNESCO (1993–99); member of Advisory Board of the International Science

Foundation; served as Chairman of the American Association for the Advancement of Science's Committee on Scientific Freedom and Responsibility; retired as Director of International Affairs, American Physical Society.

Tony **Marjoram**: Ph.D., technology for development; chartered professional engineer. UNESCO, Senior Programme Specialist in the Engineering Sciences, Division of Basic and Engineering Sciences. From 1993 to 1997, Programme Specialist in Engineering Sciences and Informatics at the Regional Office for Science and Technology in South-East Asia (Jakarta, Indonesia).

Julia **Marton-Lefèvre**: Deputy, then Executive Director of the International Council for Science (ICSU, 1978–97). After ICSU, she became Director of the Leadership for Environment and Development (LEAD) international programme, and is currently Rector of the (UN-affiliated) University for Peace.

Robert H. **Maybury**: Ph.D. in Chemistry. UNESCO Natural Sciences Sector, Paris, 1963–72; UNESCO Regional Office for Science and Technology for Africa, Nairobi, Kenya (Deputy Director), 1973–80; UNESCO Headquarters (Managing Editor, *Impact of Science on Society*), 1980–83.

James **McDivitt**: UNESCO Natural Sciences Sector, Programme Specialist (1965–67); Director, Jakarta Office (1968–78); Director, Division of Technological Research and Higher Education (1978–82).

Robert **Missotten**: Geologist, with special interests in remote sensing, GIS and Earth observation *in situ* and from space-borne sensors. Member of UNESCO staff since early 1980s; currently Secretary of the UNESCO-IUGS International Geoscience Programme (IGCP) and Head of the Earth Observation Section of the Division of Ecological and Earth Sciences.

Selim **Morcos**: Ph.D., physical oceanography of the Red Sea, Suez Canal and the Mediterranean. Professor of Oceanography at University of Alexandria, Egypt, then UNESCO Division of Marine Sciences (1972–89). Has also contributed to the submarine archaeological discoveries in Alexandria, Egypt.

Douglas **Nakashima**: Ph.D. in Human Geography. Before joining UNESCO in 1996, conducted his doctoral and postdoctoral research with Inuit hunters on their knowledge, know-how and worldviews relating to the Arctic environment. Heads the UNESCO project on Local and Indigenous Knowledge Systems (LINKS).

Annette **Nilsson**: Master's degree, Social Anthropology. Has carried out field work with the indigenous Chiquitano people in Bolivia; consultant with the LINKS project at UNESCO, 2005.

Benjamin **Ntim**: UNESCO, Programme Specialist, Engineering Sciences and Technology, Division of Basic and Engineering Sciences, 1980–2002.

Lotta **Nuotio**: Master's in Political Science, Journalist at National Radio Broadcasting of Sweden, Finnish station. Contemporary History. Research assistant for *Sixty Years of Science at UNESCO 1945–2005* project.

Bruno **de Padirac**: D.E.S.S. (French advanced degree) in International Economics. UNESCO, Programme Specialist in Science and Technology Policy Division (1976–90); Director of Cyberspace Law and Ethics Task Force (1999–2002); Director, Science Policy Studies and Information (since 2002).

John **Toye**: Professor, University of Oxford (UK). Former Director of the Globalization Division of the UN Conference on Trade and Development (UNCTAD) and the UN Committee on Trade and Development (1998–2000). He has authored seven books, most recently *The UN and Global Political Economy* (Indiana University Press, 2004), with Richard Toye.

Richard **Toye**: Ph.D. in History; Lecturer in History at Homerton College, University of Cambridge (UK); Research Assistant, United Nations Intellectual History Project (1999–2000).

Dirk G. **Troost**: Coastal scientist; worked in South-East Asia and the Caribbean, 1973–80. Joined UNESCO Natural Sciences Sector in 1980; Chief of UNESCO's Coastal Regions and Small Islands Platform since 1995.

Susan **Turne**r: Ph.D. in Geology; vertebrate palaeontologist. Member, International Geoscience Programme (IGCP) Scientific Board (2000–04); co-leader of UNESCO-IUGS IGCP 328. Since 2001, part of the UNESCO Advisory Group of Experts for Geoparks.

Indra **Vasil**: Founding Chairman of Biotechnology Action Council (BAC), 1990–2001; Graduate Research Professor Emeritus, University of Florida, Gainesville.

Gary **Wright**: B.A. and M.A. in languages and literature. United States naval officer, followed by twenty years as UNESCO Editor, eighteen of which in the Division of Marine Sciences.

Vladimir **Zharov**: D.Sc.; Professor of Physical Chemistry; over twenty years at St Petersburg University (Russia). Director of UNESCO's Division of Basic Sciences (1984–98), promoting international cooperation in the basic sciences and university science education. Contributed to preparation of the World Conference on Science (Budapest, 1999) and its follow-up.

Others

Thanks also to the following for their participation in, and support of, this project:

Massoud **Abtahi**
Association of Former UNESCO Staff Members (AFUS)
Ariane **Bailey**
F. W. G. 'Mike' **Baker**
Jens **Boel**
Etienne **Brunswic**
Siew Leng **Chan**
Brian L. **Goddard**
Eloise **Loh**
David **McDonald**
Howard **Moore**
Alfredo **Picasso**
Klaus H. **Standke**
Michiko **Tanaka**

ANNEXES

Sidney **Passman**: Physicist. Director, UNESCO Division of Scientific Research and Higher Education (1973–81); author of *Scientific and Technological Communication* (Pergamon Press, 1970).

Patrick **Petitjean**: Historian of science with the REHSEIS team (Recherches Epistémologiques et Historiques sur les Sciences Exactes et les Institutions Scientifiques) of the French National Centre for Scientific Research (CNRS) and University of Paris 7.

Ernesto **Fernández Polcuch**: M.Sc. in Science, Technology and Society. Programme Specialist since 2002 at UNESCO Institute for Statistics, after serving as consultant to UNESCO in same field and managing UNESCO's Chair on Science and Technology Indicators in Argentina.

Jean-Jacques **Renoliet**: Ph.D. in History. Author of *L'UNESCO oubliée: La Société des Nations et la coopération intellectuelle (1919–1946)* (Paris, Publications de la Sorbonne, 1999).

Jacques **Richardson**: From 1972 to 1985, Head of UNESCO Science and Society Section; Editor of *Impact of Science on Society*. Consultant to UNESCO, United Nations Development Programme (UNDP) and national governments on environmental management in West Africa and Ethiopia (1986–2003).

Badaoui **Rouhban**: Doctorate in Engineering. Joined UNESCO's Natural Sciences Sector in 1983; currently Senior Programme Specialist in Disaster Reduction in the Division of Basic and Engineering Sciences.

Saif R. **Samady**: Former Director, Division Technical and Vocational Education, Education Sector, UNESCO.

Marc **Steyaert**: Agronomic Engineer (Belgium), Third Cycle of Oceanography (France). Joined UNESCO in 1967; Programme Specialist and Head of Section (1990–94) in the Division of Marine Sciences. From 1975 to 1995, mostly concerned with the Interregional Coastal Marine Zone Project (COMAR).

Bruno J. **Strasser**: Ph.D. in Philosophy and History of Sciences; lecturer and associate researcher, Department of Philosophy, University of Lausanne, Switzerland. His work focuses on the history of the life sciences in the nineteenth and twentieth centuries and on the history of science policy. Currently, a visiting Fellow at Princeton University (USA).

Alexei **Suzyumov**: Ph.D. in Geology and Oceanology; joining Institute of Oceanology, Russian Academy of Sciences. Has taken part in several major international projects, such as Ocean Mapping (of IOC) and the Ocean Drilling Project. Joined UNESCO, Division of Marine Sciences, 1982.

M. S. **Swaminathan**: Ph.D. in Plant Genetics. Currently holds the UNESCO Chair in Ecotechnology at the M. S. Swaminathan Research Foundation, in Chennai (Madras), India, and is President of the Pugwash Conferences on Science and World Affairs. His association with UNESCO began in 1949, and with the Man and the Biosphere (MAB) Programme, in the 1970s.

Jacques **Tocatlian**: Chemist and Information Scientist. Joined UNESCO Natural Sciences in 1969, in the Division of Scientific Documentation and Information. In 1979, became Director of the General Information Programme (PGI), and later, Director of Information Programmes and Services (IPS).

INDEX

A

Abu Simbel Temples 160, 390
Academy of Sciences for the Developing World 603, 608, 613, 619
Africa, Plan of Action on Science and Technology for 632
African Academy of Sciences 608 Network of Scientific and Technological Institutions 96, 147, 151
agriculture 45, 65, 95, 107, 137, 139, 193–296, 264, 268, 310, 317, 470, 498, 508, 539, 548, 597, 603, 627 and industry 45, 137, 139, 498, 508 *see also* Food and Agriculture Organization (FAO)
agroclimatology 224, 548
Antarctic 68, 169, 201, 220, 298
Applied Systems Analysis, Institute of 271
aquaculture 139
aquatic biology 97, 140, 452, 491, 595–6, 615, 616, 628–30 biota 619
Arctic 55, 169, 201
arid lands/zones 76, 209–14, 233, 235, 238, 246, 255, 383, 515, 554 Institute 30–1, 56, 211 map of 210–11 project 33 programme 169
astronomy 53–4, 84 International Year of 624
atmosphere 113
avalanches 297, 322, 328 atlas of 328 *see also* natural disasters

B

basic and engineering sciences 24, 93–192, , 627 *see also* engineering
basins (HELP programme) 254, 255, 618 representative and experimental 240, 243
beach resources 374–5
biodiversity 285, 288, 260–96, 317, 385–6, 394–5, 416–17, 470, 499, 501, 578, 594, 600, 602, 620, 621 Convention on 414 and science and governance 388
bioethics 16–17, 98, 142, 474–5, 480, 599, 634
bioinformatics 599, 630
biological diversity 279, 282, 285, 317, 384–5, 394, 414, 549, 551 Convention on 600, 602, 620
biology 56, 61–5, 69, 97, 134–46, 176–84, 197, 199, 207, 215, 220, 223–4, 229, 282, 285, 290–1, 321, 332, 351, 354, 368, 506, 515–19, 523, 533, 543, 577, 580, 597–9, in Africa 144 and agrochemicals 137, 141 Action Council 140, 141 and biological nitrogen fixation 139 and Council for Science 134, 143, 146 and culture collections 139 expertise in 143 and food 140, 141 and Network for Molecular and Cell Biology 140–1 and Human Genome Programme 136, 137, 142 and Latin American Network of Biological Sciences

145 Microbial Resources Centres 136, 139 molecular 62 and Policy Committee on Developing Countries 146 priorities in 144 regional networks for 137 research in 134 and science Information 135 and education/training 135–7, 143 and Union of Biological Sciences 69, 134 and Universal Declaration on the Human Genome and Human Rights 138 and World Federation for Culture Collections 137 *see also* aquatic biology; cell biology; Global Impact of Applied Microbiology (GIAM); marine biology; molecular biology; neurobiology

biosphere 260–296 conferences 225, 287, 403 networks 578 reserves 24, 268–93, 317, 383, 555, 600, 621 assessments 291 breach of convention 395 cultural and natural criteria 389, 392 monitoring 317, 318 monuments 328–9 multiple functions of 275, 287 and natural heritage 393–4 networks for 288, 578 reserves 600, 621 and sacred sites 288 Scientific Advisory Panel 279 Seville Strategy 287 and traditional knowledge 381–2 *see also* Man and the Biosphere

biotechnology 59, 596, 615–16, 491, 628–30 Inter-Agency Cooperation Network on 629

Borobudur 160

brain drain 101, 111, 458, 595, 630

Brain Institute 56

Brazilian Society for Scientific Relations 35

Brundtland Report/ Our Common Future 248, 408–9, 412

Bulletin on International Scientific Relations 35

C

Canberra Manual 454

cancer 53, 58, 89

capacity-building 458, 501, 510–11, 579, 581, 583, 603, 605, 607–8, 616, 619, 626–30, 633–5 *see also* education/training, and chapters on different disciplines or topical areas

carbon dioxide 168, 338, 352, 369

Committee on Earth Observation Satellites (CEOS) 316, 319, 500

Committee on Science and Technology in Developing Countries (COSTED) 144–6, 502

CERN (European Organization for Nuclear Research) 32, 56–63, 91, 114, 117, 133

cell biology 136, 140–1, 515–19, 541 growth 61 Research Organization 61, 135

chemistry 120–5 and education 121 and exchange of scientists 124 environmental 121 and Federation of Asian Chemical Societies 124 and industry 121, 122 and International Organization for Chemical Sciences in Development 121 and Islamic Educational, Scientific and Cultural Organization 125 laboratory teaching of 122 and medicinal plants and spices 121, 124 microscience experiments in 125 and natural products 97, 121, 124 and pesticide analysis 124 priorities in 122 pure and applied 120, 122–3 technician training in 124 and Union for Pure and Applied Chemistry 120, 122, 123 workshops on 122, 124

climate change 281, 310, 333, 339, 349–51, 488–9, 499, 579, 594, 600, 625 Conference, World 347, 409–17 and desertification 301 Framework Convention on 414

coastal management 333, 352, 354, 362–3, 372–4, 381, 493 zones and small islands 197, 284, 402, 420, 424, 426, 581, 606, 631, 633 *see also* oceans

Cold War 30, 32, 33, 46, 60, 63, 78, 80, 155, 504, 552

colonialism 29, 33, 75, 86, 434, 460

communication and information 496

computer sciences 105, 524

conservation 202, 280, 285, 317 of natural resources 24, 233–96, 394, 395 Union, International 202–3, 230

INDEX

consumption 594
continental drift 221, 298, 301, 303, 306
Council for International Organizations of Medical Sciences 134, 136
creativity 464, 572, 613
cultural diversity 288, 375, 380, 385–8, 578, 601–14, 621 values 251, 269
culture 22, 86, 324, 389, 392, 398, 424, 493, 537, 552–3, 598, 604–6, 620, 622, 630, 633 *see also* cultural diversity; cultural values
Current Trends in Scientific Research 51

D

data exchange 344, 626 exploitation 624 real time 625 *see also* FRIEND
Decade for Action 'Water for Life', International 250, 552, 620 for Disaster Reduction 330 of Education for Sustainable Development 577, 606–7, 622, 631 of Ocean Exploration 337 *see also* United Nations Conference on Decade of Education for Sustainable Development
decentralization 19, 72, 75, 143, 166, 357, 491, 576, 585, 611, 613
desertification 52, 213, 225, 285, 301, 310, 317, 322, 328, 408, 523, 529 Convention to Combat 285 *see also* climate change; natural disasters
deserts 90, 212, 214, 225, 301, 310
developing countries 32, 34, 51–2, 97–9, 101–2, 105, 108–16, 121–30, 136–44, 157, 160, 168, 170, 173–4, 182, 189–90, 200, 212, 222, 228, 230, 234, 238, 240–3, 254, 258–62, 300–4, 310, 315, 317–18, 327, 354–8, 368–9, 405–6, 409, 420, 426, 434, 437, 439, 441, 443, 448, 454, 456, 458, 477, 479, 501–3, 508–9, 512, 520–1, 534, 556–7, 573–80, 594–600, 604, 606–7, 615, 627–31 *see also* capacity-building; committee on science and technology in developing countries; education/training
development 111–15 application of science and technology to 74
development, sustainable 14, 25, 195–200, 257, 342, 379–80, 382, 386, 401–28, 466, 500–1, 576, 578, 580, 586, 601, 604–7, 611–12, 622, 630–1 Conference on 410 cultural diversity for 424 economic 605 human 572 Inter-Agency Committee on 422 Island Perspective on 606 World Summit on 386, 421–5, 631 *see also* Decade of Education for Sustainable Development; United Nations Conference on Decade of Education for Sustainable Development; United Nations Conference on Sustainable Development of Small Island Developing States; United Nations Summit on Sustainable Development
Directory of Main Scientific Research Organizations 438
Directory of New and Renewable Energy Information 162
disaster management 316 mitigation 380, 488, 519, 582–4 *see also* natural disasters
discrimination 465, 467
distance learning 607
DIVERSITAS 499, 506
DNA techniques, recombinant 16, 138
drought 235, 328, 602 *see also* natural disasters

E

Earth, crust of 196 International Year of Planet 623 Observation Summit 500 Observation System of Systems 623 sciences 91, 159, 196, 219–23, 228, 263, 290, 302–3, 309, 315, 373, 411, 540, 555, 574–5, 577–80, 604, 610, 621–3 Summit 167, 248, 625, 500, 629 System Science 633
earthquakes 91, 222–3, 488, 489 *see also* natural disasters

East African Rift 228
ecohydrological services 264, 619, 620
ecological sciences 155, 195–200, 201, 225, 260–97 , 375, 373, 389, 393, 412–3, 420, 507, 540–1, 577–80, 620 see also Earth sciences
ecology 17, 196, 197, 229, 260–97, 575 Action Plan 278 assessments 262, 291–2 biosphere reserves 268, 271, 273 Coordinating Council 289–90 evaluation 291 in evolution 280 International Conference 225, 260, 287 multiple functions of 275, 287 networking 276, 288 objectives 265 poster exhibit 280 in practice 276 priorities 269 Report Series 277, 284 research themes 264 and sacred sites 288 Scientific Advisory Panel 279 Seville Strategy 287
Ecology in Practice: Establishing a Scientific Basis for Land Management 497
Economic and Social Council, United Nations (ECOSOC) 31, 48–50, 52–7, 79, 232, 322–3, 351, 361, 435, 440–1, 444
Economic and Social Commission for Asia and the Pacific (ESCAP) 304, 306, 554
economics, market and centralized 274, 454, 456
Ecosystem Conservation Group 408 and management 316
education/training 95, 97, 115, 121, 152, 203, 208, 222, 237, 241, 243, 251, 274, 278, 281, 302, 309, 318, 342–3, 352, 358, 362–3, 372–4, 381, 394, 402, 405–6, 410–11, 416–17, 423–4, 427, 434, 439, 442, 446, 453–4, 456, 458, 458, 467, 469, 471, 482, 501, 509, 511, 516–18, 526, 528, 573, 577–83, 596, 598, 603–8, 610, 615–17, 620, 622, 624, 626–9 see also Decade of Education for Sustainable Development; geoscience
Einstein, Albert 487–8
el Niño 347, 350
Encyclopedia of Life Support Systems 581, 634
endogenous technology 148 see also knowledge, local/indigenous
energy, new sources of 13 nuclear 33, 91 renewable 21, 98, 148, 150, 156, 161–4, 166–75, 185, 411, 416, 447, 519, 523, 580, 596, 602–3 solar 150, 163, 211 see also fossil fuels
engineering 147–66, 199, 630–2 see also basic and engineering sciences
environment 139, 193–427 agenda 401–2 coordination 401–27 and development 285, 401–27 and education 402, 405, 411, 418 Focal Point 412 human 230 law 406 monitoring 274, 281 protection 193–427 successes and failures 427
environmental sciences 116, 121, 193–427
ethics 17, 21, 25, 432, 476, 480, 491 in scientific knowledge and technology 504, 632
Europe, Council of 62–3
European Organization for Nuclear Research (CERN) 116–18, 133, 220, 489, 504, 508, 536, 555, 559,
European Coal and Steel Community 63 Commission 172 Culture Centre 59 Molecular Biology Organization 61–2 Space Agency 61, 218 see also CERN
evolution 42–3

F

fellowships 509, 515
Field Science Cooperation Offices 72–5, 134, 166, 491, 508, 522, 528, 585–9 Cairo (Middle East) 72 Djakarta (South Asia) 100 Montevideo (Latin America) 71 Nairobi (Africa) 73, 75–6 New Delhi (South East Asia) 72 Venice (Europe) 73 see also Regional Offices for Science and Technology
fisheries 56, 336, 352, 369, 374, 379, 383, 386
flooding 243, 320, 322, 328–9, 374, 488
flora neotropica 204, 229, 508, 559

low Regimes from International Experimental and Network Data (FRIEND) 250, 252–54
ood 141, 596 and Sustainable Development, Conference on 420 see also Food and Agriculture Organization (FAO)
ood and Agriculture Organization 48, 50, 65–7, 122, 217–18, 223–4, 234–5, 241, 251, 254, 261, 278, 334, 355, 420, 428, 493
ord Foundation 113, 536
orests 372, 402 tropical 226, 567, 624
ossil fuels 168–9, 174, 195, 594, 596
Frascati Manual 453
ree circulation of scientists 504
reshwater 233–59, 402, 407, 411, 416, 520 resources 219 see also hydrology/water sciences

G

Galapagos Islands 229, 391
gender 16, 148, 156, 266, 458, 470, 473, 541, 550, 582, 586, 596, 628, 631
genetic engineering 136, 138, 490–1 see also biotechnology
genome, human 136–7, 142
geochemical prospecting 221
geodesy and geophysics 69, 323
Geographic Information Systems 315
Geographical Union, International 218
geohazards 316
Geological Correlation 303, 307–10
Geological Correlation Programme, International 228, 230, 408, 417, 418 map of the world 217 sciences 297–300, 302, 305, 308, 520, 528
geology 197, 217, 221, 623 in the service of society 498 see also geoscience
geoparks 578, 623,
Geophysical Year, International 33, 68–70, 91, 220, 228, 298, 305
geoscience 158, 297–314, 525 basic and applied 302 and cooperative science 310 and education 302, 309 and evaluations 307, 311 financing 305 future of 310 and Geological Congress 299 and global geological research 298 and Gondwanaland 298 new resources for 303 long-term benefits of 310 and mineral exploration 306 and plate tectonics 301, 303 and rational exploration and exploitation 306 and remote sensing 306 and rural and urban planning 302 and human survival 306, 309 and Southern Continents Project 309 structure, tectonics and drifting continents 310 and Younger Scientists Programme 311 see also International Union of Geological Sciences
girls 183, 185, 189, 465, 467–9 in Africa 185
glass ceiling 465, 535, 541
global change 407, 421 Observatory 255 warming 494 (see also greenhouse gases)
Global Earth Observation System of Systems (GEOSS) 283, 316–17, 500, 578, 623
Global Ocean Observing System (GOOS) 316, 347–8, 352, 500, 579, 623–6
Global Atlantic Tropical Experiment (GATE) 350
Global Impact of Applied Microbiology (GIAM) 136
globalization 190, 376, 450, 524, 571–2, 594–5
grazing 225, 267, 277, 279, 533
greenhouse gases 168–9, 625 see also global warming
groundwater resources 235, 243, 250, 254, 528, 618

H

health 185, 426, 508, 517, 599–600, 619–20 *see also* World Health Organization (WHO)
heritage see natural heritage; World Heritage
Higher Education in the 21st Century, Conference on 502
Hiroshima 29, 48, 57, 60
HIV/AIDS 138, 140, 610, 614
human needs 95, 101, 501, 630 resources 360, 435, 537–44 welfare 41, 46, 48, 228, 252
humanism, scientific 17, 41
humanities 25, 39, 81, 198, 321, 460 *see also* social and human sciences
humid tropics 214–16 ecology 260–98
Huxley, Julian 39–43, 462, 203–6
Hydrogeologists, International Association of 237, 242, 254
hydrogeology 237, 254, 462,
hydrographic tables 365
Hydrological Decade 15, 33, 91, 233, 253–9, 498 Sciences, International Association of 235–7, 242
hydrology/water sciences 197, 219, 225, 233–59, 404, 406, 408, 418, 420, 426, 498, 511 comparative 247
Hylean Amazon Project 54, 56, 90, 205–9, 519 *see also* International Institute for Hylean Amazon (IIHA)

I

Impact of Science on Society 460–1, 463
Indian National Institute of Oceanography 70
industry 172, 189, 28, 309–10, 348, 452, 483, 498, 508, 581, 596
InfoMAB 284
informatics 98, 104–5, 113, 131–4, 489 and Federation for Information Processing 131, 132 and Intergovernmental Bureau 127, 131 programme 127, 128, 133 strategies and policies for 126–7, 131 and telematics 586
Information Programme, General 128, 161 and communication technologies 126–33, 593–4, 607, 631
information society 128, 132–3, 503–6, 586, 598 World Summit on the 133
innovation 595, 609 indicators 458
Integrated Global Observing Strategy (IGOS) 283, 316, 500, 567, 578
Integrated Global Ocean Services System (IGOSS) 346–7, 355
Intellectual Cooperation, Committee on 35 Institute 23, 29, 35–6, 38, 77, 431 Cooperation Organization 35
Inter-Academy Council 608
interdisciplinarity 25, 148, 260–96, 371–82, 401–27, 597, 599
Intergovernmental Oceanographic Commission (IOC) 33, 65–70, 91, 102, 158, 196, 217, 221, 230, 252, 283, 326–7, 332–62, 365, 368–70, 373, 407–8, 411, 416–20, 489–90, 499–500, 503, 506, 520, 533, 536, 540, 543, 574–5, 579, 583–4, 624–6 Panel on Climate Change 165 *see also* oceans
Intergovernmental Maritime Consultative Organization 334, 337, 543 (IMCO)
International Atomic Energy Agency (IAEA) 51, 57, 66, 107–8, 111–14, 118–19, 161, 241, 334, 351, 479, 491, 603
International Basic Sciences Programme (IBSP) 101–3, 120, 143, 582–3, 604, 606, 609–10, 627–8

INDEX

International Biological Programme (IBP) 229, 260–3, 266–8, 283, 403, 497, 506
International Biosciences Networks (IBN) 100, 143–6, 502, 506
International Brain Research Organization (IBRO) 135–6
International Centre for Theoretical Physics, Abdus Salam (ICTP) 99, 104, 112, 114, 488, 515, 528, 555, 575, 610, 614
International Centre for Synchrotron-Light for Experimental Science and Applications in the Middle East (SESAME) 99, 110, 116–20, 511, 555, 560
International Computation Centre 30, 33, 54, 56, 104, 131
International Council for Science (ICSU) formerly International Council of Scientific Unions 23–5, 30, 35–6, 45–6, 59, 66, 77–9, 81, 117, 129, 134, 143, 146,182, 199–200, 203, 228, 236, 242, 443, 448, 488, 494, 496–507, 600, 603, 606, 608, 613
International Geoscience Programme/International Geological Correlation Programme (IGCP) 102, 158, 196, 228, 283, 297–311, 315, 373, 408, 411, 416, 491, 498, 506, 525, 528, 540, 546, 555, 574, 578, 622–3
International Geosphere Biosphere Programme (IGBP) 282–3, 407, 499–500
International Hydrological Programme (IHP) 192, 196, 213, 217, 244–59, 263, 283, 317, 328, 371, 373, 408, 411, 416–18, 426, 491, 498–9, 519, 528, 540, 545, 551, 574, 577, 616–20
International Institute for Hylean Amazon (IIHA) 55–6, 71, 205–9 *see also* Hylean Amazon Project
International Labour Office 422
International Labour Organization (ILO) 77, 190, 204, 422, 624
International Oceanographic Data Exchange (IODE) 344, 346, 579
International Union for the Conservation of Nature and Natural Resources/World Conservation Union (IUCN) 15, 202–5, 229–30, 261, 263, 271, 278, 286, 296, 390–2, 397–8, 409, 491, 508, 559, 621
International Union of Geological Sciences 297–300, 302, 305, 308
International Union of Soil Sciences (IUSS) 217, 224
internet 95, 116, 128, 161, 187, 307, 371, 373, 376, 380, 382, 458, 479–80, 524–5, 576, 595
intersectorality 371–88, 401–27, 610, 612, 614
irrigation 212–13, 219, 234–5, 267, 279, 510

K

knowledge, fundamental 87 local/indigenous 385–8, 581, 583, 612, 635 promotion of 604 and society 598 traditional ecological 383–5 *see also* Local and Indigenous Knowledge Systems (LINKS)
Kuroshio Current 343, 565
Kyoto Protocol 489

L

laboratories, floating 220
land use 26, 218, 260–96
landslides 315, 321, 327 *see also* natural disasters
League of Nations 23, 35–6, 77, 431
Liaison Offices 45 World Centre for Scientific 30
life sciences 60–4, 97, 134, 143, 263, 385–7, 465, 471, 474, 516, 580 support system 625
Limits to Growth 260, 594
Local and Indigenous Knowledge Systems (LINKS) 197, 371, 383–8, 550, 570, 581, 612, 633 *see also* knowledge, local/indigenous

M

Man and the Biosphere 15, 24, 156, 158, 196, 225, 229, 260–96, 403, 408, 417–18, 425–6, 497–8, 577, 620, 622 Action Plan 278 International Coordinating Council 212, 230, 265, 267, 276, 289–91 Programme (MAB) 15, 24, 102, 158, 195–6, 212–13, 225, 229–31, 260, 263–93, 317–18, 328, 371, 373, 383–4, 389, 393–4, 403–5, 408–11, 416, 425–6, 470, 490, 497–9, 506, 516, 537, 540, 543, 545–7, 553, 556–7, 574, 577–8, 620–3, 634
Man, the City and Nature 414
Management of Social Transformations (MOST) 102, 419, 421
mangroves 16, 361, 363
mankind, welfare of 492, 494
maps 216–18, 225, 579 Africa 218 climatic 216 geological 217 oceanic 217 scientific 216 soil 217, 223 of UNESCO Offices worldwide 588–9 vegetation 217 world 217, 223
marginal lands 213, 218, 260–96
marine biology 65, 220, 290, 332, 339, 368 *see also* marine sciences; oceans
Marine Science Newsletter 369
marine sciences 33, 65–70, 219–21, 332–70, 373, 384, 515–16, 519, 533, 543, 552, 579 International Advisory Committee on 220 International Advisory Committee on 65, 68–70 *see also* oceans
Mathematical Union 104–5
mathematics 24, 54–5, 97, 99–100, 104–6, 113, 115, 126, 145, 176–7, 181–5, 199, 466, 469, 515, 539, 541, 580, 610 and computer sciences 105 and ICSU 104 and informatics 105 and International Centre for Theoretical Physics 104 and Computation Centre 104 cooperation in 105 and Mathematical Union 104, 105 and Latin American Centre for Mathematics 105 and South-East Asian Mathematical Society 106 training in 105 workshops 105 and World Mathematical Year 16, 105
Mediterranean 100, 142, 217–18, 267, 275, 323, 327, 366, 383, 493, 513
meteorology 54, 56, 336, 348 *see also* World Meteorological Organization (WTO)
Microbial Resource Centres 405, 629
microbiology 97–100, 134, 136–9, 207, 411, 477, 493, 541 *see also* biology, molecular
molecular biology 61–4, 97, 99, 136, 140–1, 533, 580, 599 *see also* biology
Monuments and Sites, Council of 390
Mutual Economic Assistance, Council of 434, 437

N

Nagasaki 29, 48, 57, 60
nanotechnology 16, 116, 595–7, 629–30
National Science Foundation, United States 116, 212, 458
Natural and Cultural Heritage, Convention on the Protection of the World's 285, 389–400 *see also* natural heritage
natural disasters 226, 320–31, 488, 580, 584, 600, 624, 630 and arid lands 328 *Atlas of* 328 and buildings 322, 324, 328, 330 and cultural heritage 324 data for 322, 323, 325 and desertification 312, 328 early warning systems for 312, 325–6 and earthquakes 320, 322–3 and earthquake engineering 323–4 education/training for 322, 324, 326–8 field missions on 329, 330 and floods 320, 322, 328–9 and impacts 321 information on 329 instruments for 325 mobile teams for 325 and monuments 328–9 and protective systems 320 risk assessment 321, 327 and risk mitigation/reduction 323, 326, 328 and seismology 323 and storms 322 and United Nations Environment Programme 327 and volcanic eruptions 321, 324 and warning networks 326–7, 331, 583

INDEX

natural hazards 221–3, 297, 315, 320–4, 328–9, 411, 464, 580, 583–4
natural heritage 286, 327, 389, 390–6, 400, 411
natural resources 24, 52, 83, 90, 97, 148, 195, 197, 201, 203–5, 211, 213, 216, 218, 221–6, 229–34, 260, 265, 269, 272, 276–7, 296–7, 300, 304, 383–4, 396, 403, 420, 471, 497, 513, 519–20, 577–8, 594, 602–3, 625, Research, Advisory Committee on 223, 225–6, 260, 314, *see also* environment
Nature and Resources 204, 226
nature protection/resources 33, 50, 193–427
Needham, Joseph 44, 508, 539 and UNESCO 43–47
neurobiology 97, 135, 580
networking, dynamic 599
New Partnership for Africa's Development 582, 604, 629, 631
North Atlantic Treaty Organization 63, 116
Nuclear Information System 161
nuclear weapons 49, 58, 489
nutrition 54–5, 144, 184–5, 383, 517, 600, 603

O

Obergurgl Project 271–3
Oceanographic Assembly, Joint 361–2 Data Centres, National 344 Research, Committee on 332, 334, 340, 357, 361, 363, 368 Tables and Standards 365
oceanography 56, 65, 68–9, 90–1, 196, 220, 332–70, 519–20, 524, 532 Congress on 70 Office of 67, 333, 354 physical 69, 220, 334, 360, 543, 565 *see also* Intergovernmental Oceanographic Commission (IOC)
oceans 25, 56, 65–70, 90–1, 332–70, 402, 404–5, 408, 411, 417–18, 420, 489, 624–6 and bathymetric charts 344–5 and bathythermographic data 346 and biological reference collections 366 conservation of 361 and continental margins 368 and corals 352 and cultural diversity 375 and culture 379–81 data on 336, 344–5, 347, 360 and development 357–8, 370–2 and education/training 342–3, 352, 358, 362–3, 365, 375 and electronic communication 347, 371, 376, 380 exploitation of 333 exploration and research of 337 and Geological and Geophysical Atlas of the Indian Ocean Global Atmosphere Research Programme 350 and Indian Ocean Expedition 322, 344, 348 and interaction with atmosphere 349–50 and management of 338, 340, 342, 371 mapping of 345 monitoring of 345–6, 375 and and policy 336 research 350 and resources 336, 339, 345, 351–2 and Scientific surface layers of 346 and UV radiation 338 and water export 379 and weather forecasts 346 and Weather Watch 347, 378, 381 *see also* beach resources; coastal management; Decade of Ocean Exploration; El Niño; fisheries; floating university; Global Atlantic Tropical Experiment (GATE); Global Ocean Observing System (GOOS); Integrated Global Ocean Services System (IGOSS); Kuroshio Current; pollution; salinity; sea level; small islands; Year of Oceans; World Ocean Circulation Experiment
Office for Science and Technology, United Nations (OST) 49, 436;
Organization for African Unity (OAU) 76, 172 of American States 71 for Economic Cooperation and Development (OECD) 62, 115, 434, 436–7, 450, 453–4, 456–8, 594 for Educational and Cultural Cooperation 38
Oslo Manual 458
ozone layer 169, 625

P

partnerships 77–80
peace 13, 21–2, 29, 36, 38, 40–1, 45–6, 50, 52–3, 78, 89, 92, 103, 107, 116, 121, 143, 368, 400, 432, 467, 469, 487, 492, 572–3, 586, 621, 633 science for 45, 52, 78, 432
periphery principle 32, 45–7, 71–2
pesticides 204
philosophy of Julian Huxley 40–3 *see also* science, philosophy of
physics 24, 32, 35, 55–64, 84, 97, 100, 104, 107–20, 126, 133, 145, 177–82, 184, 199, 354, 466, 487, 508, 515, 536, 539, 541, 580, 597, 610, 624 Advanced School for Theoretical Physics 112 Agenda for Scientific Cooperation 115 World Year of 487 *see also* International Centre for Theoretical Physics, Abdus Salam (ICTP)
plankton 352
plant-water relationships 219
polar regions 201, 220 *see also* Antarctic; Arctic
pollution 91, 122, 170, 174, 195, 245, 264, 267, 336–9, 347, 375, 402, 408, 482, 596 marine 351–2, 603 coastal zone 625 of oceans 336, 338, 347, 351–2, 375
population growth 195, 594
poverty 572, 593, 600, 611, 630, 632 eradication of 385, 424

R

Ramsar Conference on Wetlands 600
Regional Offices for Science and Technology (ROSTs-4) 24, 30, 33, 73–6, 84, 150, 354, 357, 528, 608, 613 Africa 73 East Asia 72 Europe 73 Latin America 73 Middle East 72 South Asia 72
remote sensing 315–19
renewable energy see energy, renewable
Reports in Marine Sciences 368
research and industrial applications 432, 452 laboratories 53–4, 57, 201, 207, 220 interdisciplinary 211, 218, 402–4, 506, 578 management of 268, 270, 274
risk assessment 222
robotics 595
John D. Rockefeller Foundation 71

S

Salam, Abdus 108, 111 *see also* International Centre for Theoretical Physics, Abdus Salam (ICTP)
saline soils 213, 219, 224
salinity 234–5 of oceans 347–8, 365
science, achievements, shortcomings and challenges of 501 American Association for the Advancement of 211 Conference on 421, 496, 501 cooperative 310 and the Information Society 503 Kalinga Prize for the Natural 88 Popularization of 461–2, 523 philosophy of 40, 81–2, 454 policy 25, 144, 421, 434–50, 455, 460, 476–9, 532, 539, 576, 580–1, 584, 604–5, 628, popularization of 598, 604 and society 429–83, social aspects and implications of 34, 44, 46, 53, 47, 50, 79, 81, 501, 598 and technology policies 434–51, 476–81, 604, 628, 631 and Technology, Advisory Committee on the Applications of 436, 439, 444 and technology for development 502 and the Use of Scientific Knowledge, Declaration on 225, 285, 312, 328 World Conference on 385, 493, 494, 598, 600–1, 627 Writers Association 461 *see also* ethics in scientific knowledge and technology; International Basic Sciences Programme (IBSP); International Council for Science (ICSU); social and human sciences; soil science

INDEX

Science and Technology Policies Information Exchange System (SPINES) 444, 447, 451, 476, 479
scientific literacy 598, 628 Publications, Network for the Availability of 503
scientists, free circulation of 504
sea level 347, 594
sediment 240, 243
small islands 197, 371–82, 402, 420, 424, 426, 631, 633
snow and ice 241
social and human sciences 264, 279, 294, 354, 419, 453, 493, 593–5, 601, 608 *see also* humanities
socioeconomics 619, 630, 632
soil erosion 53 science 217, 223–4 *see also* International Union for Soil Sciences (IUSS)
Sourcebooks of Chemical Experiments 122
space, outer 92, 315–19, 578 Education Programme 578 technologies 318, 624
Sputnik 60, 68, 97
Statistical Yearbook 453
statistics 453–9 standardization of 452 on science and technology 454
stratigraphic correlation 221
sustainable development see development, sustainable
Swedish International Development Authority (SIDA) 113

T

teaching see education/training
Technical Papers in Marine Science 368
telecommunications 115–16, 128, 133
Third World Academy of Sciences/The Academy of Sciences for the Developing World (TWAS) 102, 491, 501–4, 506, 515, 587, 603, 608, 613
traditional knowledge 381–8 *see also* Local and Indigenous Knowledge
transdisciplinarity 611
tropical forests 260–96, 624
tsunami 320, 322–3, 326–7, 488, 579, 582, 624 warning systems 326–7, 346, 536 *see also* natural disasters
tuberculosis 53, 55

U

UNESCO Courier 204, 229, 522
UNESCO-IHE Institute for Water Education 257–8, 426, 528, 575–6, 584, 587, 610, 614, 616–17, 620
Union of International Technical Associations (UATI) 242
United Nations Information System for Science and Technology (UNISIST) 129–33, 153, 161, 476, 503, 506
United Nations Organization 14, 18–19, 30, 46, 48–54, 66, 96, 173 Administrative Committee on Coordination 351, 440 Advisory Committee on the Application of Atomic Energy 51, 334 Committee on the Application of Science and Technology 51, 161 Atomic Energy Commission 49, 52 Commission on Environment and Development 409 Conference on Application of Science and Technology 33, 51, 161 Conference on Science and Technology for the Benefit of Less-Developed Areas 436 Conference on Conservation and Utilization of Resources 45, 50, 52, 79, 203 Conference on Decade of Education for Sustainable Development 424–5 Conference on Environment and Development 26, 338, 372, 411–21 Conference on Developing States 420 Conference

on Establishment of an Educational and Cultural Organization 39 Conference on Human Environment 261, 338, 391, 403, 405, 407, 501 Conference on International Organization 431 Conference on Science and Technology 435 Conference on Sustainable Development of Small Island Developing States 420–1 Conference on Convention on Law of the Sea 339, 345 Conference on World Climate 409–17 Convention on the Law of the Sea 339, 345 Economic and Social Council 31, 48, 50, 52, 54, 200–1, 337, 351, 440 Industrial Development Organization 62, 452, 493 Millennium Declaration 250 Millennium Development Goals 16, 168, 385, 572, 577, 600, 606, 611, 619, 620–1, 627, 629–30, 633 Millennium Project 597, 601, 615 Office for Science and Technology 49, 131, 436 Office for Statistics 436 and peaceful uses of atomic energy 50 Scientific Advisory Committee 49, 51 Committee on the Effects of Atomic Radiation 51, 71 Security Council 31, 49, 52 Scientific Conference on Conservation and Utilization of Resources 49–50 Summit on Sustainable Development 422–3 Technical Cooperation Unit 159 *see also* Food and Agricultural Organization (FAO); International Atomic Energy Agency (IAEA); International Labour Office (ILO); Office for Science and Technology, United Nations (OST); UNESCO; United Nations Development Programme (UNDP); United Nations Environmental Programme (UNEP); United Nations University (UNU); World Health Organization (WHO); World Meteorological Organization (WHO)

United Nations Development Programme (UNDP) 74, 76, 108, 113, 131–2, 139, 145, 149–55, 158–9, 164–6, 180, 190–1, 211, 219, 243, 323, 325, 360–1, 363–4, 443–4, 514, 532–3, 542, 594, 600, 602

United Nations Educational, Scientific and Cultural Organization (UNESCO), 'S' in 36–40, 464 Preparatory Commission for 496 ICSU Cooperation 77, 496–505 *see also* philosophy of Julian Huxley

United Nations Environment Programme (UNEP) 137, 139, 185–7, 190, 192, 200, 230, 260, 271, 278, 282, 285, 288, 296, 327, 337, 351, 363, 402, 404–10, 418, 422, 424, 427, 492–3, 500, 506, 513, 533, 600, 602, 605

United Nations University (UNU) 139, 256, 289, 502, 603, 608

university, floating 362, 366–8, 513

Upper Mantle Programme 228, 300, 303

V

volcanic eruptions 222, 231, 321, 324 and hazard mapping 316 *see also* natural disasters

vulnerability reduction 320 *see also* natural disasters

W

water 139, 196–7, 219, 225, 233–59, 387, 402, 407, 411, 416, 470, 520, 617, 619 balances 240 Centres 613, 615, 618 Conference on 248, 250 Dependencies Systems under Stress and Societal Responses 619 distribution 236 and the Environment, International Conference on 248 governance 602, 617 for Life 250 Programmes 617–20 and sustainable development 498 resources 243 resources management 317, 602–3, 619 *see also* Decade for Action 'Water for Life', International; World Water Assessment Programme (WWAP)

Water Education, Institute for 257, 258

welfare, human 46, 48, 492, 494

wetlands 16, 285, 600 *see also* water

women 22, 25, 59, 141, 148, 156, 189, 248, 266, 432, 465–73, 598, 631–2, 634 Conferences on 467, 469, 471 and Women's Decade 469 *see also* girls; glass ceiling

The World Bank 133, 190, 491, 600, 602
World Climate Research Programme 350, 498 Conference on Science 385, 471, 493–45, 598, 600–1, 627 Conservation Union 26, 202–3, 230, 271, 278, 621 Cultural and Natural Heritage 14–15, 95, 198, 216, 230, 389–400 Data Centres 503 Federation for Culture Collections 137 Solar Programme 416
World Health Organization (WHO) 48, 63, 89, 91, 134, 139, 150, 204, 241, 251, 261, 334, 351, 422, 452, 507, 539, 603
World Heritage 14–15, 198, 218, 226, 230, 285–6, 289, 317–18, 373, 389–400, 408, 523, 53, 578, 586, 614, 621–3 Centre 218, 286, 289, 373, 389, 398–400, 586, 614, 621, 623
World Meteorological Organization 66, 150, 196, 216, 224, 231, 235–7, 241, 244, 251, 234, 244, 251, 254, 256, 282, 323, 334, 337–8, 341, 346, 348–51, 355–6, 407, 409–10, 497, 499–500, 506, 543, 548, 583, 603, 624–6
World Ocean Circulation Experiment 350
World Summit on the Information Society 133, 504, 598
World Summit on Sustainable Development 421–5
World Water Assessment Programme (WWAP) 257, 552, 577, 617, 620
World Water Development Report 257, 577, 634
World Wide Web 133, 310, 609 *see also* internet

Y

Year of Astronomy 624 of Mathematics 105–6, 116 of Oceans 340 of Physics 110 of Planet Earth 498
Young Scientist Awards 602, 612
Younger Geoscientists Programme 311
youth 262, 378, 380–1, 385–7, 470–1, 622, 628, 634 *see also* girls

Achevé d'imprimer sur les presses de l'Imprimerie BARNÉOUD
B.P. 44 - 53960 BONCHAMP-LÈS-LAVAL
Dépôt légal : octobre 2006 - N° d'imprimeur : 610098
Imprimé en France